Quick reference list: philosophical topics

Quick reference list: great scientists

PHYSICS:

Concepts and Connections

PHYSICS
Concepts and Connections

ART HOBSON
University of Arkansas

PRENTICE HALL, Englewood Cliffs, New Jersey   07632

Library of Congress Cataloging-in-Publication Data

Hobson, Art, 1934–

    Physics: concepts and connections/Art Hobson.
      p. cm.
    Includes index.
    ISBN 0-02-354841-X
    1. Physics.   I. Title.
      QC23.H739   1995   94-2506
      530—dc20   CIP

Editor: Robert A. McConnin
Production Supervisor: Margaret Comaskey
Production Manager: Nicholas Sklitsis
Text Designer: Robert Freese
Cover Designer: Leslie Baker
Photo Researcher: Ken Eward

© 1995 by Prentice-Hall, Inc.
A Division of Simon & Schuster, Inc.
Englewood Cliffs, New Jersey   07632

PRINTED IN THE UNITED STATES OF AMERICA

10 9 8 7 6 5 4 3 2 1

ISBN   0-02-354841-X

Prentice-Hall International (UK) Limited, *London*
Prentice-Hall of Australia Pty. Limited, *Sydney*
Prentice-Hall Canada Inc., *Toronto*
Prentice-Hall Hispanoamericana, S.A., *Mexico*
Prentice-Hall of India Private Limited, *New Delhi*
Prentice-Hall of Japan, Inc., *Tokyo*
Simon & Schuster Asia Pte, Ltd., *Singapore*
Editora Prentice-Hall do Brasil, Ltda., *Rio de Janeiro*

For my son David

Preface

This is a liberal-arts physics textbook for nonscience college students. Its central premise is that nonscientists deserve a true *liberal-arts* physics course. Thus, this is not a watered-down version of the standard technical introductory physics textbooks for scientists. Far from being a simplified version of anything, this book is designed for a course that has a cultural sophistication not found in more technical courses. It is a cultural, rather than a technical, physics textbook. It presents physics as a human endeavor in its full philosophical and social context.

Many organizations have recommended new approaches to science education and science literacy.* The features in this book that reflect these recommendations include:

Scientific literacy. This book addresses the values, philosophical meaning, and social impact of science, and stresses the methods of science.

Modern physics. Fully half of this book is devoted to such post-Newtonian topics as relativity, quantum theory, nuclear physics, and high-energy physics.

Societal connections. This book applies the ideas of physics to such socially relevant topics as ozone depletion, global warming, technological risk, energy resources, nuclear power, and nuclear weapons.

Appropriate quantitative skills. Nonscientists should become *numerate* as well as literate in science. The abilities to interpret graphs, to think probabilistically, and to make rough numerical estimates are important for nonscientists. On the other hand, traditional algebra-based physics problems are less important for nonscientists. Just as one can appreciate a painting without being able to paint, one can appreciate the power and beauty of a scientific idea without solving equations. Thus, this book is quantitative but nonalgebraic.

---

* American Association for the Advancement of Science *Project 2061,* reported in *Science for All Americans* by F. James Rutherford and Andrew Ahlgren, (Oxford University Press, New York, 1990); National Science Board *Undergraduate Science, Mathematics and Engineering Education* (National Science Foundation, Washington, 1986); American Institute of Physics *Introductory University Physics Project* (IUPP).

Less is more. While presenting most of the great ideas of physics, this book omits many narrower topics and applications normally "covered" in introductory courses.

Unifying themes. Four story lines recur throughout the book: (1) how we know what we know in science; (2) comparisons and contrasts between Newtonian and contemporary physics; (3) the social context of physics; (4) and the unifying concept of energy.

One justification for communicating these ideas to nonscientists is simply that, as a matter of general principle, educated people should understand science's view of the modern world. But there is a more pressing, practical reason: In an age when science and technology are driving rapid cultural and physical changes on Earth, it is imperative that nonscientists contribute their understanding and their perspective to helping us figure out where we are going and where we should go. Today, science is far too important to be left to the scientists. I have written this book for you, the teachers, poets, politicians, historians, business people, journalists, and others who must help us find a rational and humane path through a time of rapid change and powerful technologies. We need your perspective, and your informed leadership.

I have tried to develop the scientific ideas slowly, with attention to the learning process.* You will find the following learning aids in the book:

Quotations in the margins lend different perspectives, and depth, to the philosophical and societal topics. Quotations, especially those on controversial topics, represent a wide range of views. Please understand that the inclusion of a particular quotation does not mean that I necessarily agree with that viewpoint, or that the scientific community necessarily agrees!

Footnotes provide additional details for students who want them. It is a perennial problem for writers of introductory textbooks, especially books for nonspecialists, to be accurate while at the same time not burdening students with details that might do more to confuse than to enlighten. Footnotes seemed to me to be the best way to handle such situations.

Dialogue questions in the main body of each chapter are my way of conducting a dialogue with the reader, to check and reinforce the reader's understanding. Readers should formulate their own answers to each dialogue question as they come to it, before reading the answer that appears in a footnote at the bottom of the same page.

How do we know is a frequently asked question in this book, and subsections bearing this title appear regularly. It cannot be emphasized too strongly that scientists have specific observation-based reasons for their theories and principles. Most principles of physics are accompanied by a discussion of how we know.

Making estimates is one of the skills that this book seeks to develop. Examples and exercises bearing this title appear frequently. Making approximate but informed estimates is one numerical technique with which educated citizens should be familiar.

---

* I have benefited, especially in my presentations of Newtonian concepts, from Arnold B. Arons' book *A Guide to Introductory Physics Teaching* (John Wiley & Sons, New York, 1990).

Summaries of ideas and terms, following each chapter, summarize and clarify the main concepts. These should be helpful when studying for exams.

Review questions, about thirty, follow each chapter. Organized by sections within the chapter, they go over the main points. Answers can be found by looking them up in the appropriate section.

Home projects, a few of them, follow each chapter. There is nothing like "hands on" experience to bring out the essential experimental nature of science. These projects could be done either at home using common household items, or in physics laboratories, or as demonstrations or desktop experiments in physics lectures.

Discussion questions, a few of them, follow each chapter. These are meant to stimulate thinking about open-ended questions, such as values questions, that have no single correct answer. They can be used for class discussion, essays, or individual thought.

Exercises, about thirty, follow each chapter. These are designed to exercise the mind, the way that jogging exercises the body. They are based on the text, but they require original thought. Most are qualitative, or "conceptual." A few are quantitative, but they require no algebra. These quantitative exercises are indicated with asterisks. Answers to all odd-numbered exercises are in the back of the book.

As you study or teach from this book, you will undoubtedly have comments, suggestions, criticisms, and/or corrections. Please send them to me!* A textbook, like learning and like life itself, is never really finished. I welcome your contributions to this book's continuing evolution.

A word to the instructor

This book takes a conceptual, nonmathematical approach to physics for nonscientists. There is a significant amount of educational research showing that all students, including science students, grasp the essential ideas of physics better if they conceptualize before they calculate. Furthermore, because the average nonscience student is deterred and discouraged by a mathematical presentation of physics, a conceptual, nonalgebraic approach is especially appropriate for these nonscientists. See the Instructor's Manual for further discussion of the conceptual approach to physics teaching.

To allow you flexibility in your choice of topics, the book is purposely too long for one semester. I usually omit the equivalent of about two chapters. Chapters and sections that may be omitted without seriously affecting the remaining material are indicated with asterisks in the table of contents. Many other omissions are possible, provided you are careful in choosing the remaining material. The flow chart following the table of contents shows connections between topics, and suggests alternative course structures. The first chapter, about scientific methodology, is one of the deletable chapters, because there is material on this topic throughout the book. However, if Chapter 1 is skipped, students should read its first section anyway because

---

* Art Hobson, Department of Physics, University of Arkansas, Fayetteville, AR 72701; phone 501-575-5918; fax 501-575-4580; internet ahobson@comp.uark.edu.

it introduces the book. Some topics, for example, interdisciplinary topics such as extraterrestrial life, could be assigned for reading only.

Selections from this book could be used for a course that focuses more intensely on a much smaller range of material, or for a short course. For example, the societal topics could be omitted, resulting in a Newtonian-plus-modern physics course. Or modern physics could be omitted along with those societal topics that depend strongly on modern physics, resulting in a Newtonian-plus-societal physics course. Or the last four chapters (nuclear physics, its societal implications, and high-energy physics) could be dropped. Instructors should consult the teacher's manual for further suggestions.

Acknowledgments

As I ponder my acknowledgments, I realize how vast is the network of people and institutions that contributed to the origin and evolution of this book. A six-month residence at the Stockholm International Peace Research Institute inspired me to further develop the interdisciplinary and social interests that led to this book. The University of Arkansas has provided a great place to work, as well as two sabbatical leaves during the past ten years that have been crucial to my professional development. The students in my general physics courses have always been a source of energy and ideas, and have provided feedback during class testing of the book. The following instructors class-tested the book: Charles Bottoms, Laurel Busse, Bernard Hoop, Mardi John, Carl Kocher, Joan Kowalski, Bill Lankford, Tom Meeks, Diane Claire Shichtman, and Robert Socolow. They, as well as their students, provided feedback. My colleagues in the Department of Physics provided help and encouragement, and have put up admirably with my questions; this is particularly true of my friends Julio Geo-Banacloche, Raj Gupta, Bill Harter, Claude Lacy, Mike Lieber, Charles Richardson, Suren Singh, and Min Xiao. Across campus, I had inspiring discussions with Lowell Nissen and Ben Anderson of the Department of Philosophy. For communications about various aspects of the book, I thank Ralph Baierlein, Al Bartlett, Timothy Burt, Malcolm Cleaveland, Martin Gardner, J. Richard Gott III, Leonard Mandel, N. David Mermin, David Renfro, Robert Resnick, F. Sherwood Rowland, John Shultz, Cynthia Sexton, Abner Shimony, and Richard Wilson. Beatrice Shube gave me advice about writing, as did Michael Hennelly and Michael Alley, especially through his fine book *The Art of Scientific Writing*. Senior Editor Bob McConnin gave me the initial encouragement to start this project; going beyond what I had perceived as the responsibilities of an editor, Bob is a real science educator who keeps unusually well-informed and maintains helpful contacts within the physics community. My entire publishing staff, and especially Ken Eward, Chris Migdol, and Margaret Comaskey, were a joy to work with. The American Physical Society's Forum on Physics and Society has played a crucial role in providing emotional and professional support; I especially thank my Forum friends Dave Hafemeister, Ruth Howes, Barbara Goss Levi, Richard Scribner, and Peter Zimmerman. The American Institute of Physics' Introductory University Physics Project has given me helpful ideas, human interactions, and professional encouragement. David Bodansky, Bruce Daniel, Paul J. Dolan Jr., Robert Ehrlich,

Norman Hackerman, Carl Kocher, William F. Lankford, Charles Leming, Theodore Lopushinsky, Mark Modera, Dwight Neuenschwander, Carl T. Rutledge, Steven Schneider, Dietrich Schroeer, Diane Shichtman, Clifford Swartz, Shelia Tobias, Carl T. Tomizuka, and James Watson have reviewed and commented on the entire manuscript. I would also like to acknowledge the following physicists and authors for their inspiration and their influence on this book: Arnold Arons, John Bell, Niels Bohr, E. A. Burtt, Albert Einstein, Richard Feynman, Werner Heisenberg, Paul Hewitt, Leon Lederman, Robert Romer, and Carl Sagan. My dear friends Dick Bennett, Marie Riley, and Rita Massey provided encouragement and stimulating interdisciplinary perspective. My daughter and son, Ziva and David, have given me support, encouragement, and cheerful discussion.

I hope that you have as much fun using this book as I had writing it and teaching the related course! Enjoy!

ART HOBSON

Contents

Flow chart of topics

For a course emphasizing modern and philosophical topics, many societal topics could be omitted. For a course emphasizing societal topics, many modern and philosophical topics could be omitted. For chapters and sections that can be omitted without affecting the continuity of the remaining material, see the table of contents.

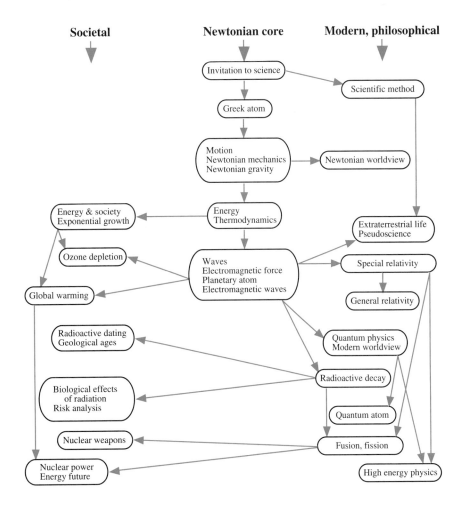

PART **I**

PRELUDE
of stars and atoms

1

THE ART OF SCIENCE
*experience, and theories
pleasing to the mind*

1.1 Stardust: *an invitation to science*

We came from the stars. We are, literally, made of atoms created and blown into space by ancient stars. This fact is only one strand in a network that connects us with the rest of the universe. **Science**—the observation and understanding of the natural world—is a path toward embracing that network.

An expanding awareness of nature is discernible in the long history of life on Earth. There is good reason to believe that our planet formed about 5 billion years ago and that the earliest simple living organisms formed over 3 billion years ago (see Chapter 15). Since then, organisms have evolved biologically to interact with their environment in increasingly complex ways. Looked at from the human perspective (an amoeba would no doubt view things differently), humankind is the latest in a sequence of increasingly aware biological organisms. We could even say that through biological evolution, the universe has become more aware of itself. Education and science can be viewed as an extension of this process. And you, as you learn about the universe, are part of that process of expanding awareness.

I hope that this book will help you discover many links between you and the universe. In writing this book, my constant criterion has been, Is this material really relevant to readers who want to participate fully in our science-related culture but who do not necessarily use science in their professional life? I have tried to use language that is meaningful to literate nonscientists. There are no extraneous technical terms and no extraneous mathematics— in particular, no algebra. The text does, however, make wide use of numbers,

The essence of education is the conquest of parochialism, that is, expanding one's horizons. You've got to look beyond the limits of your own narrow horizons.
    Walter Manger, geologist,
    University of Arkansas

proportionalities, graphs, and numerical estimates, because quantitative ideas are essential to meaningful communication today. Literate people must also be *numerate*.

We have discussed one reason for learning science: expanded awareness. A second and related reason is to help develop cultural and social values appropriate to the scientific age. Take a moment to list a dozen problems that are important to the nation or the world. A typical list might include population growth, poverty, crime, species destruction, illiteracy, global warming, urban decay, drugs, wars, air pollution, AIDS, and famine. Most or all of these problems have a science component.

Now try listing solutions to these problems. A typical list of solutions might include birth control, economic growth, education, sustainable farming, democracy, international law, environmental protection, disease control, better government, rational use of energy, more understanding among people, and concern for the environment. Most or all of these solutions have a science component.

The problems and the solutions of our times are bound up with science and its close relative, technology.* That is why we call this the scientific age.

The world needs your help. We dare not simply turn over these critical issues to "experts" or to governments. If we are to govern ourselves, we must make decisions about the issues that determine our lives, and these issues are related to science.

In his book *Of a Fire on the Moon*, about humankind's first venture to the moon, novelist and journalist Norman Mailer wrote pessimistically:

> The [twentieth] century would create death, devastation and pollution as never before. Yet the century was now attached to the idea that man must take his conception of life out to the stars. . . . A century devoted to the rationality of technique was also a century so irrational as to open in every mind the real possibility of global destruction. . . . So it was a century which moved with the most magnificent display of power into directions it could not comprehend. The itch was to accelerate. . .the metaphysical direction unknown.

Other eras have had their own problems. But if we are to resolve today's problems, we must find our metaphysical direction in this scientific age.

Science confers great power. You use this power daily when you, for instance, switch on a light, a television set, an automobile, or a computer. Every such device has powerful effects on the world, both good (light to read by) and bad (pollution from electric generating plants). The classic moral dilemma of the scientific age—a dilemma symbolized, for example, in Mary Shelley's nineteenth-century novel *Frankenstein*—is the problem of understanding and dealing responsibly with these powerful technologies. To accept technology's power without also accepting the responsibility to use that power wisely is to invite death, devastation, and pollution—the monster's retaliation against its maker.

This book focuses on that part of science called physics, and its human connections. You have heard of most of the major sciences: biology, geology, chemistry, astronomy, physics, and others. **Physics** studies the most general features of the natural world, features that are found throughout most

Technology is the application of science to achieve useful human goals. This book often uses the single word *science* to refer to science and technology.

For any man to abdicate an interest in science is to walk with open eyes toward slavery.

Jacob Bronowski, philosopher-scientist

Scientific activity is one of the main features of the contemporary world and, perhaps more than any other, distinguishes our times from earlier centuries.

Science for All Americans, a report to the American Association for the Advancement of Science

I hesitate to use the word "God," but in my studies of the universe I have come to the conclusion that there is some purpose to it. The universe has organized itself in such a way as to become aware of itself. As conscious beings, we are part of that purpose.

Paul Davies, physicist

The dangers that face the world can, every one of them, be traced back to science. The salvations that may save the world will, every one of them, be traced back to science.

Isaac Asimov, scientist and writer

or all of nature. Other sciences focus on particular facets of nature: Biologists study living organisms, and geologists study Earth's structure, for example. But physicists search for principles that apply to living organisms, Earth's structure, and all other parts of the universe. For example, most objects fall when you drop them. You can drop a rock, you can drop a frog, or you can drop many other objects, and they all fall. So the idea that objects fall when you drop them is a principle of physics.

Please pick up this book and drop it. That's physics! It's about falling and about all the other general features of nature.

The remainder of this chapter is about what scientists do when they do science. How does science operate? We answer this with a significant case study of the actual operation of science: the early history of astronomy. We begin in Section 1.2 with some commonsense observations and conclusions about the night sky. Sections 1.3, 1.4, and 1.5 present three theories about the way the heavens are organized: the ancient Greek, Earth-centered view culminating in Ptolemy's theory, then Copernicus's postmedieval sun-centered theory, and Kepler's slightly later sun-focused theory. Section 1.6 discusses what we can learn about science's operations from this history, and Section 1.7 looks at the cultural implications of all of this.

We need people who can see straight ahead and deep into the problems. Those are the experts. But we also need peripheral vision and experts are generally not very good at providing peripheral vision.
Alvin Toffler, writer and futurist

1.2 Observing the night sky

Teach me your mood, O patient stars!
Who climb each night the ancient sky,
Leaving no space, no shade, no scars,
No trace of age, no fear to die.
Ralph Waldo Emerson, *The Poet*

Science and technology are so pervasive that they are hard to fit into perspective. And yet precisely because science is pervasive, we need to understand it. This book's most important theme is the study of the nature of science itself.* How does science operate? What are its values? How valid are its conclusions? This topic is basic because science is more a method, a way of learning, than it is a body of knowledge.

The scientific method is often described as several activities that scientists sometimes practice: observing, hypothesizing, testing, and so forth. But such a list is superficial because it treats science as different from other human endeavors instead of capturing how science works in human culture. For a perspective on how science operates in real life, we will look at a historical example: the early history of astronomy. **Astronomy**, the scientific study of the stars and other objects in space, has usually been closely associated with physics.

Humans are fascinated by the heavens. The starry sky seems a more perfect place than our daily world. Life is full of clatter, the stars are serene; life is brief, the stars are forever. It is not surprising that ancient priests looked to the stars, for in addition to their beauty the stars also move in orderly, predictable patterns. Here was certain and timeless knowledge, a place, perhaps, of gods. And so for at least as long as there are records to tell the story, we have looked to the sky for the time to plant and to reap, for omens of war and peace, for a sign regarding life's meaning, for our gods (Figure 1.1).

---

* Besides this theme, three others reappear throughout the book: comparisons between Newtonian and post-Newtonian physics, the social impacts of physics, and energy.

FIGURE 1.1
Four thousand-year-old testimony to our reverence for and scientific awareness of the stars: the remains of Stonehenge, in England. Humans hauled the huge stones for more than 200 miles to make these monuments. These stones are the remains of a much larger structure used for religious purposes and to predict astronomical events, particularly solstices (longest and shortest days) and equinoxes (equal-length days and nights). Stonehenge perhaps also predicted eclipses of the moon, an impressive feat for people who did not use writing. Eclipses occur in an irregular and apparently random pattern that repeats itself only over a 56-year cycle. Even to have been aware that a repeated pattern exists required enormous dedication and attention to detail.

Although **astrology**, the belief that the positions of the stars significantly influence human affairs, has been an exploded superstition for two centuries, human scientific, religious, and aesthetic fascination with the stars may be greater today than ever (Figure 1.2).

In these electrically lit times, we sometimes fail to see the stars. On some clear night, try to follow the stars as they cross the sky. Find the moon, the seven stars of the Big Dipper, the North Star, any group of stars on the eastern horizon, and one or two stars on the western horizon (Figure 1.3). Observe all of these every 15 minutes for 1 or more hours. What happens? You should be able to see that the moon and stars move westward, that stars rise in the east and set in the west, that different stars maintain their positions relative to one another while moving as a group across the sky, that the North Star remains fixed, and that stars near the North Star move in circles around the North Star (Figure 1.4).

The Home Projects at the end of this chapter include many other astronomical observations that you can make without a telescope. You can observe that there are several small and unusually bright starlike objects that do not keep pace with the stars. When observed for a week or more, they can be seen to slowly shift their positions relative to the stars. These objects are

(a)

Figure 1.2
Our fascination with the stars seems greater than ever. (a)The Hubble orbiting space telescope. (b) Model of a four-telescope installation being built in Chile. Each telescope contains a near-perfect mirror 8 meters in diameter.

(b)

FIGURE 1.2 (continued)

(c) Many telescopes receive nonoptical signals from space. This large "radio telescope" in Arecibo, Puerto Rico, receives radio signals emitted by objects that pass overhead as Earth rotates. (d) Drawing of the "neutrino telescope" planned for installation in Arkansas. The telescope "looks" downward, through Earth, to receive subatomic particles called neutrinos emitted by objects in space. The neutrinos pass entirely through Earth at the speed of light. A few neutrinos interact with atoms near the telescope, causing these atoms to emit particles called muons that in turn emit light as they travel through the telescope's purified water. This light is detected underwater.

(c)

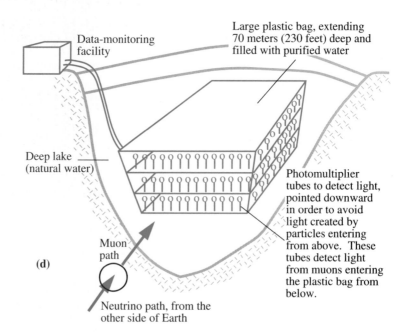

Data-monitoring facility

Large plastic bag, extending 70 meters (230 feet) deep and filled with purified water

Deep lake (natural water)

Photomultiplier tubes to detect light, pointed downward in order to avoid light created by particles entering from above. These tubes detect light from muons entering the plastic bag from below.

Muon path

(d)

Neutrino path, from the other side of Earth

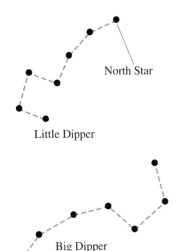

North Star

Little Dipper

Big Dipper

FIGURE 1.3
Look for these two constellations and the North Star in the night sky.

called **planets** ("wanderers" in Greek). Five planets are visible without a telescope. The moon and the sun also move at a different pace from the stars.

From such observations, most people would conclude that the stars, sun, moon, and planets all are moving in circles around Earth, with their axis of rotation fixed in the direction of the North Star. Figure 1.4 is rather convincing evidence of this idea. This is the conclusion most observers drew centuries ago, and it is surely the conclusion that observers draw today unless they learn differently in school or elsewhere.

The observations and conclusions described here are typical of science's two main processes: observing, and conceptualizing or making theories. Science is not really different from a lot of other human thought. Whenever we observe our surroundings and develop ideas based on what we observe, we are acting somewhat like scientists.

FIGURE 1.4
Time-exposed photograph showing the "star trails" near the North Star.

1.3 Ancient Greek theories: *an Earth-centered universe*

At least as early as 3000 B.C., the Babylonians and Egyptians were aware of the differing motions of the stars, sun, moon, and the five planets then known. Beginning around 500 B.C., a few Greeks sought a new kind of understanding of these motions. They desired to go beyond the observed facts, to grasp the essentials of how the system worked.

Figure 1.5 indicates the early Greek concept of the architecture of the heavens. In agreement with the observations described in the preceding section, this figure shows the heavenly objects circling a motionless Earth. Because the stars all keep pace with one another, the Greeks supposed that they all were attached to a single transparent invisible sphere centered at Earth's center and rotating around Earth once a day, carrying the stars with it. The sun, moon, and five visible planets, seven objects in all, each move along their own circle at rates a little different from that of the stars.

FIGURE 1.5

The earliest Greek conception, around 500 B.C., of the layout of the universe.

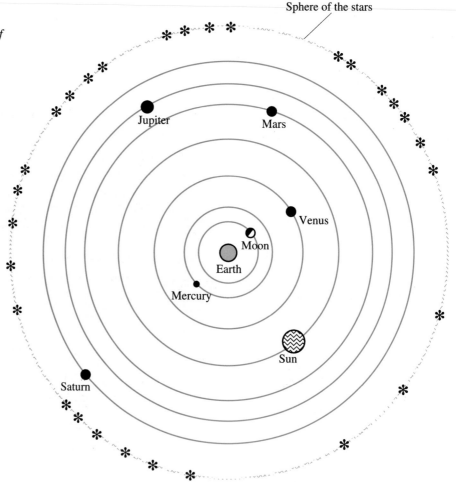

The Greeks imagined that each of these seven objects was attached to a large invisible sphere centered on Earth, one sphere for each object. Each of the seven spheres rotated at an unchanging (or **uniform**) rate around Earth roughly once each day. These spheres, and an eighth sphere carrying the stars, came to be called crystal (or transparent) spheres because although they were present in the theory, they could not be directly observed in the sky. Each sphere rotated at a slightly different, but uniform, rate about the same axis through Earth's center.

A Greek philosophical–mathematical–religious group led by Pythagoras developed this picture of the universe. The Pythagoreans formed a secretive cult that believed passionately in the importance of abstract ideas. An idea is, in a sense, eternal. A real table, for example, eventually rots and turns to dust, but the idea of "table" or "tableness" seems eternal. Pythagoras believed that the most perfect ideas were mathematical because they could be stated so precisely yet abstractly. The idea of "table" is rather imprecise—a flat rock might be considered a table, or it might just be a rock. But mathematical ideas like a straight line, a circle, or five were precise, pure. For example, a circle is all of the points on a flat surface that are at the same distance from some fixed point on the surface.

DIALOGUE 1   To check your understanding of the preceding definition of a circle, try answering this question: What do we normally call the "fixed point" and the "distance" referred to in the definition? (Dialogue questions like this will appear throughout the book. Try to answer them as you come to them, before looking at the answer at the bottom of the page.)

Number is the ruler of forms and ideas, and the cause of gods and demons.

> Pythagoras

Although this definition is precise, if you draw a circle that follows this definition, there will always be imperfections (Figure 1.6). Indeed, the Pythagoreans believed that it was the idea of a circle, rather than any particular representation of it, that was pure and eternal.

These mathematical mystics discovered that many features of the natural world could be described by mathematical ideas. The famous "Pythagorean theorem" is one example, and the simple numerical relationships between tones in the common musical intervals is another.* They believed that such ideas exist apart from human minds, that they are eternal and more real than any Earthly object, and that the universe must be based on deep mathematical principles or "harmonies" analogous to the numerical relationships between the common musical intervals. They believed that nature was ultimately mathematical and that when one studies mathematics one studies the mind of God.

Other Greeks regarded all of this suspiciously. The Pythagoreans were persecuted and eventually banished. But their thinking had a deep influence on subsequent Greek philosophers such as Plato and Aristotle and on Europe and Western civilization.

Given their general beliefs, it is not surprising that the Pythagoreans sought a precise geometric scheme for the heavens. What geometrical forms could be more fitting for the stars than the sphere and the circle? After all, the sphere is the only perfectly symmetric ("the same from all sides") shape in space, and the circle is the only perfectly symmetric shape

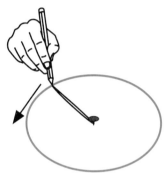

FIGURE 1.6
If you try to draw a circle, there will always be imperfections.

Let no one without geometry enter here.

> Inscription over the entrance to Plato's academy, fourth century B.C.

---

1. Remember to form your own answer before reading this! The center. The radius.
* The Pythagorean theorem states that in any triangle having a 90-degree or "right" angle, if you draw three squares, each one based on one of the triangle's three sides, the sum of the areas of the two smaller squares will equal the area of the larger square. As an example of the relationships between musical tones, if you create a musical tone by plucking a string and then precisely halve the string's length and pluck it again, the two notes you create will be exactly one octave apart. Other simple string ratios, such as 3 to 2 or 4 to 3, produce the other musical intervals that sound harmonious.

on a flat surface. And as befits the timeless stars, circular paths have no beginning or end.

In line with their elegant and fitting picture of a universe made of eight Earth-centered spheres within spheres, these early Greeks had the startling notion that Earth itself was a sphere, although not made of invisible crystal, residing motionless at the center of the crystal spheres. The Pythagorean concept of a spherical Earth (although not their idea of a motionless Earth) survives to this day.

HOW DO WE KNOW?

All good science is based on observable evidence. Science is always skeptical, always asking, What is the evidence? It is a question that we, too, will ask frequently. So we will often interrupt the flow of ideas to ask, How do we know?

How do we know that Earth is a sphere? The evidence is fairly direct today (Figure 1.7, see color insert), but what evidence did the ancient Greeks have? The Greek philosopher Aristotle, living two centuries after Pythagoras, stressed the importance of evidence. He gave many good observational reasons to believe that Earth is a sphere. For one thing, ships sink little by little below the horizon as they go out to sea, finally vanishing completely below the curve of Earth (Figure 1.8). For another, Greek travelers reported that in northern lands, the noontime sun is lower in the sky. For another, the shadow cast by Earth on the moon, as observed during an eclipse of the moon, is the shape that would be expected if both Earth and the moon were spherical.

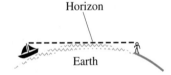

Horizon

Earth

FIGURE 1.8
One piece of evidence that Earth's surface is curved like a sphere. As a ship sails out to sea, an observer on shore sees it sink little by little below the horizon.

But there was a problem with the Pythagorean crystal-spheres "model" or "theory," concerning the motion of the five observable planets. Because of the uniformly rotating crystal spheres, the theory implies, or "predicts," that each planet moves at a uniform rate around Earth. But careful observation shows that they do not. Instead, their rate of rotation, as seen from Earth, changes. Figure 1.9 diagrams this effect for a single planet such as Mars. The diagram is drawn relative to the background stars, and so it does not show the nightly rotation of Mars and the stars. Relative to the stars, Mars generally moves from west to east, but at a variable rate. Occasionally, Mars even changes directions and moves east to west relative to the stars, a phenomenon known as **retrograde motion**. Retrograde motion is the most obvious example of the fact that each planet moves nonuniformly as seen from Earth.

The Greek philosopher Plato, convinced of the Pythagorean belief in an elegant mathematical reality lying behind the motions of the heavens, challenged his students with the problem of finding a geometric scheme that would explain the observed motion of the planets. Plato and his students constructed a theory similar to the Pythagorean theory but more complex. They retained the uniformly rotating crystal spheres centered on Earth because they believed that the universe must be based on spheres and circles. To make their theory agree with such facts as retrograde motion, they imagined several unseen spheres for each planet, all centered on Earth, all "nested" within one another, with each sphere's axis of rotation fixed to the inside of the next outward sphere, and with all of these axes pointed in dif-

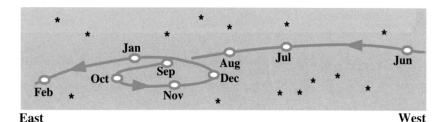

East   **West**

Figure **1.9**

The motion of a planet such as Mars, relative to the background stars. Relative to the stars, Mars usually moves from west to east. In this illustration, Mars moves more slowly during July–August than it does during June–July. It slows to a stop by October and then reverses direction during October–December and regains its normal direction during December–February.

ferent directions and the single planet riding on the outermost sphere. For example, when all of Mars's crystal spheres rotated, Mars moved in the way that it was actually observed to move in the sky. This complex theory of **multiple crystal spheres** for each planet was favored by Plato's student Aristotle and by many others.

One Greek thinker, Aristarchus, proposed that it was the sun and not Earth that was at rest at the center of the universe, that Earth and the five planets moved in circles around the sun, and that Earth spun on its axis. It was a radical idea, and few astronomers took it seriously because it seemed absurd on two different grounds. First, our senses suggest to us that Earth is nothing like the heavens. So how could Earth be a planet like the heavenly planets? Second, it seems absurd to believe that Earth moves. It's too big. What immense force could be pushing it to keep it moving? If it does move, it seems that objects such as birds and clouds that are not attached to the ground should be left behind. If Earth spins on its axis, objects should be hurled off it in the way a stone is hurled from a rotating sling. These things were not observed, and so for reasons that made sense at the time, Aristarchus's theory was rejected. It would be nearly 2000 years before a sun-centered theory would again be seriously considered.

Although the theory of multiple crystal spheres correctly described the wandering of the planets, another problem arose. The Greeks noticed that during a planet's retrograde motion, it appeared brighter than at other times, as though it were closer to Earth during this time. Yet the theory of multiple crystal spheres had each planet on an Earth-centered sphere, so that each planet maintained a fixed distance from Earth. This contradicted the observations of varying brightness.

To explain the varying brightness of the planets, the Greeks proposed a theory that was rather different from the crystal spheres but that retained the uniform circular motion favored by all Greek thinkers. The basic idea was that each planet moved around Earth in a circle within a circle. As shown in Figure 1.10, a planet such as Mars moved uniformly around a circle whose center was on another circle that was centered on Earth. The small outer circle was called the planet's *epicycle*, and the inner circle centered on Earth

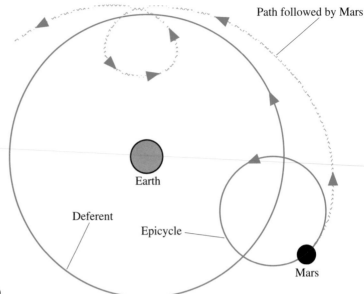

FIGURE 1.10
*The orbit of Mars around Earth,
according to the epicycle theory.*

was called the planet's *deferent*. The center of the epicycle moved uniformly along the deferent, so that Mars moved in two circles at the same time.

This produced a loop-the-loop orbit for each planet (Figure 1.10). In agreement with observation, the theory predicted that there would be occasional periods of retrograde motion (on the inside of the loops) and that the planet would be closest to Earth during retrograde motion and so should appear brightest. It was a satisfying picture, and it explained the observations. It was a good theory.

Figure 1.11 pictures the entire theory, although the figure and the preceding description are greatly simplified. First, in order to agree with the observations, each planet needed lots of epicycles—in fact, more than eighty "wheels within wheels" were needed. This theory was developed over a period of a few centuries and was finally refined and summarized in about A.D. 100 by Ptolemy, antiquity's greatest astronomer (Figure 1.12). In order to agree with the observations at that time, Ptolemy introduced two new ideas: the displacement or "eccentricity" of the centers and the "equant point" from which the motion appears uniform (Figure 1.13).* The details of these two ideas are not crucial here.

The important point is that Ptolemy devised a rather complicated, Earth-centered, mathematical theory that correctly predicted—to within the accuracy of the measurements then possible—the observed motions of the stars, sun, moon, and five known planets.

---

* None of these ideas was original with Ptolemy, but he was the first to put them all together in a consistent and quantitatively correct theory.

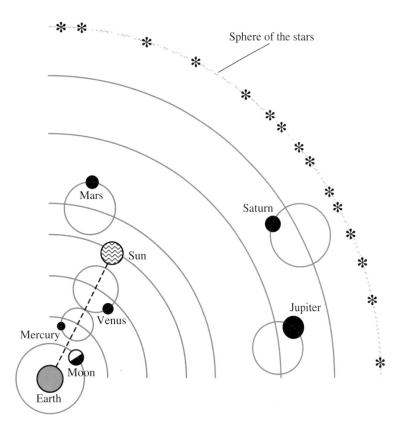

FIGURE 1.11
Ptolemy's Earth-centered epicycle theory (around A.D. 100) of the layout of the universe, according to which the five visible planets move on epicycles around Earth. The epicycles of the two innermost planets, Mercury and Venus, are centered on the line joining Earth to the sun.

Mars

Sphere of the stars

Saturn

Sun

Jupiter

Venus

Mercury

Moon

Earth

FIGURE 1.12
The ancient astronomer Ptolemy, A.D. 85–165.

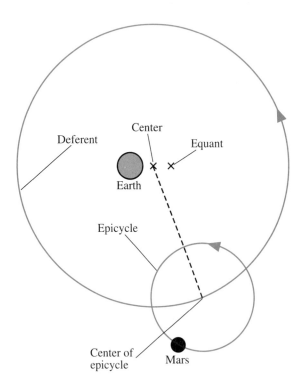

Deferent

Center

Equant

Earth

Epicycle

Center of epicycle

Mars

FIGURE 1.13
Explanation of the displaced center and the equant. Earth is displaced slightly from the center of the deferent. Furthermore, the center of the epicycle moves with unchanging speed, as seen from the "equant" and not from either the center or Earth.

Ptolemy checked his theory with many **quantitative** (numerical) measurements of the heavens. Because telescopes hadn't been invented yet, the measuring devices were long sighting rods with a scale to measure the angular position of a planet (or star). For example, one measured quantity was the number of angular degrees between the horizontal and the direction to the planet. The sighting devices were accurate to within about 0.2 degrees (recall that there are 360 angular degrees in a complete circle). Ptolemy's was the first theory to give such a complete, detailed, and quantitative account of all the known heavenly motions.

Ptolemy's theory predicted more details and was more accurate than any preceding theory. It survived, with modifications, for some fifteen centuries and was used by navigators, astronomers, and mystics such as astrologers. As the historian and philosopher of science Thomas Kuhn remarked, "For its subtlety, flexibility, complexity, and power the epicycle–deferent technique. . .has no parallel in the history of science until quite recent times. In its most developed form the system of compounded circles was an astounding achievement."

DIALOGUE 2   The ancient Greeks believed that the stars and other astronomical objects shine by means of their own light. Can they have believed this of every astronomical object?

DIALOGUE 3   Use Ptolemy's theory (Figure 1.11) to explain why the planet Venus often appears as the morning star (the last-seen "star" near the rising sun) or the evening star (the first-seen "star" near the setting sun).

1.4 Copernicus's theory: *a sun-centered universe*

The year A.D. 1543 is often taken as the birth year of modern science. In that year, an old man in Poland (Prussia, at that time), on his deathbed, signed the first printed copy of his life's work, *On the Revolutions of the Heavenly Spheres*. Nicolaus Copernicus (Figure 1.14)—astronomer, mathematician, linguist, physician, lawyer, politician, economist, and canon in a Catholic cathedral—had kept the manuscript locked up for thirty years, fearing the storm of criticism it would unleash.

FIGURE 1.14
Renaissance astronomer Nicolaus Copernicus, 1474–1543.

2. No. The moon's "phases" (new, crescent, quarter, and so on) show that it shines by means of reflected light from the sun.
3. The center of Venus's epicycle is fixed to the line joining Earth to the sun (Figure 1.11), which means that when viewed from Earth, Venus will never be seen far from the sun. Since Venus and the other planets are among the brightest astronomical objects and since Venus is never seen far from the sun, it is viewed as the morning star during those periods when it rises before the sun and as the evening star when it sets after the sun. This evening, or tomorrow morning, try to find Venus.

With this book, the sun was setting on the medieval world. During the Middle Ages (about A.D. 500 to 1500 in Europe), philosophers such as St. Augustine and Thomas Aquinas had linked Greek thought, including Ptolemy's astronomy, to Christian theology. But it was 1543 and the times were changing. During the past century the intellectual and artistic flowering known as the Renaissance had germinated in Italy and spread to all of Europe. Martin Luther had led a frontal assault on the Catholic church's authority. Christopher Columbus had made a memorable voyage.

These trends had a liberating effect on thoughtful minds. Copernicus and others were influenced by the new forms of art, new religious views, and exploration of faraway places. To begin with, Copernicus was uncomfortable with Ptolemy's theory. Not that Copernicus was a revolutionary. Quite the contrary: Copernicus's thinking resonated with the ancient views of Pythagoras. Copernicus objected to Ptolemy's theory on the grounds that with some eighty epicycles, plus all manner of eccentrics and equants, Ptolemy had strayed too far from the Pythagorean ideals. Ptolemy's system lacked the simplicity that scientists have always sought among nature's complex details. Copernicus put it this way:

> The planetary theories of Ptolemy and most other astronomers, although consistent with the data, seemed to present no small difficulty. For these theories were not adequate unless certain equants were also conceived; it then appeared that a planet moved with uniform motion neither on its deferent nor about the center of its epicycle. Hence a system of this sort seemed neither sufficiently absolute nor sufficiently pleasing to the mind.
>
> Having become aware of these defects, I often considered whether there could perhaps be found a more reasonable arrangement of circles, from which every apparent inequality would be derived and in which everything would move uniformly about its proper center, as the rule of absolute motion requires.

Note that like all thinkers since Pythagoras, Copernicus believed that the natural heavenly motions were both circular and uniform. He was adamant about this, stating that "the intellect recoils with horror" from any other suggestion and that "it would be unworthy to suppose such a thing [as noncircular motion] in a Creation constituted in the best possible way."

As an intellectual Renaissance man, Copernicus was more able than those before him had been to adopt a broad outlook and embrace a large part of nature. As Renaissance artists looked beyond Christian art and as Columbus looked beyond Europe, Copernicus looked beyond Earth itself to imagine it as an object in space, an object that he believed to be similar to other objects in space. To Copernicus, and to science since Copernicus, the ancient idea that the universe is centered on Earth seemed narrow-minded.

And so Copernicus was able to take a new point of view. He asked a broader question than had been asked before: What is the most elegant geometric scheme for the motion of the stars, sun, moon, five observed planets, and Earth that will fit the known measurements of the heavens? It was a crucial change of focus, a point of view that clashed with medieval intuitions and so had been inconceivable during the preceding 1000 years. Given this change of focus, Copernicus soon discovered "a more reasonable arrangement

If the Lord Almighty had consulted me before embarking upon the creation, I should have recommended something simpler.
Alfonso X, king of Leon and Castile, commenting on the Ptolemaic system, in A.D. 1200

In the centre of everything rules the sun; for who in this most beautiful temple could place this luminary at another or better place whence it can light up the whole at once?. . .In this arrangement we thus find an admirable harmony of the world, and a constant harmonious connection between the motion and the size of the orbits as could not be found otherwise.
Copernicus

Whether a man is on the earth, or the sun, or some other star, it will always seem to him that the position that he occupies is the motionless center, and that all other things are in motion.
Nicolas de Cusa, bishop of Brixen, 1450

of circles," a theory in which the visible planets and Earth move in uniform circular motion around the sun and only the moon circles Earth (Figure 1.15). As in previous figures, Figure 1.15 pictures only the slow wandering of the planets and moon relative to the stars. Copernicus obtained the east-to-west daily motion of the stars, sun, moon, and planets by allowing Earth to spin from west to east about a north–south line through Earth's center, rather than by allowing the sphere of the stars to rotate from east to west.

HOW DO WE KNOW?

There were still no telescopes in Copernicus's day, and data were still gathered with star-sighting devices. With properly chosen radii and rotation rates and eccentrics, Copernicus obtained quantitative agreement with the data. His theory explained many things, such as retrograde motion (see Figure 1.16 and Dialogue 5). But as Copernicus admitted, Ptolemy's theory was also "consistent with the data." Both theories agreed with the data.

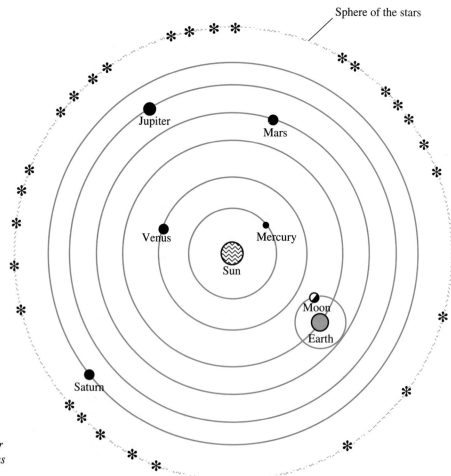

FIGURE 1.15

Copernicus's sun-centered theory of the layout of the universe, A.D. 1543. The diagram is simplified; the planets all move on epicycles, similar to those in Ptolemy's theory, but here there are far fewer epicycles. Furthermore, Copernicus used no equants.

FIGURE 1.16

The Copernican theory's explanation of retrograde motion. As Earth passes another planet, such as Mars, the other planet appears to move backward as seen against the background stars, because of the rotation of the Earth-based observer's line of sight. You can demonstrate this by following the instructions in Dialogue 5. A similar effect occurs when you pass a car moving down a straight highway. Viewed against distant background trees and houses, the slower car appears for a few seconds to move backward, because of the rotation of your line of sight.

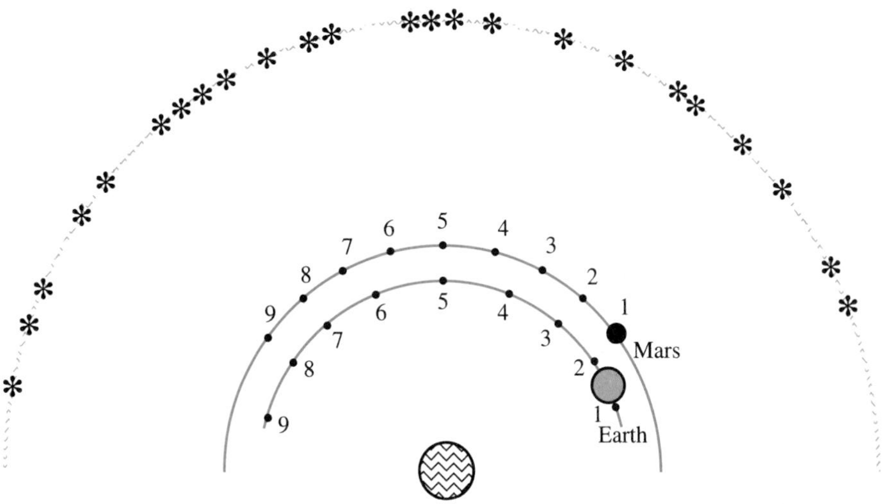

No one in his senses, or imbued with the slightest knowledge of physics, will ever think that the earth, heavy and unwieldy from its own weight and mass, staggers up and down around its own center and that of the sun; for at the slightest jar of the earth, we would see cities and fortresses, towns and mountains thrown down.

Jean Bodin, sixteenth-century
political philosopher

Despite its agreement with the data, there were good observationally based objections to the new theory, like those that had earlier confronted Aristarchus's theory. Being so large, how can Earth move? What keeps it moving? Why aren't birds and clouds left behind? Why aren't objects hurled off Earth? Copernicus did not really have an answer. Instead, he pointed out that such problems loomed even larger for Ptolemy's great spinning sphere of stars than for Copernicus's much smaller spinning Earth. Why don't the stars fly off Ptolemy's spinning sphere? What keeps Ptolemy's sphere of stars moving? In making this argument, Copernicus was assuming that the stars were subject to natural laws like those operating on Earth. Nobody had looked at it in this way before. But the new view is a natural one if you believe, as Copernicus did, that Earth is not so different from other astronomical objects.

Such questions as What keeps Earth moving? were not answered for more than a century, not until Isaac Newton and others devised a radically new view of motion, an idea known as the law of inertia. Copernican astronomy started science moving toward the physics of Isaac Newton. Newton's physics arose partly because of such questions.

The first decisive blow against Ptolemy's theory and in support of a sun-centered theory did not come until Galileo introduced the telescope into astronomy, some 70 years after Copernicus's death.*Among other things, Galileo observed that Venus goes through phases similar to the moon's phases (new moon, crescent moon, quarter moon, and the like). This means that Venus shines not by its own light but by light reflected from the sun. In Ptolemy's theory, the center of Venus's epicycle must be fixed on the line joining Earth to the sun (see Figure 1.11), in order to explain the fact that Venus is never seen far from the sun. As shown in Figure 1.17, this means that we should never be able to see a "full-Venus" phase

---

* Although Galileo did not invent the telescope, he was the first to make significant scientific use of it and the first to use it to study the heavens.

FIGURE 1.17

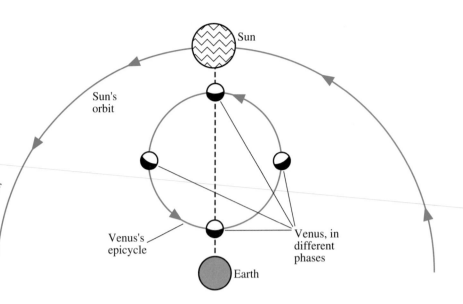

Ptolemy's theory predicted that an Earth-based observer would never see a "full" phase of Venus because Venus's epicycle lay between Earth and the sun. Copernicus's theory predicted that a nearly full Venus could be seen whenever Venus was on the far side of the sun in its orbit around the sun, as it is in Figure 1.15. But Galileo observed that the phases of Venus included a full Venus, thereby disproving Ptolemy's theory.

There is perhaps no other example in the history of thought of such dogged, obsessional persistence in error, as the circular fallacy which bedeviled astronomy for two millennia.
    Arthur Koestler, twentieth-century
        writer and historian of science

from Earth. On the other hand, the sun-centered theory predicts that we should be able to see a full-Venus phase whenever Earth and Venus are on opposite sides of the sun. Galileo observed that the phases of Venus included a full Venus.

With its eccentrics and thirty-four circles, Copernicus's new theory was still rather far from the elegant Pythagorean simplicity of eight crystal spheres. Furthermore, the observational data at that time gave no reason to prefer Copernicus's theory over Ptolemy's. Still, for Copernicus and for a few other intellectuals at the time, the new theory was an improvement because it dispensed with equants and with many epicycles. And the new planetary orbits were very close to being just simple circles around the sun, which means that for most purposes, one could use a simple circular model like Figure 1.15 to explain and predict astronomical events (Dialogues 5 and 6). Ptolemy's theory, on the other hand, would be wildly inaccurate if the epicycles were left out. Copernicus's theory was "approximately simple," so it seemed closer to the truth.

DIALOGUE 4   When we say that the sun "rises," what do we really mean, from the Copernican point of view?

DIALOGUE 5   Use Copernicus's theory to explain retrograde motion: On Figure 1.16, draw lines of sight from Earth through Mars to the background stars during the period when Earth is catching up with and passing Mars. Number the nine points (1 through 9) at which the line of sight intersects the background stars. How would Mars appear to move, as viewed from Earth against the background stars? Still using Figure 1.16, what direction would Mars appear to move during this period if you viewed it from the sun?

DIALOGUE 6   Use Copernicus's theory (Figure 1.15) to explain why
Venus often appears as the morning star or the evening star (compare
this with Dialogue 3).

1.5 Kepler's theory: *a sun-focused universe*

FIGURE 1.18
Tycho Brahe, 1546–1601.

Tycho Brahe (Figure 1.18), born three years after the death of Copernicus,
loved the night sky. A skillful fund-raiser, he obtained financing from the
king of Denmark to build a large astronomical observatory (Figure 1.19).

HOW DO WE KNOW?
Brahe's elegant sighting devices (Figure 1.20) were so accurate that his
data are sometimes still used today. Before Brahe, the best measurements
had inaccuracies (possible errors) of at least 10 arc-minutes (an arc-
minute is one-sixtieth of 1 degree). Brahe's measurements had inaccuracies of only 2 arc-minutes, a fivefold improvement.

When Brahe began his project, there were two competing theories of
the universe: Ptolemy's Earth-centered theory and Copernicus's sun-centered theory. Despite their great dissimilarity, both theories agreed with
the data known at that time. Would Brahe's measurements be able to distinguish between them and so determine which one was correct? For the
next twenty years, Brahe cataloged accurate data on the positions of the
sun, moon, and planets.

It soon became obvious that neither Ptolemy's nor Copernicus's theory
agreed with Brahe's data. Each made predictions that differed from the
observations by several arc-minutes—small differences but big enough for
Brahe to detect. This meant that both theories were flawed.

Just eighteen months before Brahe died, the 29-year-old Johannes Kepler
(Figure 1.21) managed to gain employment with the famous astronomer. Kepler was born to a ne'er-do-well father who abandoned his family when the boy
was 17 and to a mother who was later tried for being a witch. The boy was pre-
cocious above all in illness: smallpox, headaches, boils, rashes, worms, piles,
mange, and, worst of all for an aspiring astronomer, double vision in one eye
and myopia in both eyes. The hardships of his youth seem to have toughened
Kepler for the challenges to come. This philosopher-mathematician-astronomer-
astrologer was deeply devoted to the Pythagorean notion of an elegant mathematical order in the cosmos, and he harbored another devotion that could only
be described as sun worship. Regarding mathematical order, Kepler proclaimed:

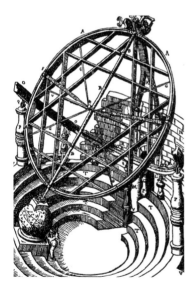

FIGURE 1.19
*Brahe's sextant for measuring the
positions of the planets.*

4. We mean that Earth has rotated eastward around its axis, to bring the sun into view.
5. The line of sight rotates backward for a brief time during passing, so when viewed from
   Earth against the background stars, Mars appears to move "backward" (in other words, opposite to its usual direction of motion). Viewed from the sun, Mars would appear to continue moving "forward" during this period.
6. The orbit of Venus is inside Earth's orbit, closer to the sun than Earth's orbit. So when
   viewed from Earth, Venus can never appear far from the sun. As explained in Dialogue 3,
   this means that Venus will sometimes be seen as the morning star and sometimes as the
   evening star.

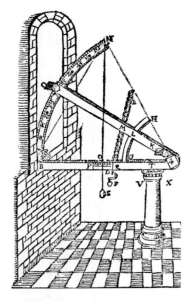

FIGURE 1.20

An instrument that Brahe used for measuring the angular altitudes of the planets. The two wooden arms are joined together. The lower arm is placed horizontally, as determined by the string and weight hanging vertically from the upper end of the scale. The upper arm is raised by turning the screw until the two sights point toward the planet. The planet's angular position above the horizontal is then read from the graduated scale. Brahe's work was done without telescopes.

Why waste words? Geometry existed before the Creation, is coeternal with the mind of God, is God Himself. . .; geometry provided God with a model for the Creation. . . .[And as for the sun:] The sun in the middle of the moving stars, himself at rest and yet the source of motion, carries the image of God the Father and Creator. . . . He distributes his motive force through a medium which contains the moving bodies even as the Father creates through the Holy Ghost.

Given his beliefs, it is not surprising that Kepler was the first professional astronomer to openly support the Copernican system, a theory whose beauty he contemplated with "incredible and ravishing delight." Kepler's words and thoughts convey the passion of the true scientist. This passion to know the universe is the driving force of science.

It is not surprising that such a man would seek out Brahe, who had made humankind's most intimate observations of the universe's architecture. Kepler deeply desired to read the mind of God, and Brahe's catalog might be the key. When he came to work with Brahe, Kepler had already spent many years pursuing the structure of the heavens, working on a beautiful scheme in which the six planets (including Earth) were attached to imaginary spheres, like the crystal spheres of the ancient Greeks but centered on the sun. He believed that the five intervals (gaps) between these six spheres were of such a size that these intervals could be exactly filled by using the surfaces of the five geometric shapes known as the *perfect solids* as "spacers" between the spheres.* He believed that the planets were arranged in such a way that this unseen collection of six spheres and five solid surfaces would precisely nest within one another, like boxes nested within other boxes, with the sun at the center (Figure 1.22).

HOW DO WE KNOW?

This idea first struck Kepler while he was teaching a class. He drew, for purposes unrelated to the layout of the planets, a blackboard diagram like that in Figure 1.23, in which an equilateral (equal-sided) triangle lies between two circles. He suddenly realized that the ratio of the sizes of the two circles was about the same as the ratio of the sizes of the orbits of Saturn and Jupiter. It was a real "eureka" moment for Kepler and no doubt an interesting moment for his students. The thought struck him that there were five intervals between the six planetary orbits and precisely five different perfect solids. Perhaps God had arranged things so that the perfect solids would just fit between the spheres of the planets! Kepler spent the next four years trying, unsuccessfully, to make this theory agree with the known data. Although he successfully pursued other theories after joining Brahe, Kepler always retained his enthusiasm for this idea, hoping to find further justification for it. But no further justification appeared. Kepler's alluring hypothesis was a scientific dead end.

---

* The perfect solids, or Pythagorean solids, are those flat-faced solid shapes whose faces are identical "regular polygons": two-dimensional (flat) figures bounded by straight lines of equal length. For example, a cube, bounded by six squares, is one of the perfect solids. A pyramid, bounded by four equilateral triangles, is another. The most complex of the five is the dodecahedron, bounded by twenty equilateral triangles. There are only five different perfect solids.

One disproof of Kepler's theory today is the fact that there are nine planets, not six. But even in Kepler's day, the only actual data supporting the theory were that it approximately predicted a few of the planetary ratios and that there were five perfect solids, the same as the number of intervals between the six planets. This could easily be, and turned out to be, a meaningless coincidence. Kepler's belief in this scheme borders on **pseudoscience**—the dogmatic and irrational belief in an appealing idea that appears scientific but that has little evidence to support it and much to dispute it.

FIGURE 1.21
Johannes Kepler, 1571–1630.

FIGURE 1.22
(a) Kepler's drawing of his model of the universe. The six planets move on six different spheres separated by the five perfect solids. The outermost sphere is Saturn's, then Jupiter's, and then Mars's. The remaining smaller spheres can be seen in (b), a detail of the central part of Kepler's illustration. The sphere of Mars is on the outside, then Earth, Venus, and Mercury, with the sun in the center.

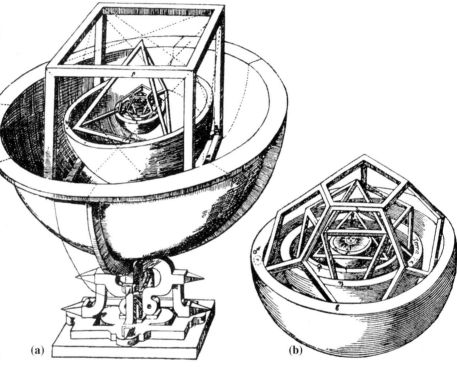

(a)   **(b)**

Kepler was a convinced Copernican when he came to Brahe. But he found that Copernicus's theory did not fit Brahe's data, despite playing with many variations on Copernicus's theory. Brahe's data for Mars proved the most difficult to fit, even though Kepler tried reintroducing the "equant" device that Copernicus had so despised. The calculations were tedious, for Kepler needed to translate Brahe's Earth-based data into sun-based data: How would the planetary positions appear if seen from the sun? Kepler spent four years on this project, filling nine hundred notebook pages with finely handwritten calculations.

The project failed. The Copernican orbit coming closest to Brahe's data for Mars was still off by 8 arc-minutes. Before Brahe, an error of 8 arc-minutes could have been ascribed to observational error. But Kepler, toughened by the confrontation with his master's hard-won data, knew that neither

FIGURE 1.23
A blackboard diagram similar to this gave Kepler the original inspiration for his planetary theory based on the five perfect solids. In this diagram, two circles are separated by a triangle.

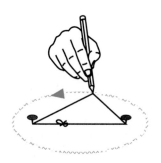

FIGURE 1.24
You can draw an ellipse with the help of a loop of string and two thumbtacks. The thumbtacks represent the two focuses.

I measured the skies, now the shadows I measure.
Skybound was the mind, earthbound the body rests.
Kepler's epitaph,
composed by Kepler

observational error nor further tinkering would make uniform circular motion agree with the observed facts. Kepler rejected the Copernican theory.

A less passionate person would have given up. Worse yet, a less tough-minded person would have found a way to fudge the data to get them to agree with the Copernican preconceptions that Kepler had believed most of his life. But the ever-fervent Kepler, writing, "on this 8-minute discrepancy, I will yet build a theory of the universe," began anew. He began studying planetary motions that, for the first time in history, were neither uniform nor circular. Copernicus, and all previous astronomers, would have been horrified.

Sixteen years later, Kepler found the precise and elegant scheme he sought. This theory is based on three geometrical principles, now called *Kepler's three laws*, describing the planetary orbits. The most significant law states that rather than moving in sun-centered circles, each planet moves in a sun-focused ellipse: an ellipse having the sun at one of its two "focuses." There is nothing at the other focus. Figure 1.24 shows how to draw an ellipse.* It is, roughly, a squashed circle. The planetary orbits are only slightly elliptical, which is why sun-centered circles come so close to fitting the observations.

The ellipse has just the kind of elegance Kepler had sought. He was elated:

> What sixteen years ago I urged as a thing to be sought, that for which I joined Tycho de Brahe. . .at last I have brought to light and recognize its truth beyond my fondest expectations. . . . The die is cast, the book is written, to be read either now or by posterity, I care not which. It may well wait a century for a reader, as God has waited six thousand years for an observer.

DIALOGUE 7   Who is the "observer" mentioned by Kepler in the preceding quotation? Kepler says that God has waited 6000 years. Why 6000?

DIALOGUE 8   Is a circle a particular example of an ellipse? Explain how to get a circle from the tack-and-string construction of Figure 1.24. Explain how to get a highly elliptical (very elongated) orbit from the tack-and-string construction.

1.6 The scientific revolution: *a dialogue between nature and mind*

The dynamic interplay between observations and theories is the essence of science. Figure 1.25 illustrates this dialogue with nature.

---

* Here is the exact definition: An ellipse is all the points for which the sum of the distances of each point on the ellipse from two fixed points (the two "focuses") is constant. The construction shown in Figure 1.24 follows this definition.
7. The observer is Brahe. In Kepler's time, most people believed the universe to be just a few thousand years old: Six thousand years is the figure that one gets from a "literal" reading of the Bible.
8. Yes, a circle is an ellipse for which the two focuses (the thumbtacks in Figure 1.24) are at the same point. For a highly elliptical orbit, move the thumbtacks far apart.

FIGURE 1.25
How we began to learn where we are in the universe.

A Summary of the Early History of Astronomy

Observations	Typical Dates	Theories
Stars, sun, moon, and planets are moving overhead.	3000 B.C.	
	500	Pythagorean theory: Earth-centered crystal spheres.
Each planet moves at a varying rate; retrograde motion.	400	Theory of multiple Earth-centered crystal spheres.
	300	Aristarchus's theory: sun-centered circles.
Heaven and Earth seem different; Earth seems motionless, apparently contradicting Aristarchus's theory.	200	
Planets are brighter during retrograde motion.	100	Theory of Earth-centered epicycles.
Detailed quantitative measurements show need for small corrections.	0	Ptolemy's theory: Earth-centered epicycles, equants.
	A.D. 100	
	1500	Copernicus's theory: sun-centered circles.
Brahe's accurate measurements disprove Ptolemy's and Copernicus's theories.	1600	Kepler's theory: sun-focused ellipses.
Galileo's telescopic observations disprove Earth-centered theories.		

The scientific method: doing one's damnedest with one's mind, no holds barred.

Percy W. Bridgman,
in *Reflections of a Physicist*

Observation is the beginning of the scientific process. But a catalog of observed facts does not add up to an understanding of nature, any more than a telephone book adds up to an understanding of a city. To understand, literally to "stand beneath," means to perceive a framework. A framework of scientific ideas is called a theory. In the development of astronomy, observations stimulated speculations that led to theories, and these theories in turn suggested new observations to check the theories and suggest new speculations. The key to science is the presence of both observation and theory.

Observation refers to the data-gathering process and has more or less the same meaning as measurement and experiment. A **measurement** is a

Physical theory without experiment is empty. Experiment without theory is blind. It is the experimentalists who keep the theorists honest.

Heinz Pagels, physicist

quantitative observation, and an **experiment** is an observation that is designed and controlled by humans, perhaps in a laboratory.

The word *theory* comes from a Greek word meaning "to see." A scientific **theory** is a framework of ideas that explains or unifies a group of observations. It is a way of understanding, a way of seeing, the observations. Theory has more or less the same meaning as principle, law, and model. A **model** is a theory that can be visualized, and a principle or law is one idea within a more general theory. The word *law* can be misleading because it sounds so absolute and certain. As we will see, no scientific idea is ever certain.

Kepler's theory makes a good example. Figure 1.26 shows the general form of Kepler's model of the solar system (the sun and its planets) extended to include all nine known planets.

Kepler's theory explains all of Brahe's data and all preceding observations, in the sense that we can deduce (predict) all these observations from Kepler's theory. The theory also unifies all these data in a few principles such as the principle of elliptical orbits. Kepler's theory represents an enormous simplification or "reduction" of many observations into a few simple ideas.

But Kepler's theory does more than describe known data. It also predicts new observations that were unknown to Kepler. For example, when new planets (Uranus, Neptune, Pluto) were discovered, Kepler's theory predicted, correctly, that they too would move in elliptical orbits. A theory having no predictive value, that needs to be patched up to account for every new observation, is not worth much. For example, Ptolemy's theory could doubtless be amended with enough new epicycles to make it agree with all of Brahe's data, but the result would be a confusing patchwork with little ability to predict, say, the orbits of Uranus, Neptune, and Pluto.

Even more important, Kepler's theory is able to suggest further developments. Isaac Newton, born a few years after Kepler's death, made important use of Kepler's theory in his theories of force, motion, and gravity. Without Kepler, Newton's more sweeping theories would have been impossible.

How valid, and how certain, is scientific knowledge? There are two common misconceptions about this. One is the belief that scientific knowledge is absolute, or certain. The other is the opposite belief, that scientific knowledge is just dubious guesswork.

Concerning the first misconception, that scientific knowledge is absolute, let us look at history. Ptolemy's theory correctly predicted the planetary observations during and after his time. So did Copernicus's theory. Both theories were useful reductions of the planetary motions to a few ideas, and both had predictive value.* Both are correct today to within some 10 arcminutes. They were, and are, good theories for many purposes.

But new observations led to improvements. Brahe's data contradicted both theories, opening the way for Kepler's theory. Kepler's theory had predictive value, predicting, for example, the phases of Venus and the shapes of the orbits of Uranus, Neptune, and Pluto. Did Kepler, then, discover the true motion of the planets?

---

* Neither Ptolemy's 80 deferents and epicycles nor Copernicus's 34 could be said to constitute a very simple way of representing the data, but these theories were certainly simpler than the volumes of data themselves.

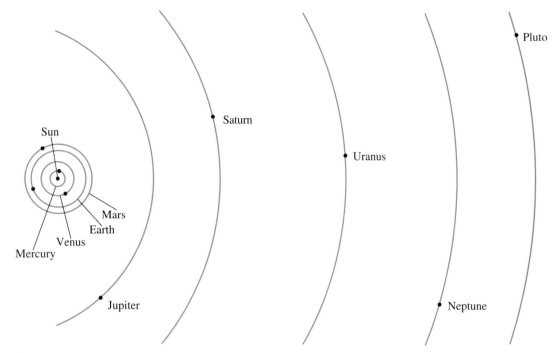

FIGURE 1.26
The arrangement of the solar system as it is now known. Uranus, Neptune, and Pluto are visible only with a telescope. The orbits are elliptical, although their ellipticity is too small to be visible in this diagram.

Not necessarily. Tomorrow night, astronomers might discover that the planets have begun severely deviating from their elliptical paths! In fact, we will see later that observations have since shown that the planets do deviate, although only slightly, from precise elliptical paths. It is always possible that new data will contradict any general theory. We cannot even be absolutely certain that the sun will rise (in other words, that Earth will continue spinning as it has) tomorrow! Good science is always provisional, nondogmatic. Its ideas dangle by the slender thread of evidence, a thread that cannot prove a theory but that can disprove it with a single contradictory observation.

In fact, if a theory is not capable of being disproved, not "falsifiable," by any conceivable observation, then it is not a scientific theory at all, because it tells us nothing about the observable universe. Scientific theories must be testable by observations that could conceivably contradict the theory. For example, a notion such as "undetectable alien creatures are living among us" is not a scientific statement, not because this notion seems odd (most scientific theories are odd), but because the creatures are said to be undetectable. Scientifically, this idea is not true and it is not even false. Being untestable, it is outside science.

Nonscientific ideas can, of course, have their own validity. "Beethoven's music is sublime," or "I love you," or "May God bless this home," can be meaningful statements, but they lie outside science.

The aim of science is not to open the door to everlasting wisdom, but to set a limit on everlasting error.
Bertolt Brecht, playwright,
in *The Life of Galileo*

Science can purify religion from error and superstition, while religion can purify science from idolatry and false absolutes.
Pope John Paul II, 1988

Extensive and highly accurate planetary observations continue to be made today, observations that can be used to test, and possibly disprove, Kepler's theory. And in fact, strictly speaking, Kepler's theory actually has been disproved by such observations! Observations made since Kepler's day show that the planets move along orbits that deviate slightly from precise ellipses. According to Isaac Newton's theories (see Chapters 3, 4, and 5), Kepler's elliptical orbits are caused by gravity acting between the sun and each planet. The main cause of the deviations from elliptical motion is gravity acting between the different planets and between each planet and its moons. These deviations cause the planets to wobble a little as they orbit the sun. Interplanetary dust and many other things also cause smaller deviations. Nevertheless, scientists have retained Kepler's theory because it is a good and useful approximation. *True* is usually not as appropriate a word to apply to theories as are words like *good* and *useful*.

Let's turn to the other misconception, namely, the belief that theories border on dubious guesswork. As one example of this belief, some people who disliked the Copernican theory on religious or other grounds argued that because the idea that Earth circles the sun was a "mere theory," it need not be taken seriously. Today, people who dislike the theory of evolution attack it on similar grounds. Our culture often disparages the word *theory* as though a theory were just a wild guess. Perhaps this is a sign of anti-intellectualism ("against thought") in our culture. At any rate, theories—coherent mental frameworks for what we observe—are what science is all about.

The fact that theories are not absolutely certain is a strength, not a weakness, of science. Absolute certainty can foster dogmatism and a rigid inability to change what needs changing. As we noted, theories can be good, useful, fruitful, or compelling, but the phrase "absolutely true" is inappropriate to scientific theories.

A plausible suggestion that is not yet well confirmed is called a **hypothesis**. Kepler's hypothesis that the five perfect solids act as spacers between the six planetary orbits turned out to be false. But because Brahe's data confirmed his hypothesis of elliptical orbits, that hypothesis was elevated to the status of a theory.

The history of astronomy shows that science thrives on creativity. Creativity shows up in the elegant tools of Brahe and the inspired theories of Pythagoras and Kepler. It is one of nature's mysteries that these beautiful inventions actually turn out to produce a consistent picture of the universe. As Albert Einstein once put it, "The most incomprehensible thing about the world is that it is comprehensible."

Science's ideas are not so much discovered as created. Guided by Brahe's data, Kepler invented elliptical orbits. Ptolemy and Copernicus created two very different models, yet in their day, both were accurate. It is possible, even today, to conceive of the sun and planets as "really" going around a stationary Earth, although such a theory would need very many epicycles indeed and would require some unusual inventions to explain the phases of Venus.* Given a choice among theories that agree with the observations, scientists opt for the theory that is simplest, least arbitrary, most fitting.

---

* In order to maintain such a contorted theory, we would also need to change the law of inertia (Chapter 3) and thus nearly all of present-day science!

Scientists generally believe in the Pythagorean ideal of a universe based on simple and elegant principles. Copernicus adopted a sun-centered theory over the hallowed Earth-centered theory because it was "pleasing to the mind." Scientists such as Kepler strove passionately to perceive such an elegant framework. When creating his theories, Einstein used to ask himself how he would have constructed the universe if he were God.

But science is more than elegant inventions. Theories must agree with the data, predict future observations, and be falsifiable. This interplay between theory and observation has led humankind to a deeper and broader understanding of nature. For example, Copernicus's theory helped inspire Brahe's observations, and both of these then led to Kepler's vastly improved theory, which in turn was the groundwork for Newton's grand theories.

All in all, the scientific method comes down to the careful observation of nature and an open-minded, creative search for general ideas that agree with and predict those observations. Scientific procedures are not really different from the procedures that usually work best in coping with daily life: careful observation, creative and open-minded thinking based on observations, and actions based on both. We will return frequently to the theme of how science operates.

DIALOGUE 9   Which theory, Ptolemy's or Copernicus's, proved fruitful for Kepler?

DIALOGUE 10   "Contrary to scientific opinion, I know that the universe was created just 200 years ago. Historical records, old-looking trees, fossils, and so forth that make Earth appear more than 200 years old are misleading. Actually, all these things were created just 200 years ago in order to fool the human race into believing the universe is more than 200 years old." Can you prove this statement wrong? Comment.

DIALOGUE 11   William claims to be absolutely certain of some idea. Can you then conclude that William's idea is right? Wrong? Crazy? Irrelevant? Is William being scientific? Unscientific?

DIALOGUE 12   "Because Darwinian evolution is only a theory, we need not take it seriously." Comment?

1.7 The Copernican revolution: *dawn of the modern age*

The scientific age has its roots in two historical developments. One is the Pythagorean idea of "natural harmonies" or "natural law," an idea that captivated influential Greek philosophers such as Plato and Aristotle and that then spread to Europe and the world. Its central premise is that the universe

---

9. Kepler was led to his theory by starting from Copernicus's theory.
10. This is a nonscientific statement because it cannot be falsified.
11. William is being unscientific.
12. Theories are precisely what we should take most seriously in science.

is organized in a framework of elegant and precise ideas that can be uncovered by observation of the natural world. The other historical development is the rejection of the "geocentric illusion" that humankind is central to the universe. Copernicus started this development, which is justly called the **Copernican revolution**. Its general idea is that we on Earth are not special; just as Earth is a planet similar to the other planets, nature is pretty much the same everywhere, differing perhaps in details at different places and times but always following the same general principles.

Others began thinking along these lines. It became apparent to them that the sun—considered by Copernicus and Kepler to be central to the universe—was a star like the other stars. We now know that the visible stars belong to a vast revolving aggregation of over 100 billion stars spread over a giant pancake-shaped region of space. Our sun is one of the stars in the outreaches of this aggregation, and it circles the center every 200 million years. But the center of this aggregation is not at the center of the universe, either. Instead, there are hundreds of billions of other similar aggregations of stars throughout the universe. And according to current theories, none of these aggregations is at the center, because the universe has no center (Chapter 11 describes this odd idea). This is the ultimate extension of Copernican astronomy.

Each of these giant aggregations of stars is called a **galaxy**. Ours is the **Milky Way galaxy**. Figure 1.27 shows a typical distant galaxy, one much like our own. If the galaxy in Figure 1.27 were our galaxy, our sun would be about two-thirds of the way out from the center toward the edge of the photograph. The cloudlike glow in the night sky that is called the Milky Way is our galaxy seen from our position within it. The glow comes from the stars in only a small, local portion of our entire galaxy. The center of our galaxy lies far beyond the visible Milky Way, in the direction of the constellation (group of stars) known as Sagittarius.

How vast those Orbs must be, and how inconsiderable this Earth, the Theatre upon which all our mighty Designs, all our Navigations, and all our Wars are transacted, is when compared to them. A very fit consideration, and matter of Reflection, for those Kings and Princes who sacrifice the Lives of so many People, only to flatter their Ambition in being Masters of some pitiful corner of this small Spot.
Christian Huygens, physicist, around 1690

FIGURE 1.27
The Andromeda galaxy, photographed through a telescope. This is the nearest other large galaxy, outside our own Milky Way galaxy. Like our galaxy, Andromeda is made of billions of stars, each one somewhat similar to our sun, whirling around a bright star-filled center. Andromeda is nearly invisible to the unaided eye and lies far beyond the visible stars of our own galaxy. Since light takes more than 2 million years to reach here from there, you are looking at 2-million-year-old history.

FIGURE 1.28

A photograph of a region of sky containing many galaxies. This is a portion of a cluster of galaxies that contains more than eight hundred galaxies, each one somewhat similar to our Milky Way galaxy. Light takes 500 million years to reach here from this cluster.

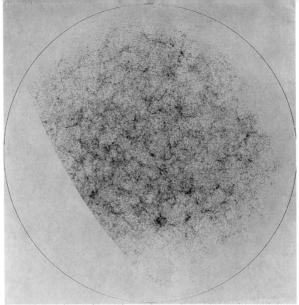

FIGURE 1.29

A computerized plot of the locations of more than a million known galaxies. On such a large-scale plot, the pattern of galaxies looks like a thin film of cobwebs. The lower left-hand part of the plot is blank because this region of the distant universe is obscured by the gas and dust of our own Milky Way galaxy.

Telescopic surveys of galaxies give some indication of the size of our universe. Figure 1.28 is a photograph of a few of the galaxies in a distant region of the sky containing hundreds of galaxies. Figure 1.29 is a computerized plot of the locations of over a million known galaxies, still just a tiny fraction of the galaxies in the universe.

The Copernican idea sowed the seeds of many revolutions. Aristotle had developed an Earth-centered physics with two sets of natural laws, one for Earth and one for the heavens. Once Copernicus announced that Earth is a planet, Isaac Newton could unify the heavens and Earth in a new physics based on principles that acted uniformly throughout the universe. And once Copernicus dispelled the geocentric illusion, Charles Darwin could conceive an evolutionary biology that unified all life and included humankind as one species among many other species. The Copernican/Newtonian conception of natural laws that apply democratically everywhere and to all people is an important component of the political transition from medieval authority to constitutional law and democracy.

Copernican astronomy was correctly perceived as revolutionary by religious and philosophical authorities. Ptolemy's system had been developed in parallel with Earth-centered Aristotelian physics, and Aristotle's thinking was a foundation of Catholic theology. The perfection of heaven, the imperfection of Earth, and humankind's centrality to God's plan for the universe were threatened by the loss of the Ptolemaic system. It is not surprising that Copernicus, prudent by nature, withheld publication of his ideas until his dying day.

Seventy years after Copernicus's death, the Catholic church pronounced his theory "false and erroneous," "altogether opposed to Holy Scripture," and "heretical." Science historians believe that during this period, science and religion fell out into two noncommunicating camps that still, today, see themselves as being at odds with each other. This situation is a far cry from the Pythagoreans who considered science and religion to be the same thing. Champions of the Copernican theory such as Kepler, Galileo, teacher and writer Giordano Bruno, and Copernicus's colleague and friend Rheticus were denounced and persecuted by religious authorities.*

The leaders of the Protestant Reformation were even more extreme in denouncing the new astronomy. The main Protestant objection was that the new theory ran counter to a "literal" reading of the Bible, in other words, a reading that interpreted the Bible as an accurate history and science text rather than, as many others argued, a spiritual work not meant as precise history or science. The Bible frequently mentions a moving sun and a fixed Earth, contrary to the Copernican theory. Even before publication of the new theory, the Protestant leader Martin Luther heard about Copernicus's ideas and violently condemned them for contradicting the Bible. In Luther's opinion, "The fool [Copernicus] will turn the whole science of astronomy upside down. But, as the Holy Writ declares, it was the sun and not the Earth which Joshua commanded to stand still."

Humanity has perhaps never faced a greater challenge; for by [Copernicus's] admission [that humanity is not the center of the universe], how much else did not collapse in dust and smoke: a second paradise, a world of innocence, poetry and piety, the witness of the senses, the conviction of a religious and poetic faith. . .; no wonder that men had no stomach for all this, that they ranged themselves in every way against such a doctrine.
Johann Wolfgang von Goethe, nineteenth-century German poet and dramatist

JOHANNES KEPLER'S UPHILL BATTLE

---

* In 1984 the Vatican stated that church officials had erred in condemning Galileo and called for increased dialogue between science and religion. Then in 1992, the pope announced that the church had wrongly accused Galileo, laying the blame on seventeenth-century church authorities who interpreted the Bible too literally.

Summary of Ideas and Terms

Science The observation and theoretical understanding of the natural world.

Physics The branch of science that studies the most general principles underlying the natural world.

Astronomy The scientific study of the stars and other objects in space.

Astrology The belief, rejected by science for over two centuries, that the stars have a significant direct influence on human affairs.

Solar system The sun and the objects that orbit the sun. It includes the nine planets and the moons that orbit many of the planets.

Uniform circular motion Motion in a circle at an unchanging or uniform speed.

Pythagoreans The earliest scientists. A society of ancient Greeks who believed that the fundamental principles of the universe were precise mathematical ideas.

Pythagorean theory of the universe The earliest Greek theory. The stars, sun, moon, and each of five planets circle Earth on uniformly spinning, transparent, Earth-centered spheres to which each is attached.

Retrograde motion A temporary change in the direction that a planet moves relative to the stars, as seen from Earth.

Multiple crystal spheres theory An extension of the Pythagorean theory, to account for the planets' "wandering." Each planet has several transparent, Earth-centered spheres, with the planet riding on the outermost sphere.

Aristarchus's theory A sun-centered theory that was rejected because it seemed to conflict with everyday observations.

Ptolemy's theory An Earth-centered theory in which the planets move in circles within circles, or loop-the-loops. It was a good theory that survived for 1500 years.

Copernicus's theory A sun-centered theory, similar to Aristarchus's. The planets, including Earth, circle the sun, and Earth spins on its axis.

Tycho Brahe's observations Highly accurate data on planetary positions that disproved both Ptolemy's and Copernicus's theories.

Kepler's theory The planets orbit the sun in ellipses with the sun at one focus. This theory agrees with Brahe's observations.

Pseudoscience The dogmatic and irrational belief in an appealing idea that appears scientific but that has little observational support.

Theory An idea, or several related ideas, that explains or unifies a group of observations. A **model** is a theory that can be visualized. A **principle** or **law** is a single idea, often within a larger theory.

Hypothesis An educated guess, a tentative theory.

Observation The fact-gathering process. A **measurement** is a quantitative observation, and an **experiment** is a controlled observation.

Copernican revolution The rejection of the idea that humankind is central to the universe.

Galaxy A large aggregation of stars. Most galaxies, such as our own Milky Way, have a disklike, pancake shape and revolve about their centers.

Review Questions

Each chapter has review questions and exercises at the end. Review questions go over many of the essentials; you can find the answers by looking in the chapter. Exercises ask you to draw on your understanding of the material to answer new questions not directly answered in the chapter. The answers to the odd-numbered exercises are in the back of the book. The many exercises allow your instructor flexibility; you will probably want to tackle only some of them.

SCIENCE AND THE NIGHT SKY

1. What two reasons does this chapter give for studying science?
2. What is physics?
3. Distinguish astronomy from astrology.
4. What astronomical objects can you normally see in the night sky? Describe their motion as seen from Earth.

ANCIENT GREEK ASTRONOMY

5. What did the Pythagoreans believe, and how did these beliefs influence the development of science?
6. According to the earliest Greek theory, the planets orbit Earth in uniform circular motion. In what way does this theory disagree with simple observations made without telescopes?
7. Give two observational reasons for believing that Earth is a sphere.
8. How does Ptolemy's theory explain the retrograde motion of the planets and the fact that planets are brighter during retrograde motion?

9. Did Ptolemy's theory agree with the quantitative observations known in Ptolemy's time? How were these observations made?

COPERNICUS'S IDEA

10. "Copernicus rejected Ptolemy's theory because it disagreed with the data, and he proposed a new sun-centered theory that did agree with the data." True or false? Explain.
11. Use Copernicus's theory to explain the retrograde motion of the planets and the fact that they are brighter during retrograde motion.
12. Why did Copernicus propose his theory?
13. State at least one plausible argument *against* the notion that Earth moves around the sun.
14. How did new telescopic evidence decisively disprove Ptolemy's theory?

KEPLER'S IDEA

15. What is pseudoscience, and in what way does Kepler's earliest theory of the planetary orbits border on pseudoscience?
16. "Kepler was attracted to Copernicus's theory because the known data supported that theory." True or false? Explain.

17. Describe Brahe's work and its effect on the theories of Copernicus and Ptolemy.
18. What aspect of Kepler's theory would have horrified all previous astronomers?
19. According to Kepler's theory, what geometric shape fits the planetary orbits?

THE SCIENTIFIC AND COPERNICAN REVOLUTIONS

20. What is the most characteristic and significant feature of science?
21. Describe several characteristics of a good scientific theory.
22. Can a scientific theory be proved (can we show that the theory is certainly true)? Can it be disproved? Explain.
23. Strictly speaking, Kepler's theory has been disproved. What has been found wrong with it? Why, then, do we still use it?
24. How does a hypothesis differ from a theory?
25. Distinguish between the Copernican theory and the Copernican revolution.
26. In what sense can evolutionary biology be said to be "Copernican"?

Home Projects

There is nothing like hands-on experience in science. Try some of these at home.

1. On a clear night, follow the stars across the sky. Look at the moon, the seven stars in the big dipper (Figure 1.3), the North Star (Figure 1.3), any group of stars that starts on the eastern horizon, and one or two stars that start on the western horizon. Try to observe them every 15 minutes for least 1 full hour. Describe your observations.
2. Observe the moon each night for a few weeks. Begin with a new crescent moon in the west, near the setting sun. Look at the moon at the same time each evening after sunset. On paper, draw its appearance (crescent, quarter, or whatever) each night and its position relative to the setting sun. After about two weeks, you will no longer see the moon just after sunset. Where is it? When does it rise and where? Continue recording your nightly observations until the moon has returned to its initial position near the setting sun. How many days did the complete process take? Explain why the moon had the appearance it did on different nights. Draw a diagram showing the relative positions of the sun, moon, and Earth; when the moon is new; when it is quarter full; and when it is full.

3. Find one of the visible "outer planets" (preferably Mars, but Jupiter or Saturn will do) that is high in the night sky at some convenient time such as after sunset. A member of your local astronomy or physics department might help you locate one. On one night, observe the planet every 15 minutes for an hour or more. During that time interval, does the planet appear to stay in step with the east-to-west motion of the stars? Does the planet move along the zodiac—the path followed by the sun and the moon? Follow the planet during a longer time by observing it once each night for a few weeks. Does the planet remain in step with the stars, or do your observations show that it moves relative to the stars? About how many days does it take for the planet to noticeably shift its position relative to the stars? Does it move east to west faster than the stars do or slower? Is it following its normal motion or its retrograde motion?
4. Find Venus. It is often either the morning or evening "star." A member of your local astronomy or physics department might help locate it. Verify that Venus is a

planet and not a star, by noting its changing position relative to the stars over a few weeks. Your observations of Venus should verify that it stays close to the sun in the sky. What is the explanation of this observation, according to Ptolemy's theory (Figure 1.11) and Copernicus's theory (Figure 1.15)?

5. Together with members of your class, lay out a scale-model solar system. Begin with a ball, about 23 cm in diameter (the size of a bowling ball), to represent the sun. Then lay out nine objects to represent the planets, as shown in the accompanying table. The model stretches over 983 meters, nearly a kilometer. We could choose a smaller scale, but then planets such as Earth and Mercury would be even smaller than a pinhead. The solar system is essentially all empty space.

To represent	—use an object of about this diameter		—placed this many paces (meters) from the preceding object
Sun	23 cm	(bowling ball)	
Mercury	1 mm	(pinhead)	10
Venus	2 mm	(peppercorn)	8
Earth	2 mm	(peppercorn)	7
Mars	1 mm	(pinhead)	13
Jupiter	2.4 cm	(chestnut)	92
Saturn	2.0 cm	(acorn)	108
Uranus	9 mm	(peanut)	240
Neptune	8 mm	(peanut)	271
Pluto	0.5 mm	(pinhead)	234

For Discussion

The discussion questions are meant to stimulate thinking about questions that have no specific correct answer. They can be used for class discussion, individual essays, or individual thought.

1. Discuss the similarities and differences between a scientific theory such as Kepler's and a work of art such as a symphony.
2. In light of the history in this chapter, speculate on what might have happened if Copernicus had never lived. Would we still believe today that Earth is at rest at the center of the universe? Would somebody else have proposed a similar theory? If so, then when might this have happened—just a few years after 1543 (when Copernicus published his theory), a century after 1543, or several centuries after 1543? When considering this, keep these historical details in mind: Aristarchus, Brahe, the Renaissance.
3. List ten ways in which science and technology have directly or indirectly improved your life. Now list ten ways in which they have worsened your life.

Exercises

Exercises indicated by asterisks are quantitative, although they require no algebra. The answers to the odd-numbered exercises are in the back of the book.

SCIENCE AND THE NIGHT SKY

1. How can you tell, from direct observation alone, whether or not a particular object in the sky is a planet?
2. Draw a diagram showing the positions of Earth, the moon, and the sun at new moon, crescent moon, nearly full moon, and full moon.
3. Are the stars in Figure 1.4 circling clockwise or counterclockwise? A time-lapse photograph made in the Southern Hemisphere, looking toward the South Pole, would also show the stars moving in a circle around a fixed point in the southern sky. Would the stars in the southern view be circling clockwise or counterclockwise?

ANCIENT GREEK ASTRONOMY

4. Describe an unaided-eye observation you could make to disprove the theory that the planets orbit Earth in a simple uniform circular motion.
5. Describe an unaided-eye observation you could make to disprove the theory that the planets orbit Earth attached to transparent spheres that rotate in a complicated fashion but that are always centered on Earth.
6. In seeking an explanation of retrograde motion, why didn't the Greeks just allow the planets to change their speed and direction of motion as the planets moved along circular paths around Earth, instead of resorting to circles within circles?

COPERNICUS'S IDEA

7. Is it possible that on some evenings, the planet Mars is the evening star? Is this very likely? (See Figure 1.15.)

8. Use Copernicus's theory to predict whether Mars goes through moonlike phases. Do we ever see a "full Mars"? A "new Mars"?

9. It is possible, but difficult, to see the planet Mercury with the unaided eye. How, then, would you go about finding it?

KEPLER'S IDEA

10. Which aspects of Kepler's theory would Copernicus have liked? Disliked?

11. Would Kepler's theory have agreed with the data available in Ptolemy's time? In Copernicus's time?

12. Did Brahe's data *prove* that planets move in ellipses? Explain.

13. Is there anything in Kepler's theory that resembles the displaced centers of Ptolemy and Copernicus?

THE SCIENTIFIC AND COPERNICAN REVOLUTIONS

14. What is the most important and characteristic feature of science?

15. Can two different theories both be true in the sense that at some particular time in history, they correctly predicted the known data? Defend your answer with a historical example.

16. "If Earth is curved, it must have a spherical shape, because a sphere is the most perfect curved solid form." Does an aesthetic argument like this have any place in science?

17. A sensationalist tabloid "news"paper carries this headline: "SCIENTISTS PREDICT THAT THE UNIVERSE AND EVERYTHING IN IT WILL DOUBLE IN SIZE AT THE BEGINNING OF THE NEXT NEW YEAR!" Is this a testable hypothesis? If so, how could you test it, and if not, why not? Is this good science, bad science, or neither?

18. What should be your attitude toward claims by proponents of astrology, dianetics, extrasensory perception (ESP), a 6000-year-old universe, creationism (belief in the literal truth of Genesis), visitations by extraterrestrials, the Bermuda Triangle, and pyramid power?

19. Aristotle, a careful observer of living organisms, wondered where the material that contributes to the growth of a plant comes from. He hypothesized that all of it comes from the soil. Based on your knowledge of biology, do you consider this hypothesis to be correct? Propose an experiment to test this hypothesis.

20. Some people believe that plants will grow better if they talk to them. Is this a testable hypothesis? If so, propose an experiment to test it.

21. "Certain people are gifted with extrasensory perception (ESP), such as the ability to move material objects with their own minds. However, ESP is so delicate that every attempt to verify it always destroys it." Is this a scientific hypothesis?

22. Isaac Newton predicted that because of its spinning motion, Earth would bulge out near the equator and be flattened near the poles. In 1735 the French Academy of Sciences sent an expedition to the Arctic to measure the exact shape of Earth. When they returned, reporting the predicted results, the philosopher Voltaire mocked them with the following couplet:

To distant and dangerous places you roam
To discover what Newton knew staying at home.

Was Voltaire's sarcasm justified? Why or why not?

23.*Since there are some 100 billion stars in a typical galaxy and since there are at least 100 billion galaxies in the known parts of the universe, how many stars are there in all? Write out this number.

24.*Figure 1.29, representing 1 million galaxies, is only a tiny fraction of the 100 billion galaxies in the known parts of the universe. How tiny? Write your answer as a decimal number.

2

ATOMS
the nature of things

We turn now from stars to atoms. One of science's key ideas is that everything is made of imperceptibly small particles. This explains an extraordinary range of observations, and it deeply influences our culture's perception of reality. As we will learn in later chapters, science profoundly changed its view of these "atoms" during the twentieth century, a development crucial to one of this book's four themes:* comparisons between Newtonian and post-Newtonian physics.

Section 2.1 presents the 2500-year-old idea that everything is made of small particles. Section 2.2 discusses the atom as chemists have seen it for the past 200 years and distinguishes atoms from molecules. Following a brief excursion in Section 2.3 into metric units and large and small numbers, Section 2.4 explores a tiny portion of the tremendous explanatory power of the atomic idea. Section 2.5 views the amazingly small size of atoms, and Section 2.6 considers the philosophical implications of the atomic idea, a topic important to our theme of comparisons between Newtonian and post-Newtonian physics. Section 2.7 looks ahead briefly to compare this chapter's so-called Greek model of the atom with two more modern models presented in later chapters. Finally, Section 2.8 explores several examples of chemical reactions, or recombinations of atoms.

2.1 The Greek atom: *the smallest pieces* _____

The ancient Greeks produced an astonishing number of profound and original thinkers. These philosophically minded people wanted to think their way

_____

* The three other themes are the methods of science, social impacts of physics, and energy.

to the bottom of everything. Because of their pro-intellectual attitude, this small group of people, during just two centuries centering on the fifth century B.C., laid the foundations for most of the Western world's great ideas. Among the things they thought about was the nature of matter: material substances such as wood, cotton, sausage, ice, water, soil, and gold. They speculated about the underlying unity that they believed lay behind the different substances. What do sausage and gold have in common? What is matter?

Leucippus and his student Democritus sharpened their focus on this question with a "thought experiment," an imagined experiment that seemed possible in principle but was difficult to carry out in practice.* Suppose, they argued, you cut a piece of gold in half. After the division, each part would still be gold. How far could you continue making such divisions? Either the divisions could go on forever, or there would be a limit at which no further divisions would be possible. That is, matter is either "continuous," divisible without limit, or else it is made of particles that cannot be divided. The first alternative seemed absurd to them. Matter, they concluded, is made of imperceptibly small, "a-tomic" (Greek for "not divisible"), particles. They called these smallest particles **atoms**.†

DIALOGUE 1   Would you classify this idea as scientific fact, experimental observation, hypothesis, theory, principle, or law? [Dialogue questions like this appear throughout the book. Try to answer them as you come to them, before looking at the answer at the bottom of the page.]

This turned out to be an influential and useful idea. We will call it:

THE ATOMIC THEORY OF MATTER
All matter is made of tiny particles, too small to be seen.

This idea is a good example of a scientific principle or law or theory: an idea or group of ideas that explains a broad range of observations. In Leucippus's and Democritus's time, the atomic idea not confirmed by observations; rather, it was an educated guess or hypothesis. When observations confirmed this idea during the nineteenth and twentieth centuries, it then became an established theory (or principle or law). A general idea like this cannot, however, be labeled as an observation or fact, no matter how often observations may have confirmed it, because we cannot observe every possible material object to prove for certain that everything really is made of atoms. Theories are never certain.

---

* The Greek philosopher Epicurus later incorporated and developed the ideas of Leucippus and Democritus. During the first century A.D., the Roman poet Lucretius elaborated the ideas of the Greek atomists in his long poem *On the Nature of Things*.

† *Atom* is used in a slightly different sense today. Today, *atom*, or *chemical atom*, refers to the smallest particle of a chemical element. This chemical atom is actually divisible, being made of smaller, subatomic parts: electrons, protons, and neutrons, and in fact the protons and neutrons are themselves made of quarks. As far as we can tell today, electrons and quarks are the ultimate, smallest particles, of which the others are made. These smallest possible parts are what the Greeks meant by an atom.

1. In Leucippus's and Democritus's time, this was a hypothesis, but today it is an established theory. It is incorrect to call a general idea, such as this one, a fact or observation.

If, in some cataclysm, all of scientific knowledge were to be destroyed, and only one sentence passed on to the next generations of creatures, what statement would contain the most information in the fewest words? I believe it is the atomic hypothesis (or the atomic fact, or whatever you wish to call it) that all things are made of atoms—little particles that move around in perpetual motion, attracting each other when they are a little distance apart, but repelling each other upon being squeezed into one another. In that one sentence. . .there is an enormous amount of information about the world.

Richard Feynman, physicist

What do your thumbnail, beer, the Brooklyn Bridge, an acorn, Saturn, the Amazon River, and the period at the end of this paragraph have in common? Each is made of atoms, the unity beneath the diversity of materials and the kind of underlying unity that the Greeks loved and that scientists seek.

HOW DO WE KNOW?

Science's power comes from its insistence on evidence. So we frequently ask, How do we know? This "show me" skepticism characterizes much of science.

How do we know everything is made of small unseen particles?

Although the ancient Greeks had no direct microscopic evidence for atoms, Democritus did have some ingenious indirect evidence. He argued, for instance, that since we can smell a loaf of bread from a distance, small bread particles must break off and drift into our noses; these bread particles were, he proposed, related to the bread's atoms. This is still an acceptable explanation of odors today (see Section 2.4). The medieval Christians rejected the atomic theory because they associated it with atheism. But scholars in Islamic countries preserved Greek culture, including the atomic theory, during the Middle Ages, and they then reintroduced these ideas to the Christian West. In this way, the atomic principle became part of the background for the physics of Galileo, Newton, and others. But even then, the idea had no direct observational support, and many scientists did not believe in atoms.

The first specific evidence for atoms was discovered around 1800 by the chemist John Dalton. Dalton discovered that when certain substances combine chemically to form other substances, they always combine in simple ratios by weight. For example, when hydrogen and oxygen combine to form water, the ratio of the weights of the two substances is always 1 to 8.

The occurrence of such simple ratios is not easy to understand based on the hypothesis that matter is infinitely divisible. But if matter is made of atoms, then there is a simple explanation: If, for example, 1 atom of hydrogen and 1 atom of oxygen have a simple weight ratio and if these atoms always combine in simple ratios to create water, the weight ratios of the hydrogen and oxygen in water will be simple numbers also. Today we know that individual atoms of hydrogen and oxygen have a weight ratio of 1 to 16 and that it always takes 2 hydrogen atoms for every 1 oxygen atom to form water. So we can see, today, why the weight ratio should be 1 to 8.

So the atomic theory explains the simple ratios that Dalton observed. But does this prove the theory? The answer is no! No, because it is possible that atoms do not exist and that there is some other explanation, or even no explanation, for the simple ratios. A general idea cannot be proved by observations, but it can be made more plausible.

A few decades later, a botanist, Robert Brown, using a microscope, observed that tiny pollen grains suspended in liquid move around erratically (Figure 2.1), even though the liquid itself had no observable motion. His first hypothesis was that the grains were alive. But lifeless dust grains suspended in liquid executed the same erratic dance, disproving

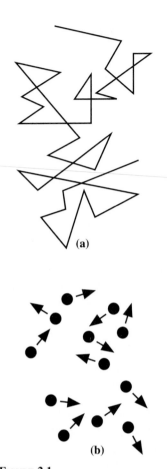

FIGURE 2.1

Brownian motion. (a) This erratic pattern is the typical path of a small particle such as a single dust grain, suspended in water, observed under a microscope. (b) Brownian motion is explained by the ceaseless, rapid, random motion of water molecules. Although the molecules are far too small to be seen even under a microscope, the effect of their impact on a dust grain can be seen in the erratic motion of the dust grain.

the hypothesis. Hypotheses and theories can be disproved, although they cannot be proved.

What unseen cause lay behind this "Brownian motion"? Toward the end of the nineteenth century, it was suggested that submicroscopic (smaller than could be seen with a microscope) motions of atoms caused Brownian motion. The idea was that atoms were in constant motion and that the dust grains were jostled by atoms (Figure 2.1).

This hypothesis got strong support in 1905 from an unknown young physicist named Albert Einstein. Einstein's contribution was on the theoretical, not the observational, side. He used an already established theory to calculate the details of the jostling of particles such as dust grains—particles that are much larger than atoms—when bombarded randomly (from all directions) by moving atoms.* He made several quantitative (numerical) predictions, such as the rate at which a collection of dust grains—if they all are bunched at one point to begin with—should spread out, or **diffuse**, because of this random jostling in a liquid. Predictions like this could be, and soon were, checked by experimental measurements. The measurements agreed with Einstein's predictions. It was difficult to dispute this quantitative evidence. Either unseen atoms really were the cause of Brownian motion, or else Einstein's calculations were extremely lucky in just happening to give all the right numbers. Since Einstein's work, scientists have not questioned the atomic theory.

DIALOGUE 2 An individual sulfur atom has twice the weight of an individual oxygen atom. When sulfur and oxygen combine to form sulfur dioxide, 1 sulfur atom is required for every 2 oxygen atoms. What is the weight ratio of sulfur and oxygen in the formation of sulfur dioxide? What is the weight ratio in the formation of sulfur *tri*oxide, having 1 sulfur atom for every *3* oxygen atoms?

2.2 Atoms and molecules

Think of all the different material substances around you: this paper, your shirt, your hair, and so forth. You could list hundreds. Nineteenth-century chemists found that most of them could be "decomposed" into a much smaller number of simpler substances but that this small number could not be further decomposed by chemical means. Any process that changes a single substance into two or more other substances is called a **chemical decomposition** of the original substance.

For example, water can be decomposed† into two quite distinct substances, called hydrogen and oxygen, neither of which are anything like water. But hydrogen and oxygen turn out to be among that small (fewer than 100) group of substances that nineteenth-century chemists could not decompose. No

---

* That theory was the kinetic theory of heat, the theory that atoms move because of their thermal energy. Some of these topics will be discussed in Chapters 6 and 7.

2. [Are you reading this before forming your own answer? If so, do you exercise by watching somebody else jog? Exercise your mind by providing your own input to the dialogues!] 1 to 1. 2 to 3.

† By passing an electric current (Chapter 8) through water.

matter how they tried to decompose hydrogen, it remained hydrogen, and the same was true for oxygen.

Apparently, these roughly 100 substances that cannot be chemically decomposed are particularly fundamental. They are called **chemical elements**, or simply elements.

By studying the weight ratios just discussed, John Dalton and others soon recognized that the different elements corresponded to different kinds of atoms. The idea was that each element was made of only one kind of atom and that elements differed because their atoms differed. That is, oxygen was made entirely of oxygen atoms, hydrogen was made entirely of hydrogen atoms, and so forth. But water is "compounded" of two kinds of atoms, hydrogen and oxygen, which is why you can decompose water but not oxygen or hydrogen.

Today, 109 elements are known. About 90 occur naturally on Earth and about 20 others are made in laboratories. The table on the inside back cover of this book gives the name and standard abbreviation for each element and numbers them. Each number, known as the element's **atomic number**, represents a particular kind of atom. Higher atomic numbers correspond roughly (with some exceptions) to heavier atoms. We will discuss the physical meaning of an atom's atomic number in Section 8.7.

Scientists found that certain groups of elements have similar chemical properties. If we list these groups vertically in order of increasing atomic number and also list the elements horizontally in order of increasing atomic number, the result is the *periodic table* shown on the inside of the back cover. This table is a nice example of the regularities that scientists find in natural phenomena. When many elements had still not been discovered, scientists used the periodic table as a predictive device, by noting the table's unfilled gaps and searching for elements with properties that just fit those gaps.

What about all the other substances, those made of more than one element? A pure substance, such as pure water (with no impurities like salt or dirt mixed in), that is made of more than one element is called a **chemical compound**.

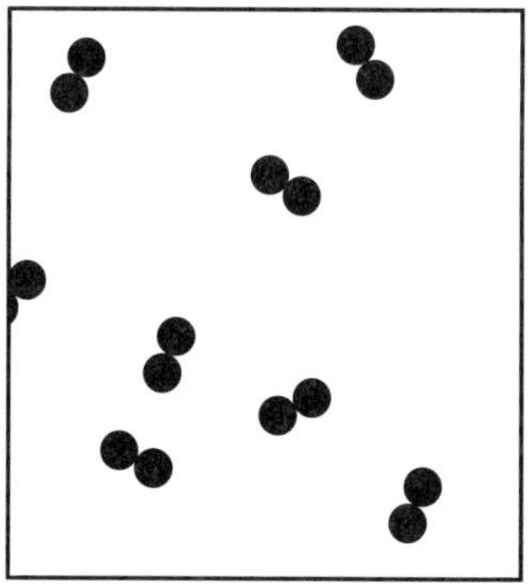

Imagine dividing a cup of water into smaller and smaller amounts. If the water is pure, you will always get just water—not something else such as salt or dirt. Working downward in size, you will eventually arrive at the smallest particle of water. In pure water, every one of these smallest particles must be a particle of water, and so they should be identical. And every particle must contain atoms of both hydrogen and oxygen, because we know that water can be chemically decomposed into these elements. This smallest particle that still has the characteristics of water is called a *molecule* of water.

This reasoning shows that every pure chemical compound must be made of tiny particles that are identical and that are themselves made of two or more kinds of atoms attached together into a single identifiable unit. Such a particle, the smallest particle of a compound that still has the characteristics of that compound, is called a **molecule** of that compound.

Some elements are made of 2-atom molecules. For example, a molecule of hydrogen gas is made of 2 hydrogen atoms (Figure 2.2).* Helium gas, on

FIGURE 2.2

A simplified drawing of hydrogen gas, magnified a billion times. Each molecule of hydrogen is made of 2 hydrogen atoms. Each molecule moves rapidly in a nearly straight line, changing direction only when it collides with another molecule or with the container wall. This and other microscopic drawings view only a tiny region within a much larger container.

---

* However, outside Earth's atmosphere nearly all the universe's hydrogen is in the "atomic" (single-atom) form rather than the "molecular" (2-atom) form, because the universe began with atomic hydrogen. Each hydrogen atom has been relatively isolated ever since, and so these atoms have not had a chance to combine into molecules.

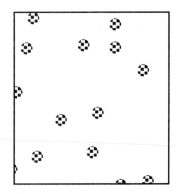

FIGURE 2.3

A simplified drawing of helium gas, magnified a billion times. Each molecule of helium is simply an unattached atom of helium.

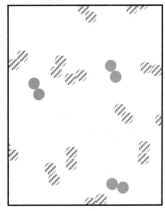

FIGURE 2.4

A simplified drawing of air, magnified a billion times. Air is a mixture mostly of nitrogen (striped), and oxygen (green) molecules, both 2-atom molecules. About 80% of all air molecules are nitrogen, and about 20% are oxygen.

the other hand, is made of individual helium atoms (Figure 2.3). The common forms of oxygen gas and nitrogen gas are also made of 2-atom molecules, and air is made primarily of these two kinds of molecules (Figure 2.4).

Chemists can deduce a compound's molecular structure by means of chemical decomposition and experiments like Dalton's. Such experiments show, for example, that the water molecule is made of 2 hydrogen atoms and 1 oxygen atom (Figure 2.5).

We represent compounds and elements by abbreviated formulas. For example, water is represented by H_2O, where the subscript 2 belongs to the symbol preceding it and indicates the number of atoms of that type in each molecule. Hydrogen gas is represented by H_2, oxygen by O_2, and helium by He.

Molecules can get pretty complicated. You might expect the molecules of life, such as your protein and DNA molecules, to be highly complex because they must carry the coded information needed to organize your body. Biological molecules are among the most varied and complicated known. Hemoglobin, the protein responsible for the red color of blood, has the formula $C_{3023}H_{4816}O_{872}N_{780}S_8Fe_4$. DNA molecules contain millions of atoms and vary from one individual to the next. DNA carries the instructions that make you you.

DIALOGUE 3 Name four other elements that have chemical properties similar to those of chlorine. Of these five similar elements, which is made of the lightest-weight atoms?

DIALOGUE 4 The simple sugar known as glucose has the chemical formula $C_6H_{12}O_6$. What elements, and how many atoms of each, does a glucose molecule contain?

DIALOGUE 5 What is the chemical formula for carbon dioxide (1 carbon and 2 oxygen atoms)? Carbon monoxide (*mono*-oxide)? Methane (carbon and 4 hydrogens)? Sulfur dioxide? Carbon tetrachloride ("tetra" means "four")?

2.3 Metric distances and powers of 10 _____

Because we will soon need to use specific units of measurement for distance and volume, this is an appropriate place to introduce many of the units of measurement that we'll be using in this book. And because we'll soon deal with some very small and very large numbers, we need to discuss a topic known as powers of 10.

---

3. Fluorine, bromine, iodine, astatine (in the same column in the periodic table with chlorine). Fluorine is the lightest.
4. 6 carbons, 12 hydrogens, 6 oxygens.
5. CO_2, CO, CH_4, SO_2, CCl_4.

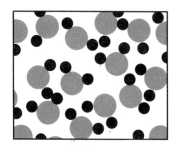

FIGURE 2.5
A simplified drawing of liquid water, magnified a billion times. As in the preceding figures, all the molecules are in rapid motion. In a liquid the molecules are in close contact and slide past one another. Each water molecule is made of 1 oxygen atom (green) and 2 hydrogen atoms (black).

Benefits to U.S. industry if it converts to metric usage. . .can be summed up in one word: Survival. Overseas countries already are refusing entry to some U.S. inch–pound goods,. . .our industries will lose the buying power of 320 million people in the European Community (EC) if we don't wake up and begin producing to the EC metric standards. . . .U.S. industry must convert to metric production if it wants to survive.
Valerie Antoine, executive director of the U.S. Metric Association

TABLE 2.2
Metric Prefixes

Mega (M)	one million,	10^6
Kilo (k)	one thousand,	10^3
Milli (m)	one-thousandth,	10^{-3}
Micro (μ)	one-millionth,	10^{-6}

TABLE 2.1
Metric Distances

Name of unit	Distance	Conversion to "English" units
Kilometer (km)	1000 m = 10^3 m	1 km = 0.62 mi, 1 mi = 1.6 km
Meter (m)		1 m = 3.3 ft = 39 in, 1 ft = 0.30 m
Centimeter (cm)	0.01 m = 10^{-2} m	1 cm = 0.39 in, 1 in = 2.5 cm
Millimeter (mm)	0.001 m = 10^{-3} m	
Micrometer (μm)	0.000001 m = 10^{-6} m	

Unfortunately, U.S. students must learn two systems of measurement: the nationally used "English" system based on feet and pounds and so forth, and the metric system that everybody else uses.* Because not even England uses it any longer, we will refer to the feet-and-pounds system as the "English" (in quotes) system.

The basic metric distance unit is the **meter** (abbreviated m). It is about 39 inches, a little over 3 feet. Table 2.1 lists other metric distances and relates them to the meter. The most important are the **kilometer** (km), the **centimeter** (cm), and the **millimeter** (mm). Table 2.2 lists four common prefixes that can be attached to any metric unit: **mega-** (M), **kilo-** (k), **milli-** (m), and **micro-** (μ). For example, the kilowatt is 1000 watts, and the megawatt is 1 million watts (we'll later see what a watt is).

A **power of 10** means 10 raised to some power. For instance, 10^2 means 10×10, which equals 100, and 10^5 means $10 \times 10 \times 10 \times 10 \times 10 = 100,000$. The superscript is the power to which 10 is raised, the number of 10s that are multiplied together.

Powers of 10 are good for handling large and small numbers. For example, the solar system's diameter (distance across) is 12,000,000,000,000 m. You can write this as $1.2 \times 10,000,000,000,000$ m or as 1.2×10^{13} m. Each multiplication by 10 moves the decimal point one place to the right, so to write out 1.2×10^{13}, you begin with 1.2 and move the decimal point 13 places to the right. The number in front (the 1.2) is usually written as a number between 1 and 10. If there is no number in front, you can think of a 1 in front; for instance, 10^5 is the same as 1×10^5.

Negative powers are used for small numbers. For instance, 10^{-2} means $1/10^2$, which equals 1/100, or 0.01, and 10^{-5} means $1/10^5 = 0.00001$. The minus sign indicates that the power of 10 is to be divided into 1. For example, the diameter of an atom is about 0.000 000 000 11 m, which can be written as 1.1×10^{-10} m. Since each division by 10 moves the decimal point one place to the left, to write 1.1×10^{-10}, you begin with 1.1 and move the decimal point 10 places to the left.

Thousand (10^3), **million** (10^6), **billion** (10^9), and **trillion** (10^{12}) all represent various powers of 10. Similarly, thousandth (10^{-3}), millionth (10^{-6}), and so forth represent negative powers of 10.

To multiply two powers of 10, just add their powers. For instance, $10^2 \times 10^5 = 10^{2+5} = 10^7$, and $10^2 \times 10^{-5} = 10^{2+(-5)} = 10^{-3}$. The numbers in front

---

* Burma also uses the "English" system.

of the power of 10 must be grouped together first, before multiplying. For example,

$$(1.5 \times 10^2) \times (3 \times 10^5) = (1.5 \times 3) \times (10^2 \times 10^5) = 4.5 \times 10^7$$
$$(1.5 \times 10^2) \times (3 \times 10^{-5}) = (1.5 \times 3) \times (10^2 \times 10^{-5}) = 4.5 \times 10^{-3}$$

To divide two powers of 10, subtract the denominator's power from the numerator's power. For instance, $10^2 / 10^5 = 10^{2-5} = 10^{-3}$, and $10^2 / 10^{-5} = 10^{2-(-5)} = 10^7$. Numbers in front of the powers of 10 are grouped together first, before dividing them. As an interesting example, the solar system's diameter divided by an atom's diameter (the ratio of the two diameters) is

$$1.2 \times 10^{13} \text{ m} / 1.1 \times 10^{-10} \text{ m} = (1.2/1.1) \times (10^{13} / 10^{-10}) = 1.1 \times 10^{13-(-10)}$$
$$= 1.1 \times 10^{23}$$

This number, 110,000,000,000,000,000,000,000, is the number of atoms you would have to line up side by side in a row in order for them to stretch across the solar system. It is difficult to calculate such numbers, or even to write them down, without using powers of 10.

Powers of ten are something that all of us need to be able to handle. You need these ideas just to begin to understand the U.S. national debt!

DIALOGUE 6   (a) Convert 6 miles to yards (there are 1760 yards per mile). To inches. (b) Now convert 6 kilometers to meters. To centimeters. Would you rather do this sort of thing in "English" or metric units?

DIALOGUE 7   The universe is only seconds old, a million trillion seconds old, in fact. Write this number in ordinary notation and in powers of 10. The diameter of an atomic nucleus is about a hundredth of a trillionth of a meter. Write this in powers of 10.

2.4 The atom's explanatory power: *the odor of violets*

What are smells? How do they reach you? When you smell violets, say, do tiny bits of violets propel themselves into your nose? Is smell an invisible wave? Magic? These are good questions to introduce us to the power of the atom idea and of careful observation.

Observation shows that we are surrounded by an invisible substance, air. You know it is there because you can feel the wind. The atomic theory tells us every material substance, anything that you can pick up or touch, is made of atoms. It is reasonable to suppose that air is a material substance, too, because you can feel it blow on you. A careful measurement would show that air has weight, further confirming our hypothesis. We conclude, from the atomic theory, that air is made of atoms. As we know from the Brownian motion experiment, the atoms in a liquid move all the time, even when

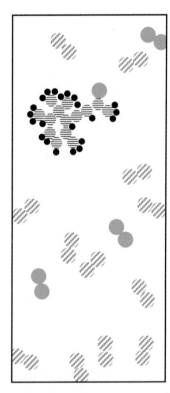

FIGURE 2.6
The odor of violets, in air. The odor-of-violets molecules are made of carbon (horizontal stripes), hydrogen (black), and a single oxygen atom (green).

6. (a) 6 miles = 10,560 yards = 380,160 inches. (b) 6 km = 6000 m = 600,000 cm.
7. 1,000,000,000,000,000,000 seconds, or 10^{18} seconds; 10^{-14} m.

FIGURE 2.7

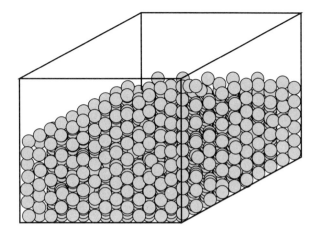

If you toss a large number of balls one by one into a box, the balls often lock into an orderly pattern. For similar reasons, the atoms in many solid materials lock into an orderly "crystal pattern" at the microscopic level.

the liquid appears to be motionless. So it seems reasonable to suppose that air molecules are in constant motion, too, even in still air.

Violets, too, must be made of tiny particles. Among a violet's various molecules, there must be some that make it smell the way it does. Chemists have learned how the violet's odor molecule is strung together (Figure 2.6). Scientifically, at least, that's what the smell of violets is—those molecules. In order for a violet's smell to spread out, odor molecules must break loose from the violet. Once in the air, moving air molecules knock odor molecules around, just like the Brownian motion of dust particles in water. This random jostling causes the odor molecules to spread out, to diffuse, in all directions in the air. Eventually, they reach your nose. Think about it the next time you smell something.

The atomic theory links human-scale or macroscopic phenomena that we can see around us to phenomena at the unseen **microscopic** level.* One case in which the microscopic perspective is especially enlightening is the states of matter. Water, for example, comes in three states: ice, liquid, and steam. We call these the **solid state**, **liquid state**, and **gas state** of water. Nearly every substance can exist in any one of these three states. For instance, iron is a solid at ordinary temperatures but becomes a liquid at sufficiently high temperatures and becomes a gas at still higher temperatures.

At the macroscopic level, the three states of matter are distinguished by their shapes. In a closed container, a solid maintains its shape; a liquid spreads out over the bottom; and a gas fills the container. How do they differ at the microscopic level? Simple macroscopic observations, the atomic theory, and some thought can lead us to a microscopic picture.

In solids, molecules must be locked into a fixed arrangement in order to maintain a fixed shape. And they also must be crowded against one another, because observation tells us that it is difficult to compress a solid into a smaller volume. The precise arrangement is determined by the ways in which the substance's molecules push and pull on one another when they get close together. If you have ever seen a large number of balls tossed one by one into a big box (Figure 2.7) or gunshot (BBs) filling a small container, you

FIGURE 2.8

An unseen microscopic order lies behind the beautiful symmetries seen in macroscopic crystals such as these hexagonal quartz crystals.

---

* *Submicroscopic* would be a better term for atoms, because atoms are too small to be detected by light-based microscopes.

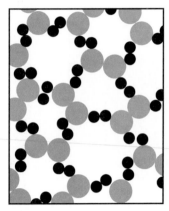

FIGURE 2.9

Ice. Compare this diagram of solid water with that of liquid water in Figure 2.5. In the solid state, atoms vibrate around their average position in the crystal pattern, but they do not migrate throughout the material, as water molecules do in the liquid state.

FIGURE 2.10

The hexagonal symmetry we see in snowflakes mirrors the hexagonal symmetry of the underlying crystal structure (compare this with Figure 2.9).

can guess that molecules tend to lock into an orderly pattern that repeats itself throughout the solid. Figure 2.7 is, in fact, a reasonable representation of the arrangement of molecules in many solid substances. Orderly molecular patterns are responsible for the regular surfaces and beautiful symmetries seen in macroscopic crystals (Figure 2.8). For example, Figure 2.9 shows the microscopic six-sided crystal pattern of ice, and Figure 2.10 shows the macroscopic snowflake crystals that are formed from it.

Most substances take up about the same volume, whether in a liquid state or a solid state. For example, when ice melts, its volume changes by only 8%. So a liquid's molecules must be in close contact with one another, somewhat as they are in a solid. Another indication that a liquid's molecules are tightly packed together lies in the fact that liquids are extremely difficult to compress (reduce their total volume).

Figure 2.11 pictures the differences among solids, liquids, and gases. The big difference between solids and liquids is that liquids have no fixed shape, indicating that their molecules are not rigidly attached to one another. At the microscopic level, a liquid is like a jumbled bowlful of marbles that can be molded into different shapes. The molecules in a liquid can slide past one another and are free to migrate throughout the liquid (Figure 2.11b). Most liquids take up more volume than do the solids of the same substance, because their molecules need room to move around. Water, however, happens to be an exception to this rule; because of its open crystal structure (Figure 2.9), ice takes up more volume than does liquid water.

Gases are easy to compress into a much smaller total volume, indicating that their molecules are widely separated. From the microscopic point of view, the widely separated molecules dart back and forth, bouncing off the container's walls or colliding and rebounding from one another (Figure 2.11c). From this microscopic picture, we would expect that a gas should press outward against any surface it contacts. This outward press, or **gas pressure**, is caused by a continual torrent of gas molecules hitting the surrounding surfaces (Figure 2.12). It is as though hundreds of baseballs were thrown at a wooden wall, pressing the wall backward. You can see the effect of gas pressure when you fill a balloon or tire with air. A balloon's elastic material is pressed outward by the trillions of gas molecules hitting the walls every second.

A complete absence of air and all other forms of matter is called a **vacuum**. A perfect vacuum is impossible to achieve in any ordinary macroscopic volume on Earth, but it is not difficult to achieve a partial vacuum, in which the container holds far less air than it would if filled with air at its normal density. A good vacuum in a laboratory still contains a trillion molecules in every cubic centimeter! But the large regions between the galaxies are nearly perfect vacuums, containing an average of only 1 molecule in every 8 cubic meters—the molecules are some 2 meters apart!

How does a warm substance differ from a cool substance, at the microscopic level? In order to answer this important question, let's consider the process of evaporation of a liquid. Imagine a pot of water from a microscopic point of view. It is reasonable to suppose that if some of the more rapidly moving water molecules are moving toward the surface, they might

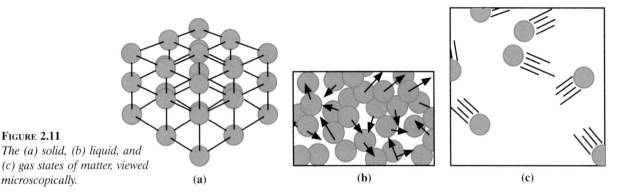

FIGURE 2.11

The (a) solid, (b) liquid, and (c) gas states of matter, viewed microscopically.

(a)   **(b)**   **(c)**

FIGURE 2.12

A gas exerts an outward push, or "pressure," on the walls of its container, because of the effect of trillions of unseen gas molecules hitting every portion of the wall. The drawing shows a microscopic portion of a box full of gas. At any particular time, a few gas molecules are hitting the inner wall, causing pressure.

FIGURE 2.13

A microscopic view of the evaporation of water. The water is evaporating into the air, which is a mixture of O_2 and N_2 molecules. Compare this with Figures 2.4 and 2.5.

escape into the air above. This is what happens during **evaporation** (Figure 2.13). This scenario leads to a further conclusion: Those liquids with faster-moving molecules should evaporate faster than do liquids with slower-moving molecules, because faster-moving molecules are more likely to break through the surface and escape into the air above.

It is an experimental fact that warmer liquids evaporate faster than colder ones do. You can easily verify this for yourself (see the Home Projects at the end of this chapter).

From the preceding two paragraphs, it is reasonable to hypothesize that the molecules in a warmer liquid move faster than do the molecules in a cooler liquid. This idea, proposed during the nineteenth century and now amply confirmed, gives us a microscopic picture of what happens when a liquid is warmed: Its molecules move faster. Most of a liquid molecule's motion is a sort of high-speed jitter around a fairly stationary average position, a jitter that increases as the temperature is raised.

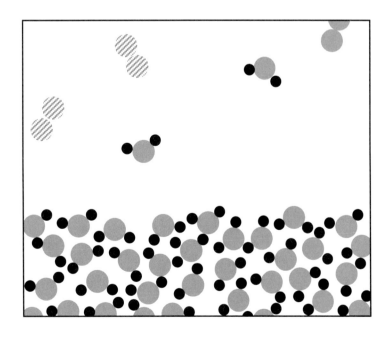

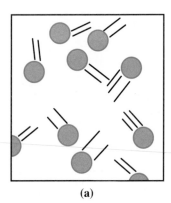

(a)

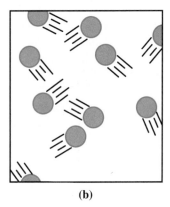

(b)

FIGURE 2.14
The microscopic meaning of "warmth": The difference between a cooler gas (a) and an identical but warmer gas (b) is that the molecules move faster in the warmer one.

We would expect that gases behave in the same way, that when they are warmed, their molecules would move around faster (Figure 2.14). Many observations verify this idea. For example, it is commonly observed that the pressure in an automobile tire increases on a hot day and also when it is driven for some time at highway speeds. The hot day, or the driving, increases the tire's temperature so that the air molecules inside move faster and hit the inside walls of the tire harder, and the tire's pressure increases. That's why it is easier to blow out a tire on a hot day or when you drive fast.

The general idea is that temperature is associated with microscopic motion. This microscopic motion cannot be directly seen macroscopically. Even in a motionless glass of water or in still air, the molecules are jittering rapidly. The speed of a typical air molecule is about 1800 kilometers per hour (1100 miles per hour)! But you do not feel this motion directly because the molecules move in all directions and result in no overall macroscopic motion of the air. The microscopic motion is disorganized, mixed up, random. You detect this random molecular motion only as temperature and gas pressure.

Random microscopic motion occurs in solids, too. Each molecule in ice, for example (Figure 2.9), jitters or vibrates randomly around a fixed position in the crystal. These vibrations increase as the temperature increases.

We can summarize this idea in the following way:

> THE MICROSCOPIC INTERPRETATION OF WARMTH
> At the microscopic level, temperature (warmth) is the disorganized, jittering motion of a substance's molecules. This motion cannot be directly observed macroscopically but is observed instead as temperature or warmth. When temperature increases, microscopic motion also increases.

The atomic theory unifies the odor of violets, diffusion, solids, liquids, gases, evaporation, crystals, chemical compounds, gas pressure, and temperature. The atomic theory unifies these and much more. It is a good theory.

DIALOGUE 8   Give several observations that confirm that we are surrounded by air.

DIALOGUE 9   Can you suggest an experiment that would show that air has weight?

DIALOGUE 10   What if, in addition to its random molecular motion, all the air molecules in some volume of air had an overall "collective" motion, all of them moving, say, eastward. Could this collective motion be observed macroscopically? What would we call it?

---

8. Trees bending in the wind, leaves blowing in the wind, air entering and leaving your nose, wind blowing a candle flame, an air-filled balloon.
9. Weigh two identical rigid containers, one containing air and one that has had some of its air removed. If the air has weight, the air-filled container should weigh a little more.
10. It can be observed as wind.

2.5 The smallness of atoms: *we all are breathing one another*

The most convincing evidence for atoms would be actually to see one. But can you see individual atoms? It is a surprisingly subtle question. The answer depends on what you mean by "seeing." Traditionally, when we say we see an object, we mean that we detect it with our unaided eyes, using light. By this criterion, atoms are certainly unseeable.

But today we also "see" things through microscopes. In a traditional microscope, a small object is illuminated with light. This light reflects from the object and passes through lenses that bend the light in such a way as to magnify the image of the small object, and a human eye detects the enlarged image. To see an object with a microscope means, then, detecting it with our aided eye, using light.

But even according to this expanded definition, it is impossible to see individual atoms. In fact, if "seeing" means "detecting using light," then nobody will ever see individual atoms. The reason lies in the nature of light itself. Light is a wave, similar in some ways to water waves on the surface of a pond. But the wavelength of light, the distance from one crest to the next, is very small—only about 0.0005 millimeters, or one-half of one-thousandth of a millimeter (a millimeter is about the width of a toothpick)! It is 10 to 100 times smaller than the smallest dust particle that your unaided eyes can see.

But atoms are much smaller still. A single atom is about 5000 times smaller than the wavelength of light. To visualize this, imagine that light has a wavelength of 5 meters. On this same enlarged scale, an atom would be a tiny speck just 1 millimeter across!

It is impossible for such relatively large waves to detect an object as small as an atom. Imagine, for example, a tiny cork afloat on the ocean's surface. Suppose the cork is camouflaged to look like water so that you cannot see it directly but are trying to detect its presence by observing the cork's effect on passing ocean waves. Large ocean waves would show no effect—the cork would just ride up and down on them, and you would see no alteration in the waves. Only by using much smaller water waves, ripples comparable to or smaller than the cork itself, could you detect the presence of the cork by its effect on the waves. It is the same for an atom and light waves: Light itself is simply too coarse to respond to individual atoms.

HOW DO WE KNOW? DETECTING THE UNSEEABLE

Before 1970, Brownian motion was probably the closest we had come to seeing atoms. The object actually seen, through a microscope, is a tiny dust particle darting this way and that, but we can infer from its sudden changes in direction that atoms are bumping it.

In 1970, scientists developed a more direct way to detect individual atoms: the scanning electron microscope. It shoots a steady stream, or beam, of tiny material particles called electrons at the object to be detected. Your television set operates with an electron beam that sprays elec-

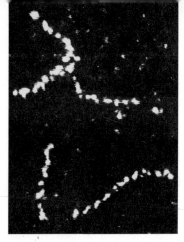

FIGURE 2.15
Chains of thorium atoms, as detected by a scanning electron microscope.

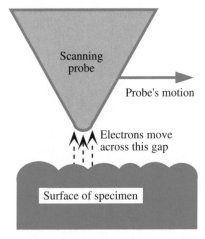

Scanning probe

Probe's motion

Electrons move across this gap

Surface of specimen

FIGURE 2.16
The tip of a scanning tunneling microscope and a small portion of the specimen being scanned.

FIGURE 2.17
A scanning tunneling microscope image of a tiny part of the surface of a silicon crystal, showing the location of individual atoms. The surface shows a few areas containing impurities.

trons across the inside of the screen's face. An electron beam is fundamentally different from a light beam because electrons are particles of matter—material substance having weight—whereas light is not made of matter. During the 1920s, physicists discovered that every particle of matter, such as an electron, has a certain kind of wave associated with it. These waves are called *psi-waves*.*

The wavelength of some psi-waves is thousands of times smaller than the wavelength of light, small enough that an individual atom can affect the psi-waves. Scanning electron microscopes use electron psi-waves to detect individual atoms. As the electron beam scans, or sweeps by, an individual atom, the beam's psi-wave is disturbed. The beam is then focused on a screen that is similar to the inside of a television screen. The pattern that the electrons make when they hit the screen can be detected using visible light.

Figure 2.15 is a photograph of a scanning electron microscope image of a string of thorium atoms. It is the first clear image of individual atoms. It was made not with visible light but with an electron beam. It is pretty direct evidence for the existence of individual atoms, but it is not what one normally means by seeing. It is more accurate to say that we can detect individual atoms.

The scanning tunneling electron microscope, developed in 1983 and also based on electron psi-waves, operates on an even smaller scale. A tiny probe, shaped like a sharp pencil tip, scans the surface of a specimen. Electrons sense the shape of the surface by moving between the probe and the surface in a psi-wave process known as *quantum tunneling* (Figure 2.16). The results are processed by a computer and displayed as a three-dimensional image of the atomic details of the surface (Figure 2.17).

We can even use the scanning tunneling electron microscope to perform the thought experiment that Leucippus and Democritus could only imagine 2500 years ago. In addition to detecting individual atoms, the probe tip can pick up individual atoms and drag them from place to place. In 1990, scientists picked up single atoms of xenon gas and rearranged them.[†] In effect, they divided xenon gas into the smallest particles of xenon, into atoms, just as Leucippus and Democritus had supposed was possible. These scientists positioned 35 xenon atoms to spell out the name of their laboratory (Figure 2.18). Democritus would have been delighted!

Atoms are among the smallest of all objects (Figure 2.19). The nucleus at the atom's center (see Section 2.7), is about 100,000 times smaller. The electron, also found within the atom, is known to be smaller than the smallest distance yet measured, which makes it at least 100,000 times smaller than the nucleus. Since it might in fact have zero size, the electron does not appear in Figure 2.19.

At the other end of the scale, the galaxies are among the largest objects known. Larger still are clusters of galaxies, forming relatively thin "sheets" of galaxies that can individually stretch across as much as 1% of the entire known universe. The largest structures ever detected are the ripples in the

---

* There is more about electrons in Chapter 8, more about light in Chapters 8 and 9, and more about psi-waves in Chapters 13 and 14.
† Xenon atoms do not combine easily with other atoms, making it easy to manipulate them.

FIGURE 2.18
Thirty-five individual xenon atoms have been manipulated into position by the tip of a scanning tunneling microscope. The distance between atoms in the pattern is about 10^{-9} m, or ten times the width of a single atom.

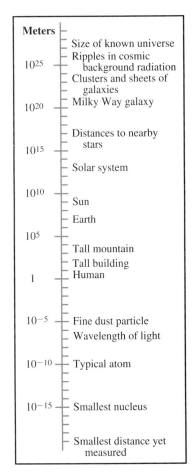

Meters	
	Size of known universe
10^{25}	Ripples in cosmic background radiation
	Clusters and sheets of galaxies
10^{20}	Milky Way galaxy
10^{15}	Distances to nearby stars
	Solar system
10^{10}	Sun
	Earth
10^{5}	
	Tall mountain
	Tall building
1	Human
10^{-5}	Fine dust particle
	Wavelength of light
10^{-10}	Typical atom
10^{-15}	Smallest nucleus
	Smallest distance yet measured

FIGURE 2.19
The range of sizes in the universe.

so-called cosmic background radiation, the faint afterglow of the great explosion that created the universe, ripples that are thought to indicate thin clouds of matter stretching across as much as two-thirds of the known universe. And standing somewhere in the middle, somewhere between the atoms and the stars, is humankind.

Atoms are pretty small. If you put a million of them side by side, the lineup would be no longer than the period at the end of this sentence. The head of a pin contains more than 10^{18} atoms. One breath of air, about 1 liter (1000 cm^3, about a quart), contains more than 10^{22} atoms.

Now, 10^{22} happens to be about the number of liters of air in the Earth's atmosphere, too, which leads to an interesting conclusion. Any particular parcel of air, such as the liter of air you will exhale in your next breath, mixes throughout Earth's atmosphere during the next few years. This means that of the air you exhaled a few years ago in any particular breath, about one or two atoms are now in every liter of air on Earth and inside the lungs of every person on Earth! And one or two atoms breathed out by every person on Earth, in any particular breath, are in your lungs now. One or two atoms from George Washington's first breath, from his dying breath, and from every other breath he ever took, are in your lungs right now. So are a few atoms from each of the breaths of all the other people who have ever lived. We all are breathing one another.

Atoms are ageless. Earth's atoms have been here since Earth formed 5 billion years ago, and very few have changed during that time.* It is only the connections between atoms that have changed. A particular oxygen atom might be part of a nerve cell in your brain today, part of an atmospheric water molecule a century from now, and part of a tree a century after that. "Your" atoms, the ones in your body, have just been borrowed from the air, from the breaths of every person who ever lived, from Earth, borrowed by the acts of breathing and eating, to be given back perhaps soon, perhaps later, to be given back entirely when you die.

MAKING ESTIMATES

About how many times have you exhaled during your life? Of all the atoms you have exhaled during your life, about how many will your class instructor inhale during his or her next breath?

These questions look difficult at first glance. Although it would be impossible to answer them precisely, it is surprisingly easy to make a rough estimate. In estimating very large numbers like the number of times you have exhaled, an estimate to the nearest power of 10 is usually good enough. In other words, is the answer closer to 10, or 10^2, or 10^3, and so on? Such estimates are called *order-of-magnitude estimates.*

We will make such estimates of all sorts of things throughout this book, but do not expect to get a single correct answer. Although different people make different estimates, all of them should be in the same ballpark.

How many exhales in a lifetime? Here is my estimate (yours may be different): By timing my breathing, I find I take about 12 breaths per minute, 12 in each minute (note how *per* is used—it always means "in

---

* Only those relatively few atoms that have been involved in radioactive decay, fission, or fusion have changed into different types of atoms, creating different elements.

each"). To get the number of breaths per year, multiply 12 breaths per minute by 60 minutes per hour, then by 24 hours per day, then by 365 days per year: $12 \times 60 \times 24 \times 365$. Since we are making only a rough estimate, we might as well round off these numbers so that they will be easy to multiply: Make it $10 \times 60 \times 25 \times 400$.

$$10 \times 60 \times 25 \times 400 = (1 \times 6 \times 25 \times 4) \times 10^4$$
$$= 600 \times 10^4 = 6 \times 10^6 \ (6 \text{ million}) \text{ breaths per year}$$

I am 58 years old, or "about 50" (I like this better than "about 60"). So the number of times I have exhaled is about

$$50 \times 6 \times 10^6 = 300 \times 10^6 = 3 \times 10^8$$

or 300 million exhales. If you are younger, your answer will be smaller.

Now, how many of my exhaled atoms will some other person—you, for example—inhale in your next breath? Recall that every liter of air contains about one atom from every one of my exhaled breaths (except for recent breaths that have not yet had time to mix throughout Earth's atmosphere). So you will inhale about one atom from every one of my exhaled breaths in your next breath. Since I have exhaled 3×10^8 times, you will inhale about 3×10^8 of my exhaled atoms in your next breath! And in your next breath, you also will inhale some 300 million atoms from the exhaled breaths of each person living on Earth and from each person who has ever lived on Earth. And all of this will only be a small fraction of the total number of atoms you will inhale on that breath. It's something to consider when you take a breath.

DIALOGUE 11 MAKING ESTIMATES   About how many millimeters thick is one sheet of paper? *Hint*: Roughly how thick is a 500-sheet package of typing paper?

DIALOGUE 12 MAKING ESTIMATES: THE SIZE OF $1 TRILLION   If you stacked up a trillion dollars in new $100 bills, about how many kilometers high would the stack be? *Hint*: Assume that they stack like typing paper, and see the previous dialogue.

2.6 Atomic materialism: *atoms and empty space*

From 1550 to 1700, the revolutionary ideas of Copernicus, Kepler, and others became widespread, and consequently, to educated people, Earth was no longer the motionless central focus of the universe. Humans became passengers on one planet among many, inhabitants of an impersonal mechanical universe (Figure 2.20). The new point of view, coupled with the new free-

---

11. A 500-sheet package of typing paper is about 4 to 6 cm (a few inches) thick, say 5 cm. So the thickness of one sheet is about $5/500 = 0.01$ cm $= 0.1$ mm (one-tenth of 1 millimeter).
12. The number of $100 bills needed is $10^{12}/100 = 10^{10}$. The thickness of one bill (previous dialogue) is 0.01 cm $= 10^{-2}$ cm $= 10^{-4}$ m. The height of the stack is $10^{10} \times 10^{-4}$ m $= 10^6$ m $= 10^3$ km $= 1000$ km (about 600 miles)!

dom of ideas in Renaissance Europe, stimulated an advancing scientific tide. The high point was Newtonian physics, the ideas about motion, force, and gravity developed by Isaac Newton (1642–1727) and others. Newtonian physics dominated science during the eighteenth and nineteenth centuries and is still influential. Its influence extended far beyond science to the general culture of the times, and it deeply affected the way that people thought about themselves, their society, and their role in the universe. Although Newtonian physics has been partly superseded today by new theories, Newtonian ideas continue to dominate our general culture.

In summing up his scientific career, Isaac Newton once stated, "If I have seen farther than others, it is by standing on the shoulders of giants." Two such giants, René Descartes (1596–1650) and Galileo Galilei (1564–1642), strongly influenced the philosophical underpinnings of Newton's physics. Together, Descartes, Galileo, and Newton were the leading founders of science as we know it today.

Although there was no evidence at that time for the atomic theory, the idea of atoms underlies much of the work of Descartes, Galileo, and Newton. It was an idea that went pretty deep, a philosophical idea. Democritus stated it in a far-reaching form:

> By convention sweet is sweet, by convention bitter is bitter, by convention hot is hot, by convention cold is cold, by convention color is color. But in reality there are atoms and empty space. That is, the objects of sense are supposed to be real, and it is customary to regard them as such, but in truth they are not. Only the atoms and empty space are real.

FIGURE 2.20
One way of picturing humankind's shift in psychological perspective, when medieval science gave way to the new science of Copernicus and Newton. The cozy pre-Newtonian universe was replaced by a vast impersonal mechanical universe. This woodcut was made during the nineteenth century, long after the transition had taken place.

This goes considerably further than the atomic theory. Democritus is saying more than that all matter is made of atoms and, more than that, everything is made of atoms. He is saying atoms are all there is. Nothing else is real. Sweetness, bitterness, hot, cold, color, and all other sense impressions are only ways of speaking. They are not real. Only atoms moving in empty space are real.

According to this view, when you say "the water is hot" or "the shirt is red," you really mean that the atoms in the water and shirt and in your body are moving in a certain way. There really is no such thing as hot or red—there are only atoms. And when you say that "the painting is beautiful," you really mean only that the painting's atoms cause your brain's atoms to move in a certain way, a way that ultimately causes your vocal chords to form the words "the painting is beautiful." It is all purely mechanical. The beauty that you think you perceive in the painting is not real, and your thinking is not real either; only the atoms are real. And so it is with all sense impressions, all feelings: Beauty, ugliness, love, hate, and so forth are really only atoms moving mechanically in your brain or in the external world.

According to Descartes, sense impressions are merely "secondary qualities," qualities that exist not in the real universe of atoms but rather only in our minds. The real universe contains only the atoms and their physical properties, such as weight and size. These are the "primary qualities." According to Descartes, science is the study of the primary realm, and so its chief task is to explain natural phenomena by means of atoms. Although Newtonian physics does not require this point of view, it is quite congenial to it.

But such a view—concerning what is real and what is less real—goes beyond what can be observed and verified, and so it lies outside science. It is a philosophical view. The philosophical view of the Greek atomists and the associated views of the founders of modern science are one version of **materialism**: the view that matter is the only reality and that everything is determined by its impersonal workings. Not that Descartes, Galileo, or Newton were materialists themselves—it was psychologically impossible for an educated person of that time to maintain a purely materialistic position. They all subscribed to nonmaterialistic religious ideas, in particular a belief in God. But in the new scientific philosophy there was little room for the God of the Middle Ages, a God who is continually and intimately involved in the world. Instead, these thinkers believed in a creator-God, a God who created the universe and set it into motion, who established the physical laws of the universe once and for all, and who then made his presence felt only by maintaining those laws.

Newtonian physics is quite compatible with the materialist position. As we will see (Chapters 3, 4, and 5), Newtonian physics explains an astonishing range of phenomena, in ways that are compatible with the idea that the universe is an impersonal clocklike machine whose parts are atoms. Since Newton, many later ideas have pointed in the same direction. We have seen in this chapter that heat is the motion of atoms. We will see in Chapter 9 that light and color are caused by vibrations of atoms. According to modern biochemistry, human sense impressions are connected with chemical changes in certain molecules in the human brain. It certainly seems that every observable phenomenon is connected with the arrangement or motion of atoms. Is it true, then, that the atoms and their motions is all there is?

All nature then, as it exists by itself, is founded on two things: there are bodies and there is void in which these bodies are placed and through which they move about.

Roman poet Lucretius, about 50 B.C.

All these things being considered, it seems probable to me that God in the Beginning formed Matter in solid, massy, hard, impenetrable, movable Particles, of such Sizes and figures, and with such other Properties, and in such Proportion to space, as most conduced to the end for which he formed them.

Isaac Newton, 1704

Lucretius is the truly Roman heroic poet; his heroes are the atoms, indestructible, impenetrable, well-armed, lacking all qualities but these; a war of all against all, the stubborn form of eternal substance. Nature without gods, gods without a world.

Karl Marx

For every good philosophical idea, there are always good arguments on the other side. One of the twentieth century's greatest scientists, the physicist Niels Bohr, put it this way: "The hallmark of a profound idea is that its converse is also profound." And so it is with atomic materialism. Respectable arguments on the other side include the following:

1. Considered as a philosophical worldview, materialism is a view that is rooted in science. But science is only one way of viewing reality. Other views—religious, aesthetic, intuitive, and psychological—have equal claim to being real.
2. Considered as a scientific principle, atomic materialism is only tentative, just as all scientific ideas are tentative. For example, we cannot know for certain that everything is really made of atoms, because we have not yet observed everything.
3. During the past century, scientists have found that Newtonian physics is incorrect in several respects.*

As we will see throughout the last half of this book, post-Newtonian physics can be interpreted to have philosophical implications quite different from those of Newtonian physics. Whereas Newtonian physics is congenial to materialism, recent theories seem neutral or perhaps even uncongenial to materialism.

2.7 Three atomic models: *Greek, planetary, and quantum*

Science's way of working back and forth between observations and theories comes with a kind of guarantee of success. If an observation agrees with a theory, that is fine for that theory. And if an observation disagrees with a theory, that is too bad for that theory, but it is fine for science, because science makes its greatest progress by repairing, or replacing, disproved theories.

The historic evolution of the atomic idea is a good example. Science's notion of the nature of atoms has changed several times. In the original Greek model of the atom, an atom was an unchangeable, single object like a small and rigid pea. Atoms were thought to come in many kinds, each with different properties. The Greek atom was also Descartes's and Galileo's and Newton's way of looking at the atom, and it got considerable experimental support from the nineteenth-century discoveries of elements, compounds, and Brownian motion.

Around 1900, experiments with electricity (Chapter 8) contradicted the Greek model of the atom. Electricity had been studied throughout the nineteenth century. Nobody suspected that an entirely new model of the atom would be needed to explain these experiments until, in 1897, scientists discovered a new, very lightweight, "electrified" particle. It was the first discovery of a particle that weighed less than an atom. It was, apparently, one part of the so-called a-tom (remember, the word means "indivisible"). This was the **electron**.

---

* More precisely, Newtonian physics is only approximately correct over only a limited range of phenomena. Outside that limited range it is not even approximately correct.

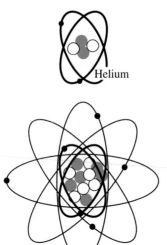

FIGURE 2.21

An atom of helium and an atom of carbon, according to the planetary model of the atom. The small black dots are electrons in orbit around the nucleus, and the green circles and white circles represent protons and neutrons in the atom's nucleus. This is not drawn to scale! The nucleus should be 100,000 times smaller than the electron orbits, and most physicists believe that the electron has no size at all.

A few years later, in 1911, physicists discovered that an atom is itself nearly entirely empty space and that nearly all of an atom's material substance resides in a tiny central core, or **nucleus**. Experiments indicated that each atom also contained electrons moving through the large empty regions outside the nucleus. Scientists developed the theory that electrons orbit the nucleus much as the planets orbit the sun. Later, scientists learned that the nucleus itself is made of two kinds of subnuclear particles, **protons** and **neutrons**. Figure 2.21 pictures an atom of the element helium and an atom of the element carbon, according to this **planetary model of the atom**. The planetary model got support from experiments in nuclear physics, chemistry, and electricity. Niels Bohr developed a version of the planetary model that explained certain previously inexplicable observations of the light that atoms can emit.

During the 1920s, new experiments involving electrons contradicted the planetary model and in fact contradicted Newtonian physics itself. An entirely new theory of matter was needed to explain the new results. The planetary model was inconsistent with this new **quantum theory** (Chapters 13 and 14), and so scientists developed a new model, the **quantum theory of the atom**. To date, no disagreements have been found between experiments and the quantum theory of the atom.

So our understanding of the atom has evolved through three different theories. At each stage, new experiments disproved the old theory, and scientists invented a broader theory that explained both the old and the new observations. But scientists did not discard the Greek and planetary models, despite their shortcomings, because they are useful within their proper range. For instance, we can use the Greek atom to explain all sorts of things like air pressure and the odor of violets. We need not resort to the planetary atom or the quantum atom to explain these things, because the atom's internal structure and its quantum nature are irrelevant to these phenomena. Restricted to its proper range, the Greek atom is a fine theory. Once again, we see that theories are best described as useful rather than true.

DIALOGUE 13   Since the quantum theory of the atom agrees with every experiment to date, can we now say that this theory is certainly correct?

2.8 Chemistry and life: *what did atoms ever do for you?*

You can get atoms to do fantastic things by connecting them in sufficiently subtle ways. It is possible, for example, for a pile of atoms to acquire additional atoms from its environment, to move itself from one place to another, to respond to external events such as the presence of particular molecules or light or warmth, and to create copies of itself. We would call such a pile of atoms "alive." Indeed, you are just such a pile of atoms. The pile of atoms that is you has an especially surprising property: It is aware of itself, and in this scientific age it is even aware that it is a pile of atoms.

The element that gives life its powerful abilities is carbon. Life emerges from the ways that atoms in biochemical molecules connect chemically with

---

13. No, general scientific theories always are tentative.

one another. Carbon, plentiful on Earth and connecting readily to a variety of other plentiful elements, is the key. Other elements that are abundant in biological molecules are oxygen, hydrogen, and nitrogen (remember "COHN"). Whenever anything happens in your body, whenever you move or breathe or sense or feel or think, some change is occurring at the atomic level.

Atoms are frequently rearranged into new molecular forms—in other words, a **chemical reaction** takes place. **Burning** is a simple nonbiological example of a chemical reaction.* It was once believed that fire was one of the substances of which things are made and that a burning object released the fire that it already contained. The prevailing theory was that this intangible and nonmaterial substance—fire—had weight and carried its weight away when any object burned. Around 1780, Antoine Lavoisier studied burning more closely. He accurately weighed all of the materials involved when an object burned, including the gases used and given off. The prevailing theory predicted that the weight should decrease, since fire carried away some of the weight. But Lavoisier showed that there was no net change in weight. This was one of the earliest experiments supporting the idea that everything is made of indestructible atoms, and it initiated the modern science of **chemistry**—the study of the properties and transformations of substances (chemical compounds).

The key to the new science of chemistry was the concept that chemical reactions are rearrangements of atoms that are themselves changeless and indestructible. It followed that the total amount of matter involved in any chemical reaction does not change, that the total amount is the same before and after the reaction. This idea, known as the **conservation** (preservation) **of matter**, or the conservation of mass, was assumed to extend to every physical process. It was and is a useful theory, especially in chemistry, but experiments have proved it wrong. Although it is a good approximation in chemical reactions, in other situations it is entirely wrong.

Most burnable substances are derived from biological materials that contain carbon and perhaps hydrogen, along with other elements. As simple experiments demonstrate (see the Home Projects), burning requires air, and something is removed from the air during burning.

Air is not a pure substance but is a mixture of molecules of many different substances. Nitrogen and oxygen dominate: Nearly 80% of air's molecules are nitrogen (N_2), about 20% are oxygen (O_2), and 1% are single argon atoms. All the other gases added together total less than 1%. These "trace gases" include all sorts of compounds. Some, such as water vapor (H_2O) and helium (He), are natural. Others, such as carbon dioxide (CO_2) and ozone (O_3), come from both natural and industrial sources. And still others, such as carbon monoxide (CO) and complex chlorofluorocarbon and hydrocarbon molecules, come almost entirely from industry.

For burning, the crucial component needed from air is oxygen. Carbon from the burning substance, the fuel, combines with oxygen from the air to form carbon dioxide. We abbreviate the preceding sentence symbolically as

$$C + O_2 \rightarrow CO_2$$

---

* Burning is one form of "combustion," which means any chemical reaction that generates heat.

The plus sign means "combines with," and the arrow means "changes into." If the fuel contains hydrogen, it too combines with oxygen to form water vapor. For example, methane gas, CH_4, is the simplest of the hydrocarbon (hydrogen and carbon) fuels. It burns in air to form carbon dioxide and water vapor*

$$CH_4 + O_2 \rightarrow CO_2 + H_2O$$

An important feature of any chemical reaction is its energy balance. For now, we use the important word *energy* to mean either of two things: heat or the ability to move things around. As you know, heat—perhaps from friction or a burning match—is needed to start a substance burning. Once it starts, the burning reaction itself creates more than enough heat to maintain itself, so excess heat is given off. Including heat, the reaction formula for burning a typical fuel such as methane is

$$CH_4 + O_2 \rightarrow CO_2 + H_2O + \text{excess heat}$$

Let us now turn to biology. Animals get their bodily material and their energy from the food they eat and the air they breathe. Your blood absorbs carbon-based molecules from food, and oxygen from air, and ferries them all over your body. When they arrive at, say, your thumb, they enter a biological cell there. Certain parts of the cell combine these substances to create the molecules needed for new cells. This is a complicated chemical process using the cell's genetic materials. Other parts of the cell use food and oxygen to create biologically useful energy. The chemical reactions that do this are called **respiration**. In a typical case, a simple sugar called glucose reacts with oxygen to create carbon dioxide and water:

$$C_6H_{12}O_6 + O_2 \rightarrow CO_2 + H_2O + \text{useful energy}$$

As you can see, this is similar to burning. Animal life is a slow burn. So it is not surprising that it generates biologically useful energy, just as burning generates heat. In respiration, part of this energy goes into making a high-energy molecule known as ATP, and the rest appears as heat. ATP is the energy carrier in animals. It can remain in storage, or it can move from place to place within the cell. It can be used for all sorts of things, such as bending your thumb. As you can see from the reaction formula, respiration generates water and carbon dioxide as waste products that are excreted in your sweat, urine, and exhaled breath.

The difference between plants and animals is in their strategies for generating useful energy. Plants use energy directly from the sun, whereas animals gather energy by eating plants and other animals that have stored it. Plants gather carbon dioxide and water from their surroundings and put them together to form high-energy carbohydrates (carbon compounded with water) such as glucose. From the formula for respiration, you can see that this process in plants is exactly the opposite of respiration in animals! Since respiration generates energy, it is not surprising that the opposite reaction consumes energy. Through evolution, plants have worked out a complicated process using a compound called *chlorophyll* that absorbs the energy in sunlight and

---

* This formula is not quantitatively "balanced." For example, there are four Hs on the left, but only two on the right. The balanced formula is $CH_4 + 2O_2 \rightarrow CO_2 + 2H_2O$. Since we are not interested here in how much of each compound enters into a reaction, we omit the numerical "coefficients."

converts it to a form that plants can use. It also comes as no surprise that plants generate oxygen, just as animals consume it. The overall reaction is

$$CO_2 + H_2O + \text{solar energy} \rightarrow C_6H_{12}O_6 + O_2$$

This is called **photosynthesis** (putting together by light).

Both the food and the fuels such as glucose that animals need can be traced, directly or indirectly, to plants. We depend on plants not only for food and fuel but also for the crucial component of the very air we breathe. Earth's oxygen comes from photosynthesis. Without plants, animals would soon be out of food, fuel, and breath.

Plants and animals complement each other quite nicely in the balance that evolution has worked out on Earth. The carbon dioxide exhaled by animals is taken in by plants for photosynthesis, and the oxygen released by plants is inhaled by animals for respiration. The two kingdoms of life recycle each other's primary waste products.

DIALOGUE 14   Is there another element, other than carbon, that might be the key element in forming living organisms elsewhere in the universe? *Hint*: In what table should you look to find an element whose chemical properties are similar to carbon's?

"OF COURSE THE ELEMENTS ARE EARTH, WATER, FIRE AND AIR. BUT WHAT ABOUT CHROMIUM? SURELY YOU CAN'T IGNORE CHROMIUM."

---

14. In the periodic table, beneath carbon, is silicon. It is quite abundant on Earth, in sand.

Summary of Ideas and Terms

Atomic theory of matter All matter is made of tiny particles, too small to be seen.

Diffusion The spreading out of a collection of particles due to their random jostling by one another or by other neighboring particles.

Chemical decomposition Any process that changes a single substance into two or more other substances.

Chemical element A substance that cannot be chemically decomposed, of which 109 are known. They are listed in the **periodic table,** along with their **atomic number**. An **atom** is the smallest particle of an element.

Chemical compound A pure substance that can be chemically decomposed. Its smallest particle is a **molecule,** two or more atoms connected into a single unit. All of a compound's molecules are identical.

Microscopic Too small to be seen with the unaided eye, as opposed to macroscopic, or visible to the unaided eye.

Solid, liquid, and gas The three states of matter. Nearly every substance can exist in any of the three states. A solid's molecules are locked closely together in a regular pattern, whereas a liquid's molecules are close together but not fixed in position, and a gas's molecules are far apart and move around rapidly. **Evaporation** is the slow transition from the liquid to the gas state, caused by the escape of faster molecules from the liquid.

Gas pressure The outward push caused by gas molecules hitting the walls of the gas's container.

Vacuum An absence of atoms and other material particles.

The microscopic interpretation of warmth. Temperature is associated with disorganized, or **random,** microscopic motions that are not visible macroscopically.

Per In each.

Newtonian physics The ideas about motion, force, and gravity developed by Isaac Newton and others around 1700.

Materialism The philosophy that only matter is real and that everything is determined by its impersonal workings. The early Greek atomists were atomic materialists. Newtonian physics is compatible with this philosophy.

Models of the atom The **Greek atom** is a tiny indestructible object, like a small and rigid pea. The **planetary atom** is made of parts, including a tiny central **nucleus,** and one or more **electrons** orbiting far outside the nucleus. The **quantum atom** is based on the post-Newtonian quantum theory.

Air A mixture of several chemical compounds: N_2 (80%), O_2 (20%), and a smattering of trace gases.

Chemical reaction A rearrangement of atoms into new molecular forms. **Burning** creates heat by combining oxygen from air with a fuel such as carbon or hydrogen. **Respiration** in animals combines oxygen and carbon-based molecules such as glucose to generate useful energy, along with carbon dioxide and water. **Photosynthesis** in plants combines atmospheric carbon dioxide and water to form carbohydrate molecules such as glucose, along with oxygen.

Review Questions

Each chapter has review questions and exercises at the end. Review questions go over many of the essentials. You can find the answers by looking in the chapter. Exercises ask you to draw on your understanding of the material to answer new questions not directly answered in the chapter. The answers to the odd-numbered questions are in the back of the book. The many exercises allow your instructor flexibility; you will probably want to tackle only some of them.

THE ATOMIC IDEA

1. What macroscopic evidence is there for atoms?
2. What light-based microscope evidence is there for atoms?
3. What experiment did the ancient Greek atomists imagine doing, and what did they believe the result would be?

4. An experiment such as the Greeks (previous question) imagined was actually carried out recently. Describe it.
5. Which is bigger, an atom or the wavelength of light? A little bigger or a lot?

ATOMS AND MOLECULES

6. From the microscopic point of view, what is the difference between an element and a compound?
7. From a macroscopic point of view, what is the difference between an element and a compound?
8. What is the difference between an atom and a molecule?
9. Why is the periodic table arranged in the way that it is?

10. If you chemically decompose water, what will you get? Will you get anything like water?

THE ATOM'S EXPLANATORY POWER

11. Describe the microscopic process by which perfume gives off an odor that you can smell some distance away.
12. How do solids, liquids, and gases differ macroscopically? Microscopically?
13. Which is easiest to compress: solids, liquids, or gases? Why?
14. Is a perfect vacuum ever attained on Earth, over a volume as large as 1 cubic centimeter? Elsewhere?
15. What is the microscopic difference between hot water and cold water?

ATOMIC MATERIALISM AND ATOMIC MODELS

16. Name several things that people ordinarily regard as real but that, according to atomic materialism, are not real.

17. Describe the philosophy of materialism.
18. Give arguments for the materialist philosophy.
19. Give arguments against the materialist philosophy.
20. Name three different models of the atom. Describe two of them.

CHEMISTRY AND LIFE

21. What is meant by a "chemical reaction"?
22. Name three different chemical reactions.
23. Is air a single substance (a single compound)? Describe its chemical composition.
24. Describe an experiment involving burning that supports the notion of conservation of matter.
25. In what types of experiments is the conservation of matter correct to a very good approximation?
26. Describe the following reactions: burning, respiration, and photosynthesis.

Home Projects

There is, literally, nothing like "hands on" experience in science. Try some of these at home!

1. The diffusion game. Draw a row of 11 squares, making each square large enough for a penny. Number and mark them as in Figure 2.22. Stack 4 pennies on square 0. Now flip another coin, and move the top penny one place to the right (to +1) or to the left (to −1), depending on whether the flip was heads or tails. Do the same with the other three pennies, flipping each time. You have now completed one "diffusion step." Take another diffusion step, moving each penny one place rightward or leftward, depending on the coin flip. Take several diffusion steps. Do the pennies diffuse outward from the center? How many diffusion steps does it take before a penny "diffuses" to your nose or to Joe's nose? Who smelled the pennies first? What are the similarities and differences between this process and the physical diffusion of an odor discussed in this chapter?

2. Does a warm liquid evaporate faster than a cold one? Check your prediction by putting identical amounts of water into identical pans. Keep one pan warm, but not boiling, on the stove and the other at room temperature. Or keep one pan in the refrigerator, but not the freezer, and the other at room temperature.
3. You may have heard that warm water freezes faster than cold water does. Try it and see, starting with identical ice trays of cold and warm water.
4. Light two identical candles, and invert a large jar over one. What happens? Explain. Now get two identical large jars, take a deep breath of the air from one, and exhale back into it. Quickly put the jars over the candles. Explain what you observe.

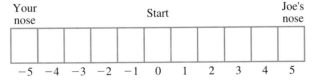

FIGURE 2.22
The diffusion game.

For Discussion

The discussion questions are meant to stimulate thinking about questions that have no specific "correct" answer. They can be used for class discussion, individual essays, or individual thought.

1. What problems would be created if the United States switched to the metric system? What would be the advantages?
2. Would you know that there is such a thing as "air" if nobody had ever told you about it?
3. In what respects do you agree or disagree with the atomic materialist philosophy described in this chapter? Defend

each of your opinions. Points to consider: What happens at the microscopic level when you have a sense impression, for instance, when you smell perfume or touch a warm object? Is that all that happens? What if, instead of a smell or touch, your sense impression had been of love, anger, beauty, or religious feelings?
4. Social commentators often note that modern society is materialistic. What does this use of the word *materialism* mean? Is this *social* materialism the same thing as the *philosophical* materialism described in this chapter? Are social and philosophical materialism at all related?

Exercises

Exercises indicated by asterisks are quantitative, although they require no algebra. The answers to the odd-numbered questions are in the back of the book.

THE ATOMIC IDEA

1. Is the atomic theory known, for certain, to be true?
2. Carbon atoms are about 25% lighter than oxygen atoms (the ratio of their weights is 3 to 4). What is the weight ratio of the carbon and oxygen that go into the formation of carbon monoxide? Answer the same question for carbon dioxide.

ATOMS AND MOLECULES

3. How many atoms are in a molecule of H_2SO_4 (sulfuric acid)?
4. Which of these is a pure compound, which is an element, and which is neither: pure water, oxygen gas, liquid mercury, H_2SO_4, U, air, He, carbon dioxide, H_2, polluted water?
5. Suppose you obtained the smallest single particle of each of the following substances. In which cases would this particle be a molecule made of more than one atom, and in which cases would it be a single unattached atom: pure water, oxygen gas in the form found in Earth's atmosphere, H_2SO_4, U, He, carbon dioxide, H_2, H?
6. Helium is an inert gas, meaning that it does not readily enter into chemical reactions with other substances. List five other substances that you would expect to also be inert gases.
7. On the simplifying assumption that oxygen and carbon atoms have the same weight, how much carbon dioxide gas is formed when 1 ton of coal burns (coal is nearly pure carbon)?

8.* Referring to the preceding exercise: In a typical large coal-fed electrical generating plant, a ton of coal is burned every 10 seconds. About how many tons of carbon dioxide enter the atmosphere every hour from such a plant?

POWERS OF 10

9.*Write as ordinary numbers: 10^9; 10^{-6}; 3.6×10^{13}; 5.9×10^{-8}.
10.*Write in powers of 10 notation: 3 trillion; five-thousandths; 730,000,000,000,000; 0.000 000 000 082.

THE ATOM'S EXPLANATORY POWER

11. Why can't you observe Brownian motion in easily visible objects such as floating bits of paper?
12. What is the chemical formula for the odor of violets (see Figure 2.6)?
13. A dog follows an escaped convict's trail by putting its nose to the ground. Explain this from a microscopic point of view.
14. If air is put into a sealed container that is then compressed (reduced in volume), what do you predict will happen to the air pressure on the container walls? Explain this from a microscopic point of view.
15. If air is put into a sealed container and heated, what do you predict will happen to the air pressure on the container walls? Explain this from a microscopic point of view.
16. If a balloon is partially filled with air (so that it isn't fully expanded), sealed, and then heated, what do you predict will happen to the balloon? Explain this from a microscopic point of view. What if the balloon is cooled instead?

17. Why is it so difficult to remove the lid from a vacuum-sealed jar?

THE SMALLNESS OF ATOMS

18. Put these in order from lightest to heaviest: water molecule, oxygen atom, raindrop, hydrogen atom, glucose molecule, electron, DNA molecule.
19.*Making estimates. According to Dialogue 11, one sheet of paper is about 0.1 mm thick. An atom is about 10^{-10} m across. About how many atoms thick is one sheet of paper?
20.*Making estimates. The average weight, per atom, of the atoms in your body is about 10^{-26} kg (2×10^{-26} pounds). About how many atoms are there in your body? Write out your answer without powers of 10.
21.*Making estimates. The smallest dust particle visible to the unaided eye measures about 0.05 mm across. About how many atoms across is this—in other words, if we line up atoms side by side, about how many would it take to make a line of atoms 0.05 mm long?
22.*Making estimates. Referring to the preceding exercise: Assume the small dust particle is shaped like a cube.

About how many atoms does it contain?

CHEMISTRY AND LIFE

23. What is the chemical reaction formula for burning hydrogen gas in air? What substance is created by this reaction?
24. For safety, gas-filled balloons are filled with helium instead of hydrogen. What does this tell you about the behavior of helium in the atmosphere?
25. Gasoline is a hydrocarbon fuel. What are the main compounds created when gasoline burns in a car engine?
26. "NO_x" (nitrogen oxide and nitrogen dioxide) is one pollutant from automobiles. What elements must combine to form NO_x?
27. Referring to the preceding exercise: Gasoline contains neither oxygen nor nitrogen. So where must these elements come from when NO_x is formed in car engines?
28. Are there any molecules in your body that you could claim are "your" molecules, unique to your body and unlike any other molecules in the universe?

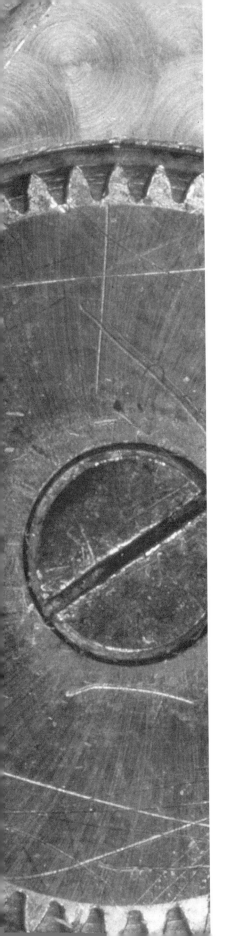

PART **II**

THE NEWTONIAN UNIVERSE
a clockwork kingdom

3

HOW THINGS MOVE
Galileo asks the right questions

When you look around, your eye falls on this book, a flower, your foot. What are these things made of? It's a natural question if you want to study your physical environment. We saw in Chapter 2 that the early Greeks, the earliest people to seriously ask such questions about their physical surroundings, answered that things are made of atoms.

You notice that the book falls, the flower sways, your foot taps. Why, and how, do things move? Again, the ancient Greeks asked first. And this is what we ask, in this and the following two chapters.

The Greek philosopher and scientist Aristotle (Figure 3.1) developed and stated the earliest real theory of motion. His theory was plausible and had some observational support, but post-Renaissance (1600–1700) European scientists such as Galileo and Newton found Aristotle's theories inadequate. They discarded Aristotelian physics in favor of powerful new ideas about motion. These ideas, known as Newtonian physics, dominated science for three centuries. Beginning in 1900, science again changed its fundamental view of motion, when the relativity and quantum theories altered most of the Newtonian principles. Comparisons and contrasts between post-Newtonian physics and Newtonian physics is one of the themes of this book.

Chapters 3, 4, and 5 discuss Newtonian physics, a group of ideas that dominated not only science but also Western culture during three centuries (see Section 2.6). Although Newtonian physics has been superseded today, it continues to be useful for understanding the way the macroscopic world around us works. Most important, Newtonian ideas have retained their powerful cultural influence.

After a look at Aristotelian physics (Section 3.1), this chapter discusses Galileo's ideas about motion itself as distinct from the causes of motion.

FIGURE 3.1
The Greek philosopher and scientist Aristotle, 384–322 B.C.

Section 3.2 presents Galileo's objections to Aristotelian physics and the experimental background for the law of inertia, the centerpiece of Newtonian physics, and Section 3.3 examines this law. Sections 3.4 and 3.5 explore the two ideas needed to describe motion, namely, velocity and acceleration. Section 3.6 applies all of this to a familiar and fundamental phenomenon, falling.

Looking ahead, Chapter 4 discusses the causes of motion, more accurately stated as the causes of the changes in motion. Chapter 5 discusses one such cause: gravity. Throughout, we focus on those aspects of Newtonian physics that are importantly related to our lives and our times.

3.1 Aristotelian physics: *a commonsense view*

Aristotle's physics agrees with most people's commonsense feelings about motion. But these plausible notions are precisely the ones that Newtonian physics discarded. Because Aristotelian physics is so ingrained in our intuitions, we must look at it closely to see why, and where, it is wrong.

In his science, Aristotle was no armchair philosopher. Keenly observant of his surroundings, he is regarded as the father of biology because of his observations of plants and animals. In observing motion, Aristotle noticed that some motions maintain themselves without assistance but that others could be maintained only by an outside agent. One example of motion that Aristotle considered to be unassisted: When you push a rock off a ledge, it falls toward the ground with no obvious assistance. Aristotle regarded motion that could maintain itself in this manner to be **natural motion**.

In addition to solid objects falling, Aristotle perceived three other sorts of natural motion: liquids falling or running downhill, air rising, and flames leaping upward. These four represented to Aristotle the four natural motions that could occur on Earth. He believed that natural motion occurred because everything on Earth was made of four different substances or elements—earth, water, air, and fire—and that each of these was striving to reach a different natural place. Although **Aristotle's elements** are different from the chemical elements (Chapter 2), the word *element* is used in a similar way to mean the substances of which everything is made.

Each Aristotelian element has its own type of natural motion and its own natural place toward which it strives: The element earth moves downward because Earth's center is its natural resting place (educated Greeks knew that Earth was a sphere). Water flows downward because its natural place is just above earth. Air rises because the atmosphere is its natural place, and fire leaps upward because its natural place is above the air.

Aristotle believed that all objects were composed of these four elements and that an object's natural motion was determined by how much of each element the object contained. So a rock sank in water because rocks were mostly made of earth; wood floated because it contained air in addition to earth; and warm air rose because it contained fire.

All these natural motions are directly downward or upward. Horizontal motions seem different. When you pull a cart along a road, throw a stone horizontally, or push a box along the floor, your activity maintains the motion: You keep the cart moving; the stone moves only because you threw it; you must push the box. Pushes and pulls are needed to keep the cart, the ball, and the box moving. Aristotle believed these pushes and pulls were needed because objects must be forced to behave contrary to their own natures, in other words, contrary to their natural motion. Aristotle classified all such forced motion as **violent motion**, meaning that an external push or pull was needed to maintain it.

He believed that all motion on Earth was either natural or violent, but he perceived in the heavens an entirely different kind of motion. He believed that the moon, sun, planets, and stars were made of a fifth element, *ether* (from the Greek word for "to kindle or blaze"), not found on Earth. Unlike the four Earthly elements, ether had no weight and was incorruptible (unchangeable, eternal). Perfect in every way, ether's natural place was in the heavens, and it naturally moved in perfect circles around Earth. This third kind of motion was called **celestial motion**.

This theory explained a wide range of observations with a few general principles. It was a fitting and useful theory and gained wide acceptance, partly because of its plausibility. It does seem to us that a rock falls all by itself, that a push is needed to maintain horizontal motion, and that motion in the heavens really is different from motion on Earth. But we will see that Newtonian physics contradicts all of these ideas.

Aristotle's theory supported other Greek ideas about Earth's place in the universe. Earth was believed to be at rest, with the stars circling it. Aristotle's physics explained why: Earth could not move because it was made mostly of the element earth, whose natural resting place was at Earth's center. The stars moved in circles because that was ether's nature. Objects on Earth could not move in the way that the stars did, because there was no ether on Earth. Aristotle's physics explained why Earth was so different from the heavens.

Medieval Christian philosophy found Greek physics and astronomy quite congenial. An Earth-centered universe, an Earth made of corruptible elements, humans made of corruptible mortal elements, a natural resting place for everything, a perfect heaven made of incorruptible ether in eternal celestial motion—all of these agreed nicely with church theology. Until about 1600, an Earth-centered astronomy and Aristotle's physics were central to the philosophy and culture of European civilization.

Beginning in about 1550, Copernicus and others introduced new astronomical theories. New theories of physics were not far behind. Descartes, Galileo, and others began rethinking Aristotelian physics around 1600.*

DIALOGUE 1   How would Aristotle's theory explain these facts: People float better in water when their lungs are full of air; an air-filled balloon rises when heated; liquid mercury sinks in water?

Odd as it may seem, most people's views about motion are part of a system of physics [Aristotle's] that was proposed more than 2000 years ago and was experimentally shown to be inadequate at least 1400 years ago.
I. Bernard Cohen, science historian

---

* Post-Aristotelian ideas had actually been brewing for many centuries, although most educated people continued to believe Aristotle's physics until Newton's time.

1. Full lungs contain more of the element air, so they stay better above the element water; heated air contains the element fire, making it rise; liquid mercury contains a lot of the element Earth.

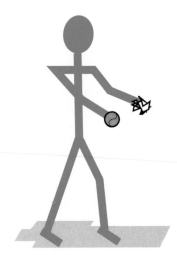

FIGURE 3.2

Hold a baseball and a wadded-up piece of paper above the ground, and drop them simultaneously. What does Aristotelian physics predict? What do you observe?

FIGURE 3.3

A feather dropped in vacuum falls as fast as a stone.

3.2 How do we know? *difficulties with Aristotelian physics*

Aristotelian physics explained a lot, but it had its weaknesses. You can demonstrate some of these for yourself. Drop a piece of notebook paper to the ground. Now crumple it into a tight ball and drop it again. Does it fall faster?* It does fall noticeably faster, and this is not easy to explain on the basis of Aristotelian physics. The crumpled paper is still the same paper, and so it has the same amount of Aristotelian earth in it. It should fall no faster than the flat sheet does.

Aristotle would argue here that the flat sheet falls slower because in plowing through so much of the element air, it carries along enough air that it now seeks the ground less strongly. To get around this argument, try dropping two objects that have the same shape—so that they plow through equal amounts of air—but that have very different weights: a block of wood and a block of iron of the same shape or a baseball and a tightly crumpled piece of newspaper of about the same size (Figure 3.2). What does Aristotelian physics predict? What do you find?

According to Aristotelian physics, the heavier object, which contains much more earth, should fall noticeably faster. But it does not. If you do this experiment carefully and from a high place such as a second-story window, you might detect that the heavier object actually does fall a little faster, but not a lot faster, as Aristotle's physics predicts. Today's theories predict that for two objects of the same shape, the lighter one will fall slower because of **air resistance**—the resistance to the motion of an object through the air due to the object's collisions with numerous air molecules. According to today's theories, air resistance also explains why a flat sheet of paper falls slowly.

It is possible to test the hypothesis that these small differences in falling are due to air resistance. Suppose that two objects fall in a vacuum, perhaps in a container from which the air has been removed. One then finds that the light and heavy objects fall at precisely the same speed. A feather dropped in a vacuum falls as fast as a stone, in clear contradiction of Aristotelian predictions (Figure 3.3)!

Aristotle's concept of violent motion also has problems. If you shoot an arrow, it can travel a great distance horizontally while hardly slowing down. A brief strong push from the bowstring starts it, but what external assistance keeps it moving once it is released from the bow? Aristotle himself had difficulty reconciling this sort of example with his own theory, and later scientists had similar difficulties.

Galileo Galilei (Figure 3.4) was a brilliant, cocky Italian who supported Copernican astronomy, issued sarcastic opinions about Aristotelian physics, and generally annoyed the authorities. His writings eventually earned him a visit by the Catholic Inquisition, which "persuaded" the now elderly man to "confess" and then confined him to house arrest for his remaining ten years of life. Even under house arrest, the irrepressible Galileo pursued his experiments and wrote a large physics book.

---

* Please do these simple experiments when they are suggested. There's nothing like observing the real thing!

FIGURE 3.4
Galileo Galilei, 1564–1642.

To focus his thinking, Galileo imagined the following experiment: Let a ball roll down an incline. Its speed will increase. Now give the ball a quick starting push and let it roll up an incline. It slows down (then stops and rolls back down). Suppose we make the inclines nearly horizontal (Figure 3.5). If you have ever let a ball roll down a very slight incline, you know that it is likely to slow down and come to rest, even though it is going downhill. Galileo understood that this slowing is due to the roughness of the incline and ball. Today we call it **friction**. Galileo's crucial step was to idealize the experiment by neglecting, at least in his mind, the effect of friction.

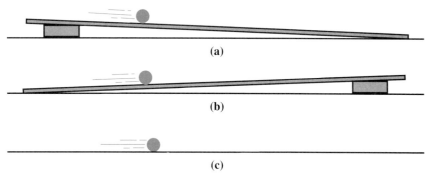

FIGURE 3.5
A smooth ball on a smooth incline always (a) speeds up going down and (b) slows down going up, even for a very slight incline. In the limiting case (c) of a perfectly smooth and level surface, the ball should keep going forever once it has started rolling.

Galileo saw that neglecting friction, the ball would speed up on any downward incline, no matter how slight, and would slow down on any upward incline. Then he took another characteristically brilliant step: He considered the "limiting case" of slight inclines, namely, a perfectly horizontal surface. On a frictionless horizontal surface, the ball could neither speed up nor slow down because the horizontal surface was intermediate between downhill and uphill. Galileo concluded that in absence of friction, a ball that once started rolling on a horizontal surface would roll forever. This radically contradicted Aristotle's theory, which stated that continued pushing or pulling was needed to maintain violent (that is, horizontal) motion.

Galileo's methods have been crucial to science ever since. They included

· *Experiments*, designed to test specific hypotheses. Although careful observation dates back at least to Aristotle, Galileo was the first to refine this process with controlled experiments to test specific hypotheses.
· *Idealizations* of real-world conditions, to eliminate (at least in one's mind) any side effects that might obscure the main effects.
· *Limiting the scope of the inquiry* by considering only one question at a time. For example, Galileo did not make Aristotle's mistake of trying to understand both why horizontal motion occurs and how it occurs. Aristotle believed that horizontal motion occurred because of external assistance (which is why it occurs), so that an unassisted horizontally moving object should

slow down and stop (that is how it occurs, says Aristotle). But Galileo looked more carefully at only the question of how horizontal motion occurred and found that in the absence of friction an object would not slow down.

· *Quantitative methods.* Galileo went to great lengths to measure, for example, the motion of bodies. He understood that a theory capable of making quantitative predictions was more powerful than one that could make only qualitative, descriptive predictions.

3.3 The law of inertia: *the roots of Newtonian physics*

With experiments and arguments such as Galileo's, scientists eventually arrived at a profound non-Aristotelian insight. It involved an extreme idealization: Suppose you could get away from the effects of friction and air resistance *and also gravity*. It isn't easy to imagine such a thing, because gravity is such an unchanging part of every person's daily existence that we don't even notice its presence. It's hard to imagine the absence of something that we aren't even aware of in the first place! French philosopher and scientist René Descartes was the first to arrive at the new insight, with contributions from Galileo and Newton.

What if you could turn off gravity? This question would have been meaningless to Aristotle. To Aristotle, there was no such thing as gravity. Things just fell, by themselves, because that was the nature of things. But the question had meaning from the enlarged post-Copernican perspective that saw Earth as one among many planets. One could imagine that on other planets, gravity might be very weak, that in outer space—far from all planets and stars—gravity might have practically no effect. One could extrapolate to a situation of zero gravity, just as Galileo had extrapolated to an absence of friction.

If you released a stone in midair and there were no gravity, the stone would not fall. It would hang in midair. And if you flicked the stone with your finger to start it moving and there were no gravity or friction or air resistance, *it would coast in a straight line, with no change in speed, forever!* This was Descartes's new insight.

This idea was an invention of the human mind, like all scientific theories. But it was not a free invention of the mind; like all scientific theories, it was conditioned on observation of the natural world. Partly, it was a redefinition of "natural motion" and "externally influenced" motion. Not only friction but also gravity were now to be regarded as external influences rather than something that bodies provide for themselves. It was a new way of looking at nature, much as Copernican astronomy was a new way of looking at the heavens.*

Descartes is saying that if it were moving to begin with, an unassisted object will keep moving all by itself. It does not need external assistance. An unassisted object that was at rest to begin with will stay at rest; it will hang

---

* Historians refer to such a shift of fundamental viewpoint as a *paradigm shift.*

in midair, for instance. It is a strange idea. At first, our intuition rejects it. Because gravity, friction, and air resistance are all around us, our intuition tells us that objects can keep moving only if they are pushed or pulled.

As a way of hanging on to the old idea that something must assist an object if it is to keep moving, we give a word to an object's ability to keep moving: *inertia*. We also apply this word to an object's ability to stay at rest. In other words, an object's **inertia** is its ability to keep moving and to remain at rest. So all objects have inertia. All objects have the ability to keep moving and to remain at rest. The idea of inertia doesn't really explain anything; inertia is simply a word that stands for the unexplainable fact that bodies do keep moving.

We summarize all of this as:

THE LAW OF INERTIA*
A body that is subject to no external influences (also called external forces) will stay at rest if it was at rest to begin with and will keep moving if it was moving to begin with; in the latter case, its motion will be in a straight line at an unchanging speed. In other words, all bodies have inertia.

It is possible to see the law of inertia in action by observing motion that is horizontal (to eliminate the effect of gravity) and nearly frictionless. For example, an object that coasts on a cushion of air or other gas moves with practically no friction. Figure 3.6a is a multiple-flash photograph of such a moving coaster, made in a completely dark room with a camera whose shutter stayed open, using a rapidly flashing light to illuminate the coaster only briefly at several equally spaced times.

As you can see by checking the meter stick next to the coaster, the coaster is moving in a straight line at an unchanging speed: It moves the same distance (as far as it is possible to measure this in the photo) in each time interval. Although careful measurement would show that air resistance slows the coaster slightly, this example is close enough to Galileo's ideal case that if you view this demonstration in a lab, it will give you an intuitive feel for the law of inertia.

Space furnishes lots of nice examples. **Space** refers to those regions of the universe outside Earth and outside other astronomical objects, where "Earth" means the solid and liquid and gaseous Earth, including the atmosphere. The atmosphere is so thin at 10 kilometers high (6 miles) that it can no longer sustain life, although air resistance still affects spacecraft at this altitude. Low-orbit satellites operate at several hundred kilometers high. The atmosphere is so thin above 100 kilometers that the drag (air resistance) on satellites is nearly negligible. Figure 3.7 puts these altitudes into perspective. As you can see, even a hundred kilometers high is still relatively near Earth. Beyond this lies space.

Of all the intellectual hurdles which the human mind has confronted and has overcome in the last fifteen hundred years the one which seems to me to have been the most amazing in character and the most stupendous in the scope of its consequences is the one relating to the problem of motion.
　　Herbert Butterfield, historian, referring to the law of inertia, in *The Origins of Modern Science*

---

* This is often called Newton's first law, even though Descartes invented it, because Newton listed it first among his three basic principles of motion. We refer to these three principles as the law of inertia, Newton's law of motion, and the law of force pairs, rather than by their more common but less accurate, less descriptive, and more boring titles: Newton's first law, Newton's second law, and Newton's third law.

FIGURE 3.6
(a) Multiple-flash photo of the motion of a glider on a smooth surface, viewed from above. (b) The glider at rest on the smooth surface.

When astronauts traveled to the moon, their spacecraft's rocket engines first boosted (pushed) them up and into orbit around Earth. Then they fired their rocket engines for a few minutes to leave Earth's orbit and start toward the moon. Then they shut their engines off and coasted for three days, covering about 400,000 kilometers (more than a quarter of a million miles) to the moon. The spacecraft became a long-distance coaster and a great example of the law of inertia.

However, this coasting was not entirely free of external influences. Although there is no significant air resistance in space, there is still significant gravity unless the spacecraft is extremely far from all large bodies such as Earth, sun, and moon. For example, at one-sixth of the distance to the moon, gravity is still 1% as strong as it is on Earth's surface—a much-reduced effect but still not negligible. Figure 3.8 demonstrates this. Keep in mind that gravity has a very long range effect: Although one escapes Earth's atmosphere at just a few kilometers high, Earth's gravity can be felt even at great distances. The spacecraft to the moon actually slowed during the first part of its journey because of the pull of Earth's gravity and then sped up during the last part because of the pull of the moon's gravity.

A big difficulty for a sun-centered astronomy, such as Copernicus proposed around 1550, was the problem of how Earth could keep moving. What is pushing this gigantic sphere around the sun? Without external assistance, it should soon grind to a halt. This problem perplexed Copernicus, and it gave his opponents powerful ammunition.

The law of inertia has the answer, but it came a century too late to help Copernicus. Earth is a coaster in space! Like the astronauts coasting to

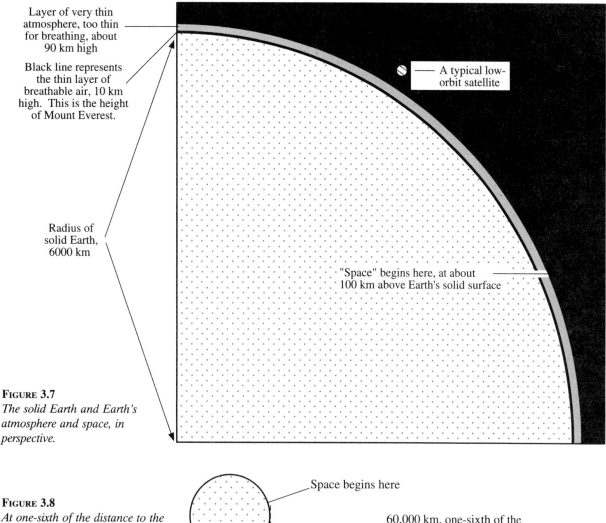

Layer of very thin atmosphere, too thin for breathing, about 90 km high

Black line represents the thin layer of breathable air, 10 km high. This is the height of Mount Everest.

A typical low-orbit satellite

Radius of solid Earth, 6000 km

"Space" begins here, at about 100 km above Earth's solid surface

FIGURE 3.7
The solid Earth and Earth's atmosphere and space, in perspective.

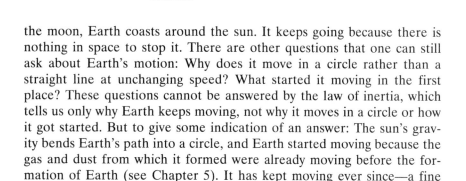

FIGURE 3.8
At one-sixth of the distance to the moon, gravity still has 1% of the effect it has on Earth's surface: much reduced but still not negligible.

Space begins here

Radius of Earth, 6000 km

60,000 km, one-sixth of the distance to the moon. At this distance, the effect of Earth's gravity is reduced to 1% of its normal value.

the moon, Earth coasts around the sun. It keeps going because there is nothing in space to stop it. There are other questions that one can still ask about Earth's motion: Why does it move in a circle rather than a straight line at unchanging speed? What started it moving in the first place? These questions cannot be answered by the law of inertia, which tells us only why Earth keeps moving, not why it moves in a circle or how it got started. But to give some indication of an answer: The sun's gravity bends Earth's path into a circle, and Earth started moving because the gas and dust from which it formed were already moving before the formation of Earth (see Chapter 5). It has kept moving ever since—a fine example of the law of inertia!

Another difficulty for a moving-Earth theory of astronomy is that it seems as though birds, clouds, and baseballs in flight should be left behind the moving Earth. Since they are not attached to the ground, how can they keep up with Earth as Earth moves through space? The answer is that birds, clouds, and baseballs have inertia, too. A bird standing on the ground is participating in Earth's 30-kilometer-per-second (70,000 miles per hour) coaster ride around the sun, so when the bird takes to the air, it is already moving at this speed around the sun, and the law of inertia says that there is no reason for it to stop. It coasts right along with Earth.

DIALOGUE 2   According to Aristotelian physics, what outside influences act on a stone while it falls? According to Newtonian physics?

DIALOGUE 3   Suppose that because of unknown causes, the sun suddenly "stood still," in other words, suddenly stopped moving across the sky as seen by an Earth-based observer. From the viewpoint of Copernican astronomy, what would this mean? Because of Earth's daily spin, a typical point on Earth's surface moves around Earth's center at about 1600 kilometers per hour (1000 miles per hour). So what would happen to people and houses at the moment the sun "stood still"?

3.4 Measuring motion: *velocity*

Sometimes **quantitative**, or numerical, methods are needed to get at nature's deeper secrets. At other times, however, quantitative details are superfluous, and **qualitative**, or nonnumerical, descriptions are preferable. For a clear understanding of motion and related ideas like force and energy, we need to think both qualitatively and quantitatively. And quantitative methods are needed to understand important practical matters such as the world's energy problems.

Scientists like to specify their measurements in terms of just a few basic quantities. To describe motion, only two are needed: distance and time. For example, suppose we want to describe quantitatively the motion photographed in Figure 3.6 using only the meter stick shown and a clock. From the photo we can read the coaster's position at each instant that the light flashes. With a clock, we can measure the time at which each flash occurs. Suppose, then, we start the clock at the first flash, when the coaster is at the 10-centimeter (cm) mark. Then, measured on this clock, successive flashes occur at 0.40 seconds (s), 0.80 s, 1.20 s, 1.60 s, 2.00 s, and 2.40 s. Table 3.1 tabulates these data. The distances are estimated to the nearest millimeter (0.1 cm) from Figure 3.6.

---

2. None. Gravity and air resistance.
3. It would mean that Earth stopped spinning. Because of the law of inertia, people and houses would then immediately start sliding across the Earth's surface, at 1600 km/hr. Eventually, friction would bring them to rest. There would also be 1600 km/hr winds across Earth's surface (ten times faster than strong hurricane winds); clouds would move at 1600 km/hr across the sky; and so forth. Most objects would slide for about 2 minutes, across a distance of about 25 kilometers (15 miles), and would be heated by friction to nearly the boiling point of water.

TABLE 3.1

Positions and Times for Coaster of Figure 3.6

Clock time (s)	Position (cm)
0.00	10.0
0.40	24.1
0.80	38.2
1.20	52.3
1.60	66.4
2.00	80.5
2.40	94.6

How can we use these data to describe how fast the coaster is moving? In other words, how can these data tell us what a speedometer attached to the coaster would read? The seven clock times in Table 3.1 divide the trip into six time intervals (spans of time), each having a duration of 0.40 s. Inspection of Table 3.1 shows that the distance traveled during any one of these time intervals is 14.1 cm. What would a speedometer read during, say, the first of these time intervals?

A speedometer tells you the distance traveled per (in each) unit of time, such as the number of kilometers in each hour or the number of centimeters in each second. To choose easier numbers, suppose you walked 6 kilometers in 2 hours. The number of kilometers in each hour would be 6 divided by 2, or 3 kilometers in each hour. This is the distance divided by the time. So a speedometer attached to the coaster should register

$$\frac{14.1 \text{ cm}}{0.40 \text{ s}}$$

We call this the coaster's **speed**, or the distance traveled during a time interval divided by the duration of that time interval.

If you divide 14.1 cm by 0.40 s, you will get 35.2 centimeters per second. This second means that the coaster would travel 35.2 centimeters in each second if it traveled at this unchanging speed for a second or more. We abbreviate this as 35.2 cm/s, where the divide sign (/) is to be read "per" or "in each."

But there is a problem with defining speed in this way. Suppose you ride your bicycle 24 kilometers in 3 hours. According to this definition, your speed would be 24 km / 3 hr = 8 km/hr. But surely you did not maintain exactly this speed during every minute of the 3-hour trip. The single value, 8 km/hr, does not describe changes of speed during the trip. Rather, the 8 km/hr is an overall speed, actually the speed you would have had to maintain in order to make the entire trip at an unchanging speed. We call this the **average speed** for the trip. It is only in the special case of an unchanging speed that the average speed is really the speed during the entire time interval.

But a car's speedometer gives a value at every instant of time. What does a speedometer read? A speedometer is based on wheel rotation rates. If you look closely at the way it operates, you will find that it actually reads only an average speed during some time interval but that this time interval is very short—so short that the speed is practically unchanging during this time interval.* This idea is called the **instantaneous speed**, the average speed during a time interval that is so short that the speed hardly changes. It is defined in the same way that average speed is, but with the understanding that the time interval is short. It is what a speedometer reads. We use the unmodified word *speed* to mean "instantaneous speed," the kind of speed you read on a speedometer, and we will always use "average speed" when that is what we mean.

---

* A car speedometer is based on the effects created by a magnet that rotates when the car's wheels rotate. It usually gives a good approximation of the instantaneous speed. But if the wheel rotation rate changes very quickly, for instance, during sudden braking, the speedometer cannot respond fast enough to measure the instantaneous speed and so gives only an average. Every measured speed is really an average speed during some time interval.

Quantitative statements, like our definition of average speed, are often easier to grasp if written as an abbreviated formula:

$$\text{average speed} = \frac{\text{distance traveled}}{\text{duration of time interval}}$$

This can be further abbreviated using symbols. We can choose the symbols to suit ourselves. We will use s, d, and t to stand for the average speed during some time interval, the distance traveled during that time interval, and the duration of that time interval. With this understanding, the formula is

$$s = \frac{d}{t}$$

Formulas like this are not essential to physics! The idea, not the formula is essential. The formula can be stated in words, without symbols. In fact, if you do use a formula, be sure that you can first state it completely in words, because otherwise you could easily use the formula incorrectly. For example, the t in the speed formula's denominator is not just any arbitrary time—it means something very specific: the duration of the time interval during which the object traveled the distance indicated in the numerator. It's easy to get carried away by formulas in physics and to think that you understand an idea when all you've done is remember a formula. Physics is about ideas, not formulas, and ideas (including quantitative ones) are expressible in words.

When you travel, it makes a difference which direction you move. Jogging at 10 km/hr northward will get you to a different place than will jogging at 10 km/hr westward. The combined concept of speed and direction of motion occurs so frequently in physics that it is useful to have a separate word for it. We use the word *velocity* to refer to this combination. The words *speed* and *velocity* are interchangeable in everyday language, but in physics, **velocity** always means both the speed and the direction of motion. The distinction might seem small, but small distinctions can be crucial.

DIALOGUE 4 A car driving through a city travels 12 kilometers in half an hour. A bicycler sprints for 1 minute at a steady 30 km/hr. Which has the higher average speed, the car or the bicycler?

DIALOGUE 5 Are the following speeds increasing, decreasing, or unchanging? (a) A car covers longer and longer distances in equal times. (b) A car takes longer and longer times to cover equal distances. (c) A car covers equal distances in equal times. (d) A car covers equal distances in shorter and shorter times. (e) A car takes equal times to cover equal distances. (f) In equal times, a car covers shorter and shorter distances.

DIALOGUE 6 Two bicyclers pass each other on a straight road, one moving north and the other south. Could they possibly have the same velocities? The same speeds?

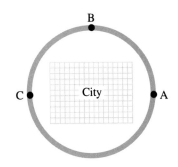

FIGURE 3.9
Illustration for Dialogue 7.

DIALOGUE 7   The automobile beltway around many cities is approximately circular. Suppose that you start driving at point A on the east side of the beltway and drive counterclockwise at an unchanging 90 km/hr (Figure 3.9). (a) As you pass point B on the north side, what is your speed? Your velocity? (b) As you pass point C on the west side, what is your speed? Your velocity?

3.5 Measuring motion: *acceleration*

The law of inertia says that a body that feels no external forces must move in a straight line at an unchanging speed, or it must remain at rest if it were at rest to begin with. But motion in a straight line at an unchanging speed is motion at an unchanging velocity, and the condition of remaining at rest is also a condition of unchanging velocity (the velocity remains zero). So we can state the law of inertia more concisely:

> **THE LAW OF INERTIA (MORE CONCISE FORM)**
> A body that is subject to no external forces must maintain an unchanging velocity.

Now suppose that there are external forces. How will they affect an object's motion? It's not too difficult to guess the answer once you understand the law of inertia: External forces must cause changes in velocity. As we will see in the next chapter, this idea is central to Newtonian physics.

So changes in velocity are crucial to understanding Newtonian physics. Any object whose velocity is changing is said to be **accelerated**. The next dialogue will exercise your thinking about this idea. Remember: An object will be accelerated only if its velocity is changing, and velocity refers to the combined instantaneous speed and instantaneous direction of motion.

FIGURE 3.10
Illustration for Dialogue 8(e).

DIALOGUE 8   During a trip, a car executes several different kinds of motion. For each case, state whether the car is accelerated: (a) moving along a straight level road at a steady 70 km/hr, (b) moving along a straight level road while speeding up from 70 km/hr to 80 km/hr, (c) moving along a straight level road while slowing down from 80 km/hr to 50 km/hr, (d) rounding a curve at a steady 50 km/hr, (e) moving uphill along a straight incline at a steady 50 km/hr (Figure 3.10), (f) rounding the top of a hill at a steady 50 km/hr (Figure 3.11), (g) stopping (coming to rest) along a straight level road, and (h) starting up from rest along a straight level road.

FIGURE 3.11
Illustration for Dialogue 8(f).

4. The average speed of the car = 12 km / 0.5 hr = 24 km/hr, which is slower than the bicycler.
5. (a) Increasing. (b) Decreasing. (c) Unchanging. (d) Increasing. (e) Unchanging. (f) Decreasing.
6. No. Yes.
7. (a) 90 km/hr, 90 km/hr toward the west. (b) 90 km/hr, 90 km/hr toward the south.
8. (a) No. (b) Yes. (c) Yes. (d) Yes, because the direction of motion is changing, so the velocity is changing. (e) No, because neither the instantaneous speed nor the instantaneous direction of motion is changing. (f) Yes, because the direction of motion is changing. (g) Yes. (h) Yes.

We have seen how to describe velocity in terms of two measured quantities: distance and time. What about changes in velocity? To answer this, imagine a car moving north along a straight level highway. Suppose it speeds up, say from 70 km/hr to 82 km/hr. Its change in speed is then 12 km/hr. Imagine how this would feel to you if you were in the car. It would make a difference to you how fast this change took place. For instance, if it took place over an entire hour, you would hardly notice it but if it occurred during one-tenth of a second, you could wind up with a whip-lashed neck!

So to describe the speedup adequately, we need to know the time interval during which it occurred. Suppose that as measured on a clock that was started when the car began its trip, the clock time was 48 s when the car was moving 70 km/hr and 56 s when the car was moving at 82 km/hr (Figure 3.12). The speed change of 12 km/hr then occurred during a time interval of 8 s. How should we combine these numbers to get something useful? Recall how we combined distance and time to get speed: Speed is the distance covered per unit of time. This suggests that in our example of a 12 km/hr speed change, the thing we want to look at is the amount of speed change per unit of time. Since the speed change occurs during 8 s, the speed change per second is

$$\frac{12 \text{ km/hr}}{8 \text{ s}}$$

FIGURE 3.12
A car speeding up.

Clock reading:	0 s	48 s	56 s

Speed: 0 km/hr (starting to move)   70 km/hr   82 km/hr

If you divide 12 km/hr by 8 s, you will get 1.5 kilometers per hour per second. These units tell us that the speed changes by 1.5 km/hr in every second. The rate at which the speed changes is 1.5 km/hr per second. Using the division sign to mean "per," we write this as 1.5 (km/hr)/s. This is the rate at which the car's speed changes, or the rate of speeding up.

This looks like a useful quantity. It is called the *acceleration* of the car. But we found that because of the law of inertia, it is changes in velocity that are important, and not only the changes in speed. An object's velocity changes not only when it speeds up but also when it slows down and when it changes its direction of motion (see Dialogue 8). So we want to define our new concept, acceleration, in such a way that an object accelerates in all of these cases. This will allow us to express the ideas of Newtonian physics in the simplest possible way.

So we define the **acceleration** of a moving object during some time interval as its change of velocity during that time interval, divided by the duration of that time interval. In symbols,

$$\text{acceleration} = \frac{\text{change in velocity}}{\text{time to make that change}}$$

Although this definition is useful in physics, it has the drawback of not entirely agreeing with the popular meaning of acceleration. The scientific meaning includes not just speeding up (the common meaning of acceleration) but also slowing down (commonly called deceleration) and changing direction.

Definitions of words like *acceleration* in physics are arbitrary, in the sense that nature does not tell us that we must define these words in any particular way. Physicists define their words for maximum convenience.

The distinction between velocity and acceleration is important and often misunderstood. *Velocity* refers to motion itself—an object has a velocity whenever it is moving. But *acceleration* refers only to changes in velocity.

The law of inertia says that in the absence of external forces, objects move with unchanging velocity. But "unchanging velocity" is the same thing as "unaccelerated." So the acceleration concept gives us yet another, very concise, way to state the law of inertia:

> THE LAW OF INERTIA (MOST CONCISE FORM)
> A body that is subject to no external forces must be unaccelerated.

You might wonder exactly how you would measure the change in velocity in those cases in which the direction changes. We will take this up in Chapter 5. Here, our quantitative examples will stick to objects speeding up along a straight line. Then the change in velocity is numerically the same as the change in speed, and the acceleration is just the change in speed divided by the time to make that change.

DIALOGUE 9   Give examples of an object that has (a) a high velocity and low acceleration, (b) a low velocity and high acceleration, and (c) an unchanging speed but high acceleration.

DIALOGUE 10   Let us return to the car speeding up along a straight road. Find its acceleration when (a) it speeds up from 70 to 82 km/hr in 4 s instead of in 8 s, (b) it speeds up from 70 to 82 in 16 s, (c) it speeds up from 70 to 94 km/hr in 8 s, and (d) it speeds up from 70 to 76 km/hr in 8 s.

3.6 Galileo's law of falling

We have seen that simple experiments demonstrate that if air resistance is negligible, any two objects dropped together will fall together. We summarize this idea as:

> GALILEO'S LAW OF FALLING
> If air resistance is negligible, then any two objects that are dropped together will fall together, regardless of their weights and regardless of the substances of which they are made.

---

9. (a) Any fast-moving object whose speed or direction of motion is changing slowly, such as a high-velocity bullet that is gradually slowing down; (b) any slow-moving object that is quickly speeding up, such as a car starting up rapidly; (c) any object turning a sharp corner at a high but unchanging speed.
10. (a) 3.0 (km/hr)/s, (b) 0.75 (km/hr)/s, (c) 3.0 (km/hr)/s, and (d) 0.75 (km/hr)/s.

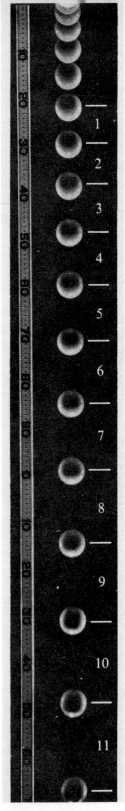

This is a pretty general statement. It applies to any two objects. Each object could be anything: a stone, shoe, frog, person, feather, balloon, even a helium-filled balloon, just as long as the air resistance is negligible. For a feather, balloon, or helium-filled balloon, you need a nearly complete vacuum before air resistance becomes truly negligible, but once you do have a vacuum, you will find that even a helium-filled balloon that floats upward in air falls just like a stone. As Galileo put it, "a grain of sand falls as rapidly as a grindstone." Scientists get excited about ideas like this that apply to such a wide variety of situations.

Galileo's law of falling tells us that we can learn about the fall of any object by studying the fall of just one particular object, because all objects fall in the same way. How does your book fall?

HOW DO WE KNOW? YOU CAN LEARN A LOT JUST BY LOOKING.

Hold this book, flat, above the floor, and let it go.

—This is a pause, for dropping your book.

What is the numerical value of the book's speed at the instant you release it? While you are holding it, the book's speed is zero, so its speed at the instant of release must also be zero. But does its speed remain zero? Obviously not. At the beginning of the motion, the speed increases from zero to something bigger than zero. So the book must accelerate, at least at the beginning.

Now drop your book from about half a meter above the floor. Pick it up and drop it again from about 2 meters above the floor. (Maybe you should do this with some other book if you want to preserve this one!) On a hard floor, you should hear the book hit noticeably harder the second time. So the book must have been moving faster at the end of the second drop. This tells us that books get going faster during 2 meters of fall than during half a meter. You could repeat this experiment for other heights, with similar results demonstrating that objects that fall farther attain a higher speed at the end of the fall. Applied to the fall of a single object, this tells us that the farther an object falls, the faster it moves. In other words, neglecting air resistance, falling objects accelerate all the way down.

The preceding two paragraphs demonstrate the power of careful observation, including the observation of everyday occurrences like a falling object. As the Yankee catcher and famous sage Yogi Berra once said, "You can learn a lot just by looking."

Figure 3.13 shows how falling can be measured quantitatively. It is a multiple-flash photo (described in Section 3.3) of a falling object. A billiard ball falls past a 2-meter stick while being photographed at several equally spaced times. The time interval between photos is 1/30 s. Directly from the photo, you can see that the portion of the fall shown covers a total distance of nearly 2 meters and takes a total of a little more than half a second. You can see the ball's acceleration: The images get farther and farther apart.

Figure 3.14 is an idealized diagram of a ball falling through a larger distance, neglecting air resistance. A real object falling farther than about 20 m is strongly affected by air resistance because air resistance becomes

FIGURE 3.13
A multiple-flash photo of a falling billiard ball. The position scale is in centimeters, and the bulb flashed every 1/30 second.

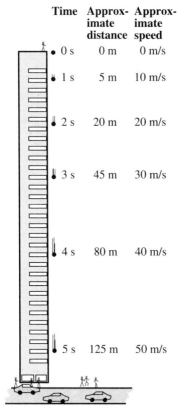

Time	Approx-imate distance	Approx-imate speed
0 s	0 m	0 m/s
1 s	5 m	10 m/s
2 s	20 m	20 m/s
3 s	45 m	30 m/s
4 s	80 m	40 m/s
5 s	125 m	50 m/s

FIGURE 3.14
A freely falling ball dropped from the top of a tall building. The effect of air resistance is neglected in this illustration.

stronger as an object moves faster. We will, however, neglect the effect of air resistance for now, in order to focus on the effects due to gravity alone. So Figure 3.14 shows how an object would fall if dropped from a high building onto an airless Earth. An object like this whose falling is influenced by gravity alone, and not by air resistance or other influences, is said to be in **free fall.** The ball in the figure is in free fall for more than 5 seconds.

Suppose you measure distances downward from the release point and that a speedometer attached to the ball measures the ball's speed as it falls. At the instant of release, you start a clock that tells you the time elapsed since the ball was dropped. Figure 3.14 pictures the ball's fall and gives some of the data you would get in this idealized experiment. The data show the approximate positions, measured downward from the top, and the approximate speeds, at clock times of 1 s, 2 s, 3 s, 4 s, and 5 s after release. In order to emphasize the pattern, the actual experimental data are rounded.*

You can see, in three different ways, that the ball accelerates: First, the drawing shows the stretching out noted in Figure 3.13. Second, the speed is greater at each clock time, so the ball is moving faster and faster. Third, if you look closely at the distance data, you will see that the distance intervals covered during each successive time interval grow larger and larger. For instance, the distance intervals covered during 0 to 1 s is 5 m; the distance covered during 1 to 2 s is 15 m; the distance covered during 2 to 3 s is 25 m; and so forth. These distance intervals (5 m, 15 m, 25 m) stretch out, so the ball must be accelerating.

How big is this acceleration? Recall that the acceleration during any time interval is the change in velocity during that time interval divided by the duration of that time interval. Since this is just straight-line motion at increasing speed, we can replace *velocity* in this prescription with *speed*, as discussed in the preceding section. Now look at each successive 1-second time interval. All the changes in speed are precisely the same! During 0 to 1 s, the speed change is 10 m/s. During 1 to 2 s, it is also 10 m/s, and so forth. So calculating the acceleration for any one of these time intervals,

$$\text{acceleration} = \frac{\text{change in speed}}{\text{time to change}} = \frac{10 \text{ m/s}}{1 \text{ s}}$$

which is 10 meters per second per second, or 10 (m/s)/s, usually abbreviated as 10 m/s^2. This means that the speed changes by 10 m/s in every second. The units can be confusing because both times are measured in seconds. Time must appear twice because acceleration is a rate of change of speed and speed is a rate of change of position. Two rates of change are involved.

Since we get 10 m/s^2 for every time interval, the acceleration is the same all the way down. And Galileo's principle of falling tells us that it must be the same for every freely falling object. The numerical value of this **acceleration due to gravity** is, more accurately, 9.8 m/s^2. When accuracy is important, you should use 9.8 rather than 10 m/s^2. The precise measured value varies by as much as 0.5% at different places in the world because of

---

* The rounding-off process introduces a 2% error. More precise distances are 4.9 m, 19.6 m, 44.1 m, 78.4 m, and 122.5 m. More precise speeds are 9.8 m/s, 19.6 m/s, 29.4 m/s, 39.2 m/s, and 49 m/s. Using these more accurate numbers, the calculated acceleration is (9.8 m/s) / (1 s) = 9.8 m/s^2.

variations in Earth's shape and in the material of which Earth's crust is made. For example, it is 9.832 m/s² at the North Pole, 9.803 in New York City, and 9.782 in Panama, near the equator.

There is a pattern in the speed data. At clock times of 0 s, 1 s, 2 s, 3 s, 4 s, and 5 s, the speed (in m/s) is 0, 10, 20, 30, 40, 50, and so on. Just by looking at this string of numbers, it is easy to guess what should come next: 60. The reason this can be guessed is that there is a recognizable pattern here. Patterns are what scientists look for in nature. Kepler (see Chapter 1) searched for years through Brahe's data before finding the elliptical pattern in the data. Your expectation that the sun will rise tomorrow is also based on a pattern in nature.

One way to describe this pattern is with proportionalities. One quantity is **proportional to** a second quantity if doubling the first means you must double the second, tripling the first means you must triple the second, and so forth. In general, whatever you multiply the first quantity by, you must also multiply the second quantity by. Looking at the speed data, doubling the time (from 1 to 2 seconds, say) results in doubling the speed (from 10 to 20 m/s), and similar proportionalities hold throughout the data. So, for free fall:

$$\text{speed is proportional to the time;} \qquad s \propto t$$

The symbol $\propto$ means "is proportional to."

Similarly, speed is proportional to the clock time for any object that starts from rest and moves in a straight line with unchanging acceleration. For example, an object dropped onto the surface of the moon falls freely (there is no air resistance on the moon, because there is no air or other atmosphere) with an unchanging acceleration of 1.7 m/s². So a falling object reaches a speed of 1.7 m/s at the end of 1 s of falling, 3.4 m/s at the end of 2 s, 5.1 m/s at the end of 3 s, and so forth. As on Earth, the speed is proportional to the time; for instance, the speed at the end of 3 s is three times greater than the speed at the end of 1 s.

The pattern in the position data is not as easy to recognize. To make it easier to find, let's express the position data in multiples of the position at 1 s, in other words, in 5-meter units. The data tell us that at 2 s, the position (the total distance fallen) is four of these units, at 3 s it is nine units, at 4 s it is sixteen units, and at 5 s it is twenty-five units. So the positions, measured in 5-meter units, are 0, 1, 4, 9, 16, and 25. What is the next number in this sequence? In other words, what is the pattern?

To recognize the pattern, note that each of the numbers is a perfect square: 0^2, 1^2, 2^2, 3^2, 4^2, 5^2. What comes next? 6^2, or 36 of our 5-meter units! To get the distance in meters, multiply this by 5: $5 \times 36 = 180$ m.

This pattern can also be described quantitatively with proportionalities. The distances, in meters, are **proportional to the square of** the time. This means that doubling the time multiplies the distance by four, tripling the time multiplies the distance by nine, and so forth. In general, whatever number you multiply the time by, you must square this number and then multiply it by the distance. For instance, if you multiply the time by 3 (from 1 to 3 seconds, say), then you must multiply the distance by 3^2, or 9 (from 5 to $9 \times 5 = 45$ m). So for free fall:

$$\text{distance fallen is proportional to the square of the time;} \qquad d \propto t^2$$

Again, this is a proportionality that is valid for any case of unchanging acceleration. For instance, distance is proportional to the square of the time for an object falling freely onto the moon.*

DIALOGUE 11   How much farther does a freely falling object fall in three times as much time? In four times as much time? Would your answers be the same if the object were falling freely onto the surface of the planet Mars?

DIALOGUE 12   Answer "true" or "false" to each of the following statements regarding free fall: (a) The total distance covered keeps increasing during the fall. (b) The distance covered during each second keeps increasing. (c) The speed keeps increasing. (d) The change in speed during each second keeps increasing. (e) The acceleration keeps increasing.

---

*   With further analysis, we could arrive at two general formulas for the speed and position of any object that starts from rest and moves in a straight line with unchanging acceleration:

$$s = at, \qquad d = (\tfrac{1}{2}) at^2$$

where a represents the acceleration. For freely falling objects on Earth, $a = 9.8$ m/s².

11. Nine times as far. Sixteen times as far. Yes, both answers would be the same on Mars.
12. (a) True. (b) True. (c) True. (d) False. (e) False.

Summary of Ideas and Terms

Natural motion Unassisted motion. According to Aristotelian physics, falling is one form of natural motion, and horizontal motion is always unnatural, or violent. According to Newtonian (and current) physics, any motion at a constant velocity is natural motion, and falling is not natural motion.

Aristotelian elements Earth, water, air, fire, and ether.

Difficulties with Aristotelian physics Contrary to Aristotelian predictions, heavy and light objects often fall at the same speed, and horizontally moving objects would move forever if there were no external forces.

Inertia An object's ability to stay at rest or to maintain an unchanging velocity when acted on by no external forces.

Law of inertia A body that is subject to no external forces maintains an unchanging velocity. More briefly: All bodies have inertia.

Space The universe outside Earth and its atmosphere and outside other astronomical bodies.

Average speed An object's average speed during a time interval is the distance traveled during that time interval divided by the duration of that time interval, $s = d/t$.

Instantaneous speed, or **speed** The average speed during a time interval that is so short that the speed hardly changes. Speedometers measure this.

Velocity The combined instantaneous speed and instantaneous direction of motion.

Acceleration An accelerated object is one whose velocity is changing. The acceleration is quantitively the change in velocity during a time interval divided by the duration of that time interval.

Galileo's law of falling Neglecting air resistance, any two objects dropped at the same speed fall at the same speed, regardless of their weights.

Free fall Falling that is influenced only by gravity and not by air resistance or other influences.

Acceleration due to gravity The acceleration of any freely falling object. On Earth this is about 10 m/s^2 or, more precisely, 9.8 m/s^2.

Proportionality One quantity is proportional to a second quantity if the first is always multiplied by whatever the second one is multiplied by. One quantity is **proportional to the square of** a second quantity if the first is always multiplied by the square of whatever the second is multiplied by.

Quantitive description of free fall For an object that starts from rest and then falls freely: $s \propto t$, $d \propto t^2$. These proportionalities are correct for any motion that starts from rest and maintains an unchanging acceleration in a straight line.

Review Questions

ARISTOTELIAN PHYSICS

1. Describe the four kinds of motion that Aristotle considered to be natural on Earth.
2. Give two examples of violent motion.
3. According to ancient Greek thought, in what fundamental way does Earth differ from the heavens?
4. Give an example of a motion that contradicts Aristotelian physics.
5. According to Aristotelian physics, why does a stone fall when it is released above the ground? According to Newtonian physics?
6. Describe at least one principle of Aristotelian physics that seems intuitively plausible but that Newtonian physics rejects.

THE LAW OF INERTIA

7. What does it mean to say that an object has inertia?
8. What does the law of inertia say about the velocity of a body that is subject to no external influences? What does this law say about such a body's acceleration?

9. Which of the following describes how high you must go before you will first reach "space": Anywhere above the ground, about 100 meters, about 1 kilometer, about 100 kilometers, beyond the moon, beyond the solar system?
10. Is there air in space?
11. Give at least one example that is very nearly unassisted motion at a constant velocity.

VELOCITY

12. Describe how you could use a clock and a meter stick to measure a moving object's speed.
13. When we say "5 centimeters per second," what does the "per" mean?
14. What is the difference between speed and average speed? In what circumstances are they the same?
15. Can you give an example in which the speed is unchanging but the velocity changes? If so, give one.
16. Can you give an example in which the velocity is unchanging but the speed changes? If so, give one.

ACCELERATION

17. A car speeds up along a straight line. Describe how you could use clocks and meter sticks to measure its acceleration.
18. How is acceleration related to velocity?
19. If an object's position is changing, can we be certain that it has a nonzero velocity? Can we be certain that it has a nonzero acceleration?
20. If an object is moving in a circle at an unchanging speed, is it accelerated?
21. If an object is slowing down, is it accelerated?

FALLING

22. An object is released above the ground and falls freely. At which of these positions during the fall is its velocity greatest? The top, the midpoint, or a point near the bottom? At which position is its acceleration greatest?
23. What is the meaning of the phrase "acceleration due to gravity"? What is its approximate value on Earth?
24. What does "speed is proportional to time" mean?
25. In twice the time, does a freely falling object fall (from rest) twice as far, at twice the speed, or both?
26. Which of these increases as an object falls freely: distance (from the starting point), speed, velocity, or acceleration?
27. Which of these increases for an unassisted (unforced, or isolated) moving object: distance (from the starting point), speed, velocity, or acceleration?

Home Projects

1. Drop two sheets of notebook paper to the ground at the same time, one of them open and the other crumpled up tightly. Drop the crumpled sheet and a stone at the same time. Drop a pebble and a large stone at the same time. Explain the results.
2. Get a small piece of notebook paper, lay it on top of a large book, and drop the book. Now hold the book and the paper in different hands and drop the two at the same time. Try to explain the result, and compare your explanation with the answer in the footnote.*
3. Measure the speed of an ant. Use a ruler (metric, preferably) and a watch with a second hand. Try to find an ant that moves in a straight line.
4. Check the accuracy of your car's speedometer, using the kilometer or mile markers posted along many highways. Have a friend measure the time, using a watch with a second hand. Calculate the speed. It is not safe to do this alone—have a friend along to measure the time.
5. Measure various speeds, such as your top bicycling speed, running speed, and hopping-on-one-foot speed. Measures the distances with a tape measure or by pacing after you have measured your pace size.
6. Measure the time for your car to accelerate from 0 to 50 km/hr (30 mi/hr). For safety, have a friend along to do the timing. Find the car's acceleration in (km/hr)/s or (mi/hr)/s. Convert your result to m/s^2 by using the facts that 1 km = 1000 m, 1 mi = 1610 m, and 1 hr = 3600 s. An acceleration of 1 g means 9.8 (about 10) m/s^2. What is your car's acceleration in g's?

For Discussion

1. "Meteoroids" are pebble-sized to boulder-sized rocks, similar in many ways to Earth rocks, moving through space. A typical meteoroid moves at a nearly unchanging speed in a nearly straight line. How might Aristotelian physics have been affected if Aristotle had known about this? How might Newtonian physics have been affected if Descartes, Galileo, and Newton had known about this? (Meteoroids were not known to science until about 1800.)
2. Many intuitively plausible beliefs conflict with science. For example, the belief that the sun moves around a stationary Earth conflicts with current astronomy. List several other intuitively plausible beliefs that conflict with either Copernican astronomy or Newtonian physics or both.

---

* When the sheet of paper is placed on top of the book, it keeps up with the book because the book sets the surrounding air into motion at the book's velocity. Since the sheet of paper is falling in *moving* air, the sheet no longer needs to push through the air as it falls. This effect is called *drafting*. Bicycle racers know about this.

Exercises

ARISTOTELIAN PHYSICS

1. You roll a ball. It soon rolls to a stop. How would Aristotle interpret this? How would Galileo interpret it?
2. What explanation would Aristotle's physics give for these facts? Dry wood floats and water-logged wood sinks; an air-filled balloon falls more slowly than does a partly deflated balloon; and smoke rises.

THE LAW OF INERTIA

3. Most meteoroids—pebble-sized to boulder-sized rocks in space—have been moving for billions of years. What, if anything, keeps them moving?
4. If you ride on a smooth fast train at an unchanging speed and throw a baseball upward inside the train, will the baseball then get left behind, and will it come down toward the rear of the car? Explain.
5. If a ball is moving at 20 m/s and no forces ever act on it, what will its speed be after 5 seconds? After 5 years?

VELOCITY

6. Can you drive your car around the block at a constant velocity?
7. Mary passes Mike from behind while bicycling. As she passes him, do the two have the same velocity? Do the two have the same speed?
8. Figure 3.15 represents a multiple-flash photo of two balls moving to the right. The figure shows both balls at several numbered times. The flash times are equally spaced. Which ball has the greater acceleration? The greater speed? The greater velocity? Does either ball pass the other, and if so, when?

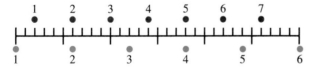

FIGURE 3.15

9.*In the preceding exercise, suppose that the large divisions on the measuring rod are centimeters and that the time intervals each have a duration of 0.20 s. Find the speed of each ball.
10.*Find the average speed of a jogger who jogs 3 km in 15 min. Give your answer in km/hr.
11.*You wish to travel from downtown New York City (NYC) to downtown Washington, DC (DC), a distance of 330 km. You consider two options: train and plane. The high-speed train takes 1.5 hr plus 30 min in stations. The airplane flies the 330 km (airport to airport) in just

30 min, but the drive to the NYC airport takes 30 min; you must arrive 30 min before departure time; the plane waits 15 min for takeoff; and it takes 45 min to get your luggage and drive into DC. Find the train's track speed, the plane's flying speed, the total travel time for each option, and the overall average speed for each option.
12.*You drive from New York City to Washington, DC, by car. You drive in traffic for the first hour at an average 50 km/hr. You cover the next 250 km in 3.0 hr and then drive the remaining 30 km into Washington in 30 min. Find your total time and average speed.

ACCELERATION

13. A French TGV train cruises on straight tracks at a steady 290 km/hr (180 mi/hr). What is its acceleration?
14. Is the "motion sickness" that some people get in a car actually due to motion per se or to something else? Describe one form of motion that would not make people sick.
15. When you drive a car, might you depress the accelerator pedal without actually accelerating? Might you be accelerating without having your foot on the accelerator? Explain.
16. One car goes from 0 to 30 km/hr. Later another car goes from 0 to 60 km/hr. Can you say which car had the greater acceleration? Explain.
17. Figure 3.16 represents a multiple-flash photo of two balls. Describe each ball's motion. Does either ball pass the other? When? Do they ever have the same speed? When?

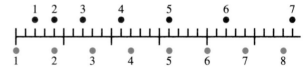

FIGURE 3.16

18. In each of the following cases, is the motion accelerated or not accelerated? (a) A rock falling freely for 2 m. (b) A meteoroid (a rock in space) that is so far from all planets and stars that gravity is negligible. (c) An artificial satellite orbiting Earth at a steady 30,000 km/hr. (d) The moon. (e) An ice-skater coasting on smooth ice, neglecting friction and air resistance.
19. Can a slow-moving object have a large acceleration? Can a fast-moving object have a small acceleration?
20. How many devices in a car are designed to cause acceleration?
21.*A bicycler increases her speed along a straight road from 3 m/s to 4.5 m/s in 5 s. Find her acceleration.

22.*A car accelerates from 0 to 100 km/hr in 10 s. Find its acceleration. Drag racers can reach 400 km/hr from rest in 5 s. How big is this acceleration?

FALLING

23. Multiple choice: Two metal balls are dropped from a cliff at the same time. They are the same size, but one weighs twice as much as the other. The time to reach the ground will be (a) about twice as long for the heavy ball; (b) about twice as long for the light ball; (c) about the same for both; (d) considerably longer for the heavy ball, but not necessarily twice as long; (e) considerably longer for the light ball, but not necessarily twice as long.

24. Figure 3.14 shows that a falling object has a speed of 10 m/s at $t = 1$ s. So it seems that it should move 10 m in 1 s. Yet the data say that the object moves only 5 m in 1 s. What is wrong here?

25. Figure 3.17 represents a multiple-flash photo of a falling ball. Neglect air resistance. At which point, A or B, is the ball's acceleration larger? At which point is its velocity larger?

26. By how much does a freely falling object's speed increase during its third second of fall (from $t = 2$ s to $t = 3$ s after release)? During its fourth second of fall?

27.*An astronaut on another planet drops a rock off a cliff. How much faster is the rock moving at the end of 3 s, as compared with 1 s? How much farther (measured from the release point) does the rock fall in 3 s, as compared with 1 s?

28.*Neglecting air resistance, would the answers to the preceding exercise be different if the rock were dropped on Earth? Neglecting air resistance, would the distance fallen in 3 s on Earth be likely to be the same as the distance fallen in 3 s on the other planet?

29.*Along a straight highway, a car speeds up from rest. Its acceleration is unchanging. How much farther (measured from the starting point) does it go in 10 s, as compared with 1 s? How much faster it is moving in 10 s, as compared with 1 s?

30.*On the planet Mars, a free-falling object released from rest falls 4 m in 1 s and is moving at 8 m/s at that time. How fast would such an object be moving after 2 s? Three seconds? How far would such an object fall in 2 s? Three seconds?

FIGURE 3.17

4

WHY THINGS MOVE AS THEY DO

Nature and nature's laws lay hid in night;
God said, "Let Newton be," and all was light.

Alexander Pope, eighteenth-century British poet

The world changed in 1687. In that year, Isaac Newton published his *Mathematical Principles of Natural Philosophy* (Figure 4.1). To take just one example, Newton's work firmly established the law of inertia as the new way of thinking about what kinds of motion are natural. First proposed by Descartes and Galileo, this law became the foundation of Newton's new theory. After Newton, the inertial view of motion was the norm for physical science. In one stroke, this view undermined our intuitive notion of how things move, a 2000-year-old view first made explicit by Aristotle.

The Newtonian world is surprisingly simple. Using only a few key concepts and principles, Newton was able to give clear, quantitative explanations for motion on Earth and in the heavens. Newton's theory explains the orbits of planets and moons, Kepler's laws of planetary motion, the motion of comets, the motion of falling objects, weight, ocean tides, and Earth's slight equatorial bulge. It lies behind most technology, from the motion of an automobile to the stresses on a bridge. It was a unification and deepening of our understanding of nature, such as the world had never before seen.

Newtonian principles guided not only physics, astronomy, and technology but also all of the other sciences and much of society. Newtonian concepts such as force, inertia, mechanical interactions, precise predictability, determinism, and universal law, found their way into scientific fields like chemistry and biology. And thinkers in fields that had not previously been considered scientific—fields such as history, economics, law, politics, sociology,

FIGURE 4.1
Isaac Newton, 1642–1727.

and psychology—thought in terms of Newtonian-like natural laws governing those fields. Historians of science argue that the concepts of inalienable human rights that inspired the American and French revolutions were Newtonian in origin.

Reigning supreme for more than 200 years, Newtonian physics worked so well that it came to be regarded as the absolute truth. Indeed, the very word *understand* came to mean "to explain in terms of Newtonian physics."

Newton's physics starts from only a few concepts and principles. We have already discussed two Newtonian concepts, velocity and acceleration, and two principles, the law of inertia and the law of falling. Newton's other key concepts are force and mass, examined in Sections 4.1 and 4.2. His other key principles are the law of motion (Sections 4.3 and 4.4), the law of force pairs (Section 4.5), and the law of gravity (Chapter 5). As an example, we use these ideas in Section 4.6 to look at the motion of a device that has had a remarkable social impact: the automobile.

4.1 Force: *why things accelerate*

The preceding chapter used the word *force* synonymously with *external influence*. Now we need to be more specific. Like velocity and acceleration, there is no predetermined way that we must necessarily define "force." We want to define this word in the most useful way.

One example of an object that experiences no external influences is a body that is isolated in distant space, far from all planets and stars. The law of inertia predicts that such a body will be unaccelerated. But what about a body that is not isolated, for example, a stone in space with a comet nearby? We often find it convenient to use examples of objects in space, because Earthly effects such as friction and gravity tend to obscure the basic principles of motion. It was precisely such confusion that kept the scientific world thinking in Aristotelian terms for 2000 years. An air coaster riding on a cushion of air on a horizontal surface (Figure 3.6) is another convenient example, because friction is negligible and gravity does not affect the horizontal motion. A smoothly rolling ball is a pretty good example, for the same reasons.

How can we tell whether or not a nearby comet "influences" a stone in space? Physicists answer this question by using the law of inertia as a definition of what we mean by an "influence": We say that the comet influences, or *exerts a force on,* the stone if the comet causes the stone to accelerate.* This is the way we use the word *force*. That is, we say that a body exerts a force on another body whenever the first body causes the second body to accelerate.

There are many misconceptions about the word *force*. The operational term here is *exert a force on*. This term is a verb. Force is an action. Force is not a thing: A body cannot be a force. And force is not a property of a body: A body cannot possess force. Instead, a body can "exert a force on" another

---

* If you are saying to yourself that this reasoning sounds circular, you are right, for the law of inertia is partly a way of defining natural motion and force. But experiments, like the one shown in Figure 3.6 of Chapter 3, are needed to demonstrate the consistency of the definitions. The law of inertia is part definition, part natural law.

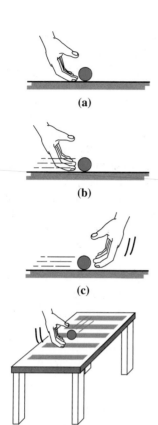

FIGURE 4.2
A push (a) starts a ball moving and (b) speeds a ball up. (c) A brief pat from in front slows a ball down. (d) A brief push from the side changes the direction of motion of a moving ball.

body. It would be best to use *force* only in a pure-verb sense, as in "the comet forces the stone," but unfortunately this is not done.*

Some examples (Figure 4.2): If a ball is at rest on a table and you push it with your hand, the ball will start moving. If a ball is already moving across a table and you push it briefly from behind, the ball will speed up. If a ball is moving across a table and you "pat" it briefly and lightly from in front, toward the rear, it will slow down. If a ball is moving across a table and you push it briefly and lightly sideways, perpendicular to the direction of motion, it will change directions. In all four cases, your hand push accelerated the ball. In fact, a little experimentation with a smooth hard ball rolling on a smooth hard surface (better yet, with an air glider) shows that you cannot push the ball without accelerating it.† Since we defined a force as any influence that causes an acceleration, we conclude that every hand push is a force.

The hand push is probably the most familiar example of a force. When you use the word *force*, it is useful to keep in mind that a force is similar to a push.

Pulling is another example. Starting with the ball at rest on the table, you can clasp it and pull it toward you, accelerating it into motion. So pulls, as well as pushes, are examples of forces. Instead of pulling the ball with your hand, you could attach a string and pull the string, which pulls on the ball to accelerate the ball. So strings, when they pull, exert forces.

Only a single force is exerted on the ball in these examples. Now suppose that you and your friend Ebenezer both push on a ball in opposite directions (Figure 4.3). If you and Ebenezer adjust your pushes, you can get them to balance so that the ball remains at rest. Yet you both are pushing on the ball. Even though the ball is unaccelerated, we will still say that you exert a force on the ball and that Ebenezer does too. In cases like this, involving more than one force, we say that a body exerts a force on another body if, in the absence of the other forces, the first body would cause the second to accelerate.

You could tap a ball with a hammer instead of pushing or pulling it with your hand. Since the hammer tap accelerates the ball, it exerts a force on the ball. Try tapping a motionless or moving ball from various directions yourself, and observe it carefully. Exactly when is it accelerating?

—This is a pause for finding a ball and a hammer or other tapper and trying this.

The ball accelerates only during the brief fraction of a second that the hammer is actually touching it. So the hammer exerts a force on the ball only during this fraction of a second. After the tap, the ball moves at a constant velocity, so there is no force exerted on it. Remember that the moving ball does not "have force" or "carry force along with it." To exert a force is to perform a certain action. In this example, the hammer exerts a force on the ball, and only during the brief time of contact.

---

* Scientists like to turn their verbs into nouns, so rather than the verb *to force*, physicists persist in using *to exert a force on*. This bad habit has caused centuries of confusion for students, but it seems impossible to break.
† We are assuming that only a single push is exerted at a time. Two simultaneous pushes could cancel each other, as we will discuss later.

FIGURE 4.3
Both you and Ebenezer are exerting a force on the ball, but the ball is not accelerating. You exert a force on the ball whenever you would cause it to accelerate if no other forces were acting. In the case pictured, Ebenezer's force on the ball prevents your force on it from causing it to accelerate.

Friction and air resistance also are forces. If you briefly shove a book and let go so that it slides across a table, the book will slow down as it slides. We know that some force must cause this slowing, because the law of inertia says that the speed would be unchanging if there were no force on the book. If we glide an air coaster across a table, the air coaster will not slow down. So the force that slows the book must result from the contact between the book and the tabletop. It must be exerted by the tabletop on the bottom of the book. This force, by one surface on another surface, is called **friction**.

A fast bullet moving horizontally through air slows down a little, so there must be a force on the bullet. Since there is no slowing when a bullet moves through a vacuum, we conclude that this force is due to the air around the bullet. It is caused by air molecules that the bullet hits as it travels and is called **air resistance**. The effect is similar to friction: The bullet slides through the air in somewhat the same way that a book slides across a table.

As we know from the preceding chapter, an apple falling freely to the ground accelerates all the way down. Since the apple accelerates, there must be a force on it. It is commonly called **the force of gravity**. But remember that forces are always actions by one object on another object. Gravity is a force on the apple, but what is this force exerted by? The answer must be that this force is exerted by Earth. The experimental evidence for this is that no matter where you go on Earth, a falling apple always accelerates downward, toward Earth's center. The direction of its fall is determined by the position of Earth's center.

You can think of a gravitational force as a pull, although not a human, muscular pull. Forces do not require living creatures to exert those forces. The gravitational force is one example: It is a pull by Earth on nearby bodies. There is an interesting difference between the gravitational force and the other forces we've looked at. Hand pushes, hand pulls, hammer taps, string pulls, friction, and air resistance all are contact forces: forces exerted by an object that is touching another object. The gravitational force by Earth on a falling apple is different, because Earth is not actually touching the apple while it falls. Air is touching the apple, but we could imagine removing the air, or we could take the apple above Earth's atmosphere and then drop it, and the apple would still fall. The gravitational force acts at a distance.

As a final example, if a magnet is brought near another magnet, the second magnet will accelerate toward or away from the first, depending on the orientation of the two magnets. So the first magnet exerts a force on the second, a force that can apparently be either attractive (a pull toward the first magnet) or repulsive (a push away from the first magnet). This **magnetic force** is another force that acts at a distance.

4.2 Connecting force, mass, and acceleration

The basic idea of Newton's theory of motion is that forces cause accelerations. It is a nonintuitive, surprising idea, as we have seen. Our Aristotelian intuitions would like us to believe that forces cause *velocities,* that outside influences are needed to keep a thing moving. The Newtonian view is that objects maintain their velocity without assistance and that outside influences (forces) only *change* an object's velocity. This is science's view today. Although observations during the past century contradicted some fundamental Newtonian principles, no exceptions to the law of inertia have been observed.

What are the details of the relationship between force and acceleration? We'll first discuss this relationship in broad terms and then go through it more carefully.

If you put a smooth ball on a smooth table and tap it once with a hammer, it will accelerate. If you tap it lightly, it will move slowly after the tap, and if you tap strongly, it will move rapidly. Since the time durations for the taps are about the same, the ball must have had a larger acceleration during the strong tap than during the light tap. We conclude that stronger forces cause larger accelerations.

If you tap two different balls, one heavier than the other, you will find that different objects respond differently to similar forces. If you give an equally strong tap to each ball, you will find that the heavy ball accelerates into slower motion than does the light ball. So the heavy ball gets a smaller acceleration during the tap.

It is useful to extend the inertia idea to this situation. Recall that an object's inertia is its ability to maintain its velocity. Because a heavy ball is harder to accelerate than a light one, the heavy one has a greater tendency to maintain its velocity. So we say that the heavy ball has more inertia than the light one. We use the word *inertia* to mean a body's degree of resistance to acceleration.

Summarizing, we conclude that a body's acceleration is larger when a larger force is exerted on it and a body's acceleration in response to a given force is smaller when the body is heavier. This idea, stated only loosely so far, is called *Newton's law of motion*.

Now for a more careful look at this idea. Let's imagine that all experiments are done in distant space, far from all outside influences, including gravity. Objects that are outside the influence of gravity are "weightless"; in other words, if you were in distant space and you put an object in your hand, the object would not be "heavy"—it would not weigh down on your hand. We will look at the reason for this in the next chapter. Here we want to consider objects that are removed from gravity's influence, because gravity has little to do with the basics of Newton's law of motion, and it tends to obscure the discussion.

Experiments show that if you tap any object in space, it will accelerate just as it does when you tap it on Earth in the absence of friction. This is an important and perhaps surprising point. Objects in distant space still respond to a hammer tap the same way they would respond on Earth if you removed the friction. A light tap causes a small acceleration, and a strong tap causes a large acceleration, and the accelerations are quantitatively the same as they are on Earth (neglecting friction). Even though an object in distant space has no weight, just as much force is needed to accelerate it.

So if you take a heavy stone and a light stone into space and tap both of them with similar taps, they will accelerate the same way they accelerated on Earth: The heavy one accelerates a little, and the light one accelerates a lot. But now we cannot say that this is due to their different weights, because neither stone weighs anything in space, and in fact the words *heavy* and *light*

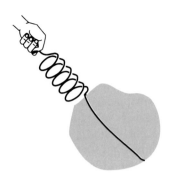

FIGURE 4.4
Imagine pulling on a stone in distant space by means of a spring attached to the stone.

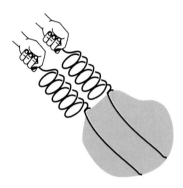

FIGURE 4.5
What happens to the stone of Figure 4.4 when you double the force acting on it?

don't even mean anything in space, where everything has zero weight. How can we describe the difference between the two stones in space without resorting to the concept of weight? One answer is that the first stone (the heavier one on Earth) has a larger amount of matter in it, more material. Microscopically, what we mean here is that the first stone contains more atoms or that its atoms are individually larger (larger atomic numbers—gold as opposed to hydrogen, for instance). So the stone containing more matter gets a smaller acceleration during the tap.

To put this another way, the stone with the most matter has the most inertia. Furthermore, an object's inertia when it is in space is quantitatively the same as it is on Earth.

To make these ideas quantitative, we'll look at some quantitative experiments. It is more convenient now to imagine that springs (Figure 4.4), rather than hammer taps, exert the forces.

Imagine pulling on a stone in space by means of a spring (Figure 4.4). The spring stretches and the stone accelerates, showing again that forces cause acceleration. If you pull in such a way that the amount of stretch doesn't change, then surely the strength of the force will not change either. Newton performed experiments like this, only he couldn't do them in outer space. You might be able to do a similar experiment with air coasters in your physics lab. Such experiments show that when we exert an unchanging force on an object, the object always has an unchanging acceleration. Furthermore, if the force is doubled by attaching two springs instead of one, the acceleration will be doubled too (Figure 4.5). A tripled force causes a tripled acceleration, and so forth: An object's acceleration is proportional to the total force exerted on it. We will call this total force—exerted by all the springs or other objects that might be exerting a force—the **net force** exerted on the object.

We can summarize this in the abbreviated form

$$\text{acceleration} \propto \text{net force};\qquad a \propto F$$

where F stands for the net force exerted by all other objects on the accelerated object.

We've seen that objects containing more matter get smaller accelerations when pulled by the same force. To study this quantitatively, we need to know how to quantitatively measure the amount of matter in an object. This presents problems because we want to be able to do all of this in distant space where objects don't weigh anything. How could we, for example, compare the amounts of matter in a piece of steel and a piece of plastic foam, in outer space, where neither one weighs anything?

To deal with this kind of problem, we must return to the concept of inertia. Recall that objects having more matter also have more inertia. An object's inertia can be measured even when the object is in distant space and has no weight, for instance by tapping it and seeing how much it accelerates. And its inertia in space is the same as on Earth; that is, its inertia is the same everywhere. So it makes sense to use inertia as our way of quantitatively comparing the amount of matter in different objects.

When the concept of inertia is made quantitative in this way, it is called *mass*. That is, *the **mass** of an object is its amount of inertia.* In the next section we will find that this definition of mass agrees with our intuitive notion that two objects have the same mass if they balance on a balance beam.

FIGURE 4.6
The U.S. National Standard Kilogram no. 20, an accurate copy of the International Standard Kilogram kept at Sèvres, France. It is stored inside two bell jars.

To have a measurement scale for mass, we choose one particular object, called the *standard kilogram*, and define its mass to be 1 **kilogram**, abbreviated kg (Figure 4.6). There are good duplicates of it in most physics laboratories. Any other object that has the same inertia as the standard kilogram has a mass of 1 kilogram. And any object having the same inertia as 2 kilograms bundled together is said to have a mass of 2 kilograms, and so forth: We can define the mass of any object in this way, by comparing its inertia with that of 1 or more kilograms (or with half a kilogram or some other fraction).

Now imagine using identical forces to pull two stones through space, with one stone having twice as much mass as the other (Figure 4.7). How do their accelerations compare? When we do this experiment, we find that the stone containing twice as much mass has half as big an acceleration. And three times as much mass yields one-third as much acceleration; four times as much mass yields one-fourth as much acceleration; and so forth. The acceleration is divided by whatever you multiplied the mass by. We express this by saying that an object's acceleration is *proportional to the inverse* of its mass. The inverse of a number is 1 divided by that number. Briefly,

$$\text{acceleration} \propto \frac{1}{\text{mass}}; \qquad a \propto \frac{1}{M}$$

We can put our two proportionalities, $a \propto F$ and $a \propto 1/M$, together to read

$$\text{acceleration} \propto \frac{\text{net force}}{\text{mass}}; \qquad a \propto \frac{F}{M}$$

This says that an object's acceleration is proportional to the net force on it divided by its mass.

We have not yet mentioned any units for force. The unit of force is called, appropriately, the **newton**. The newton is defined as the amount of force that can give a 1 kg mass a 1 m/s² acceleration. When the mass is 1 kg and the force is 1 newton, it must work out that the acceleration is 1 m/s². This means that we can write the proportionality as an equality:

$$\text{acceleration} = \frac{\text{net force}}{\text{mass}}; \qquad a = \frac{F}{M}$$

This formula gives the acceleration in m/s², provided that we express the net force in newtons and the mass in kilograms.

In the "English" system, force is measured in the familiar (in the United States) "pounds." One newton is a little less than a quarter of a pound. Think of a quarter pound of margarine. The "English" unit of mass happens to be called the *slug*. Sounds terrible. You've probably never heard of it, and you can immediately forget it. We can only hope that the United States will someday be able to forget the entire "English" system!

4.3 Newton's law of motion: *centerpiece of Newtonian physics*

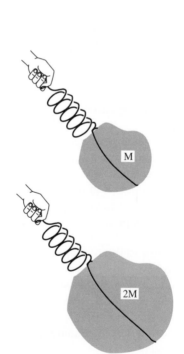

FIGURE 4.7
What happens when a stone having twice the mass is pulled by the same force?

We need to discuss two other facets of Newton's law of motion. The first is net force, which we have described as the overall or total force. If just one force is exerted on an object, that single force is also the net force on the object. But if two or more forces are exerted on the same object, they all will affect the object's motion, and so we need to include all of them in the net force. If all the individual forces act in the same direction—as was the case for the two springs attached to a stone (Figure 4.5)—their effects will add up, and the net force will be the sum of the individual forces. But what if the individual forces are in different directions?

For example, what if two forces act in opposite directions on the same object? If the two forces are of equal strength (Figure 4.3), you know from experience that the object will not accelerate. So the net force must be zero. This suggests that we should subtract the two forces when they are in opposite directions. This suggestion is correct. If two forces act in opposite directions, the net force will be in the direction of the larger force and is equal to the difference between the larger and the smaller force.

In this book, we will not be concerned with finding the net force when the individual forces are neither in the same nor in opposite directions. The main point to keep in mind here is that if two or more forces act on an object, the net force does not represent any one of these individual forces. Instead, the net force represents the overall effect of all the individual forces acting simultaneously.

For instance, suppose that you push your book along a tabletop with a force of 10 newtons and that the tabletop exerts a 3-newton frictional force on the book (Figure 4.8). Then the net force on the book is 7 newtons. These 7 newtons represent the net, overall effect of the external environment pushing and pulling on the book. It is this 7-newton net force, and not just your 10-newton pushing force, that accelerates the book.

The second point we need to consider is the direction of the acceleration. Recall that an acceleration is a change in velocity. Since velocity has a direction, it is not surprising that acceleration should have one too. So far, the only accelerations we have discussed quantitatively were ones in which an object was speeding up along a straight line. In this case, the velocity has an unchanging direction, and the velocity changes by becoming larger, so that the change in velocity is forward (Figure 4.9). That is, the direction of the acceleration is forward.

What about an object moving along a straight line and slowing down? In a case like this, the velocity keeps getting smaller, so the change in velocity is backward (Figure 4.10). This means that the acceleration is backward, opposite to the velocity. In the next chapter, we will discuss the direction of the acceleration in cases in which the object does not move along a straight line.

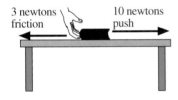

FIGURE 4.8

How strong and in what direction is the net force on the book?

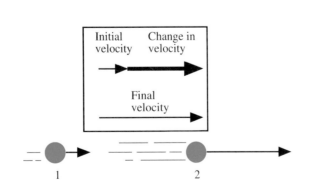

FIGURE 4.9

When an object speeds up along a straight line, its change in velocity is along the direction of the motion.

FIGURE **4.10**

When an object slows down along a straight line, its change in velocity is opposite to the direction of the motion.

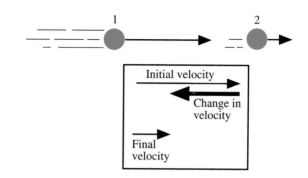

Since an object's acceleration is determined by the net force on it, it seems plausible that the direction of this acceleration should be the same as the direction of the net force. Indeed, this is true, and this idea is part of Newton's law of motion. Simple experiments verify this idea: If you give a motionless ball a brief hammer tap, it will accelerate into motion along the direction in which you tapped, so the acceleration is along the direction of the force. If the ball is already moving and you tap it from behind, it will speed up; the acceleration is forward, again along the direction of the force. And if you tap a moving ball lightly from in front, the ball will slow down; this is a backward acceleration, which again is along the direction of the force. We save the case of a sideways force, which is neither forward nor backward, for the next chapter.

In summary:

NEWTON'S LAW OF MOTION*
An object's acceleration is determined by the net force exerted on the object by its environment and by the object's mass. Quantitatively, the acceleration is proportional to the net force divided by the mass. Furthermore, the direction of the acceleration is the same as the direction of the net force. In symbols,

$$\text{acceleration} \propto \frac{\text{net force}}{\text{mass}}; \quad a \propto \frac{F}{M}$$

If force is measured in newtons, mass in kilograms, and acceleration in m/s², the proportionality becomes an equality:

$$\text{acceleration} = \frac{\text{net force}}{\text{mass}}; \quad a = \frac{F}{M}$$

DIALOGUE 1   You push your 2 kg book along a tabletop, pushing it with 10 newtons of force. If the book is greased so that friction is negligible, what is the book's acceleration? Another, ungreased, book also has a mass of 2 kg, and you also push it with a 10-newton force, but now there is a 4-newton frictional force. Find the net force on this book, and the book's acceleration. What object exerts the frictional force?

---

* Often called Newton's second law.

DIALOGUE 2   You are in distant space. A giant boulder, many times larger than you, is at rest in front of you. You tap the boulder lightly with a small hammer. Describe the motion of the boulder during and after the tap. If the boulder were on Earth, at rest on level ground, would the experimental result be the same? If not, why not?

DIALOGUE 3   You are in distant space. You hold two blocks of metal. They look and feel identical when touched, but you have been informed that one is made of "lightweight" aluminum and the other of "heavy" lead. How can you determine which one is which? Can you determine how much more massive the lead block is than the aluminum block (that is, determine the ratio of the two masses)? Explain.

4.4 Weight: *gravity's force on a body*

We have omitted three complications from our study of motion: friction, air resistance, and gravity. Now we put gravity back into the picture.

As we know (Section 4.1), Earth exerts a gravitational force on objects that are falling to the ground. We use the word *weight* for this force. An object still has weight even when it is not falling, for instance, when it is at rest on the ground. This very plausible fact is one of the consequences of Newton's law of gravity (next chapter), which says that an object's weight is the same regardless of whether it is moving or at rest.

As we will discover in the next chapter, the sun, moon, planets, stars, and all other astronomical bodies exert gravitational forces too. It is useful to extend the meaning of the word *weight* to include all such possibilities. In other words, *the **weight** of an object refers to the net gravitational force exerted on the object by all other objects.*

Weight and mass are related concepts, but they certainly are not the same thing. An object's weight is the force on it due to gravity, while its mass is its quantity of inertia. Like all forces, weight is measured in newtons, while mass is measured in kilograms. An object's weight depends on its environment; for instance, an object's weight is different when it is near the sun than when it is near Earth. But an object's mass is a property of the object alone, and not of its environment.

As one example, the acceleration of a stone falling onto the moon's surface is only 1.6 m/s^2, one-sixth of the acceleration due to gravity on Earth. Since Newton's law of motion says that acceleration is proportional to net

---

1. First book: $a = F/m = 10/2 = 5$ m/s^2. Second book: $F = 10 - 4 = 6$ newtons; $a = F/m = 6/2 = 3$ m/s^2. The table (or tabletop) exerts the frictional force.
2. It accelerates during the tap. After the tap, it moves at a slow, unchanging velocity. On Earth, the boulder would not accelerate because when you tap the rock, the ground exerts a friction force in the other direction, making the net force (caused by the hammer tap and ground) zero.
3. Exert a force on each block, for instance by giving each a small tap with your finger. The one that accelerates more is aluminum. To find the ratio of their masses, exert the same force on each block, and measure their accelerations. The ratio of the masses is the inverse of the ratio of the accelerations; for instance, if the aluminum's acceleration is three times larger, then its mass will be three times smaller.

force and since gravity is the only force exerted on the stone, the force exerted by gravity on this stone must be only one-sixth of the force exerted by gravity on the same stone on Earth. In other words, the stone's weight on the moon is one-sixth of its weight on Earth. But the stone's mass is the same in both places, since its inertia doesn't change. The stone's mass would be the same even in distant space, where its weight would be essentially zero.

If a stone and a baseball are held above Earth, dropped, and allowed to fall freely (no air resistance) to the ground, Galileo's law of falling tells us that their accelerations will be the same (9.8 m/s^2). Suppose that the stone and the baseball happen to have the same mass. Then Newton's law of motion tells us that the forces on the two objects are the same because their accelerations are the same and their masses are also the same. But this force is just the force of gravity; in other words, it is the weight of each object. So their weights are equal.

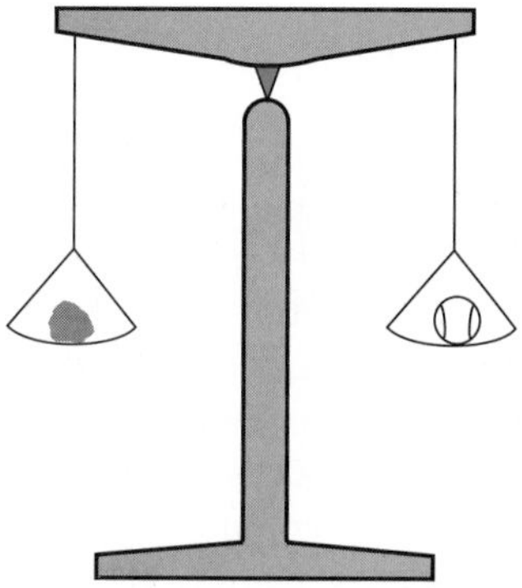

This is a fairly plausible conclusion and also a very important one. It says that *two objects of equal mass also have equal weight*. So you can compare masses by comparing weights, for instance on a balance beam (Figure 4.11). Any object that balances a kilogram has a mass of 1 kilogram, for example.

Since weight is a force, it can be measured in newtons. In "English" units, weight is measured in pounds.

The **metric ton**, or **tonne** (it is always spelled this way), is 1000 kilograms. It is useful for larger masses. The similar "English" unit—the "English ton" or simply "ton"—is 2000 pounds. It is not a unit of mass but, instead, a unit of force. On Earth, the mass of a ton is about 900 kilograms, so a ton is a little less massive than a tonne. As you can see, things can get confusing if we try to keep up with both metric and "English" units, so we will confine ourselves to metric.

As a common example, let's think about an object at rest on a horizontal surface, such as this book resting on a table. Suppose the book weighs 12 newtons; in other words, suppose that the gravitational force by Earth on the book is 12 newtons. This force has a downward direction. But the book is obviously not accelerating downward through the table. Since the book's acceleration is zero, Newton's law of motion tells us that the net force on it must also be zero. So there must be an upward force of 12 newtons acting on the book, to balance the downward weight. The table must exert this force, because the book will fall if we suddenly remove the table.

It may seem strange that an inanimate object could exert a force. Why should a table push on a book? The tabletop doesn't seem to be doing anything. But if we believe Newton's theory, then we must believe that this force exists, because Newton's theory implies it. Once again, Newton's theory is counterintuitive.

A microscopic view is enlightening. The upward force is exerted by the atoms in the table on the atoms in the book.* When the book presses against the tabletop, the tabletop is squeezed down and slightly deformed. That is, the book squeezes the atoms in the tabletop, in the way that a coiled spring can be squeezed. And like a squeezed spring, the atoms then push upward against the book (Figure 4.12). The direction of this force by the tabletop is directly away from the surface, perpendicular to it.

FIGURE 4.11
You can compare masses by comparing weights, for instance on a balance beam.

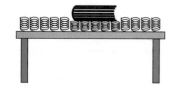

FIGURE 4.12
An explanation of the perpendicular force by a table on a book.

---

* More precisely, this force is an electric force by the electrons in the table's atoms on the electrons in the book's atoms. Electrons repel one another strongly when they get very close together.

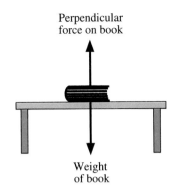

Perpendicular
force on book

Weight
of book

FIGURE 4.13
The forces exerted on a book at rest on a table.

A force similar to the upward force by the table on the book is exerted when any object touches a solid surface. We will call any such force a **perpendicular force**,* because it is always perpendicular to the surface (the table, in our example).

Figure 4.13 shows the forces exerted on the book. Each force is represented by an arrow. A **force diagram** like this can help in analyzing forces and motion. When you draw a force diagram, show every one of the individual forces acting on whatever object is of interest. Show each force as an arrow pointing in the direction in which that force is exerted on the object, and give each force a descriptive name.

As another example, suppose that a rocket at liftoff weighs 150,000 newtons and has a mass of 15,000 kilograms and that the rocket engines exert a 180,000-newton "thrust" force on the rocket (Figure 4.14). (We'll learn more about the thrust force in the next section.) What net force accelerates the rocket, and how big is the rocket's acceleration?

Figure 4.14 shows that the net force is 30,000 N (newtons) upward. It is only this 30,000 N that actually accelerates the rocket. Notice that the force diagram does not show the net force. It shows only the individual forces. Adding another arrow, labeled "net force," pointing upward, would only confuse things. To find the acceleration, just divide the net force by the mass (Newton's law of motion): $30,000 / 15,000 = 2$ m/s^2.

DIALOGUE 4   (a) Suppose an engine fails and the rocket develops a thrust of only 165,000 newtons. Find the net force and the acceleration. (b) Suppose the thrust is only 150,000 newtons and that the rocket is still on the launch pad. Find the net force and the acceleration. Does the launch pad also exert a force on the rocket, and if so, how much force? (c) What if the thrust were only 120,000 N? (d) What if the rockets didn't exert any thrust at all?

DIALOGUE 5   (a) Would it be easier to lift this book on Earth or on the moon? (b) Neglecting friction, would it be easier to set this book into horizontal motion at 5 m/s on the Earth or on the moon? (c) Neglecting friction, would it be easier to set this book into horizontal motion at 5 m/s on Earth or in distant space?

DIALOGUE 6   A standard kilogram in your physics lab weighs about 10 newtons (2.2 pounds). Find its mass and weight on the moon, where the acceleration due to gravity is one-sixth of its value on Earth. Find its mass and weight in distant space.

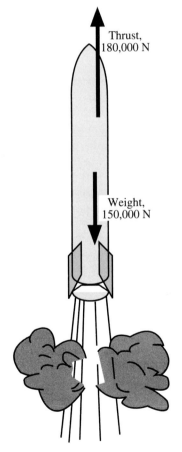

Thrust,
180,000 N

Weight,
150,000 N

FIGURE 4.14
The forces exerted on a rocket during liftoff.

---

* Many textbooks call it a *normal force*, because "normal" sometimes means "perpendicular."
4. (a) 15,000 N, 1 m/s^2. (b) net force = 0, $a = 0$. The launch pad exerts no force. (c) The rocket is at rest, $a = 0$. Now the launch pad must exert a 30,000 N force in order to make the net force zero. (d) The rocket is just resting normally on the launch pad; $a = 0$, the launch pad exerts 150,000 N (upward), net force = 0.
5. (a) On the moon. (b) The same in both places. (c) The same in both places.
6. On the moon: mass = 1 kg, weight = $1/6 \times 10 = 1.7$ N. In space: mass = 1 kg, weight = 0.

4.5 The law of force pairs: *you can't do just one thing*

Try these: Slap a tabletop with your hand. Grasp the edge of a table and pull hard on it. Now push hard against the table. Find two marbles, place one at rest, and "shoot" the other marble at it.

When you slap a table, it slaps back, as you can feel when it stings your hand. This slap by the table is a force, because it accelerates your hand (by stopping your hand).

When you pull on a table, the table pulls you toward it. And when you push on the table, the table pushes you away. The table pulls back and pushes back. These are forces exerted by the table on you.

When the marbles collide, the shooter marble exerts a force on the second marble, as you can see from the fact that the second marble accelerates into motion. But the second marble exerts a force on the first marble, too, as you can see from the fact that the first marble's velocity changes.

When you slap, pull, or push a table, the table slaps, pulls, and pushes back on you. When one marble exerts a force on a second marble, the second exerts a force on the first. These experiments indicate that whenever one object exerts a force on a second object, the second exerts a force on the first: Forces always come in pairs.

Do things still work out this way even if the two objects are not touching? If you can find a pair of magnets, you can investigate this. Place the magnets on a smooth surface, and hold them at rest with their "poles" near each other but not touching. When you release them from rest, they both move. Both are accelerated. So each exerts a force on the other. Even for two separated objects, forces come in pairs.

Touch your friend's face. Your hand, too, is touched, by your friend's face. Forces are never alone; they always come in pairs. You cannot touch without being touched.*

The harder you slap on a table, the harder it slaps back. This gives you quantitative information about these **force pairs**: When one member of a force pair grows bigger, so does the other. In fact, quantitative experiments show that the two members of any force pair have the same strength. If one of them is, say, 3 newtons, the other one will be 3 newtons too.

FIGURE 4.15
When you pull or push on a table, it pulls or pushes on you in the opposite direction.

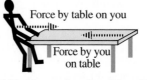

(a) Pulling on the edge of the table.

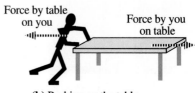

(b) Pushing on the table.

---

* Thanks to Paul Hewitt, author of *Conceptual Physics*, 6th ed. (Boston: Scott, Foresman, 1989), for this nice way of putting it.

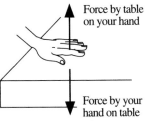

(a) Slapping a table.

Force by table
on your hand

Force by your
hand on table

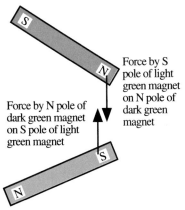

Force by S
pole of light
green magnet
on N pole of
dark green
magnet

Force by N pole of
dark green magnet
on S pole of light
green magnet

(b) Opposite poles of two magnets
attracting each other.

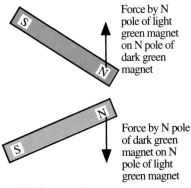

Force by N
pole of light
green magnet
on N pole of
dark green
magnet

Force by N pole
of dark green
magnet on N
pole of light
green magnet

(c) Like poles of two magnets
repelling each other.

FIGURE 4.16
Force pairs.

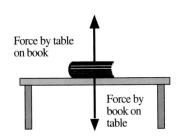

Force by table
on book

Force by
book on
table

FIGURE 4.17
Another force pair.

The directions of the two forces in a force pair are not the same, however. In fact, our examples show that they are opposite in direction. For instance, when you pull a table toward you, the table pulls you toward it (Figure 4.15).

This idea about force pairs is one of the essentials of Newton's physics. We will summarize it this way:

THE LAW OF FORCE PAIRS*
Forces always come in pairs: Whenever one body exerts a force on a second body, the second exerts a force on the first. Furthermore, the two forces are equal in strength but opposite in direction.

Figures 4.16 and 4.17 show several examples of force pairs. These diagrams are not complete force diagrams. Figure 4.14, for example, is a complete force diagram for a rocket—it shows *all* the forces acting on a *single* object, namely, on the rocket. On the other hand, each illustration in Figures 4.16 and 4.17 shows forces on two objects and generally does not show all of the forces on either one of the two objects.

In Figure 4.16a, b, and c, the arrow representing each force is attached to the object on which that force is exerted. The two arrows in each force pair point in opposite directions and are drawn with the same length to indicate that they have equal strengths.

Figure 4.17 is the example of a book on a table (Figure 4.13). Although Figures 4.13 and 4.17 look similar, the difference between them is important: In Figure 4.13, the downward force is the gravitational force by Earth on the book; all the forces in this figure are exerted on the book. In Figure 4.17, the downward force is by the book on the table. Microscopically, this force is exerted by the atoms in the book's lower surface on the atoms in the tabletop. This force is repulsive, away from the book. If you lay a book on the palm of your hand, you can feel this downward force by the book's pressing on the atoms of your hand.

Figure 4.18 illustrates an interesting point. Since Earth exerts a gravitational force on an apple, the law of force pairs says that the apple must exert a gravitational force on Earth! Furthermore, the strengths of these two forces are equal: If Earth exerts a 2-newton force on the apple (in other words, if the apple weighs 2 newtons), then the apple also will exert a 2-newton force on Earth. This might seem surprising. Why haven't we noticed this force, by objects such as apples, on Earth? If an apple exerts a force on Earth and if forces cause accelerations, then why doesn't Earth accelerate toward the apple?

The answer is that Earth has such a large mass that the acceleration of Earth caused by an apple is extremely small. It is so small that you don't notice it, in fact so small that accelerations of the entire planet in response to objects on the surface of the planet have never been measured. Large astronomical objects, however, can noticeably accelerate the planet. Our planet accelerates in response to the motions of the moon; the gravitational force by the moon on Earth pulls our planet back and forth far enough that this motion can be measured.

---

* Often called "Newton's third law."

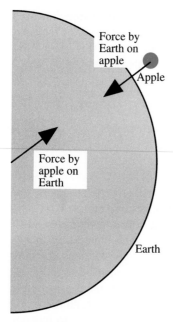

FIGURE 4.18
Earth and a falling apple: Which one exerts the larger force? (Answer: They are the same.)

DIALOGUE 7   The following is a list of forces. In each case, describe the other member of the force pair that includes the listed force: (a) The force a pitcher exerts on a baseball while throwing it. (b) A bat hitting a baseball. (c) A baseball hitting a catcher's mitt. (d) An apple's weight. (e) Your weight. (f) Your arms pulling on a box as you slide it across the floor. (g) A rope pulling forward on a water skier.

DIALOGUE 8   A 12-newton book rests on the palm of your hand. Describe the two forces that act on the book. How strong is each force, and in what direction does each act? Are these two forces equal but opposite to each other? Are these two forces part of one force pair?

DIALOGUE 9   Referring to the preceding dialogue, you accelerate the book into upward motion. Describe the individual forces that act on the book during the acceleration. What can you say about the strengths of each of these two forces? Are they equal but opposite to each other?

DIALOGUE 10   (a) A large truck collides head-on with a small car. Which force is bigger: the force by the truck on the car or the force by the car on the truck? (b) A moving car hits a stationary car from behind. Which is bigger, the force by the moving car on the stationary car or the force by the stationary car on the moving car?

4.6 Drive forces: *Newton meets the automobile*

An interesting dilemma arises when we consider any "self-propelled" object. For example, animals seem to be able to accelerate themselves into motion. The dilemma arises from Newton's law of motion, which states that forces by the environment cause an object to accelerate, implying that things cannot accelerate themselves.

Simple experiments demonstrate that you cannot get anywhere by pushing or pulling on yourself. Try pulling on your nose. You might pull your nose out of joint, but you won't go anywhere because your nose pulls back on your hand, and both pulls are on your body, so they result in zero net force on your body (Figure 4.19). If you want to accelerate, something in your environment must exert a force on you.

FIGURE 4.19
You can't get anywhere by pulling on your nose.

---

7. (a) The baseball pushing backward on the pitcher's hand. (b) The ball hitting the bat. (c) The mitt stopping (exerting a backward force on) the baseball. (d) The apple exerting an upward gravitational force on Earth. (e) Your body exerting an upward gravitational force on Earth. (f) The box pulling backward on your arms. (g) The skier pulling backward on the rope.

8. The weight of the book and the force by your hand. The weight is 12 N downward, and the hand force is 12 N upward. They are equal but opposite. They are not members of a single force pair.

9. All the answers are the same as they were for Dialogue 8, except that the strength of your hand force is now larger than 12 N (in order to give a net upward force).

10. (a) The law of force pairs says that the two forces are the same size (same strength). One is just as big as the other. (b) Ditto.

What happens when you walk? Stand up for a moment and take one step, noting carefully the sensations in your legs, especially along the bottom of the foot that is accelerating you. —The foot pushes backward against the floor.* You can demonstrate more convincingly that your foot actually pushes backward when you walk or run, by accelerating rapidly from rest on a dusty dirt road. You push dust backward, showing that your feet exert a backward force on the road.

The law of force pairs tells us that if you push on the ground, the ground will push on you. And since your foot pushes backward on the ground, the ground must push forward on your foot. This is the force that propels you forward! It is one example of a **drive force**: the forward force exerted by the environment on a self-propelled object[†] that accelerates it or keeps it moving against external forces that resist the object's motion.

The automobile is a good example of many Newtonian principles. More important, motor vehicles are one of the technologies that have drastically reshaped the social fabric of the modern world. Like all powerful technologies, the car has important social pros and cons. Motor vehicles give us unheard-of freedom to move across Earth's surface; they pervade our physical and economic environment; they have restructured our cities and our society; they loom large in all questions relating to energy resources and pollution; and they are the leading cause of death of Americans under 35 years of age. Since the social impact of science is one of the themes of this book, cars will come in for a lot of discussion.

Try listing, or drawing, the forces exerted on a car while it is driving along a straight level road. (See Figure 4.20 for the answer.)

The car's weight is a force by Earth on the car in the downward direction. A perpendicular force is exerted upward by the road on the car. These two forces act vertically. Since a car on a level road has no acceleration in the

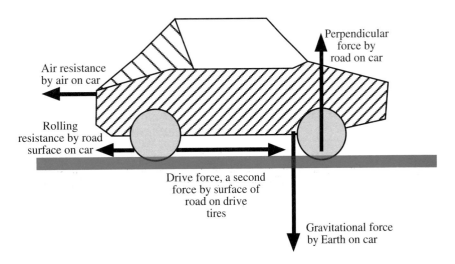

FIGURE 4.20
The five forces on an automobile.

---

* —If you are barefoot. Otherwise, your foot pushes backward against your shoe, which in turn pushes backward against the floor.
† "Self-propelled" means that the energy (Chapter 6) to propel the object comes from within the object itself.

vertical direction, the net vertical force must be zero. So these two vertical forces must cancel out each other.

The horizontal forces relate more directly to the car's motion. Two **resistive forces** are exerted backward on the car. **Air resistance** is exerted by the air through which the car moves. The second resistive force is exerted by the road on the tires and is called **rolling resistance** because it acts backward, just as the frictional force on a sliding object acts backward. But this force is not really a frictional force, because the tires roll rather than slide along the road. Rolling resistance is caused by the squeezing or compression of the tire where the rubber meets the road. The force that the road must exert to cause the compression turns out to be partly backward, retarding the car's motion. Another way to look at it is that this continual compression and decompression around the surface of the tire represents a loss of energy (Chapter 6) by the car. This is most pronounced in flexible, air-filled tires. Hard tires rolling on a hard smooth surface, such as steel wheels on steel tracks, reduce rolling resistance to a minimum. This is one reason trains are more energy efficient than cars.

These four forces act even on a car that is coasting with its engine shut off. If these are the only forces on the car, then the net force is backward, and Newton's law of motion tells us that the acceleration is backward, so that the car slows down.

When a car is driving instead of coasting, there is a drive force on it. This force is not exerted by the engine on the car, because a car cannot push itself. Instead, the engine causes the drive wheels to turn; the drive wheels push backward against the road; and the *road* pushes forward against the drive wheels. This is the drive force.

Figure 4.20 shows the five forces acting on a car driving down a straight level road. If the car moves at an unchanging velocity, there is no acceleration. Newton's law of motion tells us that the net force must then be zero, which means that the five forces shown in the figure must balance. In this case, the forward drive force must equal the sum of the strengths of the two resistive forces. Note that "no net force" does not mean "no drive force." Even with no acceleration, a drive force is needed. The car speeds up whenever the drive force is larger than the sum of the two resistive forces and slows down whenever the drive force is smaller than the sum of the two resistive forces.

Most other drive forces are similar to this: The driving object pushes backward on its surroundings, and the surroundings push forward on the object. A bicycle's drive tire pushes backward on the road, and the road pushes forward on the tire. A swimmer pushes backward on the surrounding water, and the water pushes forward on the swimmer. A motorboat's propeller pushes backward on the water, and the water pushes forward on the propeller. A prop-driven airplane's propeller pushes backward on the surrounding air, and the surrounding air pushes forward on the propeller, just like water pushing forward on a motorboat's propeller. A jet airplane pushes air backward, too. As a jet engine moves through air, the air flowing into its front end heats by combustion with jet fuel, and the heated gas expands and rushes out of the back end at high velocity. Air has been pushed backward, but by a different mechanism than in a prop-driven airplane.

One nice thing about space travel is that there are no resistive forces in space, so you don't need a drive force to keep going. Your spaceship keeps going because of its own inertia. But if you want to give your spaceship an acceleration—for instance, a change in direction—then you have a problem. It is difficult to get the surroundings to exert a drive force on your spaceship because there is nothing around to push against!

A similar problem would arise if you were stranded in the middle of a smoothly frozen pond. If the ice were absolutely smooth, you could not walk off it because your feet would slip. You could not get the ice to push on you because, with no friction, you could not push on the ice. How could you get off? You could fan the air, pushing air backward in the way that a swimmer pushes water backward. That would work. But suppose that, as in space, there were no air? Then there would be nothing in your surroundings to push against.

However, you would probably have something along with you that you could push against. Your physics book, for example. Or a shoe. You could take off a shoe and throw it away! While throwing, you would push on the shoe so it would push in the other direction on you, and your body would accelerate away from the shoe. If you now let go of the shoe (if you hung onto it, the shoe would slow your body to a stop), you would have acquired a velocity. So you would slide along the pond. If you wanted to slide faster, you could either throw your shoe away faster or throw away your other shoe too.

This is the principle of **rocket drive**. Rockets take along their own material just to have something to push against. Shoes would work (Figure 4.21), but they aren't very practical. The rocket fuel for the U.S. space shuttle's main rocket engines is hydrogen and oxygen, stored as low-temperature liquids. When combined, their combustion produces steam, which streams rapidly out of the back end of the engine, pushing the shuttle forward.

FIGURE 4.21
Shoe power.

DIALOGUE 11   A car weighing 10,000 newtons moves along a straight level road at an unchanging 80 km/hr. Air resistance is 300 newtons, and the rolling resistance is 400 newtons. Find the net force on the car and the strength and direction of the perpendicular force and the drive force.

DIALOGUE 12   Magnetic forces can levitate railroad trains a short distance above the tracks, making friction practically negligible. Suppose such a "maglev" train runs inside an evacuated (emptied of most air) tunnel, from New York City to Chicago. Assume that friction and air resistance are negligible. During what parts of the trip would a drive force or other external horizontal force act on the train? Discuss the direction of the external horizontal force.

---

11. Zero, 10,000 newtons upward, 700 newtons forward.
12. There would be a forward drive force during a relatively brief acceleration period while leaving New York; no drive force would be needed during most of the trip; and a backward force would slow the train to a stop in Chicago. *Note*: Although there are plans for maglev trains, there are no plans for evacuated tunnels.

Summary of Ideas and Terms

Force A body exerts a force on another body whenever the first body causes the second body to accelerate. A force is an action by one body on another; it is not a thing or a property of a body. Every force is similar to a push or a pull.

Newton A unit of force. The amount of force that can accelerate a 1 kg mass by 1 m/s^2.

Net force The total, overall force on an object. The net force due to two forces acting in the same direction is the sum of the two. The net force due to two forces acting in opposite directions is the difference between the two and acts in the direction of the stronger force.

Inertia A body's ability to stay at rest or to maintain an unchanging velocity. Quantitatively, a body's inertia is its degree of resistance to acceleration, or its **mass**.

Matter Material substance. Anything made of atoms.

Mass A body's mass is its amount of inertia and also its quantity of matter. We find a body's mass by comparing its inertia with the inertia of a standard kilogram.

Kilogram A unit of mass. The mass (or inertia) of the object known as a "standard kilogram." Any object that has the same inertia as the standard kilogram has a mass of 1 kilogram, abbreviated as kg.

Metric ton, or **tonne** One thousand kilograms or about 2200 pounds.

Proportional to the inverse A quantity is proportional to the inverse of another quantity if the first is proportional to (equal to some number times) 1 divided by the second quantity.

Newton's law of motion An object's acceleration is proportional to the net force exerted on it by its surroundings and is proportional to the inverse of its mass. The direction of the acceleration is the same as the direction of the net force, $a \propto F/m$. In the appropriate units (m/s^2, newtons, kilograms), $a = F/m$.

Weight The weight of an object is the net gravitational force exerted on it by all other objects.

Resistive force Any force that acts on a moving body in a direction opposite to the body's motion. **Friction** is the resistive force by a surface on an object sliding across that surface. **Air resistance** is the resistive force that air molecules exert on an object moving through the air. **Rolling resistance** is the resistive force by a surface on a rolling object.

Perpendicular force The force, perpendicular to a solid surface, that is exerted by any solid surface on any object touching it.

Force pair The two forces that two bodies exert on each other.

Law of force pairs Forces always come in pairs: Whenever one body exerts a force on a second body, the second exerts a force on the first. The two forces are equal in strength but opposite in direction.

Drive force The force, exerted on a "self-propelled" (self-energized) object, that accelerates it or keeps it moving against resistive forces.

Review Questions

FORCE

1. How can we tell whether or not a body is exerting a force on another body?

2. Can an object have force? Can an object exert a force? Can an object be a force? Can an object feel a force?

3. List at least six specific examples of forces. Try to list examples that are significantly different.

4. What is a resistive force? Give two examples.

5. Give two examples of forces that act at a distance.

NEWTON'S LAW OF MOTION

6. What does Newton say that forces cause? What does Aristotle say?

7. What do we mean when we say that one object has "more inertia" than another object does?

8. When you move an object from Earth to the moon, does its inertia change? Does its weight change? Does its mass change? Does its amount of matter change? Does its acceleration differ while falling freely? Does it respond differently to a force of 1 newton?

9. Forces of 8 newtons and 3 newtons act on an object. How strong is the net force if the two forces have opposite directions? The same directions?

10. Is an object's acceleration always in the same direction as its velocity? If not, give an example in which it is not. Is an object's acceleration always in the same direction as the net force on the object? If not, give an example in which it is not.

11. As you increase the net force on an object, what happens to its acceleration? What if you double the net force? As you increase the mass of an object (for example, by gluing additional matter to it), what happens to its acceleration? What if you double the mass?

WEIGHT

12. What is weight? Is it the same as mass? If not, what is the difference?

13. Describe a simple way to determine, in a lab, whether two objects have equal masses. Would this method work in distant space? What would work in distant space?

14. Find the gravitational force on a 1-newton apple. Would it still be a 1-newton apple if we took it to the moon?

15. Draw a force diagram showing the forces on an apple at rest on a table. Find the net force on the apple.

16. Draw a force diagram showing the forces on a rocket during liftoff. Which force is largest? What is the direction of the net force?

17. Where is it easiest to lift your automobile: on Earth or on the moon? Where is the automobile's mass larger?

LAW OF FORCE PAIRS

18. Describe several experiments demonstrating that forces come in pairs.

19. Do you exert a gravitational force on Earth? How do you know? What direction is this force?

20. Describe the other member of the force pair for each of the following forces: the perpendicular force on a book, the weight of an apple, the force by a bat against a baseball, the force by a baseball hitting a catcher's mitt.

DRIVE FORCES

21. Describe four examples of drive forces.

22. Draw a force diagram showing the forces on a car driving along a straight level road. How would this force diagram be altered if the car were coasting? What if the car were braking?

23. How does propeller drive differ from jet drive?

24. How does a car's drive force compare with the resistive forces when the car maintains a constant speed? When the car is speeding up? Slowing down?

25. When a car moves at constant speed along a straight road, is the drive force zero? Is the net force zero? Is the acceleration zero? Is the velocity zero?

26. What is the main difference between rocket drive and other drive forces such as propeller drive and automobile drive?

Home Projects

1. Experience the meaning of an unchanging force and an unchanging acceleration. Pull a smoothly rolling wagon along a smooth level sidewalk, with a spring or a strong rubber band stretched by an unchanging amount. You will have to move faster and faster—you must accelerate! Try increasing the mass by putting an object, or a friend, in the wagon. For the same amount of stretch, is the acceleration larger or smaller with the larger mass?

2. Place a spool of thread on a table, as shown in Figure 4.22a. According to Newton's law of motion, what should the direction of the spool's acceleration be when you pull on the thread? Try it. Now turn the spool over, as shown in Figure 4.22b. Intuitively, what do you believe will happen when you pull on the thread? What does Newton's law of motion predict? (*Hint*: What is the direction of the net force?) Try it. You might not have expected this result. The experiment does agree, however, with Newton's law of motion. Nature, and Newton's theory of motion, can be surprising!

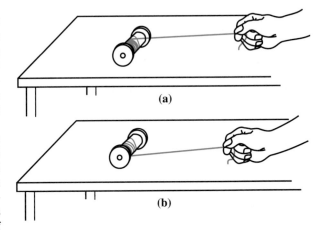

FIGURE 4.22
Try this.

Exercises

FORCE

1. Is any force exerted on you when you speed up along a straight line? When you slow down along a straight line? How do you know?

2. Is any force exerted on you while you move in a circle at unchanging speed? How do you know?

3. A smooth ball rolls on a smooth table. Initially, no horizontal forces are exerted on the ball. Then you bring a magnet near the rolling ball, but you are not sure whether the magnet actually exerts a magnetic force on the ball. How can you tell whether or not the magnet is exerting a horizontal force on the ball?

4. Does a high-velocity bullet contain force? Does a stick of dynamite contain force?

NEWTON'S LAW OF MOTION

5. You place your book on a table and hit it horizontally with a hammer, strongly but briefly. Do not neglect friction. Describe the motion of the book, beginning from just before you hit it with the hammer. Describe the direction and strength of the net force on the book during the entire motion.

6. If you exert a force on an object and then exert three times as strong a force on the same object, what (if anything) can you say about the object's acceleration during the exertion of each force?

7. A ball weighing 8 newtons is thrown straight upward. Disregarding air resistance, find the direction and strength of the net force on the ball as it moves upward. What is the direction of the ball's acceleration? Are the net force and the acceleration in the same direction in this case? Can they ever be in different directions?

8. An object moves with unchanging velocity. Does it then have no forces acting on it? Explain. What about an object at rest—does it have any forces acting on it?

9. When you stand on the floor, does the floor exert a force on your feet? In which direction? Why, then, don't you accelerate in that direction?

10. You push on a solid concrete wall. Is your push the only horizontal force on the wall? How do you know? What can you say about the net force on the wall?

11.*A car weighing 8000 newtons moves along a straight level road at a steady 80 km/hr. The total resistive force on the car is 500 newtons. Find the net force on the car, the acceleration of the car, and the drive force.

12.*The car of Exercise 11 has a mass of 800 kg. The driver pushes down on the accelerator, increasing the drive force to 2100 newtons. All other forces on the car remain unchanged. Find the net force on the car, the acceleration of the car, and the directions of both.

13.*The driver of the car of Exercises 11 and 12 lets up on the accelerator, reducing the drive force to only 100 newtons. All other forces are unchanged. Find the net force on the car, the acceleration of the car, and the directions of both.

14.*A freely falling apple has a weight of 1 N. Earth's mass is 6×10^{24} kg. How strong is the force by Earth on the apple? How strong a force does the apple exert on Earth? How big is the apple's acceleration? Find the acceleration that the apple would cause Earth to have if the apple was the only object exerting a force on Earth.

WEIGHT

15. Roughly, what is your weight in newtons?

16. Which has the greater mass, a tonne of feathers or a tonne of iron? Which has the greater weight? Which has the larger volume?

17. Would you rather have a hunk of gold whose weight is 1 newton on the moon or one whose weight is 1 newton on Earth—or wouldn't it make any difference?

18. Would you rather have a hunk of gold whose mass is 1 kg on the moon or one whose mass is 1 kg on the Earth—or wouldn't it make any difference?

19. Find the strength and direction of the net force on an apple weighing 2 newtons, neglecting air resistance, in each of the following cases: The apple is held at rest in your hand. The apple is falling to the ground. The apple is moving upward, just after you threw it upward.

20. An apple is accelerated upward by your hand. Which is larger, the apple's weight or the upward force by your hand? What if you accelerate the apple downward while it is in the palm of your hand? What if you lift the apple at an unchanging velocity? What if you lower the apple at an unchanging velocity?

21. Would it be easier (in other words, would it require less thrust and less rocket fuel) to lift a rocket off the moon's surface than off Earth's surface? Why?

LAW OF FORCE PAIRS

22. "Planet Earth is pulled upward toward a falling boulder with just as much force as the boulder is pulled downward toward Earth." True or false? Why?

23. "Planet Earth is pulled toward a falling boulder with just as much acceleration as the boulder has as it moves toward Earth." True or false? Why?

24. A large truck breaks down on the highway and receives a push back into town by a small car. The car speeds up while pushing the truck. While speeding up, the car exerts a forward force on the truck. But does the truck exert any force on the car? If so, is this force weaker or stronger than the force that the car exerts on the truck?

25. A car collides head-on with a large truck. Which vehicle exerts the stronger force? Which has the larger force exerted on it? Which experiences the larger acceleration?

26. When a rifle fires, it accelerates a bullet along the barrel. Explain why the rifle must recoil.

27. A 2-newton apple hangs by a string from the ceiling. Describe the two forces on the apple. How strong is each of these forces? Do these forces form a single force pair? If not, then for each force, describe the other member of that force's force pair.

28. A horizontally moving bullet slows down. Is anything exerting a force on it? How do you know? Is it exerting a force on anything? How do you know?

29. I push you away from me. Do you also push (exert a force on) me? Which force is stronger—the force by me on you or the force by you on me?

30. Since the law of inertia states that no force is needed to keep an object moving at an unchanging velocity, why does a car need a drive force to keep moving?

31. While driving your car on a straight level road, you slam on your brakes. Draw a force diagram of the car during braking. What is the direction of the net force? Draw a force diagram for a car that is coasting without braking. In which of the two cases is the net force stronger?

32. Why is it easier to pedal a bicycle with hard high-pressure tires, as compared with soft balloon tires?

33. When you hold your foot on a car's accelerator pedal, is the car necessarily accelerating? Could it be accelerating? Could it have a forward acceleration? Could it have a backward acceleration?

34. There are three acceleration devices on any car. What are they, and what kinds of accelerations does each one give to the car?

35. If a jet airplane were above Earth's atmosphere, could it then accelerate? What about a rocket-driven plane?

5

THE UNIVERSE
ACCORDING TO NEWTON

And from my pillow, looking forth by light
Of moon or favouring stars, I could behold
The antechapel where the statue stood
Of Newton with his prism and silent face,
The marble index of a mind for ever
Voyaging through strange seas of Thought, alone.

William Wordsworth, Prelude (Book III), 1850

The principle of inertia might be history's most fruitful scientific idea. Besides unifying natural motion on Earth and in the heavens, undermining Aristotelian views, promoting the idea of universal natural law, and leading to Newton's law of motion, it also led Newton to look in a new way at one specific kind of force: gravity. This is the subject of most of this chapter.

Gravity is perhaps the most universal force, because it is all around us on Earth and also accounts for most motions observed in the heavens. It is with us so much that we are hardly aware of it, and this historically made it difficult to conceptualize. Prior to the inertial view of motion, it was believed that every body containing the element "earth" had a natural tendency to seek out Earth's center, in the way that a thirsty person seeks out water. External influences—forces—were not needed to explain why objects fall: They fell because that was what they "wanted" to do.

The inertial view is that bodies "want" to maintain their velocity. It was Descartes who, more than anyone else, first conceived of this new view of motion. It was a conceptual shift comparable to Copernicus's shift to a sun-centered view. Newton then built on Descartes's idea.

If you believe that bodies have inertia, then you must ask why an apple, held at rest in midair and then released, falls. The inertial view made this question possible. Newton's answer applied to more than apples. The answer reaches beyond the moon and demonstrates convincingly that in the heavens and on Earth, the same forces are at work. We live in one natural universe, not two.

Section 5.1 presents the general idea of Newton's law of gravity, and Section 5.2 gives the specifics along with several examples. Section 5.3 applies these ideas to understanding the origin of the sun and Earth, and the future final gravitational collapse of the sun. Section 5.4 tells the tale of the final violent collapse of stars that are more massive than the sun and the exotic objects that result from the collapse. Sections 5.5 and 5.6 return to our theme of comparing Newtonian and post-Newtonian physics: Section 5.5 looks at some broad philosophical implications of Newtonian physics, particularly the general worldview known as the mechanical universe. Section 5.6 places Newtonian physics within a broader scientific perspective, noting the limitations of Newton's theories in light of the more recent relativity and quantum theories.

5.1 The idea of gravity: *the apple and the moon*

At age 22, Isaac Newton had just completed his bachelor of arts degree at Cambridge University in England. He had been invited to remain there, but the school then closed for 18 months because of an epidemic of the plague, and so the graduate returned to his family's farm. During those 18 months Newton laid the foundations for both a theory of gravity and a theory of light and developed an early form of calculus.

Some say that greatness is partly a matter of timing. Newton lived at a time that was ripe, scientifically and culturally, for a new view of the universe. The scientific foundations had been laid by Copernicus, Brahe, Kepler, Descartes, and Galileo. We have seen that the inertial view of Descartes and Galileo leads naturally to Newton's law of motion. The ideas surrounding the law of motion, combined with the astronomical ideas of Copernicus and Kepler, then led Newton to the law of gravity. Newton truly stood, as he himself said, "on the shoulders of giants." But by the same token, it took the supreme inventive genius of a Newton to put all this together.

As Newton recounted it late in life, he invented the central idea of his law of gravity during that stay on his family's farm. The idea came to him when an apple fell from a tree while he could see the moon in the sky. Beyond the fact that both are more or less round, it is difficult to think of two more dissimilar objects than an apple and the moon. One is on Earth, the other is in the heavens; one quickly rots, the other seems eternal; one falls to the ground, the other stays aloft. Yet where others saw difference, Newton saw similarity. Great science often comes from just such unifying insights.

Let us trace Newton's thinking. Figure 5.1 shows an apple falling toward the ground, accelerated by Earth's gravitational pull. The directions of the apple's velocity and acceleration and of the gravitational force all are downward toward Earth's center, as shown. But the moon's motion is quite dif-

FIGURE 5.1
The apple and the moon.

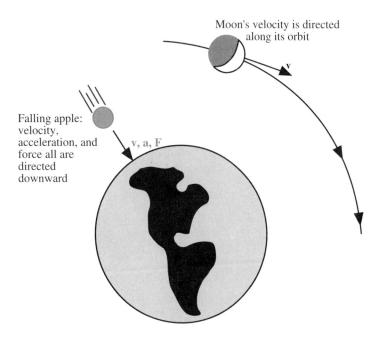

ferent (Figure 5.1). The direction of its velocity is around Earth rather than toward its center. But we are interested in the forces on each, and according to the inertial view, forces cause accelerations, not velocities. So the forces on the two could be similar, despite the dissimilarity of the velocities. How do the forces compare?

Aristotle would say that no force is needed to make the moon move in a circle because that is its natural motion, but the new inertial view is that in order for the moon to deviate from a straight-line motion, a force must act on it. What is the direction of this force? If the moon were at point A in Figure 5.2 and if no force acted on it, the moon would keep moving in a straight line from A to B. But instead it moves around to point C. The moon

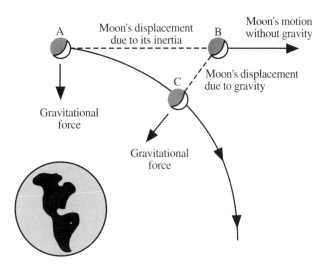

FIGURE 5.2
The moon is held into its orbit by an inward-directed force.

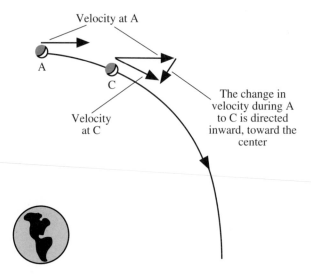

Velocity at A

A

Velocity
at C

C

The change in
velocity during A
to C is directed
inward, toward the
center

is doing two things simultaneously: Its inertia keeps it moving forward and would carry it from A to B, but at the same time it is being pulled inward, toward Earth's center, so that it also "falls" from B to C. The moon's departure from its natural straight-line path can be described as a continual falling—similar to the fall of an apple. The force required to pull the moon inward—so that it arrives at C rather than B—is directed inward, toward Earth's center, just like the force on the apple.

Newton was the first to understand that an inward-directed force must act on the moon, to hold it in its circular orbit. He hypothesized that this force had the same source as the force that pulled an apple downward: Earth's gravitational attraction.

Since the force on the moon is inward, Newton's law of motion tells us that the moon's acceleration must be inward too, even though its velocity is horizontal. Figure 5.3 shows in another way that the moon's acceleration is directed toward the center of its orbit: In order for the moon to maintain a circular motion, its velocity must keep changing toward the inside. The change in velocity is always toward the center, so the acceleration is toward the center. Although the velocity of the apple and the velocity of the moon are quite different, both their accelerations are directed toward the center.

Newton offered another argument showing that the force on the moon is inward. If you pick up an apple or anything else and throw it horizontally, it will follow a curved path as it falls to the ground (Figure 5.4). If you throw the apple faster, it will go farther before falling to the ground. And if you throw it fast enough, it might "fall" around a large part of Earth's surface before striking the ground (Figure 5.5).

If the apple is launched at such a high speed that the curvature of its path just matches Earth's curvature, it will fall all the way around. The required speed is about 8 km/s, or 29,000 km/hr. At this speed, the apple will fall all the way around Earth without getting any closer to the ground, because the

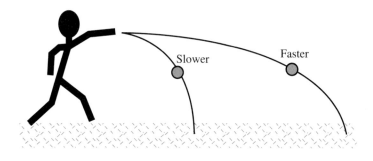

FIGURE 5.4

If you throw an apple horizontally, the faster you throw it, the farther it will go.

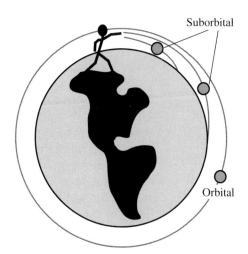

FIGURE 5.5

If you throw an apple fast enough, it will fall around a large part of Earth's surface or even go into orbit. A diagram like this appears in Newton's notebook.

downward curvature of its orbit just matches the downward curvature of Earth. Once it makes one circuit, it will make another (ignoring air resistance). It goes into orbit. This is what the moon does, and it is what any other orbiting satellite does. The force that shapes the moon's path is gravity—the same gravity that pulled the apple to the ground that day on Isaac Newton's family's farm.

It was an imaginative leap, in more ways than one. It was difficult to believe that anything at all was pulling on the moon, much less that it could be the same force that pulled on an apple. Most difficult of all was the notion that the gravitational force could reach out across nearly half a million kilometers of empty space (the distance was known, roughly, in Newton's time), to pull on the moon. It is easy to see that things exert forces on one another when they are in direct contact. But a force that acts across so much empty space seems like magic.

DIALOGUE 1   Figures 5.4 and 5.5 show five possible paths for an apple that has been thrown horizontally. Assume that air resistance is negligible. For each path, draw three arrows—labeled **f**, **a**, **v**—attached to the apple that show the directions of the gravitational force on the apple, the acceleration of the apple, and the velocity of the apple.

DIALOGUE 2   A 2-newton apple falls from a tree. While it is freely falling, what are the strength and direction of the force on it? What are the numerical value and the direction of the apple's acceleration?

DIALOGUE 3   Continuing Dialogue 2, suppose we throw the apple horizontally, as shown in Figures 5.4 and 5.5. While the apple is in each of the five positions shown in the figures, what are the strength and direction of the force on it (neglect air resistance)? Would the apple's mass still be the same as it was in Dialogue 2? Since the net force and the mass are the same as they were in Dialogue 2, what must the numerical value and the direction of the apple's acceleration be?

5.2 The law of gravity: *moving the farthest star*

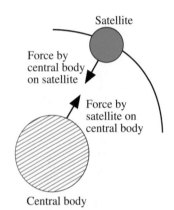

Satellite

Force by central body on satellite

Force by satellite on central body

Central body

FIGURE 5.6
An astronomical object with a satellite. The two bodies could be Earth and the moon, the sun and Earth, the sun and Mars, Earth and an artificial satellite, or whatever. Each of the two bodies pulls on the other.

What makes planets go around the sun? At the time of Kepler some people answered this problem by saying that there were angels behind them beating their wings and pushing the planets around an orbit. . . .The answer is not far from the truth. The only difference is that the angels sit in a different direction and their wings push inwards.

Richard Feynman, physicist

Since Earth's gravitational pull holds the moon into its orbit, it is reasonable to suppose that all satellites—bodies in orbit around larger astronomical bodies—are held in their orbits by the gravitational pull exerted by the larger body. Since the planets are satellites of the sun, Newton's insight regarding the moon also resolves the old question of why the solar system moves as it does! The planets keep moving forward because of their inertia, and the sun's gravitational pull bends their orbits into ellipses (elongated circles). Similarly, the four moons of the planet Jupiter (which Galileo discovered before Newton's work) are held in their orbits by Jupiter's gravitational pull. And because of the law of force pairs, every satellite must pull back on the larger body. The gravitational force is exerted between an astronomical body and a satellite: The central body pulls on the satellite, and the satellite pulls equally strongly but in the opposite direction on the central body (Figure 5.6).

But why would gravity act only between astronomical bodies and their satellites? For instance, since there is a gravitational force between the sun and Earth and also between the sun and Mars, it seems logical that there should be a gravitational force between Earth and Mars. Such a force between planets had not been noticed yet in Newton's day, but Newton supposed that this was only because it was so much smaller than the force by the sun on the planets. Likewise, there should be a gravitational force between any two stars. Even though the force between such widely separated objects would be small, Newton supposed that the sun must pull on even the farthest star and that every astronomical body exerted a gravitational pull on every other astronomical body.

But why should gravity be restricted to astronomical bodies? Why shouldn't a gravitational force be exerted between smaller objects on Earth—

---

1. The arrows labeled **f** and **a** should point directly toward the Earth's center, and the arrow labeled **v** should point along the path of motion.
2. 2 newtons downward. 10 m/s^2, downward.
3. 2 newtons downward. The mass would still be the same. Since the mass and force are the same as in Dialogue 2, Newton's law of motion tells us the acceleration is the same: 10 m/s^2, downward.

Force by
apple on
book

Force by book
on apple

FIGURE 5.7
*Even ordinary-sized objects exert
gravitational forces on one another.
Your physics book exerts a force on
an apple, and vice versa. It's a
small force, but forces like this
have been measured.*

*Pick a flower on Earth and you
move the farthest star!*
                    Paul Dirac, physicist

oranges, rocks, books, and so forth? Your physics book, for instance, should exert a gravitational pull on an apple, and vice versa (Figure 5.7). We do not notice any such attractive force between objects like apples and books, but that is only because the force between such objects is very small.

So Newton reasoned that the gravitational force is universal, that is, that it is exerted between every pair of objects throughout the universe. This is the central idea of Newton's theory of gravity.

Newton understood the importance of quantitative methods. Although his basic insight was qualitative, its expression in a quantitative form led to powerful explanations and predictions. Quantitatively, we expect that the gravitational attraction between two objects must be stronger when the objects' masses are larger, because an apple's weight is larger when its mass is larger (double the mass, for example, by replacing the one apple by two apples joined together, and you double the weight). And since widely separated objects do not attract each other very strongly, the gravitational force should get smaller when the distance between the objects gets larger.

Newton put all this together (see How do we know?, p. 122, for how he did this) and came to the following conclusions:

NEWTON'S LAW OF GRAVITY
Between any two objects there is an attractive force that is proportional to the product of the two objects' masses and proportional to the inverse of the square of the distance between them:

$$\text{gravitational force} \propto \frac{(\text{mass of 1st object}) \times (\text{mass of 2nd object})}{\text{square of distance between them}}$$

$$F \propto \frac{m_1 \times m_2}{d^2}$$

If mass is expressed in kilograms, distance in meters, and force in newtons, then this proportionality becomes

$$F = 6.7 \times 10^{-11} \frac{m_1 \times m_2}{d^2}$$

Let's look at some examples, beginning with your own weight. According to the previous chapter, your weight is the gravitational force exerted by Earth on you. The law of gravity tells us that this force is proportional to your mass times Earth's mass, which means that the force is proportional to each of the two masses separately. So doubling your mass would double your weight, tripling your mass would triple your weight, and so forth—which seems quite plausible. But the law of gravity also tells us that if you imagined that somehow Earth's mass were doubled (without, however, changing its size), this also would double your weight; halving Earth's mass would halve your weight; and so forth. You can reduce your weight without dieting or exercising: Simply reduce Earth's mass!

What if you altered both masses? For instance, suppose you tripled your mass while simultaneously doubling Earth's mass. Since the force is proportional to the product of the two masses, this would multiply your weight by 6. As another way of looking at this, take it in steps: If you first tripled your mass, this would triple your weight; if you then doubled Earth's mass, this threefold weight would be doubled, giving an overall multiplication by 6.

What happens when the distance between Earth and you is changed? In fact, exactly what do we mean by the "distance between the objects" in a case like this? Does the "distance from Earth to your body" mean the distance from the near side of Earth (the ground beneath your feet), from the far side, from the center, or from some other point? And to what point in your body should we measure the distance?

Newton worked through a lot of mathematics to answer this—in fact he needed to invent "integral calculus" to answer it. Newton's answer was that the distance between the "centers" of the two bodies is the correct distance to use when applying the gravitational force law to two extended bodies. In the case of a body such as Earth that has an obvious center, distance is measured from that center. For other bodies, such as your own, the distance should be measured from the body's "balance point"—the point at which the body would be balanced under the force of gravity. But because your body is so small compared with the distance from Earth's center to your body, it matters little which point you choose within your body.

If you travel away from Earth, your distance from the center will increase. Since the gravitational force is proportional to the inverse of the square of this distance, the increased distance makes the force decrease— another way to reduce your weight! For instance, your weight at the top of Mount Everest, nearly 10 km above sea level, is 0.3%—3 parts in 1000— less than at sea level. If your weight is normally 600 newtons (135 lbs), it will be 598 N (134.5 lbs) at 10 km high. Your weight reduction is greater at an altitude of a few hundred kilometers, where low-orbit artificial satellites travel. For example, at a 200 km altitude, your weight would be reduced by 6%, so a person normally weighing 600 N would weigh only 560 N. Now you're really losing weight (but not mass!).

Let's consider the situation inside an orbiting satellite. Judging from Figure 5.8, you would feel weightless in an orbiting satellite, at any altitude. But we just saw that if the satellite is in low orbit, your weight is actually only a little less than normal. Why, then, would you feel entirely weightless? To answer this, let's imagine a somewhat similar situation (Figure 5.9): Suppose you are in an elevator and the elevator cable breaks. The elevator is then in free fall, and so are you, because all bodies fall with the same acceleration. After the cable breaks, your feet no longer press down against the floor. If you try to press your feet against the floor, you will simply push yourself away from the floor. You are apparently weightless, even though Earth still exerts its usual gravitational force on you. But because we have defined weight as "the gravitational force on an object," you are not really weightless.

You would feel weightless in an orbiting satellite for the same reason that you would feel weightless in a freely falling elevator: As we saw in the preceding section, the satellite falls freely around Earth. You are falling around Earth too, regardless of whether you are inside the satellite or outside in

It is only by bringing into the open the inherent contradictions, and the metaphysical implications of Newtonian gravity, that one is able to realize the enormous courage—or sleepwalker's assurance—that was needed to use it as the basic concept of cosmology. In one of the most reckless and sweeping generalizations in the history of thought, Newton filled the entire space of the universe with interlocking forces of attraction, issuing from all particles of matter and acting on all particles of matter, across the boundless abysses of darkness.

Arthur Koestler, author and philosopher, in *The Sleepwalkers*

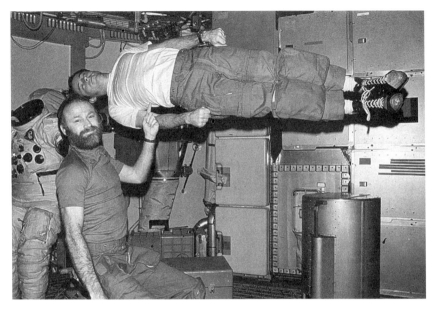

FIGURE 5.8
Space travelers feel weightless when they are in orbit and at any other time that they are "falling" freely through space. (a) Balancing. (b) Floating. (c) Spacewalking.

(a)

(b)   **(c)**

FIGURE 5.9
Falling freely in a freely falling elevator.

space. Since both you and the satellite are just falling around Earth, you have the sensation of weightlessness: You float around inside the satellite, and your body behaves as though it were removed from the effects of gravity. But you are not really weightless.

Let's move to still higher altitudes. Suppose you are 6400 km—1 Earth radius—above the ground. What is your weight? The proportionalities in the law of gravity make this an easy question. If you rise 1 Earth radius above the ground, the distance from Earth's center to your body will have doubled from its normal value, so the square of the distance is multiplied by 4. And since the force is proportional to the inverse of the square of the distance, the force is divided by 4. Your weight is now one-fourth of normal.

Figure 5.10 is a graph of your weight at various distances, in Earth radii, measured from Earth's center. No values are shown for distances less than 1. The gravitational force by Earth on your body never will reach precisely zero, no matter how far you are from Earth. But very far away, the force becomes very small. For example, at 10 Earth radii, your weight is 1% of your normal weight. Figure 5.10, and our entire discussion about the weight of your body, applies to the weight of any object. It could be an apple, an orbiting satellite, or the moon.

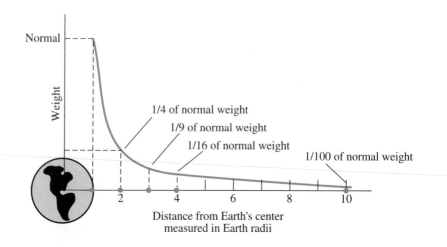

Normal

Weight

1/4 of normal weight

1/9 of normal weight

1/16 of normal weight

1/100 of normal weight

2   4   6   8   10

Distance from Earth's center
measured in Earth radii

It's a good thing for us that the force of gravity declines at larger distances in just the way it does. If the gravitational force declined a little faster, the planets would not move in ellipses but would instead spiral into the sun, and we would not be here to ask about things like gravity. If, on the other hand, the gravitational force declined a little more slowly, the gravity from distant stars would dominate the gravity from Earth, and again we would not be here.

The law of gravity can be used to calculate the gravitational attraction between any pair of objects, from apples and books to stars and moons. For example, the force between 1 kilogram and another kilogram 1 meter away is found by putting 1 kg, 1 kg, and 1 m into the gravitational force formula. The answer is 6.7×10^{-11} newtons, or 0.000,000,000,067 newtons! It is no wonder that the gravitational force between ordinary objects is difficult to detect. The delicate experiments needed to measure such tiny forces could not be performed until about a century after Newton's work. When they were performed, they did verify Newton's predictions.

HOW DO WE KNOW?

Since the forces between ordinary objects on Earth were too small for Newton to measure, how did he verify his hypothesized law of gravity? Because an object's weight is proportional to its mass (for instance, two identical apples tied together surely have twice the weight of one, so doubling the mass doubles the weight), Newton reasoned that the force of gravity must be proportional to each of the two masses.

But what about the dependence on distance? Newton knew that the distance to the moon is about 60 times larger than Earth's radius (Figure 5.11). Newton's hypothesized law of gravity then implies that an object at the moon's distance should experience a force that is 3600 (the square of 60) times smaller than the force on the same object on Earth. So (since Newton's law of motion says that acceleration is proportional to force), the acceleration of an object at this distance should be 3600 times smaller than the acceleration of an object falling to Earth.

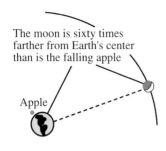

The moon is sixty times farther from Earth's center than is the falling apple

Apple

FIGURE 5.11
The moon is 60 Earth radii away from the center of Earth.

In other words, Newton's hypothesis implies that the moon's acceleration should be $(1/3600) \times 9.8$ m/s^2, or 0.0027 m/s^2. But from the observed facts that the moon is 60 Earth radii from Earth and that the moon takes 27 days to complete a circle around Earth, Newton could calculate directly from observation that the moon's acceleration (due to its circular motion) actually is 0.0027 m/s^2. Newton's hypothesis agreed with observation.

DIALOGUE 4   Suppose that you were in distant space, far from all planets and stars, and that you placed an apple and a book at rest in front of you, separated by about 1 meter. What effect, if any, would gravity have on the apple or the book? What effect would your own presence have on this experiment? How could you reduce your own influence on the experiment?

DIALOGUE 5   Do you weigh less in a high-flying jet plane? Can you decrease your mass in this way? How can you decrease your mass? When people say that they are "overweight" and want to reduce, what do they really want to lose—weight or mass?

DIALOGUE 6   Calculate your weight at an altitude of 2 Earth radii. At an altitude of 3 Earth radii. Calculate your weight at a distance of 10 Earth radii from Earth's center. At a distance of 100 Earth radii.

5.3 Gravitational collapse: *the birth and death of the solar system*

Gravity builds stars, and it builds planets. The star-birth story starts with an extremely diffuse gas, made mostly of widely separated hydrogen atoms, that is spread throughout the universe. In some regions, this material is gathered slightly more densely into great gas clouds, which are the spawning grounds for stars (Figure 5.12, see color insert). Because of the gravitational pull between all bits of matter, all gas and dust in space tends to aggregate. If by chance there is a place within a gas cloud where matter is gathered more thickly than it is elsewhere, this place can become a center of gravitational attraction for other matter. If enough material falls into this center, it can become a new star. In this manner, stars create themselves. This falling together of matter due to gravitational attraction is called **gravitational collapse**.

---

4. The apple and the book will move very slowly toward each other, speeding up as they get closer. Your own body will cause the apple and book to move toward you. You can reduce this effect by viewing the apple and book from a larger distance away. (If undisturbed, a 1 kg apple and a 1 kg book started 1 m apart would come together in about 26 hours).
5. Yes. No. You can reduce your mass by eating less, exercising, or cutting off one of your ears (not recommended!). Mass.
6. One-ninth of your normal weight; 1/16 of normal; 1/100, or 0.01 (1%) normal; 1/10000, or 10^{-4}, of normal.

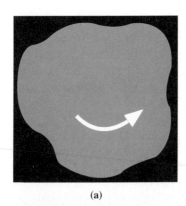

(a)

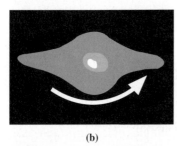

(b)

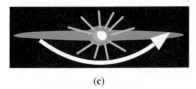

(c)

FIGURE 5.13
In (a), a slowly rotating gas cloud begins to contract because of its own gravity. In (b), the cloud rotates faster and flattens, and a hot central region begins forming. In (c), the sun forms in the cloud's center, surrounded by a rotating disk of gas. This disk will evolve into the sun's planets.

This is how the sun and Earth were born. The solar system, including the atoms in your body, began as a diffuse, cold cloud of gas and dust in space, a cloud thousands of times larger than our present solar system. About 5 billion years ago, partly by chance, matter gathered more thickly in one part of the cloud than elsewhere. This clump of matter drew surrounding material toward it, making the clump more massive, which increased its pull so that even more material was gathered, which made the clump even more massive, and so forth. This self-reinforcing process continued until a large part of the original cloud had collapsed around that first clump of matter, forming a giant ball of gas larger than our solar system. Our ball of gas continued pulling inward on itself. Atoms accelerated inward. As gas collected, new gas was pulled inward at higher and higher speeds. Atoms at the center collided at greater and greater speeds. In other words, the gas at the center heated up.

Every gas cloud has a slight overall spinning motion, simply because of the net effect of its chaotic flowing and swirling. As our gas cloud contracted, this spinning increased, just as a figure skater spins faster and faster as she brings her outstretched arms into her sides. Because of continued contraction, this spinning became rapid enough to flatten the outer regions of the gas ball into a thin disk, much as a wad of dough can be flattened into a pizza shape by spinning it (Figure 5.13). Some of the gas in the outlying flattened region eventually rotated fast enough to go into orbit around the larger central ball of gas. Because it was orbiting, this material was left behind as the center collapsed. The flattened outer region continued orbiting while cooling, condensing, and aggregating into clumps. Some of the clumps eventually became Earth and the other planets.

As the warming sun became hot enough to glow, light streaming outward from the warming sun swept away the dust and gas that had filled the solar system. And then there was light on Earth (Figure 5.14, see color insert).

The large central ball of gas continued collapsing and heating until the center reached million-degree temperatures. New things happen at such high temperatures: Atoms collide so violently that their electrons are stripped off, leaving a gas made of bare nuclei and electrons. The violently colliding nuclei occasionally stick together, a process known as **nuclear fusion** (Chapter 16). Nuclear fusion creates lots of heat, and the pressure from this heat then prevents the ball of gas from collapsing further. A balance is created between the inward pull of gravity and the outward pressure of the hot gas, much like the balance in a balloon between the tension of the inward-pulling rubber and the pressure of the outward-pushing air in the balloon.

When it initiated nuclear fusion about 5 billion years ago, the sun turned itself on and became a normal, self-sustaining star.

All stars turn themselves on in roughly the same manner that the sun did. It's a process that is going on all the time, all over our galaxy and all over the universe. The starry sky is not a static scene. Stars are born, have a lifetime, and die, like everything else.

Once the sun stopped collapsing, it settled into a middle age that has been going on now for nearly 5 billion years. The long-term stability of this period made it possible for atoms on one of the sun's planets to gather into highly complex and structured molecular forms that eventually evolved into life-forms and then, much later, into ourselves (Chapter 12). Humanlike creatures have existed on Earth for only about the past 4 million years, or about a thousandth of the sun's long history. Like the rest of the solar system, we came from the universe.

The sun's fuel for the fusion process is the hydrogen gas from which the sun is mostly made. Over billions of years, the supply of hydrogen must gradually deplete until, about 5 billion years in the future, it will no longer support nuclear fusion. Then the sun will enter old age. With no heat source, the pressure that has kept the sun big will vanish. The core will begin to collapse gravitationally, and this collapse will again heat the sun but now this "gravitational heating" will cause the outer portion of the sun to expand far beyond its present size. The sun will expand beyond the orbits of the two innermost planets, Mercury and Venus, and perhaps as far as Earth. Its heat will vaporize (turn into gas) any life remaining on Earth. Enveloped in the sun's gases, Mercury, Venus, and perhaps Earth will slow down and spiral into the sun's center, where they will be vaporized.

These nuclear reactions will eventually run their course. Then gravity will reassert itself for the final time. Without a nuclear heating source, there will be little to stop the sun from falling inward on itself. Certainly the interatomic forces that hold up solid matter against outside pressures on Earth are far too puny to stand up against the enormous inward pull of gravity in the final collapse of a star.

The sun will squeeze itself far inside its present boundaries and far inside the volume it would have if it were made of ordinary solid material, squashing its atoms out of recognizable existence until all that remains at the microscopic level will be a solid, densely packed expanse of bare nuclei and unattached electrons. At this point, the collapse will be permanently stopped by an effect known as *quantum exchange forces* between the electrons.*

The sun's burnt-out "corpse" will be hot, solid, and small—about Earth's size. The sun's entire mass will be packed into one-millionth of its present volume! The sun's material will be extraordinarily compact, with many tonnes packed into each cubic centimeter. A thimbleful of this material would collapse a strong steel table on Earth! Gravitational heating will heat the sun enormously during its final collapse, but once the collapse ends there will be no further source of heating. This remnant of the sun will glow brightly for a while and then slowly dim like a dying ember.

A star the size of Earth? When the first astronomical body of this type was discovered, in 1862, astronomers found these figures too staggering to be believed and thought there must be an error in their observations. But two other small stars were soon discovered, and now more than two hundred are known. Because of their white-hot glow and their size, these

---

* Quantum exchange forces have no explanation within Newtonian physics. Quantum exchange forces between electrons are far stronger than the ordinary repulsive electrical forces between electrons. These ordinary electrical forces, which can be included in Newtonian physics, are what maintains the solidity of normal solid matter.

compact stars are called **white dwarfs**. They are believed to be remnants of collapsed stars.

HOW DO WE KNOW?

Theories of the evolution of stars predict the evolution of the sun as just sketched. Observations of stars in the various evolutionary stages described and observations of Earth's oldest rocks, the moon, moon rocks, meteorites, other planets, other moons, and the sun itself all support these theories.

The natural place to look for star births is among thick gas clouds in space. When astronomers searched the dense gas cloud known as the Orion nebula, they found thousands of newly minted stars (Figure 5.12). Just as the theory predicts, nearly all of these new stars were wrapped in disks of dust and gas, disks that are expected eventually to coalesce into planets.

DIALOGUE 7   If the sun were to collapse tomorrow to become a white dwarf, but without any explosions or expansion that would directly affect Earth, would the gravitational force by the sun on Earth be any different from what it is now? Would Earth's orbit be any different? Would you expect life on Earth to be affected?

5.4 Gravitational collapse: *the death of more massive stars*

A star's life cycle is determined primarily by how much mass it has. A star needs at least 10% of the sun's mass in order to get hot enough to initiate nuclear fusion and become a star in the first place. All stars massive enough to initiate nuclear fusion go through a middle age that is similar to the sun's present state. Then when its hydrogen fuel has been used up, it enters its final phase. Stars having masses up to three times the sun's mass go through a final phase similar to the sun's, ending their lives as white dwarfs.

But a quite different fate awaits more massive stars—those with more than about three times the sun's mass. Like the sun, these more massive stars use up their hydrogen fuel and then contract at the center. But because of the star's larger mass, the contraction is stronger, and its center gets hotter. The high temperatures initiate a wide range of nuclear reactions that eventually turn the star's small central "core" into solid iron. Iron continues forming until the inner core becomes so massive that it cannot hold itself up. The solid iron core then suddenly collapses during a time in-

My suspicion is that the universe is not only queerer than we suppose, but queerer than we can suppose.
John B. S. Haldane, British geneticist, 1856–1928

7. No, because the sun's mass would be unchanged. Earth's orbit would not be affected, because the gravitational force by the sun on Earth would not change. Life on Earth would cease, however, because the sun's radiation (output of light and other elements) would be radically altered.

terval of just 1 second. This unimaginably violent process jolts the entire star, ripping it apart. Most of the star explodes into space, an event called a **supernova explosion**.

Only 10 to 20% of the original star remains after the explosion. No further nuclear reactions can occur in this remnant, so there is little to oppose the inpull of gravity. The remainder of the star enters its final collapse.

If the original star's mass (before collapse) was between 3 and 5 solar (sun) masses, then exchange forces between electrons will be able to stop the collapse, just as they will be able to stop the final collapse of the sun. The star then ends its life as the sun will, as a white dwarf.

But if the original star was more massive, perhaps 10 or 20 solar masses, the final collapse will be stronger, and the electron exchange forces will be too weak to stop it. There is only one other known force that can stop the collapse. It is called the *neutron exchange force*. It is similar to the electron exchange force, except that it occurs between neutrons (Chapter 2). Viewed at the microscopic level, the collapse not only squashes atoms out of existence; it also squashes electrons out of existence by forcing them to merge with nuclei. This turns each nucleus into a collection of neutrons, and it turns the entire star into an object that resembles a giant nucleus made of neutrons. This object is called a **neutron star**.

Nuclear physicist J. Robert Oppenheimer, who later became the leader of the team that developed the atomic bomb (Chapter 16), predicted neutron stars in 1938. None was discovered until 1967, when Jocelyn Bell, a sharp-eyed astronomy graduate student in England, discovered a source of radio waves in space that sent out "beeps" or "pulses" every 1.3 seconds. Project scientists thought at first that she might have discovered a radio beacon from an extraterrestrial civilization. But another was soon discovered, and by now more than 500 are known, with a wide range of pulsation rates. There is little question about their identity: They are neutron stars.

A neutron star is an exotic object. Although more massive than the sun, it is only a few kilometers across with a billion tonnes of material packed into each cubic centimeter. On Earth, a tiny speck of this material would weigh as much as a large, fully loaded highway truck! Like a spinning ice-skater bringing in her arms, a neutron star spins faster and faster during its final collapse. Indeed, the entire star spins at speeds that are unimaginable for such a massive object. The fastest stars spin hundreds of times every second. This spinning combines with magnetic effects to create the rapid pulses of visible light and radio signals observed from Earth, the signals that Bell discovered in 1967. As seen from Earth, the entire star appears to flash on and off many times every second. Figure 5.15 is a sequence of photographs of a neutron star, showing two visible flashes. The supernova explosion that created this neutron star was seen and recorded on Earth in 1054. For a few days the light from the explosion was brighter than the planet Venus. Today, this star is called the Crab nebula because of the crab-like shape of its dispersed, nebulous halo of gases blown into space by the explosion.

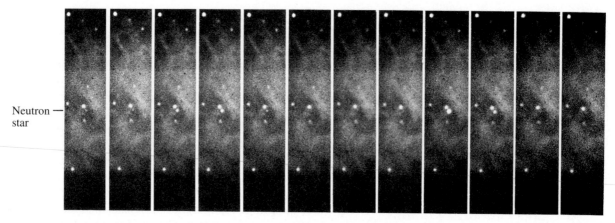

Neutron star

FIGURE 5.15

A sequence of photographs of the neutron star at the center of the Crab nebula. Portions of a surrounding gas cloud, the remnant of the supernova explosion that created the neutron star, can be seen. This sequence, which lasted a total of about 1/20 second, includes two flashes. The first flash occurred during frames 3 and 4, and the second occurred during frames 9 and 10.

Gravitational forces are enormous on a neutron star's surface. The star's radius is only about 10 kilometers, which is 10^5 (100,000) times smaller than the sun's radius. Yet the star is more massive than the sun. Newton's law of gravity tells us that the weight of an object on the surface of a star is proportional to the inverse of the square of the star's radius. If a collapsing star's radius becomes 10^5 times smaller, the weight of an object on its surface will become $10^5 \times 10^5$ times larger, in other words, 10^{10} (10 billion) times larger.

What about stars that are even more massive than those that collapse to form neutron stars? When a star whose original mass (before its collapse) exceeds about 30 solar masses runs out of fuel, the ensuing collapse is so strong that no known force can stop it. According to current theories, it collapses into a single point! Its matter—its atoms and subatomic particles—is squeezed out of existence. The star retains its mass, however, and so it retains its gravitational influence on the space around it.

Gravitational forces become unimaginably large near such a star. If you had the misfortune to get too close, you could not escape, because gravity won't let anything escape. If you were within a few hundred meters of the central point and you tried to throw an object away from the center, you would need to throw it faster than the speed of light in order for the object to escape the star. But as Chapter 10 explains, objects cannot be thrown faster than the speed of light. Nothing can get out of the immediate vicinity of the star's central point, not even light itself.* Such an object is called a **black hole**.

HOW DO WE KNOW? OBSERVATIONS OF BLACK HOLES

We can detect black holes by means of their gravitational influence on things around them. The first probable black hole, Cygnus X-1, was discovered in 1972. This object is thought to be a double star, two stars orbiting around each other, held together by their gravitational attraction.

---

* More precisely, quantum theory allows black holes to emit subatomic particles, but this effect is negligible for collapsed stars. This effect is expected to be important to low-mass black holes, although such small black holes have never been observed and may not exist.

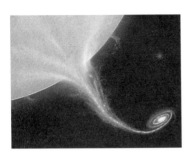

FIGURE 5.16

In this artist's conception, a black hole pulls matter from a companion star and accelerates it into a hot, X-ray–emitting disk before slowly swallowing it.

One star is a visible giant star, and the other is an unseen compact (in other words, far smaller than a normal star) object. By observing its gravitational effect on the visible star, the compact object's mass can be deduced to be some 15 solar masses. Since theories indicate that a compact object of more than 3 solar masses can only be a black hole, astronomers believe that Cygnus X-1 is a black hole. Satellites in orbit around Earth detect X rays from Cygnus X-1 that further confirm that it is a black hole. Apparently the invisible object's gravitational pull is drawing gases from the visible star and accelerating them down into and around the black hole, a process that tears apart the gas atoms and causes them to emit X rays that we can observe (Figure 5.16).

It seems difficult to believe that objects as exotic as black holes actually exist. Scientists don't go out of their way to invent strange ideas like this. To the contrary, they always look for the most natural, least strange, explanation. For example, people once found it strange that Earth could orbit the sun, but astronomers such as Copernicus found that this was the most natural idea that could account for the data. In the same way, astronomers today find that a black hole is the most natural explanation of Cygnus X-1. If it is not a black hole, then this object is not compact, or it does not have a mass larger than 3 suns, or compact objects of greater than 3 solar masses do not always collapse to become black holes. Astronomers find it easier to believe that Cygnus X-1 is a black hole than to believe any one of these three options. Science is conservative, always preferring the least-odd conclusion possible.

Using X rays to identify possible black holes and then studying these candidates by more direct means, several objects have now been discovered that are believed to be black holes resulting from the gravitational collapse of a star.

Much more massive black holes probably exist at the centers of many galaxies, including our own Milky Way galaxy. In some galaxies, confirmation of a central black hole comes from telescopic observations of the speeds of individual stars near the center. These stars are orbiting the center, indicating that something at the center is holding them into orbits. Measurements of the stars' orbital speeds enables the mass of the central object to be deduced from the law of gravity; faster speeds indicate a larger mass at the center. In some galaxies, the orbital speeds are high enough to indicate that the central object must be a black hole with a mass of millions of suns.

Such observations lead astronomers to conclude that many galaxies have black holes at their centers and that the distant and powerful galaxy-like objects called *quasars* are powered by massive black holes.

DIALOGUE 8   Suppose that Earth collapsed to half of its present radius. How much would you weigh? What if Earth collapsed from its present 6000 km radius to only 60 km?

8. Four times your present weight. The 60-km radius is 0.01 (1/100) of the original radius, so your new weight is 100 × 100 times larger, or 10,000 times your present weight.

5.5 The Newtonian worldview: *it's a mechanical universe*

During the sixteenth and seventeenth centuries, the new sun-centered astronomy and inertial physics ushered in not only new science but also a new philosophical and religious view that we will call the **Newtonian worldview.*** It is one of the more significant consequences of Newtonian physics. This worldview retains its influence today, even though Newtonian scientific ideas have been superseded by theories developed during the twentieth century. To appreciate this worldview properly, we will look at it in its historical context.

The pre-Newtonian or "traditional" worldview was centered on medieval religion, the Earth-centered astronomy of the ancient Greeks, and Aristotle's physics. Central to the traditional view is the idea of purpose, or future goals. Purpose is important to all of us in our individual lives, as many of our actions are directed toward some future goal. It is natural for people to extend this notion of purpose to human existence in general and to the physical universe. During most of human history, and certainly during the Middle Ages in Europe, popular culture and religion and science have been united in the belief that there is a purpose not only for individual human actions but also for humankind as a whole, for Earth, for the planets, for falling objects, and for everything else that happens. Furthermore, it was believed that human purposes are closely connected to the universe's larger purposes, that the universe exists for humans.

Ancient Greek astronomy supported this traditional view. Earth was thought to be motionless at the center of things, with the sun, planets, and stars circling it for our benefit. Aristotle's physics also supported the traditional view. The five Aristotelian elements all had goals that explained their natural motions. For instance, heavy objects moved downward because they wanted to get to their natural place—their goal—at Earth's center. Everything had its natural place in a hierarchy of places and purposes.

Astronomy and physics since the Middle Ages have fundamentally contradicted Earth-centered astronomy and Aristotelian physics. In the sixteenth century, Copernicus placed the sun, not Earth, at the center. Kepler replaced the planets' "natural" circles with ellipses. Descartes declared that there was only one kind of natural motion, namely, motion at an unchanging velocity, the same in the heavens as on Earth. According to the law of inertia, bodies move not because they have any goal but simply because there is nothing to stop them. The hierarchy of natural places, the notion that Earth is special, the centrality of humankind, and the scientific basis for universal purposes were swept away.

Galileo emphasized that he sought only to describe how things behave, not why they behave as they do. He was not concerned with a physical phenomenon's purpose. Analysis—the new technique of separating phenomena

---

* Newton was only one of many scientists and philosophers who contributed to these beliefs. Whereas Newton was the main scientific contributor to the new worldview, Descartes and Galileo contributed more to its philosophical formulation.

Without the living animal, I do not believe that odours or tastes or sounds are anything else than names—they have truly no other existence than in use; I say that I am inclined sufficiently to believe that heat is of this kind, and that. . .if the animate and sensitive body were removed, heat would remain nothing more than a simple word.

Galileo

I am induced by many reasons to suspect that they [the phenomena of nature] all may depend upon certain forces by which the little parts of bodies. . .are either mutually impelled to one another and crowd together according to regular figures, or are repelled and recede from one another.

Newton

I shall express what I call general nature by "cosmic mechanism," i.e., [composed] of all the mechanical affections (figure, size, motion, etc.) that belong to the. . .great system of the universe.

Robert Boyle, physicist and chemist, 1627–1691

Mathematical and mechanical principles are the alphabet, in which God wrote the world.

Robert Boyle

The universe is like a rare clock, such. . .that the engine being once set moving, all things proceed according to the artificer's first design, and the motions. . .do not require the peculiar interposing of the artificer, or any intelligent agent employed by him, but perform their functions. . .by virtue of the general and primitive contrivance of the whole engine.

Robert Boyle

[Men are] engines endowed with wills.

Robert Boyle

into their simplest components and studying those components—was one of Galileo's tools in this endeavor. This led to a focus on the simplest and smallest components of matter: atoms. And so *atomism*—the idea that nature can be reduced to the motions of tiny material particles—underlay the thinking of Galileo and others who developed the new physics.

Newtonian physics represents a return to the materialism and atomism of Leucippus and Democritus (Chapter 2). For example, in a view remarkably similar to Democritus's view, Galileo stated:

> I find myself impelled. . .to conceive that in its own nature [any piece of matter] is shaped in such and such a shape, that in relation to others it is large or small, that it is in this or that place, in this or that time, that it is in motion or remains at rest. . . .But that it must be white or red, bitter or sweet, sounding or mute, of a pleasant or unpleasant odor, I do not perceive my mind forced to acknowledge. . . .Hence I think that these tastes, odors, colors, etc., on the side of the object in which they seem to exist, are nothing else but mere names, and hold their residence solely in the sensitive body [that is, in the human or animal observer]; so that if the animal were removed, every such quality would be abolished and annihilated.

Newton noted:

> It seems probable to me that God in the beginning formed matter in solid, massy, hard, impenetrable, movable particles, of such sizes and figures, and with such other properties, and in such proportion to space, as most conduced to the end for which he formed them; and that these primitive particles being solids are incomparably harder than any porous bodies compounded of them, even so hard as never to wear or break in pieces. . . .If at any time I speak of light and rays as coloured, I would be understood to speak not philosophically and properly, but grossly, and according to such conceptions as vulgar people in seeing all these experiments would be apt to frame. For the rays to speak properly are not coloured. . . .[C]olours in the object are nothing but a disposition to reflect this or that sort of rays more copiously than the rest; in the rays they are nothing but their dispositions to propagate this or that motion into the sensorium [the brain], and in the sensorium they are sensations of those motions under the forms of colours.

In the Newtonian worldview, the fundamental physical reality is atoms. All other phenomena, such as tastes and colors, are, in Galileo's words, "mere names." Sense impressions and human feelings can be reduced to motions of atoms. For example, a napkin's "redness" is due to the motions of atoms in the napkin and in the observer's eye and brain. Things are not really "red" in themselves.

In the traditional view, human experience—feelings, colors, and so forth—had primary significance. Since humans were the universe's central focus, human experience was a fundamental reality. A red napkin did not merely appear to be red, it actually was red. This shift in viewpoint—from human experience to impersonal atoms—is central to the shift from the traditional to the Newtonian worldview.

Newton, Galileo, and Descartes were participants in the religious culture of their time and firm believers in God. What place could be found for God and religion within the new science? Descartes reconciled the new science with traditional religion by assuming that there was another reality in addition to the physical world. The other reality was a spiritual world parallel to, but independent of, the physical world. The material world was made of matter and operated according to the inflexible laws of nature, in just the same way that an impersonal machine operated. In this material universe, there was no color or human feeling, although God was needed to maintain the physical laws and to start things moving. The other world, the spiritual one, was the realm of human thoughts and feelings and communication with God. This idea—that there are two independent realities—is called **dualism**.

By putting all human qualities into a nonphysical spiritual realm, dualism lent support to the notion that physical reality consisted of atoms following the impersonal laws of nature. In the physical universe, the true realities, or **primary qualities**, were assumed to be impersonal characteristics such as position, time, shape, volume, mass, and the motions of atoms. Human sense impressions such as color and beauty were assumed to be **secondary qualities** that were not part of the physical world but were merely reflections of the primary qualities. Science and philosophy relegated human concerns to a shadowy secondary role in a physical universe.

Galileo's introduction of scientific experimentation fit this trend nicely. Traditionally, scientists had observed the universe as it existed around them. Because Galileo was the first to conduct prearranged experiments, he was also the first to isolate natural phenomena from influences that tended to cloud the main point of his experiments. For example, Galileo sometimes removed—insofar as possible—friction, air resistance, or gravity. Naturally, Galileo also wanted to remove all human influence from his experiments in order to study nature in itself, without human influences. This idea—that experimental results should not be influenced by humans, that they should be **objective**—has proved crucial to scientific progress.

It was assumed that objectivity actually was possible, at least in principle. If everything happened because of the motions of atoms and if these motions were determined by physical laws, then what was real did not depend on humans. The physical universe would do whatever it would do, regardless of human thoughts or feelings. It should be possible to observe the workings of the universe, uninfluenced by humans. In other words, it was assumed that humans did not interact with the physical universe in any essential way.

According to this view, there is little room for God in the day-to-day workings of the material universe. In the traditional view, God is continually and actively present throughout the universe, continually endowing all things with purpose. Descartes and Galileo believed that God was needed to establish the laws of nature and to start the universe moving but that once started, the whole thing would run itself.*

In the new view, God is, at most, an outside observer.

---

* With a few exceptions, Newton also believed that God did not intervene in the universe. On certain occasions, namely, in situations for which Newton himself could not find a scientific explanation, he believed that God momentarily intervened. This view—that every phenomenon that cannot be explained by science at any particular time requires an intervention by God—has been referred to as a "God of the gaps" philosophy. But as science closes the gaps, such a philosophy becomes less and less tenable.

A machine, especially a delicate and finely tuned machine such as a clock, is an excellent analogy for the Newtonian worldview. Once the owner has started them, clocks run themselves according to their own operating principles. The founding fathers of physics envisioned the universe as a clockwork mechanism whose operating principles were the laws of nature and whose parts were atoms. Once started, the universe ran itself. No further intervention was needed. Because of the machinelike quality of Newtonian physics, physicists refer to Newton's theories of motion and gravity as Newtonian *mechanics.*

According to Newtonian physics, every physical system is entirely predictable, like a perfectly operating clock. For example, Newtonian physics can predict precisely how far a freely falling object will fall in any specified amount of time.

This clocklike predictability of nature has surprising implications. To understand them, imagine a simple, self-contained collection of tiny particles that move and interact in accordance with Newtonian physics. Suppose we specify the precise positions and velocities of every one of these particles at one particular time, such as noon today. Then, according to Newton's theory of motion, the entire future behavior of this collection of particles can be precisely predicted. That is, the precise position and velocity of every one of the particles can be predicted, for all time.

This in itself is not too surprising. For example, it is not surprising that given its starting position and velocity, we can predict where a stone will be as it falls. But remember that according to the Newtonian worldview, everything is made of atoms and that these atoms obey precise laws of nature, including Newton's theory of motion. Not only is a stone made of atoms; the entire universe, including living organisms like you and me, is made of atoms that, according to the Newtonian worldview, follow these precise laws.

This means that the future of the entire universe is predictable, that the future is entirely determined by what all the atoms of the universe are doing right now or at any other time. Once started, the clockwork universe is required to do precisely what it has done and what it will do for all time in the future. Everything that happens, including the fact that you are reading this book right now, was predetermined by the behavior of the atoms at the instant the universe began. Since, according to the Newtonian worldview, every thought or feeling that enters your head is reducible to the motion of atoms within your brain and elsewhere, all of your thoughts, feelings, and actions are entirely predetermined and predictable. You never choose to scratch your nose, for example—the laws of nature choose for you. You might believe that you choose, but this too, this believing that you choose, was chosen for you by the laws of nature.

This loss of free will was expressed most directly by the French scientist Pierre-Simon Laplace (1749–1827):

> All events, even those which by their insignificance do not seem to follow the great natural laws, are indeed consequences of these laws as surely as is the rising and setting of the sun. We must therefore consider the present state of the universe as the result of its earlier state and as the cause of that which is to come. An intelligence which at a given instant knew all the forces acting in nature and the positions of every object in the universe—if endowed with a

brain sufficiently vast to make all the necessary calculations—could describe with a single formula the motions of the largest astronomical bodies and those of the lightest atoms. To such an intelligence, nothing would be uncertain; the future, like the past, would be an open book.

Beginning in the seventeenth century and continuing into the twentieth century, these ideas influenced all educated people in Europe and elsewhere. Newtonian physics was so successful that the associated philosophy, the Newtonian worldview, was accepted with little question. These ideas were absorbed with little realization that they even were being absorbed.

There are reasons today to question both the Newtonian worldview and Newtonian physics (see Chapter 2 and the next section). Nevertheless, it would be surprising if these views, which dominated for so long, did not still operate in important ways today. Influential worldviews tend to be absorbed automatically, thoughtlessly, as part of the cultural air of the times. It seems likely that the Newtonian worldview remains active in our culture today, especially among people who have never heard of Isaac Newton, let alone Descartes or Galileo.

It is for you, the reader, to determine to what extent the Newtonian worldview is valid, whether it retains a significant influence, and what difference it might make.

It's a material world.

Madonna

5.6 Beyond Newton: *limitations of Newtonian physics*

Newtonian physics was tested repeatedly during the eighteenth and nineteenth centuries, and it stood up in quantitative detail to every challenge. In fact, Newtonian physics was so powerful and accurate that scientists began to accept it as true in an ultimate, absolute sense. To some, the word *understand* came to mean "to explain in terms of Newtonian physics."

But science is never absolute. Its theories can never be proved, but they can be disproved. Every scientific idea hangs by the slender thread of experimental test. No matter how many tests have been carried out in the past, there are always new testing grounds.

During the last two decades of the nineteenth century, experimental results began appearing that proved impossible to reconcile with Newtonian physics. Accordingly, Newtonian physics was found to have limitations; in other words, it was found to be incorrect in certain situations.

During the first few decades of the twentieth century, physicists invented three new theories to account for nature's behavior in those situations in which Newtonian physics was incorrect. These three theories are called **special relativity**, **general relativity**, and **quantum theory**. To date, at least, scientists have found no exceptions to any of these three theories.

There are four types of situations in which Newtonian physics has been found to give incorrect predictions: when speeds are very high, when gravitational forces are very strong, when distances are very large, and when distances are very small. Briefly, special relativity deals with high speeds; general relativity deals with strong gravitational forces and with large distances; and quantum theory deals with small distances.

Experiments show that Newton's law of motion and deeper Newtonian views concerning time and space break down at high speeds. The disagreement between Newtonian theory and experiment is not noticeable at slow speeds, but the errors become progressively worse as the speeds increase. The non-Newtonian effects cannot be observed for automobiles, bullets, jet planes, or even orbiting satellites moving at some 10 km/s. But at 30,000 km/s, Newtonian predictions are off by 1%. At 290,000 km/s, nearly the speed of light, typical Newtonian predictions are incorrect by 1500%—a factor of 15! Scientists didn't notice these non-Newtonian effects for two hundred years because they had never studied such fast-moving objects, at least not in enough detail needed to observe the effects. The theory of special relativity gives correct predictions at all speeds, both low and high. The predictions of special relativity become indistinguishable from those of Newtonian physics whenever the speeds are considerably less than the speed of light.

Similarly, experiments show that both Newton's law of gravity and Newton's law of motion—as well as Newtonian views regarding time and space—are incorrect for objects subjected to very strong gravitational forces and also over very large distances. For example, non-Newtonian effects are measurable, but small, for the orbit of the innermost planet, Mercury, which feels strong gravitational forces by the sun. Non-Newtonian gravitational effects are quite significant near neutron stars and black holes and for physical systems that range over large portions of the observable universe. The theory of general relativity gives correct predictions for all these situations and is regarded as the correct theory of gravity. The predictions of general relativity become indistinguishable from those of Newtonian physics whenever gravitational forces are not too strong and distances are not too large.

Finally, experiments show that Newton's law of motion and deep Newtonian views concerning predictability, cause and effect, and distant interactions are incorrect for very small objects, objects whose size is comparable to that of one molecule or atom. Newtonian predictions are significantly in error when applied to individual atoms or molecules. However, when large numbers of atoms come together to form a macroscopic object such as a baseball, Newton's theory does give correct predictions for that object's overall motion. Quantum theory gives correct predictions for objects of all sizes, from small to large. For ordinary macroscopic objects, quantum theory's predictions are indistinguishable from those of Newtonian physics.

Figure 5.17 is one way of indicating, graphically, these limits of validity of Newtonian physics.* The vertical axis represents speed. Special relativity predicts that objects cannot move faster than the speed of light, 300,000 km/s, so these speeds are forbidden. The horizontal axis shows distance, or the size of individual objects. Quantum theory predicts that objects cannot be both small and slow moving, in other words, that small objects must move rapidly. So there is another forbidden region in the small-size, low-speed corner of the diagram.

---

* Thanks to Douglas Giancoli, the author of several physics textbooks, including *Physics: Principles with Applications* (Englewood Cliffs, NJ: Prentice Hall, 1991), for suggesting diagrams of this type.

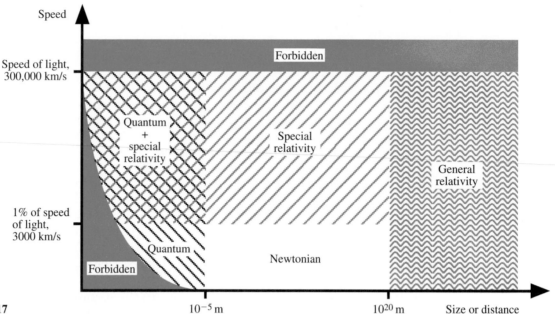

Speed

Forbidden

Speed of light,
300,000 km/s

Quantum
+
special
relativity

Special
relativity

General
relativity

1% of speed
of light,
3000 km/s

Quantum

Forbidden

Newtonian

10^{-5} m $\qquad$ 10^{20} m $\qquad$ Size or distance

FIGURE 5.17

Newtonian physics breaks down for very small objects, very large objects, and very fast objects. Newtonian physics also breaks down for strong gravitational forces, such as those near a neutron star or a black hole. The diagram is only schematic and approximate.

In summary: Newtonian physics is correct for common phenomena on Earth, but it breaks down for phenomena that are far from this "normal" range. The quantum and relativity theories apply throughout the entire range of the phenomena observed to date, and these theories reduce to Newtonian physics when applied to common phenomena on Earth.

Summary of Ideas and Terms

Newton's law of gravity Between any two objects is an attractive force proportional to the product of the two objects' masses and proportional to the inverse of the square of the distance between them: $F \propto m_1 \times m_2/d^2$.

Satellite A body in orbit around a larger astronomical body. Inertia keeps satellites moving, and the gravitational force exerted by the central body holds satellites in their orbits.

Weightlessness Bodies are weightless only when they are far from all other bodies. Bodies in orbit around Earth are not weightless, but they seem weightless because they are falling freely around Earth.

Gravitational collapse Matter falling together because of gravity.

Star birth Stars form from collapsing gas clouds. Gravitational collapse heats the collapsing gas, which initiates nuclear fusion in the center, which stops the collapse.

Nuclear fusion The process in which two nuclei stick together to form a single nucleus. This fuels the stars.

Supernova explosion An explosion of an entire star.

White dwarf A planet-sized, compact star. When stars of about the sun's mass run out of fuel, they flare up and then collapse to become white dwarfs. When slightly more massive stars (3 to 5 solar masses) run out of fuel, they explode, and the remaining remnant collapses to become a white dwarf.

Neutron star A compact star a few kilometers in diameter, resembling a giant nucleus made of neutrons. It spins rapidly, sending out radio beeps and light flashes. When massive stars (10 to 20 solar masses before collapse) run out of fusion fuel, they explode, and the remaining remnant collapses to become a neutron star.

Black hole Any object that has gravitationally collapsed into a single point. Nothing can escape from its vicinity. When very massive stars (greater than 30 solar masses before collapse) run out of fuel, they explode, and the remaining remnant collapses to become a black hole. Giant black holes having millions of solar masses probably exist at the centers of many galaxies.

Newtonian worldview The philosophical view, accompanying Newtonian physics, that the impersonal, clocklike workings of material atoms, as prescribed by natural laws, determine everything else. Human sense impressions are secondary qualities caused by motions of atoms in human brains.

Dualism Descartes's idea that there are two realities, physical and spiritual. In the physical realm, the real or **primary qualities** are objective, impersonal phenomena such as the motion of atoms. Human sense impressions are considered to be **secondary qualities**, caused by the primary qualities.

Objectivity An experiment is objective if its outcome is not influenced by humans. The Newtonian worldview assumes that perfect objectivity is possible, at least in principle.

Newtonian determinism Newtonian physics, when coupled with the idea that everything is made of atoms, implies that the universe's future behavior is entirely determined by its present state, like the workings of a perfect clock.

Limitations of Newtonian physics Newtonian physics gives incorrect predictions for (1) fast-moving objects, (2) strong gravitational forces or large distances, and (3) small objects. Special relativity, general relativity, and quantum theory do give correct predictions for each of these three classes of phenomena, respectively.

Review Questions

THE IDEA OF GRAVITY

1. What is the direction of a falling apple's velocity? Of its acceleration?
2. What is the direction of the moon's velocity? Of its acceleration?
3. Does Earth exert a force on the moon? What is its direction? How would the moon move if this force did not act?
4. In what ways are the moon and a falling apple similar? In what ways do they differ?

THE LAW OF GRAVITY

5. Does this book exert a gravitational force on your body?
6. What would happen to this book's weight if you managed to double Earth's mass? What if, instead, you doubled the book's mass?
7. In order to use Newton's law of gravity to calculate your weight, what data would you need?
8. If you were orbiting Earth in a satellite 200 km above the ground, would you be weightless? Would your weight be as large as it is when you are on the ground? Would you feel weightless? Explain.

9. What caused the sun to get hot? What keeps it hot today?

10. Describe the process that formed the planets.

11. Since gravity pulls inward on the material in the sun and since the sun is made only of gas, why doesn't the sun collapse?

12. Are there places in our galaxy where stars are being born?

13. Name the process and also the substance that fuels the sun.

14. What will happen to the sun after it runs out of fuel?

15. Name and describe the object into which the sun will evolve after it runs out of fuel.

16. What causes different stars to evolve differently?

17. All stars eventually evolve into one of three types of objects. Name them. What kinds of stars evolve into each of the three types of objects?

18. Describe a neutron star.

19. Describe a black hole. Since nothing can come out of a black hole, how can we detect it?

20. List some ways in which ancient Greek astronomy and Aristotelian physics support the traditional philosophical and religious worldview.

21. List some of the ways in which Copernican and Newtonian science are less supportive of the traditional worldview.

22. According to the Newtonian worldview, is a red napkin really "red"? Explain.

23. List several ways in which, according to the Newtonian worldview, the universe is similar to a clock.

24. For what kinds of phenomena is Newtonian physics incorrect? Why did it take so long to discover such exceptions?

25. List the three theories that give correct predictions for the situations in which Newtonian physics is incorrect.

Home Project

Discover the experience of having reduced apparent weight in an elevator. As the elevator comes to rest while going upward or starts into motion while going downward, you should feel light. Is your weight really less when you experience this? This sensation of decreased weight is due to the elevator's downward acceleration, just as an orbiting astronaut's sensation of weightlessness is due to the satellite's downward acceleration while the satellite orbits Earth. What, if anything, can be done to give you the sensation of complete weightlessness inside the elevator? What motions of the elevator make you feel heavier than normal? What is the direction of the acceleration in these cases?

For Discussion

1. Give at least one argument for the proposition that the Newtonian worldview is a valid view of reality. Now give at least one argument on the other side. What is your own view and why? List some ways in which the Newtonian worldview has influenced our culture, either in the past or today.

2. In view of the discussion of determinism and free will in Section 5.5, do you have free will? Support your view with at least two specific reasons (also see Section 2.6).

3. Discuss the meaning of ethics in a supposedly deterministic universe. If all actions really were predetermined, could we then talk meaningfully of what a person "should" do or denounce people for immoral actions? Could there be any right or wrong?

Exercises

1. Does Earth's gravity pull more strongly on a block of wood or on a block of iron having the same size? Which one falls faster when dropped (neglect air resistance)?

2. Does Earth's gravity pull more strongly on a block of wood or on a block of iron with the same mass?

3. What is the magnitude (strength) and direction of the gravitational force on you right now?

4. When you crumple a sheet of paper into a tight ball, does its mass change? Does its weight change?

5. Are you in orbit around (falling around) Earth's center? Is there anything around which you are in orbit?

6. The moon is in orbit around two objects simultaneously. Which two? (Actually, there is a third—our galaxy's center.)

7. Do you exert a gravitational pull on people around you? Do they exert a gravitational pull on you?

THE LAW OF GRAVITY

8. Which is larger, the gravitational force by Earth on the less massive moon or the gravitational force by the moon on Earth?

9. How strongly and in what direction does Earth pull on a 1-newton apple? How strongly and in what direction does the apple pull on Earth?

10. If you somehow decreased Earth's mass, with all other factors unchanged, would this affect your weight? How?

11. If you somehow increased Earth's radius, without changing its mass, would this affect your weight? How?

12. List at least three bodies that have a detectable (measurable) gravitational effect on Earth's motion.

13. The giant planet Jupiter is more than 300 times more massive than Earth. It seems, then, that an object on Jupiter's surface should weigh 300 times more than it weighs on Earth. But it actually weighs only about 3 times as much. Explain.

14. If you were in a freely falling elevator and you dropped your keys, they would hover in front of you. Are the keys falling? Are the keys weightless?

15. If gold were always sold by weight, could you make money buying gold at one altitude above the ground and selling it at a different altitude? Where would you want to buy—at a high altitude or a low altitude?

16.*Suppose that the gravitational force between an apple and an orange placed a short distance apart is one-trillionth (10^{-12}) newton. What would the force be if the distance were doubled? Halved? Tripled? Quartered?

17.*Referring to the previous exercise, what would the force be if the mass of the apple were doubled? Tripled? What if the mass of the apple were tripled and the mass of the orange were quadrupled (multiplied by 4)?

18.*Referring to the previous exercise, what would the force be if the mass of the apple were doubled, the mass of the orange were doubled, and the distance between them were doubled?

19.*Making estimates. Earth's mass is about 100 times the moon's mass, and Earth's radius is about 4 times larger than the moon's radius. From this information, use the law of gravity to estimate how much more an object weighs on Earth as compared with its weight on the moon.

20.*Making estimates. Estimate the gravitational force, in newtons, that you exert on a person standing near you.

GRAVITATIONAL COLLAPSE

21. Suppose that the sun collapsed tomorrow to become a black hole but that it collapsed "quietly," with no explosive or other direct effects on regions outside the sun. Would Earth's orbit be any different after the collapse than it is now? What important feature affecting life on Earth would be different?

22. Will Earth ever collapse to become a black hole? Why? Will the sun?

23. The orbits of all nine planets lie approximately in the same flat plane. Why?

24.*If Earth collapsed to one-tenth of its present radius, how much would you then weigh?

25.*If Earth expanded to ten times its present radius, how much would you then weigh?

BEYOND NEWTON

26. What theory or theories would be needed to predict the behavior of an atom moving at half the speed of light?

27. According to the most widely accepted scientific theory of the creation of the universe, the universe during the first few moments (much less than 1 second) of its existence was extremely hot, was full of densely packed matter, and was very tiny—smaller than an atom. What theory or theories would be needed to explain what was happening during these first few moments?

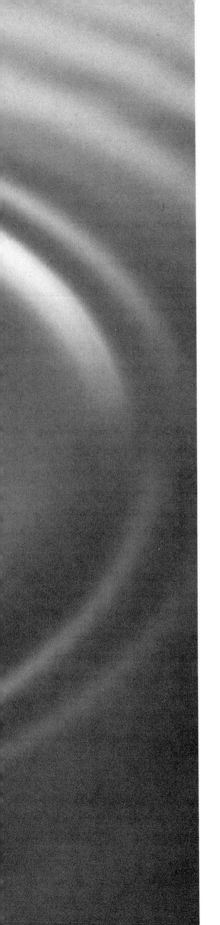

PART **III**

TRANSITION TO THE NEW PHYSICS

6

CONSERVATION OF ENERGY
you can't get ahead—

Human cultures are defined largely by their uses of nature's energy resources. "Civilization" itself is nearly synonymous with the organized use of solar energy. Although humanlike creatures evolved a few million years ago, it was not until the organized use of solar energy to grow food, just a few thousand years ago, that humans settled into villages and developed a sense of history, progress, and civilization.* The earliest village culture was an *agri*culture.

Despite the impact of energy on our lives, it is not easy to say what "energy" means. This is not surprising, for it took science two centuries following Newton's work to understand energy. Energy is an idea, a powerful and useful idea, to be sure, but still "only" an idea, not something substantial and material like the chair you are sitting on. Nevertheless, we will see in later chapters that to modern physicists, nothing is more real than energy.

The power of scientific ideas lies in their ability to explain and unify a variety of natural phenomena. At this, energy excels. And unlike many scientific principles, the principles of energy appear to be correct in all situations. Whether we observe huge cosmological structures, subnuclear particles, or an automobile on Main Street, the laws of energy are correct in every detail. In view of the limitations of even such a grand theory as Newtonian physics (see Section 5.6), the sweep of the energy principles is remarkable.

---

* Specifically human forebears, or *hominids*, evolved in Africa and are dated by radioactive and other methods at 3 million to 5 million years old. Anatomically modern humans, *Homo sapiens*, evolved from these forebears and are dated at 100,000 years in Africa. The agricultural revolution—the transition from nomadic hunter-gatherers to farmers—occurred first in the Middle East between 9000 and 5000 B.C.

Maybe it is more than coincidence, then, that energy principles have a vital social context. We have noted that the earliest human culture was based on solar energy. Our more recent industrial culture has been energized by the chemically altered remains of ancient life known as **fossil fuels**: coal, oil, and natural gas.

The coal-fueled steam engine was at the heart of the **industrial revolution** that began about 1750. Steam engines were first used to pump floodwater out of mines and later to drive industrial machines, to move coal, and to power railroad locomotives, ships, and the first automobiles. The economic and social consequences of these steam engines were profound. Because the new machines of production were expensive, they were accessible mainly to the rich. Workshops away from home grew into factories. Even unskilled workers could tend the new machines, unlike traditional craftsmen whose skills required a long apprenticeship. So surplus farm workers and children could be employed. Consequently, in Europe and North America the nineteenth century was marked by increased productivity, the capitalistic organization of industry, and a shift of population from farms to cities. The political ideologies of the twentieth century grew out of the economics of the industrial revolution. The industrial revolution continues. It is spreading all over the world and to new industries such as computers.

As the industrial revolution was transforming the planet, it stimulated scientists to understand the principles of energy. This chapter develops one of science's grandest principles, conservation of energy. The first four sections develop the concepts of work and energy needed to understand this principle, and Section 6.5 presents the law of conservation of energy. Everything that happens in the universe involves an energy transformation of one sort or another. Section 6.6 studies several examples of energy transformations. Section 6.7 looks at a related idea of great practical importance in our energy-consuming times: power, or the *rate*, per unit of time, of transforming energy. The next chapter explores the second grand principle of energy, the second law of thermodynamics.

Energy is one of our organizing themes and the central physical idea in this book. It will be used throughout the book to analyze all sorts of physical systems.

DIALOGUE 1   Since there has never been an observed violation of them, are the principles of energy then true for certain?

6.1 Work: *using a force to move something*

Because energy is based on the notion of work, we will begin with this idea. In common language, you do work whenever you exert yourself to perform a task. The physicist's definition is a refinement of this common usage. In physics, work is done whenever an object is pushed or pulled through a distance. For instance, you do work on this book when you push it across the

---

1. No! All scientific ideas are tentative. See Chapter 1.

FIGURE 6.1
Is either person doing any work on the brick wall?

table. The agent doing the work need not be a person. For instance, a magnet does work on a paper clip when the magnet pulls the clip toward the magnet. Earth does work on this book when the book falls. More precisely, object A (a person or any other thing) does work on object B if A exerts a force on B while B moves. You do work on this book when you lift it. Earth does work on the book when the book falls. Notice that work is always done by a specific object on a specific object. And notice that both force and motion are needed in order for work to be done.

DIALOGUE 2   Is either person in Figure 6.1 doing any work on the brick wall? Can you "work up" a sweat while pushing on a wall, without doing any physics-type work?

DIALOGUE 3   A giant meteoroid (a rock in space) moves at high speed through the solar system. Being far from the sun and other large bodies, the meteoroid feels little gravitational pull. If we neglect this small gravitational pull, is any work being done on the meteoroid? What keeps it moving?

It is of practical importance to be quantitative about energy. For example, is it enough to know merely that your car uses fossil fuel, or is it important to know how much fuel?

Let's think about the amount of work you do in various situations. If the word *work* is to agree roughly with common language, the work you do in lifting a load should be larger for larger loads. To see how much larger, compare lifting one book with lifting two stacked identical books (Figure 6.2). The effect is twice as big in the second case, so the work done should be twice as big. This means we want work to be proportional to force.

Now compare pushing a book across one table with pushing it all the way across two adjoining tables (Figure 6.3). Again the effect is twice as big in the second case. So work should be proportional to the distance moved.

Work should be proportional to both force and distance. Thus, we define the amount of **work** done by one object on a second object as the product of the force (exerted by the first object on the second) multiplied by the distance that the second object moves while experiencing that force:*

$$\text{work} = \text{force} \times \text{distance}$$
$$= Fd$$

FIGURE 6.2
It takes twice as much work to lift two books.

FIGURE 6.3
It takes twice as much work to push one book twice as far.

2. No, because neither person is moving the wall. Yes, you can work up a sweat pushing on a wall without doing any actual work on the wall. (At a microscopic level, your muscles are doing work on one another, although this microscopic work is done in your body, not on the wall.)

3. No work, because there are no forces. Nothing is needed to keep it moving (law of inertia).

* This definition assumes that the motion is in a straight line in the direction of the force. If the motion and the force are not in the same direction, then in the formula we must use only that part or "component" of the force that is along the motion. We won't need this refinement here.

FIGURE 6.4
Scottish physicist James Prescott Joule. His experiments, in the 1840s, helped unravel the confusion surrounding heat and so led to the first clear understanding of energy.

For instance, if you push on a book with a 3-newton force while pushing it 2 meters, you have done (3 newtons) × (2 meters) = 6 newton-meters of work. Note that the unit of work is the newton-meter, a measurement that has been used so often that it has been renamed the **joule** (J) (rhymes with school), in honor of James Prescott Joule (Figure 6.4).

Although our definition of work is related to the way that the word is used in ordinary language, the scientific reason for this particular definition is that it is useful in connection with the upcoming idea of energy.

From the following dialogue you can see that the work done in lifting an object is the object's weight multiplied by the height lifted.

DIALOGUE 4   Suppose you slowly lift a book weighing 12 N from the floor to a shelf 2 m above the floor. While you are lifting it, (a) how large is the net force on the book? (b) how large is the force by your hand against the book? and (c) how much work did you do on the book?

6.2 Work and energy: *a simple example* _____

Do this two-step experiment, and observe carefully: First, place your book (or some other nonbreakable object) on your outstretched hand on the floor; slowly lift the book from the floor to some height; hold it there a few seconds; and then slowly lower it back to the floor. Second, repeat the same lifting process to the same height, but this time suddenly remove your hand from the book so that it falls to the floor. We are going to look closely at this.

You do work on the book to lift it. But in lowering back to the floor, the book does work on you, because the book pushes downward against your hand all the way down. We could say that the raised book has an ability to do work and that the book actually does this work when it is lowered.

Let's look more closely at the "ability to do work." We just saw that raised objects have this ability. And so do moving objects. For example, suppose that you throw your book, horizontally, at a wall. One way to get work out of the moving book would be to stick a thumbtack partly into the wall, directly in line with the book's motion, so that the book will hit the tack and drive it in farther (Figure 6.5). A moving book, or any other moving object, has an ability to do work.

In the second step of our experiment, the book was again raised, and in the raised position it again had the ability to do work. But then you dropped it so that it simply fell, without doing work on your hand. As the falling book lost height, it gained speed, and so it retained an ability to do work. Just before hitting the floor, the book still had some ability to do work; only now this ability resulted from the book's speed rather than from its height. We

FIGURE 6.5
One way to get work out of a moving book.

---

4. (a) Zero. (b) 12 newtons. (c) 12 × 2 = 24 J.

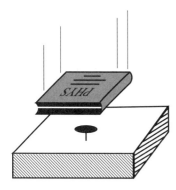

Figure 6.6
One way to get work out of a falling book.

could get this work out of the falling book by sticking a tack partly into the floor and letting the book drive in the tack farther (Figure 6.6).

Raised objects and moving objects have an ability to do work. When you lift the book, you give it the ability to do work. The work you do is "stored" in the raised book. You could get back this work at any time by letting the book push your hand back down slowly to the floor. Or you could just drop it, in which case the book would gain speed and so retain the "stored work" right up until it hits the floor.*

Physicists have a word for the ability to do work. It is called **energy**. You can think of an object's energy as stored work, work that you can get from the object. In our example, we see that both raised objects and moving objects have energy. It is useful to distinguish these different forms of energy. We say that a raised object has **gravitational energy**, because this energy is caused by the gravitational pull by Earth on the object, and that a moving object has **kinetic energy** (*kinetic* is related to the Greek word for "motion"). As we will see, there are many other energy forms.

The fact that you gave energy to the book when you raised it is one example of the **work–energy principle**, which states that work done on an object by outside agents such as your hand increases its energy.

As it falls, the book loses gravitational energy but gains kinetic energy and so retains its overall energy. This retention of overall energy when there is no outside agent (such as your hand) is one example of the **law of conservation of energy**.

These are the essential ideas about work and energy, presented in the context of a simple example. The rest of this chapter expands on these ideas.

6.3 A quantitative look at energy

Let us now look quantitatively at the simple experiment described in the preceding section. As we know (Section 6.1), the amount of work you do to lift an object is its weight multiplied by the height raised. Once the book is raised, it has gravitational energy because it can do work in pushing your hand back to the floor. How much gravitational energy does it have?

We define the book's energy quantitatively as *the amount of work it can do.*† It is measured in joules (or newton-meters), just as work is. The raised book's energy is the amount of work it can do in slowly pushing your hand back down to the floor. So the amount of work the book can do in this manner is just the book's weight multiplied by the

---

* For now, we'll worry only about what happens up until the book hits the floor. Later, we'll study the process of its actually hitting the floor.

† But we'll find out in the next chapter that for the particular energy form known as *thermal energy*, there are limits on the "efficiency" of actually doing this work.

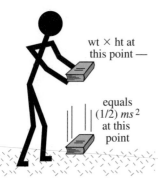

wt × ht at
this point —

equals
(1/2) ms^2
at this
point

FIGURE 6.7
An amazing thing: The gravitational energy at the beginning precisely equals the kinetic energy at the end.

1

2

FIGURE 6.8
The total energy is conserved all the way down. The loss in gravitational energy between points 1 and 2 during the fall is precisely balanced by the gain in kinetic energy between these two points.

distance to the floor. The gravitational energy of this or any other raised object is

$$\text{gravitational energy} = \text{weight} \times \text{height}$$
$$\text{GravE} = \text{wt} \times \text{ht}$$

When you slowly lower the book, it uses up its energy in pushing your hand back to the floor. But when the book falls, its gravitational energy is transformed into kinetic energy. How much kinetic energy is possessed by a moving object such as the book? In other words, how much work can a moving object do because of its motion? It is possible to work out the answer to this question, starting from Newton's laws, although we won't work it out here. The answer turns out to be

$$\text{kinetic energy} = (\tfrac{1}{2}) \times (\text{object's mass}) \times (\text{square of object's speed})$$
$$\text{KinE} = (\tfrac{1}{2})ms^2$$

This formula tells us that a more massive object has more kinetic energy and that a faster object has more kinetic energy, as we might expect. It makes sense that the formula should involve the object's mass rather than its weight, because kinetic energy is possible even in the absence of gravity.

Now we note an amazing thing, also provable from Newton's laws: If we neglect air resistance (which we'll deal with later), the amount of gravitational energy the book has at the beginning of its fall will precisely equal the amount of kinetic energy it has at the end (Figure 6.7). The book's total ability to do work, its total energy, is quantitatively unchanged during the falling process. Its energy is simply changed in form—transformed—from gravitational to kinetic, but its total energy remains the same. Energy is precisely "conserved."

Furthermore, since energy is conserved for any distance of fall, it must be conserved at halfway down, at three-quarters of the way down, and at every other point during the fall. The loss in gravitational energy during any portion of the fall precisely equals the gain in kinetic energy during that portion (Figure 6.8).

DIALOGUE 5   (a) How much gravitational energy (relative to the surface) does a diver have when she stands 2 m above the water, as compared with when she stands 4 m above the water? (b) One car moves at twice the speed of a second identical car. How much more kinetic energy does the faster car have?

DIALOGUE 6   A book having a mass of 1.2 kg and a weight of 12 N is raised to a height of 2 m and then dropped. (a) How much gravitational energy does it have at 2 m high? How much total energy? (b) Neglecting air resistance, how much total energy does it have just before hitting the floor? How much gravitational energy? How much kinetic energy? (c) How much total energy does the book have at one-quarter of the way down from the top? How much gravitational energy? How much kinetic energy?

6.4 Energy: *the ability to do work*

Now we will expand on the ideas introduced in the preceding two sections, extending them to a wide variety of situations.

Any system* having the ability to do work is said to have **energy**. A system's energy is quantitatively the amount of work it can do.†

Energy can be measured in joules, just as work can. But work and energy are different ideas. A system does work, but it has energy. Work is a process, and energy is a property, an attribute, of a system. Think of a system's energy as the amount of work the system could do, regardless of whether it ever actually does this work: A raised boulder has energy, even though it might be tied up and left that way forever and never do any work.

If you think of energy as stored work, it becomes apparent that there are many forms of energy beyond the gravitational and kinetic forms already discussed. We'll discuss eight forms of energy, beginning with the two we already know something about.

Kinetic energy is the energy of motion. It is the work that a system can do as a consequence of the system's motion, in other words, the work a system can do while coming to rest.

Gravitational energy§ is energy caused by gravitational forces. It is the work a raised system can do while Earth (or any other object that can pull gravitationally) pulls it back to its initial position. There is a quirk about gravitational energy: Its numerical value depends on the level chosen as the

---

5. (a) Half as much, because gravitational energy is proportional to height. (b) Four times as much, because kinetic energy is proportional to the squared speed.

6. (a) GravE = weight × height = 12 × 2 = 24 J; TotalE = 24 J. (b) TotalE = 24 J because of the conservation of energy; KinE = 24 J, because there is no longer any gravitational energy. (c) TotalE = 24 J. But now the height is only 1.5 m, so GravE = wt × ht = 12 × 1.5 = 18 J (it's lost one-quarter of its gravitational energy). Because total energy is conserved, KinE = 24 − 18 = 6 J.

7. (a) Suppose you can lift to about 2 m: GravE = wt × ht ≈ 10 × 2 = 20 J. (b) Suppose you can jog a kilometer in 6 minutes (a mile in 10 minutes). Your speed is then $s = d/t = 1000$ m / 360 s ≈ 3 m/s. If your mass is about 70 kg (weight 150 pounds), your kinetic energy is $(1/2)ms^2 = (1/2) × 70 × 3^2 ≈ 300$ J. (c) You probably walk about half as fast as you jog, perhaps a kilometer in 12 minutes. Half the speed means one-fourth the kinetic energy, so the energy is about $(1/4) × 300 = 75$ J. *Note:* Your own estimates will probably differ from these.

* The words *system* and *process* often come up in science. A *system* means a part of the universe, for instance, a collection of objects, and a *process* means any sequence of events.

† As mentioned earlier, in the case of thermal energy there are limits on the efficiency of actually doing this work.

§ Also known as *gravitational potential energy*. *Potential energy* is energy resulting from a system's position or configuration. Except for kinetic and thermal energies, all our listed energy forms are potential energies. In the interest of deleting excess words, we won't use the word *potential*.

initial or "reference" level, simply because the amount of work you can get from a raised object depends on how far down it must go before you consider it no longer "raised." For instance, this book's gravitational energy is only a few joules relative to the floor of your room, but it may be thousands of joules relative to sea level (depending on your present height above sea level). So we sometimes need to be specific about the agreed-upon reference level when discussing gravitational energy.

If you stretch a rubber band or bend a ruler, it can snap back when released. There is energy in the stretched or bent system because it can do work while snapping back. For instance, a stretched rubber band can do work in slowly pulling your fingers together. This energy, resulting from the ability of a stretched or deformed system to snap back, is called **elastic energy**.

A pot of hot water has more energy than does a pot of cold water of the same size. How do we know? Well, if the hot pot is boiling, it can rattle its lid, and this requires work, so the boiling-hot water must have the greater ability to do work. If the hot pot is below the boiling temperature, one way to get work from it would be to find another liquid, not water, that boils at a temperature below that of the hot pot's temperature. Then let the hot pot warm up the other liquid to boiling so that this other liquid rattles its lid.

The energy that a system has due to its temperature (or warmth—see Chapter 7) is called **thermal energy**.*

It is enlightening also to consider thermal energy from a microscopic point of view. As discussed in Chapter 2, temperature is associated with random microscopic jittering motions that are not directly visible macroscopically. For example, the molecules in a pot of water jitter around faster and faster as the temperature rises. A higher temperature means that the molecules have greater kinetic energies. But these microscopic kinetic energies do not appear as visible kinetic energy at the macroscopic level, because there is no overall, organized motion of the molecules. From the microscopic point of view, thermal energy is this random kinetic energy that cannot be directly observed macroscopically.†

To prevent confusion, we will reserve the term *kinetic energy* for macroscopic kinetic energy and use *microscopic kinetic energy* when referring to thermal energy from a microscopic point of view.

Put one hand on a warm object and the other on a similar cooler object, for instance, on pots of warm and cool water. The only difference between the two pots is that the warm one has faster molecules. From a microscopic point of view, you are not really experiencing warmth and coolness, you are only experiencing rapidly jittering and slowly jittering molecules. Does that seem amazing to you? It does to me. How could one pot feel warmer when the only "real" difference is that its microscopic parts are shaking around faster? The idea of warmth has been replaced by, or "reduced to," motion. In a way, the notion of warmth has vanished,

---

* Thermal energy is sometimes called *internal energy* or *heat energy*.
† Thermal energy also includes "microscopic potential energy" because of the forces between molecules, a point that is important to understanding melting and other "phase transitions."

a classic example of science's reduction of a wide assortment of phenomena to a few basics. One goal of physics is to reduce, or unify, phenomena in just this way.

This reduction of sense impressions to the mechanical motion of atoms is precisely what Democritus (Section 2.6) was talking about when he proclaimed, "By convention hot is hot [and] cold is cold. . . .The objects of sense are supposed to be real—but in truth they are not. Only the atoms and the void are real." But Democritus's view does more than simply interpret sense impressions as mechanical motions; it renders the sense impressions invalid, or nonexistent, and claims that only the mechanical motions are valid, or real. As we pointed out in Chapter 2, one can accept the scientific usefulness of the atomic viewpoint without necessarily accepting Democritus's philosophical view.

It took science a long time to comprehend thermal energy. During the eighteenth and nineteenth centuries, Joule and others eventually demonstrated that heat is a form of energy, in other words, that thermal energy actually is energy. Historically, thermal energy was difficult to understand because it does not fit comfortably into the mechanical framework of Newtonian physics and because, as we will see when we consider the second law of thermodynamics (next chapter), it is fundamentally different from the other energy forms.

We will run quickly over the remaining four types of energy, returning to them in more depth in the appropriate later chapters.

The electromagnetic force (Chapter 8) is one of nature's fundamental forces. The energy that results from electromagnetic forces is called **electromagnetic energy**, or sometimes simply electric energy.

There is energy in a light beam, as you can tell from the fact that light (sunlight, for instance) can warm things, and you can get work out of warm things. The energy carried by a light beam is one form of **radiant energy**. There are other forms of radiant energy, some of them perhaps familiar to you: radio, microwave, infrared, ultraviolet, X ray, and gamma ray. All of these are similar to visible light. We will learn more about them in Chapter 9.

Chemical reactions can do work, as you can see from a wood fire used to boil water. This energy results from the molecular structure of the wood. This energy, produced by a system's molecular structure, is called **chemical energy**.

Whereas chemical energy results from molecular structure, **nuclear energy** results from nuclear structure, from the way protons and neutrons are arranged into nuclei. Nuclear energy can be obtained from nuclear reactions (Chapter 15), just as chemical energy can be obtained from chemical reactions.

DIALOGUE 8 Name the type of energy possessed by each of the following: a raised book, gasoline, a stretched spring, sunlight, a speeding train, hot steam.

---

8. Gravitational, chemical, elastic, radiant, kinetic, thermal.

6.5 The law of energy: *energy is forever*

When you drop an object, its energy is conserved (is unchanged) as it falls. At any rate, we have seen that if you neglect air resistance and observe the fall only until the object hits the floor, the gravitational energy lost during any portion of the fall is always exactly balanced by the kinetic energy gained during that portion. As we discussed earlier, it is possible to prove this remarkable result from Newton's laws. Now we expand on this idea.

It is possible to prove, still based on Newton's laws, that any system that experiences only gravitational forces conserves its total energy, just as the falling object does. Any gains or losses in kinetic energy are exactly balanced by the losses and gains in gravitational energy. Furthermore, elastic forces fit into this Newtonian picture. It is possible to prove that any system that obeys Newton's laws and that possesses only kinetic, gravitational, and elastic energy must conserve its total energy.

A perfectly elastic rubber ball that falls from some height, is squeezed elastically during its impact with the floor, and rebounds back to the initial height (this is what "perfectly elastic" means), is one illustration of such a system. Another example is the solar system: To a good approximation, only gravitational forces are important to determining the motion of the sun, planets, and moons in our solar system. In these cases, Newton's laws guarantee that the total energy remains unchanged despite the changes that occur within the system.

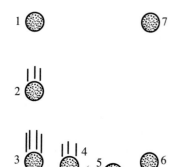

FIGURE 6.9
A bouncing ball. What forms of energy does a perfectly elastic ball have at each of these seven times during its fall to the floor and its rebound?

DIALOGUE 9   What form or forms of energy does a falling elastic ball (Figure 6.9) have at each of the following times? (a) Just as it begins falling. (b) When it is halfway down. (c) As it first touches the floor. (d) After it touches the floor while it is still being further squeezed against the floor. (e) When it is at its most-squeezed position on the floor. (f) Just as it leaves the floor during its rebound. (g) When it is back at its high point.

This is as far as Newton's laws can take us. Even without going further, it is a remarkable result. There is this abstract idea, the "ability to do work," that stays the same at every moment of some complicated process such as the rebound of an elastic ball or the motion of the solar system. But even more remarkably, experiments show that the energy principle goes far beyond Newton's physics. Energy is conserved in every physical process yet observed, including processes lying far outside the range of validity of Newton's laws and processes involving the non-Newtonian forms of energy. We call this:

---

9. (a) Gravitational. (b) Gravitational and kinetic. (c) Kinetic and a little gravitational (measured relative to its lowest position 5 in Figure 6.9). (d) Elastic and kinetic and a very little gravitational (because it is still just a little above the floor). (e) Elastic. (f) Kinetic and a little gravitational. (g) Gravitational.

THE LAW OF CONSERVATION OF ENERGY (THE LAW OF ENERGY)
The total energy of all the participants in any process must remain unchanged throughout that process. That is, energy cannot be created or destroyed. Energy can be transformed (changed from one form to another), and it can be transferred (moved from one place to another), but the total amount always stays the same.

This statement is about as true as any general idea ever gets in science: It is correct in every situation yet observed. It holds even in situations in which Newton's physics is not even remotely correct, such as near black holes, close to the speed of light, and for subatomic particles.

This law says that something, namely energy, remains quantitatively the same. This can be useful, because once you have calculated or measured the total energy at one moment during some process, you will know it at any other moment without having to calculate or observe what happened between the two moments. In Dialogue 9, for instance, if you knew the rubber ball's initial height, you would also know its initial energy, and so you would know that the ball has precisely this amount of energy at every moment during the bounce, without having to calculate the forces and motions involved in the bouncing. For another example, the number of joules of chemical energy consumed from a car's fuel tank must equal the total number of joules appearing as the car's kinetic and gravitational energy, exhausted thermal energy, chemical energy of pollutants, and so forth, during that time. Energy conservation is like the motion of a bishop in chess. If a bishop is on a black square, then after any number of moves it will still be on a black square. You don't need to know the details of the moves to predict this, because a bishop's square color is conserved.

Some of the energy principle's scientific appeal lies in its elegance and its expression of timelessness. It is perhaps reassuring that amid nature's chaos and change, energy is forever.

Physicists have discovered several other natural quantities that do not change, that are conserved. Although we will not study the other conservation principles, it is interesting to list them, beginning with energy conservation. Throughout every physical process, the participants maintain or conserve each of the following:

· Total ability to do work, or "total energy."
· Total motion through space, or "total linear momentum."
· Total rotational motion, or "total angular momentum."
· Total electric charge.
· Certain subatomic properties associated with microscopic interactions.

These conservation laws are among the most far-reaching physical principles known. Conservation of energy is the most important of the conservation laws.

There is a useful alternative way of stating the energy principle in terms of work and energy. Whenever work is done, it is done by some system on some other system. The system doing the work must lose some of its ability to do work; in other words, it must lose energy. Since total energy is conserved, this energy cannot just vanish but must instead go into the system on which work is done. So work is an energy transfer from the system doing the work into the system having work done on it. We call this

THE WORK–ENERGY PRINCIPLE
Work is an energy transfer. Work reduces the energy of the system doing the work and increases the energy of the system on which work is done, both by an amount equal to the work done.*

HOW DO WE KNOW?
Why do we believe energy is conserved? As we stated, in the realm where Newton's physics is correct, the conservation of energy can be proved from Newton's laws for processes involving only kinetic, gravitational, and elastic energy forms.

The important question of including thermal energy in the energy principle will be discussed later. For now, we will just assume that thermal energy can be included in the energy principle, and we will describe how a thermal energy measurement process known as *calorimetry* can be used to demonstrate the conservation of the other non-Newtonian energy forms (electromagnetic, radiant, chemical, nuclear).

The idea of calorimetry is to measure a system's nonthermal energy by transforming it into thermal energy in some second system whose thermal energy can be measured. For example, the chemical energy in a gram of coal can be measured by combusting the coal and using the thermal energy to warm, say, a kilogram of water. The measured rise in temperature of the water tells us the amount of thermal energy deposited in the water, which is then the amount of chemical energy in the original coal (and in the oxygen used during combustion).

A historical example illustrates how calorimetry can test the conservation of such non-Newtonian energy forms as nuclear energy. Early in the twentieth century, nuclear physicists investigated "beta decay," a process in which a nucleus spontaneously creates an electron and spits it out of the nucleus. This process alters the original nucleus. We represent this process as

original nucleus → altered nucleus + ejected electron

(Figure 6.10). If energy is conserved, the energy of the original nucleus should equal the energy of the altered nucleus plus the energy of the ejected electron.

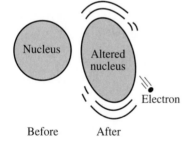

FIGURE 6.10
The beta decay process, before and after. A nucleus spontaneously creates and ejects an electron. Physicists were once uncertain about whether energy was conserved in this nuclear process.

---

* There is also another worklike process by which energy can be transferred, called *heating*. Heating is a thermal energy transfer due to a temperature difference, and it can be thought of as microscopic work. When expanded to include not only ordinary work but also microscopic work (heating), the work–energy principle is called the first law of thermodynamics. Thus conservation of energy, the work–energy principle, and the first law of thermodynamics all are equivalent ideas, stated in different forms. We won't need heating or the first law of thermodynamics in what follows.

But nature had a surprise. Measurements showed that the energies did not balance! The energy was larger before than after. Being reluctant to conclude that energy was actually not conserved, physicists hypothesized that some undetected particle was also ejected along with the electron. It was thought that when this other particle's energy was included, the energies would balance.

Although the hypothesized particle had not been detected, it was thought that its energy could be directly measured by surrounding the nucleus with a large cylinder of lead. The lead cylinder was used as a "calorimeter," or thermal energy measurer. The unseen particle would surely be slowed down and stopped inside a sufficiently large cylinder of lead and so deposit its energy in the lead. This energy would then show up as a small temperature rise in the lead, and by measuring this rise one could deduce the amount of energy deposited in the lead.

But nature sprang another surprise: There was no measurable temperature increase in the lead. Perhaps energy was *not* conserved in beta decay. This is where the matter stood between 1914 and 1930. Physicist Niels Bohr suggested in 1929 that perhaps energy conservation did not apply to the nucleus. But others did not accept this suggestion, and in 1930 physicist Wolfgang Pauli hypothesized that the new particle was so penetrating that it could not be detected and measured even in a thick lead calorimeter. This particle was supposed to be so penetrating as to pass right through the lead without depositing any energy. Pauli proposed that energy is conserved in beta decay but that calorimetry is unsuccessful in detecting part of the energy.

This set off a search for such a particle. Before long, other indirect evidence (other than beta decay) was found, and the particle was dubbed the *neutrino*. We now know that Pauli was right and that energy is conserved during beta decay. The neutrino was finally detected directly, 25 years after Pauli's hypothesis.

DIALOGUE 10   Farswell Slick invites your investment in a business venture to manufacture his remarkable "supertranspropulsionizer." His diagrams show an elaborate array of superconductors, lasers, liquid-helium coolants, and computer controls (Figure 6.11). Slick informs you that this ultimate propulsion system will accelerate spaceships to nearly lightspeed for interstellar travel. Amazingly, no fuel supply is needed, either on board or outside the spaceship. The principle involved, he explains, is "superconducting optical self-propulsion" (SOS). With SOS, the device operates in a continuous cycle that both accelerates the spaceship and "feeds back" some of its laser light to maintain, for as long as may be desired, the operation of the transpropulsionizer itself. Should you invest?

FIGURE 6.11
"A remarkable device," Farswell Slick remarks. Would you buy a supertranspropulsionizer from this man?

10. Don't invest. And don't bother investigating his design. It violates conservation of energy. Work must be done by the transpropulsionizer to accelerate the spaceship, so the device transfers energy from itself. During a complete cycle of operation, energy must be supplied to the device in order to bring it back to its initial (at the beginning of the cycle) condition. So the device needs an external fuel supply. You can't get something for nothing.

6.6 Transformations of energy

Here are some familiar processes, described in terms of energy and its transformations:

FALLING   Once again, drop your book to the floor (it's coming in for a lot of rough treatment in this chapter!). We have studied this process up until its impact with the floor. Where is the energy after impact? It might appear to have vanished, but the law of conservation of energy says that energy never vanishes. Let us proceed by elimination: The energy is not kinetic, gravitational, or elastic. There are no obvious electromagnetic, radiation, chemical, or nuclear effects. There is only one candidate left: thermal energy. So the impact must warm the book or the floor. This temperature rise may be hard to detect, but you can demonstrate the same effect by giving a nail a hard hammer blow while driving it into a board. Feel the nail before and after the blow. Try several blows.

It is useful to summarize the book's fall in a simple **energy-flow diagram**:

GravE (at the top) → KinE (just before impact) → ThermE (after impact)

Let's add the effects of the atmosphere (air resistance) to this. The participants in the process are now the book, Earth (its gravitational pull), and the air. Since air resistance slows the book, the falling book has less kinetic energy than it did before. But this energy is not lost—you cannot lose energy. It must be transformed into thermal energy. The air and book must warm a little as the book falls.

Until the work of Joule and others around 1850, scientists had long believed that the work going into forces such as air resistance and friction, work that produces "heat," was lost. The key to finally uncovering the principle of conservation of energy was discovering that these processes are simply another, more subtle, type of energy transformation and that the "heat" produced is a form of energy, a form we now call *thermal energy*. We discuss in the next chapter why it was so hard to understand thermal energy: Thermal energy is only partially recoverable as work. So although it is a form of energy, thermal energy is degraded in its ability actually to do work.

HOW DO WE KNOW? JOULE'S EXPERIMENTS

We predicted that air warms when you stir it up with a falling object. James Prescott Joule (Figure 6.4) did an experiment like this in the 1840s, using water instead of air. He placed a paddle wheel in a tub of water, stirred the water with the paddle wheel, and measured the temperature rise in the water.* He quantified the experiment by allowing a falling weight attached to a cord to turn the paddle wheel. The weight's energy loss is then its weight multiplied by the distance fallen. Joule found that the water's rise in temperature was precisely proportional to the work done, in other words, to the gravitational energy lost.

---

* When you stir hot water in the open air, the water *cools* because of evaporation. In Joule's experiment the stirring was done inside a closed container, thereby preventing any significant evaporation.

These ideas were murky in Joule's day because it was not understood that "heat" (thermal energy) is a form of energy. Joule clarified matters by showing that work is precisely, quantitatively convertible to thermal energy. Experiments such as the paddle-wheel experiment measured the amount of work needed to create a particular amount of thermal energy, as measured by a temperature rise. Joule found that a given amount of work always created the same amount of thermal energy. This break-through extended the energy principle beyond the Newtonian energy forms to thermal energy.

Joule showed that a particular amount of work always produces a 1°C rise in the temperature of 1 kilogram of water. This amount is 4184 joules, or about 4200 joules. This amount of energy is the **dietitian's Calorie**.* Al-though the Calorie is often used to measure thermal energy, Joule's work showed that it is really a general energy unit, equivalent to 4200 joules.

Back to the falling book (isn't it surprising what we can learn just from the fall of a book?): Just before impact, all the energy has converted to kinetic energy of the book and to the thermal energy of the air and book. Since air resistance has only a small effect on the motion, the thermal energy must form only a small fraction of the total. It is this small reduction in kinetic energy that shows up as thermal energy in the air. Finally, the impact converts the preimpact kinetic energy to the thermal energy of floor, book, and air.†

The energy-flow diagram, including the effect of the air, is now

GravE → KinE (nearly 100%) + ThermE (air, small %) → ThermE

ANIMAL ENERGY, AND EFFICIENCY   The energy that enables you to do useful work is obtained from foods and is stored in your body as chemical energy (Chapter 2). Dietitians measure this stored chemical energy in Calories (= 4200 J). For example, once you have eaten and respirated it (Chapter 2), a 70-Calorie slice of bread gives you 70 Calories of energy which is stored in chemical form and can then provide 70 Calories of work and thermal energy.

When animal chemical energy is used to do work, for example, when you lift a book, only a small fraction (often about 10%) of this chemical energy actually is transformed into useful work. We say that the process is "ineffi-cient" because so little of the input energy appears as useful work. A "highly efficient" process, on the other hand, is one in which most of the input en-ergy is transformed into useful output energy and the "wasted" fraction is small.

It is useful to define this idea quantitatively: The **efficiency** of any energy transformation is the useful energy output divided by the total energy input.

$$\text{efficiency} = \frac{\text{useful energy output}}{\text{total energy input}}$$

---

* The dietitian's Calorie is always spelled with a capital C. Physicists use the word *calorie* (lowercase c) to denote the energy needed to raise 1 gram of water by 1°C.
† And you can hear the impact. A small fraction of the energy is transformed into the energy of sound, which actually is a form of kinetic energy of the air.

In other words, the efficiency is the fraction of the input that appears as useful output. It is usually expressed as a percentage.

The efficiency of transforming animal chemical energy into work done on the external environment is about 10% for many human muscular activities. Most of the remaining 90% of the chemical energy is transformed into thermal energy in your body. For example, when you slowly lift your book, the energy transformation is

ChemE → GravE (of book, 10%) + ThermE (of your body, 90%)

SLIDING   Now give your book a quick hard push so that it slides across your tabletop, sliding to rest (Figure 6.12). Neglecting friction during the brief push, the energy transformation during the push is

ChemE → KinE (of book) + ThermE (of your body)

The frictional force by the table slows the sliding book until it comes to rest. The book's kinetic energy has then been transformed into thermal energy.

At the atomic level, the frictional force comes from the electrical forces exerted by the atoms of the tabletop on the atoms in the bottom surface of the book. As the book slides, these atomic forces set the atoms into more vigorous motion, and this increased microscopic kinetic energy is observed as a higher temperature.

FIGURE 6.12
What energy transformations occur when you briefly push a book and then let it slide?

6.7 Power: *the quickness of energy transformation*

What is the difference between running and walking up a flight of stairs? Your gravitational energy increases by the same amount in both cases. The work you do is the same. And yet your body knows there is a difference between running and walking upstairs. The difference is that you do the work in less time, quicker, when you run.

We need a word for this notion of how quickly work is done. We call it **power**. Quantitatively, "how quickly a task is done" is the rate of doing that task, the amount done per second. So we define power as the rate of doing work, or the amount of work done divided by the time to do it:

$$\text{power} = \text{rate of doing work}$$

$$\text{power} = \frac{\text{work done}}{\text{time to do it}}$$

Power can also be thought of as the rate of transforming energy, because work is an energy transformation.

For example, suppose you run up one flight of stairs and then walk up a second identical flight of stairs in twice the time it took to run up the first flight. You do the same work for each flight, but your power output during the first flight is double your power output during the second flight because you went up the first flight in half the time. The difference between the two processes is a power difference, not an energy difference.

The unit of power is the "joule per second" (J/s). It differs by an all-important "per second" from the unit of work or energy. Power is such a popular idea that its unit is given a special name. The joule per second is called the **watt** (W), in honor of the eighteenth-century developer of the steam engine, James Watt. The kilowatt (kW) is 1000 watts, and the megawatt (MW) is 1 million watts.

MAKING ESTIMATES

What is your power output while running up a flight of stairs? At an efficiency of 10%, what is your (chemical) power input, in watts and in Calories/second?

SOLUTION   Suppose that your weight is 500 newtons (110 pounds), the vertical height of one flight of stairs is 4 meters, and you run up one flight in 5 seconds (try it!). The work done to lift yourself and the power output are

$$\text{work} = \text{weight} \times \text{height} = 500 \text{ N} \times 4 \text{ m} = 2000 \text{ J}$$
$$\text{power} = \text{work/time} = 2000 \text{ J/5 s} = 400 \text{ W}$$

This is a large power output for a human being, as you will discover if you do the experiment (see Home Project 2). To produce this output at 10% efficiency, you must convert chemical energy at a rate of 4000 W, or 4000 J/s. In Cal/s (1 Cal = 4200 J), this is $4000/4200 \approx 0.95$ Cal/s, your **metabolic rate** in this example. If you could maintain this rate for an hour (3600 s), you would "burn up" (transform) $3600 \times 0.95 = 3400$ Calories, the energy content of a sizable steak.

Think of several devices you use every day: automobile, lightbulb, electric blender, toaster, and so forth. These devices can be understood as energy converters, as they convert energy from one form to another form that you use directly. An important feature is often the rate at which the energy is converted. For example, to get a certain lighting level from a lightbulb, the bulb must convert a certain number of joules per second to visible light. So lightbulbs and other devices are rated in power units (watts). In the United States, a popular power unit for automobile engines and other heat engines (next chapter) is the horsepower. One horsepower is about 750 watts.

Table 6.1 gives the power consumed (transformed into nonelectric forms) by typical household electrical appliances. The numbers give the power consumed only during the time that the appliance is turned on and consuming electric energy. Average power, perhaps over one day or over one year, is often quite different. For example, a toaster is used for only a short time each day, so its average power consumption is far less than 1200 watts. Although refrigerators appear toward the bottom of the table, they consume power during a large portion of each day and so are among the largest household energy consumers.

The performance of many devices depends on their power output. For example, a bulb's brightness is determined largely by its power output. But it is the bulb's total energy consumption that contributes to the electric company's fuel costs and to the consumption of energy resources.

Although your total energy use determines the amount of fossil or other resources that your home electricity consumes, your impact on the need for new power plants is determined by your power use at so-called peak times. Each electric plant has a

TABLE 6.1
Power Consumption of Household Appliances While the Appliance Is Turned on and Consuming Electric Energy

Appliance	Power (W)
Cooking range	12,000
Clothes dryer	5,000
Water heater	4,500
Air conditioner, window	1,600
Microwave oven	1,400
Dishwasher (incl. hot water)	1,200
Toaster	1,200
Hair dryer	1,000
Refrigerator, frostless	600
Refrigerator, not frostless	300
TV, color	350
Radio-phonograph, stereo	100

maximum power output, its "power rating," which is usually several hundred megawatts. The plant's actual power output is largest at times such as hot afternoons when everybody is running an air conditioner. If the power peak approaches the plant's rating, the plant will cut its output by reducing the supply to all customers, and so lightbulbs and other appliances dim in a "brownout."

When the peak demand becomes excessive, electric companies often turn on temporary sources of supply, usually smaller oil-burning plants, or they might construct new permanent plants. But there are other possibilities not involving new sources of electric power. The peak demand can be reduced with energy-efficient electrical devices that provide the same services (the same amount of light, for example) with less energy. Another option is storing the energy produced by the power plant during off-peak times, for later use during peak times. One example is pumped storage, in which a plant's excess capacity during off-peak periods is used to pump water uphill into a reservoir. The water in the reservoir is then used for hydroelectric power during peak periods (Figure 6.13). Some electric companies charge more at peak-power times, which creates incentives for people to use less power at those times and so reduces the need for new power plants.

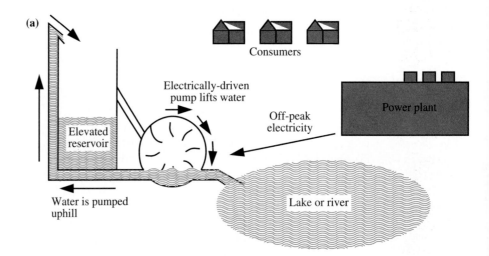

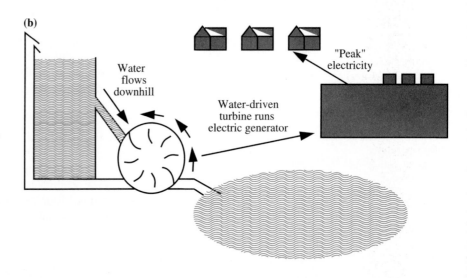

FIGURE 6.13

The principle of pumped storage. (a) During periods of low consumer demand for electricity, the plant's excess power capacity runs a pump that lifts water uphill. (b) During peak periods, this water flows back downhill to produce electricity. What energy transformations occur during peak and off-peak periods? Pumped storage moves some of the plant's electrical production from peak periods to off-peak periods, reducing the need for new plants.

You can help reduce the need for new power plants by reducing your power consumption at peak times, for instance, by not using your dishwasher on a hot afternoon. Better yet, using high-efficiency devices decreases the need for both new power plants and new energy resources.

The most useful unit for measuring your home's electric energy consumption is the **kilowatt-hour** (kW·h), the amount of energy transformed when a power of 1 kilowatt operates for 1 hour. Since 1 kilowatt is 1000 joules/second and 1 hour is 3600 seconds,

$$1 \text{ kW·h} = 1000 \text{ J/s} \times 3600 \text{ s} = 3.6 \times 10^6 \text{ J}$$

If a known power in kilowatts operates for a known number of hours, it is easy to figure the number of kilowatt-hours of energy consumed: Just multiply the number of kilowatts by the number of hours.

DIALOGUE 11   (a) A clothes dryer is equivalent to how many 100 W lightbulbs? (b) If drying a load takes an hour and if you dry 15 loads in a month, how many kW·h of electric energy will you consume? (c) How much does this cost, at 10¢ per kW·h?

DIALOGUE 12   In many U.S. homes, the refrigerator is the largest energy consumer. How can this be so when according to Table 6.1, other common items such as the cooking range and the clothes dryer operate at much higher power?

"WAKE UP, DR. ERSKINE — YOU'RE BEING TRANSFERRED TO LOW ENERGY PHYSICS."

11. (a) Dryers operate at 5000 W (Table 6.1); 5000 W / 100 W per bulb = 50 bulbs! (b) 5 kW × 15 hours = 75 kW·h. (c) ($0.10 per kW·h) × 75 kW·h = $7.50.
12. Table 6.1 shows only the rate of energy use when the device is in use. The refrigerator is in use for a much longer total time every month.

Summary of Ideas and Terms

Fossil fuel The remains of ancient life, used as an energy resource, such as coal, oil, and natural gas.

Industrial Revolution The beginning of the fossil-fueled industrial age, around 1750.

Work Object A does work on object B if A exerts a force on B while B moves. Unit: newton-meter = joule. Work = Fd.

Energy The ability to do work. The energy of a system is the amount of work that system can do. Units: joule, Calorie, kilowatt-hour.

Forms of energy:
- **Kinetic energy,** due to motion: KinE = $(1/2)ms^2$.
- **Gravitational energy,** due to gravitational forces: GravE = wt × ht.
- **Elastic energy** due to the ability of a deformed system to snap back.

- **Thermal energy,** due to temperature. Or microscopic energy, the kinetic (and other) energy of molecules that cannot be observed at a macroscopic level.
- **Electromagnetic energy,** due to electromagnetic forces.
- **Radiant energy,** carried by a light beam, radio wave, microwave, infrared, ultraviolet, X ray, or gamma ray.
- **Chemical energy,** due to molecular structure.
- **Nuclear energy,** due to nuclear structure.

The law of conservation of energy The total energy of all the participants in any process must remain unchanged throughout that process.

The work–energy principle Work is an energy transfer.

Efficiency Useful energy output/total energy input.

Power The rate of doing work. Units: joule/second, watt (= J/s), horsepower. Power = work/time.

Review Questions

INTRODUCTION

1. What type of culture was the first village culture?
2. What type of energy technology fueled the industrial revolution?
3. What is a fossil fuel? Name three kinds.

WORK

4. Is work done whenever a force is exerted? Explain.
5. Is work done whenever an object moves through a distance? Explain.
6. To what two quantities is work proportional?
7. You slowly lift a 3 N grapefruit, by 2 m. How much work did you do and on what object?
8. A 3 N grapefruit falls 2 m to the floor. Was work done during the fall? By what object on what other object?

ENERGY

9. Explain the difference between energy and work.
10. List the eight physical types of energy.
11. Explain thermal energy from a microscopic point of view.
12. Give one example of each of these energy forms: elastic, thermal, chemical, kinetic, radiant, gravitational.
13. If you double the speed, how is the kinetic energy affected? If you double the height, how is the gravitational energy affected?

14. Choose the correct answer(s): One joule is the same as 1 (watt-meter, newton-meter, meter per second squared, newton-second, kilowatt-hour).

THE LAW OF ENERGY, AND ENERGY TRANSFORMATIONS

15. Has the law of conservation of energy been found to be correct in all situations observed so far? Is the same true of Newton's laws?
16. Choose the correct answer(s): For a system that returns to its initial state, during one complete cycle you can't get more (acceleration, force, energy, power, speed) out of the system than was put in.
17. What energy transformations occur when this book falls to the floor? When you lift this book?
18. Give an example of each of these energy transformations: kinetic energy → thermal energy, kinetic energy → elastic energy, elastic energy → kinetic energy.
19. What do we mean when we say that the energy efficiency of a lightbulb is 10%?
20. An apple in a tree has 90 J of gravitational energy (relative to the ground). It falls. If you neglect air resistance, what can you say about the amount of kinetic energy the apple has just before it hits the ground? What if you do not neglect air resistance?

21. Explain the difference between energy and power.
22. Choose the correct answer(s): The (watt, newton per second, joule, calorie, joule per second, meter per second, horsepower, kilowatt-hour) is a unit of power. Which are units of energy?

23. You lift a 2 N rock by 4 m in 3 s. What is your work output? Your power output?
24. Which do you pay for (primarily) in your monthly electric bill, energy or power?

Home Projects

1. Rub one hand hard back and forth on a wooden table-top for 30 seconds. Using your other hand, compare the temperature of the rubbed part with another part of the table. (This doesn't work on a metal surface because thermal energy quickly travels away from the rubbed region in a metal.)
2. Time yourself as you run up a staircase. Measure the vertical height of the staircase, and find your total energy output (1 kg weighs 10 N on Earth). What energy conversion occurred? Find your power output in watts. How many 100 W bulbs could this light up if converted to electric energy?
3. Try a variation of Joule's experiment: Draw two glassfuls of room-temperature tap water; blend the water of one of them at high speed for a few minutes; and pour it back into its glass. Dip your finger into both glasses. Notice how much work is needed to produce even a small temperature rise: There is a lot of stored work (energy) in hot liquids.
4. Measure a tennis ball's elasticity by dropping it and measuring the height of rebound as a fraction of the initial height. What fraction of the initial energy was transformed into thermal energy? What fraction was retrieved as gravitational energy? Measure this fraction for several different initial heights. Does it vary significantly? Now try it for several different balls: new tennis ball, old tennis ball, Ping-Pong ball, and so forth.

For Discussion

1. Energy-awareness audit. Note the ways you use energy during one day, listing direct uses such as automobile travel, heating, and electric appliances. Make a second list of indirect uses such as energy-intensive foods (frozen foods, meats), packaging, throwaway containers, and nonrecycled items. List as many items as you can think of. Describe the ways in which each item uses energy. (Example: Frozen foods use energy for growing, packaging, refrigeration, and transportation.)
2. Compare and discuss the energy audits of different class members. Are there ways to lower your energy consumption without lowering your quality of life? Are there ways that you or society can reduce energy consumption while enhancing the quality of life?

Exercises

1. Does Earth do gravitational work on you as you walk downstairs?
2. Give an example of a system that has both kinetic and gravitational energy.
3. In order for you to get out of bed with the least amount of work, would it be better for your bed to be on the floor or about a meter high? Explain.
4.* Your left hand lifts a 2 N apple by 1.5 m, and your right hand lifts a 4 N grapefruit by 0.5 m. Which hand did more work? Which hand exerted the larger force?
5.* Making estimates. About how much work would it take to lift the U.S. population by 1 km?

6. Which of the eight physical types of energy was the basis for the earliest human culture? Which was the basis for the Industrial Revolution?
7. Name the type of energy possessed by each of the following: Jill at rest at the top of a sliding board, Jill sliding off the bottom of the sliding board, dynamite, sunlight, coal, hot air.

8. Explain, in microscopic terms, why the air pressure inside a tire increases on a hot day. If the air in a balloon is heated, will the balloon expand or contract? Why?

9. (a) Where would an apple have greater gravitational energy, at 100 km high or at 1000 km high? (b) Would the gravitational energy of an orbiting satellite be increased or decreased by moving it from an orbit that is 6000 km high up to an orbit that is 12,000 km high (see Figure 6.14)? (c) At which point, 6000 km high or 12,000 km high, does a satellite have the larger gravitational force on it?

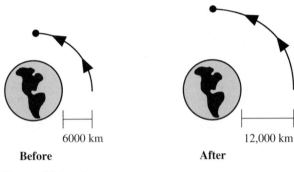

6000 km   12,000 km

Before   **After**

FIGURE 6.14
What happens to the gravitational energy of the satellite when it is moved to a higher orbit?

10.*If you triple your height above the ground, how is your gravitational energy (relative to the ground) affected? What if you halve your height?

11.*If you triple your speed, how is your kinetic energy affected? What if you halve your speed?

THE LAW OF ENERGY,
AND ENERGY TRANSFORMATIONS

12. You squeeze an elastic spring and clamp it in the squeezed position. You then drop the clamped spring into acid, dissolving the spring. What happened to its elastic energy?

13. What is the main energy transformation for each device (input and useful output)? (a) An automobile while speeding up, (b) a bicycle while speeding up, (c) an electric blender, (d) a toaster, (e) a lightbulb.

14. What energy transformation occurs when you climb a rope?

15. You throw a baseball horizontally, and Jill catches it. Neglect air resistance. Describe the energy transformation that occurs (a) during the throw (while the ball is in your hand) and (b) during the catch.

16. You throw a ball upward and then catch it at the same height. How does the ball's final speed compare with its initial speed, (a) neglecting air resistance and (b) including air resistance? Defend your answers.

17. Does an automobile use more gasoline when its lights are on? When the air conditioner is on? (*Note*: The battery does not run these devices while the engine is running.)

18. Does a stone gain energy as it drops? Does it gain kinetic energy? What kind of energy does it lose?

19. Imagine a 100% efficient automobile. Would it emit any exhaust? Would its engine be hot?

20. Figure 6.15 is a graph of a roller coaster's height above the ground versus the length of track it covers. The coaster is powered up to its high point at 100 m from the starting point. From the high point, the coaster coasts freely all the way to the end. Assume that the coaster starts from rest at the high point and encounters no friction or air resistance. (a) Between 200 m and the finish, where is it moving slowest? Fastest? (b) Is it moving faster at 1000 m or at 1100 m? (c) At what distances does the coaster have roughly half of its initial gravitational energy?

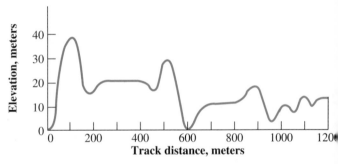

FIGURE 6.15
Elevation versus track distance for a roller coaster.

21.*In the previous exercise, suppose the coaster has 160,000 J of gravitational energy at the high point. (a) How much kinetic energy does it have at the low point? (b) How much total (gravitational + kinetic) energy does it have at the low point? At the finish? (c) How much gravitational energy does it have at those points where it is 10 m high? How much kinetic energy does it have at these points?

22.*Jack, who weighs 300 N, sits in a child's swing. You pull the swing back so that it is 2 m above its low point and release it. (a) What form of energy and how much does Jack have when he is pulled back and held at rest? (b) What form of energy and how much does Jack have as he swings through the low point (neglect air resistance and friction in the moving parts)? (c) Still assuming no air resistance or friction, how high should Jack be when he has swung forward to his high point?

23.*In a crash test, a 1000 kg automobile moving at 10 m/s crashes into a brick wall. (a) How much energy goes into demolishing (and heating) the wall and the auto? (b) What if the auto were moving twice this fast?

POWER

24. You start a bowling ball rolling by swinging it with your arm and releasing it. Then you start a second identical bowling ball rolling, at the same speed as the first, by hitting it sharply with a sledge hammer. Which process, the arm swing or the hammer blow, imparts more kinetic energy to the ball? Which process has the greater power output?

25. What other unit(s) could automobile engines be rated in, instead of horsepower?

26. Why are electricity rates often higher in summer than in winter?

27.*Which process has the larger power output: 2 J of work performed in 0.1 s or 1000 J of work performed in an hour?

28.*A drive force of 500 N acts on an automobile while it travels 60 km. (a) How much work did the drive force do on the automobile? (b) If the trip took 50 minutes, what was the automobile's power output in watts? (c) If the auto's energy efficiency is 10%, what was its power input (its rate of converting the gasoline's chemical energy into other forms) in watts? How many 100 W lightbulbs could this light up?

29.*A cyclist delivers power to her bicycle at a rate of 150 W while her metabolic rate is 1000 W. What is her body's bicycling energy efficiency?

30.*How much does it cost to run a nonfrostless refrigerator for a month? Assume that the refrigerator consumes power for 8 hours each day and the electricity costs 10¢ per kW·h. What if it is frostless?

31.*Making estimates. Use Table 6.1 to estimate the number of kilowatt-hours of electric energy that a typical single-family home consumes in one month. Don't forget lightbulbs (they aren't in Table 6.1).

7

THERMODYNAMICS
—and you can't even break even

The crucial breakthrough in understanding energy was the discovery that "heat" (thermal energy) actually is a form of energy that can be exchanged for work just as other forms of energy can. This discovery establishes that energy is conserved in processes involving thermal energy.

Because of the central role of thermal energy, the study of the general principles of energy is called **thermodynamics**. Its laws have no known exceptions and so are among the most general scientific principles known. The law of energy is the first of those basic principles and is often referred to as the **first law of thermodynamics**.* This chapter presents another great thermodynamic principle, the **second law of thermodynamics** or the "second law."

There are three different ways of stating the second law. In its most straightforward form, it is a familiar observation about thermal energy flow (Section 7.1). Like many everyday observations, this one has profound consequences. Section 7.2 discusses one of these, namely, another form of the second law that highlights the special nature of thermal energy. Unlike other energy forms, there is a restriction on the transformations of thermal energy: It can be transformed into other forms only with limited efficiency. This leads to discussion of a device that is central to society's use of energy: the heat engine (Sections 7.2 and 7.3). Sections 7.4 and 7.5 apply these ideas to our continuing study of the physics and the social implications of a significant heat engine: the automobile. Section 7.6 looks at another significant heat engine: the steam-electric power plant. Because these topics bring up the question of the use and depletion of energy resources, it is natural

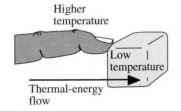

FIGURE 7.1
When you touch a piece of ice, thermal energy flows from your higher-temperature hand to the lower-temperature ice, so the ice gets warmer and your finger gets colder.

---

* Textbooks usually state the first law of thermodynamics in a form that looks different from the law of conservation of energy. Nevertheless, the two ideas are logically equivalent to each other.

at this point to discuss exponential growth and its implications for resource depletion (Section 7.7). Finally, Section 7.8 turns to yet a third, and the most fundamental, way of expressing the second law: the law of entropy.

7.1 Heating

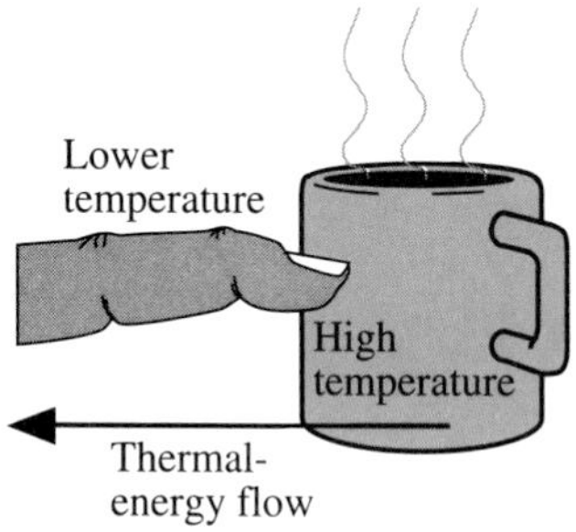

Touch a piece of ice (Figure 7.1). Energywise, what happens? Your hand gets cooler, and the ice begins to melt. So thermal energy must have flowed from your hand to the ice. Now touch a hot cup of coffee (Figure 7.2). Your hand gets warmer while the coffee cools, so thermal energy must have flowed from the coffee to your hand. Notice that in each case, thermal energy flowed from the high-temperature object to the low-temperature object. There is a general principle operating here, a principle that you experience whenever you touch an object that feels hot or cold: Thermal energy flows spontaneously from hot to cold. Any such flow of thermal energy from a high to a low temperature is called **heating**.

Heating has a one-way nature: It flows "downhill" from a high to a low temperature. It won't flow spontaneously (without external help) "uphill" from a low to a high temperature.

Like a lot of simple ideas, this notion of the one-wayness of heating has profound consequences. It is one way of stating a new basic principle of physics called the second law of thermodynamics, or the second law. We call it:

> THE SECOND LAW OF THERMODYNAMICS, HEATING FORM (LAW OF HEAT FLOW)
> Thermal energy flows spontaneously (without external assistance) from a higher-temperature object to a lower-temperature object. It will not spontaneously flow the other way.

Now we need to be more quantitative about temperature. **Temperature** is a quantitative measure of warmth. We can measure temperature by using any material that expands as it warms and contracts as it cools. Most materials do this. Such an expansion and contraction can be used as the basis for a temperature-measuring device, or **thermometer**. One choice is liquid mercury, which expands inside a glass tube.

The standard (metric) temperature unit is the degree Celsius (°C). The Celsius scale assigns 0°C and 100°C to the freezing and boiling points of water. The United States still uses the Fahrenheit scale, related to the Celsius scale, as shown in Figure 7.3.

Temperature and thermal energy are related but different. For instance, a cool lake contains much more thermal energy than does a hot cup of coffee, even though the lake has a lower temperature, because the lake is so much larger.

DIALOGUE 1   To develop a feel for Celsius temperatures, answer the following questions in °C. What is the approximate temperature of a nice day? Of a hot day? What temperatures are below freezing? Above boiling?

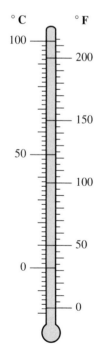

FIGURE 7.2
When you touch a hot cup of coffee, thermal energy flows from the high-temperature cup to your lower-temperature hand, so your finger gets warmer and the coffee cools.

FIGURE 7.3
The Celsius and Fahrenheit scales, compared.

---

1. 20–25°C, 35–40°C, below 0°C, above 100°C.

7.2 Heat engines: *using thermal energy to do work*

Drop a book on the floor. Slide it across the table. Smack it with your hand. Imagine tearing out a page and burning it up. (As you may have noticed, books come in for a certain amount of rough treatment in physics courses.)

Consider the energy transformations during these processes: Thermal energy is created during each one! It's hard to think of a physical process that doesn't create at least a little thermal energy. It is easy—almost inevitable—to create thermal energy.

But processes that consume thermal energy as the input and then convert it to other forms are less common. Can you think of any? One example is the automobile engine, which uses thermal energy to do work. Any repetitive (or cyclic) device that uses thermal energy to do work is called a **heat engine.**

One significant feature of an automobile engine is that a lot of thermal energy is ejected after the engine has done its work: The automobile's radiator ejects excess thermal energy, and its tailpipe exhausts residual hot gases. So *not all the thermal energy input is actually used to do work*. Some of the thermal energy going into the process remains unused at the end of the process. This turns out to be a general feature of every heat engine. A heat engine's thermal energy output is called its **exhaust.**

So the energy transformation for any heat engine is

ThermE (input) → Work (which could then produce any form of energy) + ThermE (exhaust)

Figure 7.4 is one way of representing a heat engine's energy flow.

Because of the social significance of heat engines, we want to quantify their performance. One of the most important quantities is a heat engine's efficiency. The energy efficiency of any device is its useful energy fraction—its useful energy output divided by the total energy input—often expressed as a percentage (Chapter 6). Since the work we get out of it is usually the useful output from a heat engine, we define the **efficiency** of any heat engine as its work output divided by the thermal energy put into it:

$$\text{efficiency} = \frac{\text{work output}}{\text{total energy input}}$$

The notion that heat engines always have an exhaust can be stated very simply in terms of efficiency: Heat engines are always less than 100% efficient.

This special feature of thermal energy—that you can entirely create it but you can't entirely consume it—has been found to be true every time we have checked experimentally. It is, in other words, a fundamental principle of nature. But as we will see, it turns out not to be another *new* principle of nature. It has a one-way quality about it that is reminiscent of the heating form of the second law. Perhaps it is not surprising then that it turns out to be the second law of thermodynamics, only put into words that are different from our previous statement. We call it:

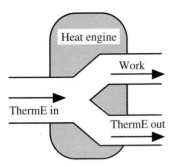

FIGURE 7.4
Energy flow for a heat engine.

THE SECOND LAW OF THERMODYNAMICS, HEAT ENGINE FORM (LAW OF HEAT ENGINES)

Any cyclic process that uses thermal energy as the input to do work must also have a thermal energy output or exhaust. In other words, heat engines are always less than 100% efficient.

HOW DO WE KNOW?

Why do we believe the law of heat engines? Rather than appealing directly to experiment, we offer a theoretical argument based on the law of heat flow, which is in turn based on experiment. It is a type of argument that logicians call an argument by contradiction or a reduction to an absurdity.

Let's temporarily suppose that (in violation of the law of heat engines) there is a heat engine that can convert thermal energy entirely to work. We could then use that heat engine to extract thermal energy from, say, a pot of warm water and convert this energy entirely to work. This work could produce thermal energy (by friction, for example) in some hotter system, such as a pot of hotter water. The net result would be to transfer thermal energy "uphill," from a lower to a higher temperature, without any other change taking place. But this is absurd—it's exactly what the law of heat flow says we cannot do. In other words, any violation of the law of heat engines implies that the law of heating is also violated. But we know, directly from many experimental observations, that the law of heating is not violated. So it follows that the heat engine form cannot be violated either.

That is how we know that heat engines are less than 100% efficient. It follows logically from the common observation that thermal energy flows from hot to cold.

The flow of thermal energy from a high to a low temperature is essential to heat engines. In fact, a heat engine may be described as a device that uses the natural hot-to-cold flow of thermal energy, by shunting aside some of the flowing thermal energy to do work (Figure 7.5). Keep in mind that Figure 7.5 is only a graphical way of showing the energy flow and is not a picture of an actual heat engine.

The second law tells us that we can never get back all the input thermal energy in the form of useful work. Even if we could eliminate all such effects as friction and air resistance that create thermal energy and hence reduce an engine's work output, every heat engine must be less than 100% efficient. Effects like friction only make an engine even less efficient.

Since heat engines are driven by the flow of thermal energy from hot to cold, you must have a difference in temperatures before you can have a heat engine. You cannot turn thermal energy into work if you have only a single temperature to work with. The ocean, for example, contains a lot of thermal energy, but you cannot use it to do work unless you have a colder system for the ocean's thermal energy to flow into.* Heat engines always operate between two systems with different temperatures, as Figure 7.5 shows.

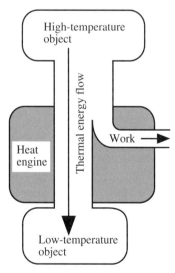

FIGURE 7.5

Heat engines use a portion of the thermal energy that flows naturally from a high to a low temperature and convert it to work.

---

* This means that temperature differences between different depths of ocean water could be used to run a heat engine; see Section 17.6.

How efficient can the very best heat engine be? It's an important question, because most of the world's primary energy (energy that comes directly from nature) passes through heat engines. In the United States, fully 60% of the primary energy consumed each year passes through either a transportation vehicle or a steam-electric power plant. Both of these devices are heat engines, subject to the second law.

Since a heat engine operates because of the ability of thermal energy to flow from hot to cold, we expect its efficiency to be influenced by both its hot **input temperature** (the temperature at which thermal energy is put into the engine) and its cooler **exhaust temperature**. Since temperature differences drive heat engines, we expect a higher efficiency for larger temperature differences between the input and the exhaust and a lower efficiency for smaller temperature differences. Nineteenth-century physicists found a quantitative formula that can be used to predict the best or ideal efficiency of a heat engine operating at any predetermined input and exhaust temperatures.* As examples, Table 7.1 lists the predicted ideal efficiencies for several specific types of heat engines, along with the actual efficiencies obtained by these heat engines in practice.

Table 7.1 shows how important the second law is to society, as even the ideal efficiencies of these heat engines are 60% or less. Friction and other nonideal processes reduce this even further. Less than half the natural energy resources used in these heat engines are employed to do work. This doesn't necessarily mean, however, that all of the remainder is wasted, because the exhausted thermal energy might be warm enough to be used

TABLE **7.1**

Heat Engine Efficiencies. Typical input and exhaust temperatures, ideal (best possible) efficiencies, and actual efficiencies.

			Efficiency (%)	
Engine type	T_{in} (°C)	T_{ex} (°C)	Ideal	Actual
Transportation				
Gasoline automobile/truck	160	25	30	10–15
Diesel auto/truck/locomotive	500	25	60	15–20
Steam locomotive	180	100	18	10
Steam-electric power plants				
Fossil fuel	550	40	60	40
Nuclear powered	350	40	50	35
Solar powered	225	40	38	30
Ocean-thermal	25	5	7	???

* The formula is

$$\text{efficiency} = \frac{\text{input temperature} - \text{output temperature}}{\text{input temperature}}$$

In this formula, the temperatures must be measured in degrees *Kelvin* (K), a new temperature scale. The temperature in K is found by adding 273 to the temperature in °C. 0 K (or −273°C) is known as absolute zero, because it is the lowest possible temperature—the temperature at which all microscopic motion is at its absolute minimum.

directly for heating. Indeed, many countries' steam-electric power plants make use of such **cogeneration** of both electricity and useful thermal energy.

Table 7.1 shows the importance of "burning hot" and "exhausting cool." For example, the fossil, nuclear, and solar generating plants have progressively lower input temperatures, but they all exhaust to cooling water that is at about 40°C. As you can see, the efficiency declines as the difference between T_{in} and T_{ex} declines.

Ocean-thermal generation of electric power is a plan to use some of the ocean's thermal energy by exploiting temperature differences at different ocean depths. In the tropics, the ocean's temperature drops rapidly from 25°C at the surface to 5°C at 300 m beneath the surface. This small temperature difference could be used to run a heat engine with an efficiency of less than 7%. Because the primary energy resource would be solar energy falling on the ocean, the low efficiency would be of little concern. But because the intake point for cool water must be 300 m below the ocean's surface, the engineering problems would be formidable. A small demonstration plant of this sort has been constructed (see Chapter 17, Figure 17.15).

7.3 Energy quality: *things run down*

The second law tells us that thermal energy is less useful than other types of energy are, because unlike the other forms, not all its energy can be used to do work. In this sense, thermal energy is lower-quality energy. A moving bullet or a raised rock can convert its kinetic or gravitational energy entirely to work, but thermal energy can be only partially used to do work. So whenever you transform other energy forms into thermal energy—say by friction or combustion—you reduce the energy's quality. Energy is not lost in such processes, but the quality of the energy is. Thermal energy, once created, is impossible to retrieve totally as useful work.

So there is an irreversibility about any process that creates thermal energy. Once a system creates thermal energy, that system will never by itself (spontaneously) be able to return to its previous condition. To return, it would have to convert spontaneously all the created thermal energy to its original form, and the second law prohibits this. The system can return to its initial state only with outside help.

A good example is a rock swinging back and forth on a string tied to a hook (Figure 7.6). Air resistance and friction (between the string and the hook) gradually bring the rock to rest. The complete system (rock, string, hook, and surrounding air) loses no energy. But it can't return to its initial condition because thermal energy is created, and this cannot be entirely reconverted to kinetic or gravitational energy. Something is lost when systems run down like this, but it cannot be energy because energy is conserved. Instead, energy *quality* is lost. The system's energy gradually decreases in quality until all of it has turned into thermal energy at a temperature only slightly warmer than the surroundings. Now the original energy is almost entirely useless.

FIGURE 7.6

A rock swinging on a string tied to some fixed point overhead. The rock "dies down," illustrating the irreversibility of natural processes implied by the second law.

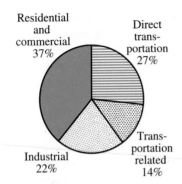

FIGURE 7.7

The fraction of total U.S. primary energy consumed by each economic sector.

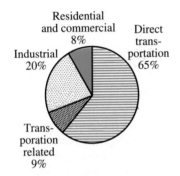

FIGURE 7.8

The fraction of U.S. petroleum consumed by each economic sector.

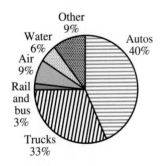

FIGURE 7.9

The fraction of U.S. direct transportation energy consumed by each transportation mode.

When we use Earth's energy resources, we don't "use up" energy. Instead, we degrade energy from highly useful forms such as the chemical energy of oil to less useful forms such as thermal energy.

One of the two great laws of energy says that energy is conserved, and the other says that energy becomes less useful. You can't get ahead, and you can't even break even.

DIALOGUE 2   (a) Assume that the energy flows in Figure 7.5 are proportional to the width of each pipe. Estimate this engine's efficiency. (b) Suppose this engine consumes 400 J of thermal energy and converts 300 J of this to exhaust. Find the engine's efficiency.

DIALOGUE 3   A large coal-fired electric generating plant burns about 1 tonne (1000 kg) of coal every 10 seconds. According to Table 7.1, how much of the tonne actually goes into producing electric energy? How much goes into exhaust?

7.4 The automobile

Few technologies shape our culture as strongly as the automobile. It brings great freedom while affecting our quality of life, family structure, self-perceptions, physical environment, health, work, community structure, resource use, economy, and even issues of war and peace. We cannot go into all these far-flung issues, but as part of our science-and-society theme, we will look at the automobile's role in energy-related matters.

Transportation consumes much of the total energy used by the United States (Figure 7.7) and most of the nation's oil (Figure 7.8). Most of this is used by cars and trucks (Figure 7.9).

MAKING ESTIMATES

Calorimetry experiments show that upon combustion, 1 gallon (3.8 liters) of gasoline releases (converts to thermal energy) about 130 million joules of energy. Use this figure to estimate the rate, in watts, at which a typical car consumes chemical energy while moving at highway speeds without acceleration. *Hints*: Typical gasoline consumption is 25 miles/gallon (10 kilometers/liter), and highway speeds are about 50 mi/hr (80 km/hr).

SOLUTION   During 1 hour at 50 mi/hr, a car travels 50 miles and so consumes 2 gallons. Since 1 gallon contains 1.3×10^8 J, the chemical energy in 2 gallons is 2.6×10^8 J. This is consumed in 1 hour. Since there are 3600 (3.6×10^3) seconds in 1 hour, the chemical energy consumed per second is

$$2.6 \times 10^8 \text{ J} / 3.6 \times 10^3 \text{ s} \approx 0.7 \times 10^5 \text{ J/s} = 70{,}000 \text{ J/s}$$

or 70,000 watts or 70 kilowatts.

---

2. (a) Something like 30%, judging from the widths of the "pipes." (b) Work = 400 − 300 = 100 J. Efficiency = 100/400 = 0.25 = 25%.
3. Forty percent of 1000 kg is 400 kg; the remaining 600 kg goes into exhaust.

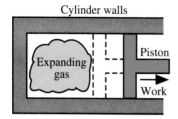

Cylinder walls

Expanding gas | Piston | Work

FIGURE 7.10
The conversion of thermal energy to useful work using a piston.

FIGURE 7.11
General Motors' experimental Sunraycer converts solar energy to electricity to power an electric vehicle.

So a typical car, moving without acceleration, consumes natural energy resources at a rate of 70 kilowatts. This is equivalent to the electric power going into seven hundred 100-watt bulbs burning the whole time the car is moving. As another comparison, the average rate at which a typical U.S. household consumes electric energy is about 1 kilowatt, so while a car is moving, its energy use is equivalent to that of 70 households. If the car is accelerating, all of these figures should be multiplied by about 5.

Most of the energy used for transportation goes into heat engines, where it first burns to produce thermal energy and then is partially transformed into useful work. Cars and trucks are powered by **internal combustion** engines. Heat engines of this type burn a fuel–air mixture. The mixture's high combustion temperature gives it a high pressure, which enables the hot gases to push strongly on a **piston**, a movable metal plate connected to a rod (Figure 7.10). The piston does the work that turns the drive wheels. Combustion is "internal" because the combustion occurs directly inside the gases that do the work, in contrast with **external combustion**, which occurs in a fuel that then provides thermal energy to a second substance such as steam that then does the actual work.

The most convenient and abundant automobile fuel has always been **gasoline**, a form of petroleum. But there are many other possibilities. As gasoline's pollution problems loom larger and as petroleum's availability declines, other, less-polluting fuels may become more popular. Table 7.2 (see also Figures 7.11 through 7.14) lists and describes possible automotive fuels.

We estimated that a car moving at highway speeds without acceleration consumes chemical energy at a rate of 70 kilowatts, or 70,000 joules in every second. Figure 7.15 is an energy flow diagram showing where this 70 kW goes. Energy diagrams like this are helpful in analyzing the connections between technology and society.

TABLE **7.2**
Fuels for Automobiles and Trucks

Fuel	Source of fuel	Description
For internal combustion		
Gasoline	petroleum	liquid, widely used
Diesel fuel	petroleum	liquid, widely used
Compressed natural gas	natural gas	high-pressure gas
Liquefied natural gas	natural gas	low-temperature liquid
Liquefied petroleum gas (propane)	petroleum or natural gas	high-pressure liquid
Methanol (wood alcohol)	wood, natural gas, or coal	liquid
Ethanol (grain alcohol)	grain or sugar crops	liquid
	agricultural trash	liquid
	municipal trash	liquid
Hydrogen[a]	any electric power plant	hydrogen produced from water
	solar electricity	hydrogen produced from water
For noncombustion		
Storage batteries[a]	any electric power plant	recharge by plugging in
	solar electricity	recharge from solar cells

[a] Since electricity for hydrogen production or for storage batteries can come from any source, the ultimate energy resource can be wind, hydroelectric, nuclear, coal, photovoltaics, and the like.

FIGURE 7.12
Toyota's battery-powered EV-50 has a range of 240 km (150 mi) and a top speed of 110 km/hr (70 mph).

FIGURE 7.13
The Honda Civic NGV (Natural Gas Vehicle) is fueled by compressed natural gas. It has a range of 430 km (270 mi), and lower hydrocarbon, carbon dioxide, carbon monoxide, and nitrogen oxide emissions than gasoline-powered vehicles.

FIGURE 7.14
Two high-pressure cylinders in the rear of the Honda Civic NGV are the only indication that it is powered by compressed natural gas.

On the average (all these figures are average), 1 kW is lost to evaporation in the carburetor. It isn't really lost, as energy is never lost. But 1 kW of hydrocarbons goes into the atmosphere, where its chemical energy essentially cannot be recovered and where it contributes to chemical pollution. The remaining 69 kW go to the engine, which produces about 17 kW of work and exhausts the remaining 52 kW as thermal energy and unused chemical energy.

About half of the exhaust energy is removed by the car's radiator, and the other half goes out through the exhaust pipe as hot gases. These gases carry the automobile's pollution. Gasoline is a hydrocarbon, made of hydrogen (H) and carbon (C) atoms. Both the hydrogen and the carbon combust with oxygen from the atmosphere, so the exhaust gases are mostly CO_2 and H_2O. Neither one is toxic (poisonous to humans), but CO_2 contributes to environmentally harmful global warming (Chapter 9).

The tailpipe exhaust carries various other molecules, mainly CO, NO, NO_2, and unburned hydrocarbons. These are toxic pollutants, and their unused chemical energy also represents an energy inefficiency. The carbon monoxide and unburned hydrocarbons are the result of incomplete combustion (partial combustion) of the fuel. The two oxides of nitrogen (collectively called NO_x) form from atmospheric oxygen and nitrogen, which combine under the influence of the engine's high temperatures. Cars and trucks are major polluters, producing about two-thirds of the nation's CO pollution, one-third of its hydrocarbon pollution, and half of its NO_x pollution.

The automobile's main loss of useful energy occurs in the engine. The engine's best possible, or ideal, efficiency is 30%, and its actual efficiency is 17 kW/69 kW = 25%. The second law is the reason for the 30% limit. This

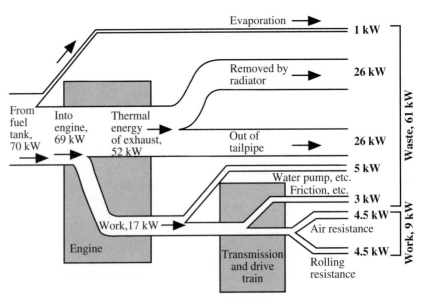

FIGURE 7.15
Typical energy flow in an unaccelerated car at highway speed.

loss, a necessary consequence of the second law, is the automobile's largest single inefficiency. Several other losses combine to bring the engine's efficiency down to 25%. These include incomplete combustion, the formation of NO_x, friction in the engine, and thermal losses through the engine's wall.

But this is not the end of the story. Of the 17 kW produced by the engine, 5 kW are used to run internal devices like the water pump and air conditioner. The remaining 12 kW go to the transmission and drive train that couples the engine to the drive wheels. This coupling is about 75% efficient, so 9 kW finally arrive at the drive wheels, and the other 3 kW produce thermal energy.

The 9 kW are the part that actually carries the car down the road. On average, about half of it goes into overcoming air resistance. Air resistance increases with speed and dominates at higher speeds. The other half of the 9 kW go into rolling resistance (Section 4.6), which dominates at lower speeds. The overall energy efficiency is 9/70 = 13%, or about one-eighth.

Nearly all of the chemical energy input, both the energy lost and the energy used, ends up as thermal energy at about atmospheric temperature. The original chemical energy has been permanently reduced in quality by passing through the automobile and has for all practical purposes lost its ability to do useful work.

DIALOGUE 4   When a car accelerates, the drive power needed (at the drive wheels) is about 45 kW. What input power from the gasoline is then needed? How does gasoline consumption during periods of acceleration compare with its consumption during periods of constant velocity?

7.5 Transportation efficiency

Our definition of efficiency—useful energy output divided by total energy input—doesn't really capture the automobile's purpose. Although its purpose is to move people, its energy goes mostly into moving the car itself rather than people. Gasoline mileage, the common measure of automobile efficiency, suffers from the same defect. Neither of these measures really captures people-moving efficiency, in other words, transportation efficiency. On the other hand, gasoline mileage is useful for comparing different cars with one another (Table 7.3, Figure 7.16).

The efficiency of using energy to move people can be directly measured by using as the output the number of passenger-kilometers delivered. For example, if a bus moves 20 passengers a distance of 3 km, it has delivered 20 passengers × 3 km = 60 passenger-km. Similarly, the efficiency of using energy to move goods can be directly measured by using as the useful output the number of "tonne-kilometers" delivered. For example, if a truck moves 5 tonnes a distance of 80 km, it has delivered 5 tonnes × 80 km = 400 tonne-km.

FIGURE 7.16
General Motors' Ultralite experimental automobile has a fuel efficiency of 40 km/l (100 mi/gal).

---

4. Forty-five kW is five times the previous output (9 kW), so the input must be five times larger, or 350 kW. The gasoline consumption rate is five times larger.

TABLE 7.3

Fuel Efficiencies of Four-Passenger Automobiles

	mi/gal	*km/l*
Averages, all cars		
U.S. fleet average	20	8
World fleet average	25	10
Averages, new cars		
U.S.	26	11
European, Japanese	30	13
Cars produced today		
Ford Escort	53	22
Honda Civic	57	24
Suzuki Sprint	57	24
Prototypes		
Volvo LCP 2000	71	30
Volkswagen E80	85	36
Toyota AXV	98	41
General Motors Ultralite	100	42
Renault VESTA	124	52

SOURCE: D. Bleviss, *The New Oil Crisis* (New York: Quorum Books, 1988).

Table 7.4 compares the passenger-moving and freight-moving efficiencies of several transportation modes. Walking and bicycling come out far ahead because no energy is put into moving a heavy vehicle. The energy input for walking and bicycling is the metabolic (chemical) energy from food. Bicy-

TABLE 7.4

Passenger-Moving and Freight-Moving Efficiencies of Several Transportation Modes, per Gallon of Gasoline or of Gasoline's Energy Equivalent in Food or Other Fuels

Passenger-moving efficiency	*Pass-mi per gal*	*Pass-km per l*	*Pass-km per MJ*
Auto (average occupancy = 1.2)	24	10	0.29
Carpool auto (occupancy = 4)	80	32	1.0
Commercial airline	50	21	0.62
Transit bus	50	21	0.62
Intercity rail	250	100	3.0
Person walking	400	170	5.0
Person on bicycle	1200	500	15.0

Freight-moving efficiency	*Ton-mi per gal*	*Tonne-km per l*
Airplane	13	5
Truck	65	25
Rail	260	100
Waterway	290	110

SOURCE: *Transportation Energy Data Book*, Oak Ridge National Laboratory

TABLE **7.5**

Mass-Moving Efficiencies of
Animals and Machines, in Gram-
Kilometers per Joule

	g-km/J
Human on bicycle	1.5
Salmon	0.6
Horse	0.5
Jet transport	0.4
Human walking	0.3
Automobile	0.3
Typical bird	0.2
Humming bird	.05
Fly, bee	.02
Jet fighter	.01
Mouse	.005

SOURCE: S. Wilson, "Bicycle
Technology," *Scientific American,*
March 1973.

cling is more efficient than walking because wheels keep rolling—wheels take advantage of the law of inertia.

For fun, take a wider view and speculate on transportation efficiencies throughout the animal kingdom. Which animal or machine is most efficient at moving itself and any passengers that might be along? Fruit flies? Horses? Jet planes? To compare fruit flies and horses fairly, we must incorporate the fact that the horse's energy goes into moving a much larger mass. So the useful output should be measured as total body (or vehicle) mass times distance moved. Table 7.5 gives several such mass-moving efficiencies, in gram-kilometers per joule of energy. For vehicles such as the bicycle and the automobile, the total mass of the occupant(s) plus the vehicle is included as the output.

Again, bicycling comes out far ahead, because animals don't have wheels, and among the vehicles used for transportation, the bicycle is the only one that is not a heat engine. Its wheels take advantage of the law of inertia, but it is not up against the second law's restrictions on heat engines.

Animals with wheels would have a big energy advantage. They have not evolved on Earth's rough surface, but tumbleweeds take advantage of rolling in order to spread their seeds. One could speculate that on another planet with surfaces created by smooth lava flows, wheeled animals might evolve and be abundant!

DIALOGUE 5   You wish to move 130 tons of freight a distance of 100 miles. About how many gallons of gasoline (or gasoline equivalent) will you need to move it by truck? By air? By rail?

DIALOGUE 6 MAKING ESTIMATES   About how much gasoline does a 55 mph speed limit save every year, for an average American car and for all American cars? Use the following data: There are 141 million cars in the United States; a typical car travels 15,000 miles in a year, about half of this being highway driving; American cars average 20 miles per gallon; and fuel mileage is about 15% worse at 65 mph (roughly the speed expected without a 55 mph limit) than it is at 55 mph.

---

5. 130 tons × 100 miles = 13,000 ton-mi. Trucks use 1 gallon of gasoline for each 65 ton-mi (Table 7.4), so 13,000 ton-mi requires (13,000 ton-mi)/(65 ton-mi per gal) = 200 gallons. For the other two questions, it's easiest to use proportions: Table 7.4 shows that air transport is one-fifth as efficient as truck transport, and so it requires five times more gasoline, or 1000 gallons. Rail transport is four times more efficient than truck transport, and so it requires one-fourth as much gasoline, or 50 gallons.

6. For one car in one year: It drives 7500 miles on the highway. Compare driving this at 55 mph (getting about 20 miles per gallon) with driving it at 65 miles per hour (where fuel efficiency is 15% worse, or only 17 miles per gallon): 7500/20 = 375 gallons, 7500/17 = 441 gallons, so an extra 441 − 375 = 66 gallons is needed at the higher speed. For the entire U.S. auto fleet, $141 \times 10^6 \times 66$ gallons = 9.3×10^9 gallons (nearly 10 billion gallons). To put this into perspective, this is 45 days worth of the United States' oil imports. Saving it could reduce oil imports by as much as 12%.

7.6 The steam-electric power plant

We turn now from the automobile to society's other major heat engine, the steam-electric power plant.

Figure 7.17 is a schematic diagram of the operation of a typical coal-burning electric power plant. Coal is the most widely used primary energy source for electricity. Other plants that use oil, natural gas, nuclear energy, or solar energy to turn water into steam in a boiler operate much as a coal-burning plant does in producing energy from steam.

FIGURE 7.17
A schematic diagram showing the operation of a coal-fueled steam-electric generating plant.

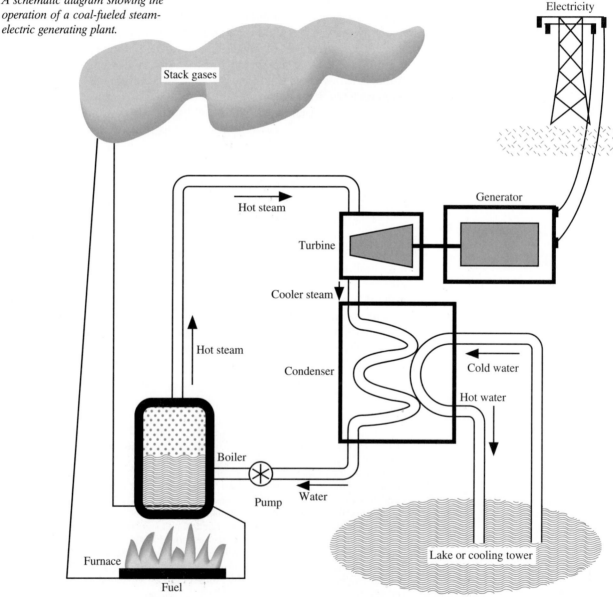

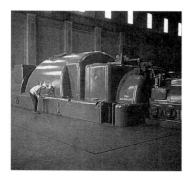

The coal combusts externally in a furnace, and its thermal energy is transferred into water inside a boiler. Most combustion products escape through the smokestack or stack (little smoke goes out today), but some pollutants are removed first. In the boiler, water turns into high-pressure steam at over 500°C, far above the normal boiling temperature. The steam moves through pipes to a large waterwheel-like **steam turbine** (Figure 7.18). The turbine is designed to turn when it feels a higher pressure on the front (upstream) side than on the back (downstream) side. Like the piston in a car, the turbine is the key device that transforms thermal energy into work. The turbine turns an **electric generator**, a device containing wires and magnets that generates (creates) electricity when the turbine makes it rotate.

The second law says that thermal energy can be converted to work only if that energy is allowed to flow from a high to a low temperature. So the exhaust side of the turbine must be cooled. If it were not cooled, the two sides would soon be the same temperature, and then there would be equal pressures on both sides, so the turbine would not turn! In generating plants, the low-temperature material that cools the exhaust side of the turbine is either a body of water or the atmosphere.

It is most efficient to remove so much thermal energy from the exhaust side of the turbine that the steam transforms back into liquid water, because this greatly reduces the pressure against the back side of the turbine.* The steam is then sucked forcefully through the turbine, from very high pressure on one side to a near vacuum on the other. The transition from steam to water is called *condensation*, and the cooling device is called a *condenser*. Once condensed, pumps move the water back around to the boiler, where the cycle begins again.

The plant is a heat engine. It converts thermal energy to work. Thermal energy flows in at the boiler and out at the condenser, and work is done by the turbine (compare Figure 7.5).

Figure 7.19 shows the energy flow. A large plant generates about 1000 megawatts (10^9 watts) of electric power, enough for a large city. This electrical output requires about 2500 megawatts, or 2500 million joules per second, of energy input from coal or another fuel. This means that in every second, 2500 million joules of chemical energy must be converted to thermal energy. Calorimetry experiments show that the combustion of 1 kilogram of coal produces about 25 million joules of thermal energy, so the plant requires 100 kilograms of coal every second.

Figure 7.19 follows the 2500 megawatts of input power through the plant. Three hundred megawatts go out through the stack, accompanied by gases such as sulfur oxides (which cause acid rain) and CO_2 (which causes global warming), and small incombustible solid particles called *ash*. Modern plants remove about 90% of the sulfur oxides and 99% of the ash, which then present a significant solid-waste disposal problem. The turbine converts the thermal energy of steam to useful work at a rate of 1000 megawatts. This work drives a generator that creates

---

* In practice, not all of the steam is condensed; some of it is removed from the turbine's exhaust and reused without condensation, to heat the water on its way to the boiler. Steam also is "reused" by being reheated and passed through lower-pressure turbines several times before finally going to the condenser.

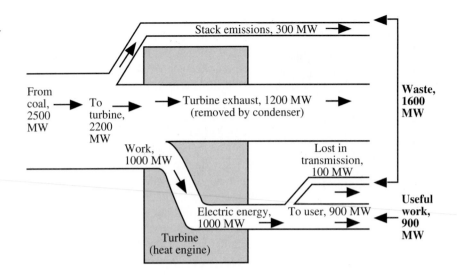

FIGURE 7.20

Cooling towers and stacks at a coal-fueled generating plant. Cool air is sucked into the bottom of the cooling towers, where it cools hot water from the plant. The tall cooling tower promotes a rapidly rising hot-air column.

1000 megawatts of electric power. The plant's best possible, or ideal, efficiency is about 60% (Table 7.1), and its real efficiency is 1000/2500 = 40%. The loss due to the second law is the plant's main inefficiency.

Twelve hundred megawatts of exhaust go into the condenser's cooling water. If the cooling water comes from a lake or river, the exhaust warms the water, an effect called *thermal pollution*. Many plants use the atmosphere as the coolant, by employing large evaporative coolers known as *cooling towers* (Figure 7.20).

Some of the 1000 megawatts of generated electricity is transformed into thermal energy during its transmission over electric power lines that might extend hundreds of miles. About 100 megawatts are lost in this way, so the efficiency of delivering electricity to the end user is 900/2500 = 36%.

DIALOGUE 7   How efficient is the boiler? The electric generator?

DIALOGUE 8 MAKING ESTIMATES   (a) Make an order-of-magnitude estimate of the amount of coal (in tonnes) that this 1000 MW generating plant uses in one day (recall that it uses 100 kg/s). How many large highway truckloads of coal is this, at about 50 tonnes per truck? (b) Make an order-of-magnitude estimate of the amount of carbon dioxide that this plant puts into the atmosphere every day. *Hint*: The oxygen atom's mass is 33% larger than the carbon atom's mass; for order-of-magnitude purposes, treat their masses as roughly equal. For your estimate, assume that coal is 100% carbon.

---

7. 2200/2500 = 88%. For the generator, 1000 MW of input power (from the turbine) produces 1000 MW of electric power (a little less, really), so the generator is essentially 100% efficient.

8. (a) The number of seconds in a day is 60 s/min × 60 min/hr × 24 hr/day ≈ 10^5 s. One hundred kilograms enter the plant every second, so the amount entering in a day is about $10^5 × 100 = 10^7$ kg, or 10,000 tonnes. This is 10,000/50 = 200 large truckloads every day! (b) Since an O and a C atom have roughly the same mass, the CO_2 molecule is about three times as massive as the C atom. Treating coal as though it were pure carbon, the mass of CO_2 created is three times the mass of coal consumed: 30,000 tonnes!

The world's most important arithmetic is the arithmetic of the exponential function.

Albert Bartlett, physicist

7.7 Resource use and exponential growth* ___

Because energy production and consumption are prominent examples of a pattern known as exponential growth, this is an appropriate point to discuss this important concept.

We begin with a familiar example. Suppose you invest $100 at a rate of return or **growth rate** of 10% per year. When will you double your money? You earn $10 during the first year, so you have $110. So in the second year, you earn $11, one dollar more than you earned the first year. Now you have $121, so you earn $12.10 during the next year. Each year you earn more than you did the previous year, because your *percentage* increase is the same each year.

Figure 7.21 graphs your account and compares it with a second graph that also increases by $10 during the first year but then follows a straight line. The second graph illustrates **linear** (straight line) **growth**, which increases by a fixed amount, instead of a fixed percentage, each year. As you can see, there is a big difference between $10 per year and 10% per year. When a quantity grows by a fixed percentage in each unit of time, its growth is said to be **exponential.**[†]

If you continue adding on 10% each year, your investment account will reach about $195 after 7 years. In 7 years, you will roughly double your money.

The arithmetic of growth is the forgotten fundamental of the energy crisis.

Albert Bartlett

What will happen if you keep your money deposited for another 7 years? Since the same arithmetic applies, it will double again, to nearly $400.[§] In the next 7 years it will double again, to nearly $800. And so forth. Exponential growth has a fixed, unchanging, **doubling time**.

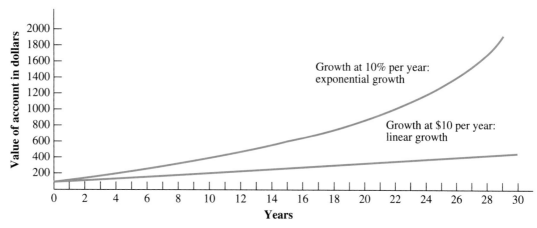

FIGURE 7.21
The 10% investment account. The initial investment is $100. Would you rather have exponential growth at 10% per year or linear growth at $10 per year?

---

* This section draws on the work of University of Colorado physicist Albert A. Bartlett.
† It is called *exponential* because the account is worth 100×1.1 after 1 year, 100×1.1^2 after 2 years, 100×1.1^3 after 3 years, and so forth. The number of years is in the *exponent*.
§ If you aren't convinced of this, think of it in this way: You begin the second 7-year period with about $200. Think of this as *two* $100 accounts. During the second 7-year period, each of these accounts must grow to about $200, so your total is $400.

MAKING ESTIMATES: THE EXPONENTIAL SALARY

Suppose you take a job requiring you to work every day for 30 days and your employer offers you just 1¢ for the first day and then a doubled salary every day for the 30 days. About how much will you earn on day 30?

SOLUTION   Starting with 1¢ on day 1, you will earn $5.12 on day 10 (1, 2, 4, 8, 16, 32, 64, 128, 256, 512¢). You aren't rich yet.

To simplify matters, round off the $5.12 to just $5. Now continue doubling every day until day 20: $10 on day 11, $20 on day 12, and so forth. On day 20, your earnings will be $5120. Now you are getting rich.

Round this off to $5000, and continue with days 21 to 30: $10,000 on day 21, and so forth. On day 30 alone, your earnings are more than $5 million! And this is an underestimate, because we rounded downward twice. Furthermore, your total accumulated earnings are much more.

Exponential growth can be surprising, because our intuitions tend to be linear.

Biological populations often increase at a fixed percentage per year. If there are no offsetting deaths due to starvation and other consequences of the population size, the number of newborns each year should be roughly proportional to the number of potential parents in the population that year. This means that the percentage increase (the number of newborns divided by the total population) should be roughly the same from year to year. So the growth is exponential. The next dialogue is an example.

DIALOGUE 9 POPULATION GROWTH IN A FINITE

ENVIRONMENT   Bacteria reproduce themselves by means of division. If you start with 1 bacterium, it will divide into 2, they will divide into 4, then into 8, and so forth. Since each population doubling occurs in the same time interval, it is an exponential process. Suppose that some strain of bacteria has a dividing time of 1 minute. You put 1 bacterium into a bottle at 11 A.M., and at noon you note that the bottle is full of bacteria. When was the bottle half full? If you were one of the bacteria, when might you have noticed that you were running out of space?

When you have a finite resource, such as space in a bottle, and you consume it exponentially, it is easy to use nearly all of it before you realize there is a problem (Figure 7.22).

Continuing with Dialogue 9, suppose that at 11:55 A.M. some visionary bacteria, realizing they have a problem, launch an all-out search for new bottles. By 11:58 A.M., this program has been successful in discovering a huge, new reserve: three new bottles! It took the bacteria an entire hour to fill the first bottle. When will the new bottles be full? The answer is at 2 minutes past noon.

---

9. Work backward from noon: The bottle must have been half full at 1 minute before noon. There is no single answer to the second question, but Figure 7.22, obtained by working backward from noon, gives a rough idea. A few forward-looking bacteria might begin to be concerned when the bottle got to be a few percent full, at a few minutes before noon.

Pressures resulting from unrestrained population growth put demands on the natural world that can overwhelm any efforts to achieve a sustainable future. If we are to halt the destruction of our environment, we must accept limits to that growth. . . .The United Nations concludes that the eventual total [world population] could reach 14 billion. . . .But, even at this moment, one person in five lives in absolute poverty without enough to eat, and one in ten suffers serious malnutrition. . . .We must stabilize population.

From *World Scientists' Warning to Humanity*, a declaration signed by nearly 1700 leading scientists from 71 countries, including 104 Nobel laureates—a majority of the living recipients of the Nobel Prize in the sciences. The declaration was sponsored by the Union of Concerned Scientists and published in 1993.

The world is projected to add at least 960 million people during this decade, up from 840 million in the eighties and 750 million in the seventies.

World Population Data Sheet, 1990

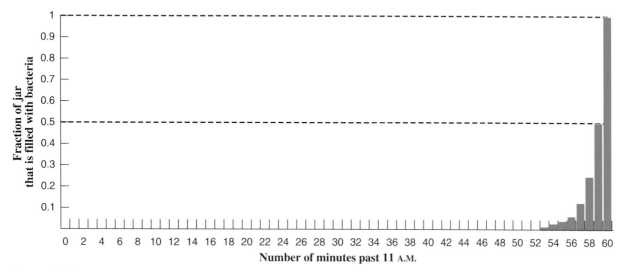

FIGURE 7.22

Bacterial growth in a jar. A finite resource, consumed exponentially, runs out surprisingly rapidly toward the end. On the scale of this graph, growth is imperceptible until 11:53 A.M. when the bottle is 1% full.

When a finite resource is consumed exponentially, new discoveries cannot change things much unless the discoveries grow exponentially too. It is the consumption pattern that dominates. Exponential growth eventually overwhelms everything else.

There is a simple and useful quantitative relation for exponential growth. Any increase in the growth rate must decrease the doubling time, so we might expect to find a relation between these two. It turns out that they are inversely proportional. The relation is, approximately,

$$\text{doubling time} \approx \frac{70}{\text{growth rate}}$$

$$T \approx \frac{70}{P}$$

where T stands for the doubling time and P is the growth rate (the percentage growth per unit time, expressed in percent). This relation can also be turned around to read

$$P \approx \frac{70}{T}$$

Either quantity, the doubling time or the growth rate, is equal to 70 divided by the other quantity. We can determine the doubling time whenever we know the growth rate, and vice versa. For instance, the 10% savings account has a doubling time of $T = 70/P = 70/10 = 7$ years.

For a significant historical example involving energy, we consider the growth of electric power in the United States. As you can see from examination of Figure 7.23, electric power production grew exponentially between 1935 and 1975, as it roughly doubled every 10 years during this period. The percentage growth rate between 1935 and 1975 was then $P \approx 70/T = 70/10 = 7\%$ per year. What would the consequences have been if this growth rate had continued for very many years past 1975? In 1975, all of the electric energy in the United States could have been

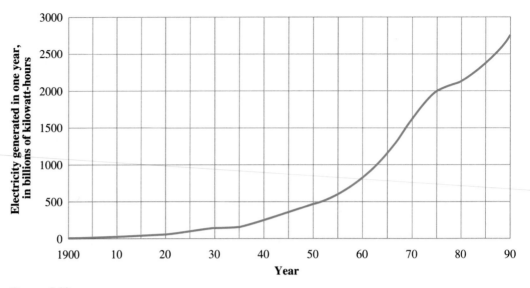

FIGURE 7.23
The history of electric power production. The annual electric energy produced, in billions of kW·h, is graphed from 1900 through 1990. The growth was roughly exponential between 1935 and 1975.

provided by about 400 large plants. If the 10-year doubling time had continued, 800 plants would have been needed in 1985, 1600 in 1995, 3200 in 2005, and 6400 in 2015. What would such numbers mean? Sixty-four hundred U.S. power plants would mean an average of more than 125 in every state, with everybody living within a few miles of a large power plant!

The point is that expansion at a fixed growth rate cannot be sustained forever. Whether it is bacteria or electricity, the space in the bottle will eventually run out. In fact, electric power production increased by 50% during the 15-year period between 1973 and 1988, for an average growth rate of $50/15 \approx 3\%$ per year. Although growth continued, the growth rate declined.

What happened in 1973 that caused the dramatic shift in growth rate? That year the Mideast oil embargo raised oil prices, which raised all energy prices, which depressed energy consumption.

U.S. oil production illustrates what can eventually happen when a finite resource is consumed exponentially. Like many industries, oil production grew exponentially during its early years, maintaining a growth rate of 8% per year between 1870 and 1930. But this could not be maintained, because for one thing, the United States' recoverable oil resources (the oil that is economically feasible to recover) would be gone by now. The growth rate declined, and then around 1970 U.S. oil production in the forty-eight contiguous states began to drop.

The consumption of a nonrenewable resource such as oil must ultimately level off and then decline. A **nonrenewable resource** is a natural resource

And in each of these decades (the 1950s and 1960s), more oil was consumed than in all of man's previous history combined.
President Jimmy Carter, 1977

Figure 7.24

A typical bell-shaped curve showing the life history of consumption of a nonrenewable resource. Exponential growth must slow as the resource is depleted. Consumption eventually levels off and declines as the resource nears exhaustion.

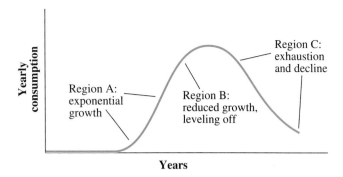

Figure 7.25

A typical life history of a renewable resource such as hydroelectric power. Exponential growth slows as consumption reaches its natural limits. Consumption eventually levels off at some sustainable value.

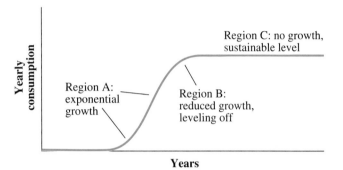

that can be used up and cannot then be readily replaced within a reasonable time span such as a human lifetime. Figure 7.24 shows the bell-shaped graph that is typical for the yearly production of such a nonrenewable resource. U.S. oil production is following this pattern and is in the early stage of region C on the graph. World oil production is not as far along on this curve and might be in the early stages of region B. In any case, resource depletion is inevitably driving the world toward the end of the oil age, although other problems such as global warming or toxic pollution might end the oil age even sooner.

Renewable resources such as wood or solar energy can follow a different kind of history (Figure 7.25). A **renewable resource** is one that is continually replaced, like raised water for hydroelectric power, or can be replaced within a few decades, like wood. In their early stages, renewable and non-renewable resource use follow the same pattern: Both grow exponentially and then level off because of natural limits. But for a renewable resource, decline need not follow. Consumption can be sustained indefinitely, provided that it proceeds at less than the replacement rate.

World population is an important example of growth. Figure 7.26 shows that the population doubled, from 200 million to 400 million, between A.D. 200 and 1300. The next doubling took only 450 years. The doubling after that took 150 years, and the next took less than 75 years. The decreasing doubling time means that the growth rate is faster than exponential. A 75-year doubling time represents a yearly increase of about 1% ($P = 70/T = 70/75 \approx 1$). World population is nearing 6 billion and is currently growing

Figure 7.26
The population explosion. The world's population has grown faster than exponentially.

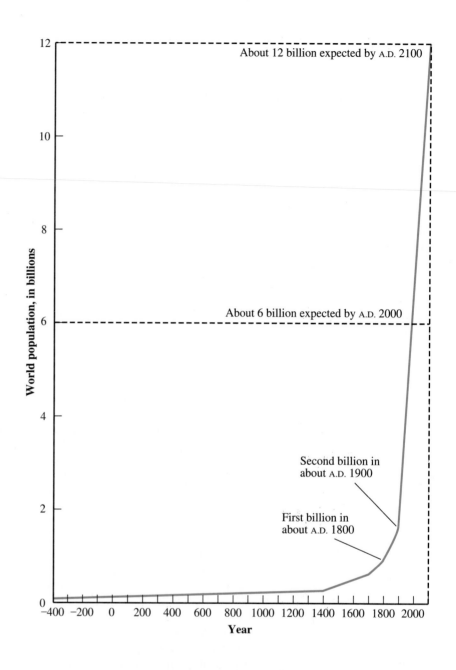

at 1.5% per year. The predictions are that it will reach 8.5 billion by 2025 *if* we are successful in holding down the birthrate. It we are not successful, this number will, of course, be even larger.

DIALOGUE 10   Between 1870 and 1930, U.S. oil production had an 8% growth rate. What was its doubling time? It reached 1 billion barrels (1 barrel = 42 gallons) per year by 1930. What would the production be this year if it had maintained this growth rate?

When you ask the question of an ecologist, "What is the most important environmental issue," they will be unanimous in saying it is exponential population growth.
  Gaylord Nelson, former U.S. senator and founder of Earth Day in 1970

DIALOGUE 11   During the 1980s, U.S. car and truck miles traveled increased by 4% per year, but the length of highway increased by only 0.1% per year. Find the doubling time for vehicle miles traveled and for miles of highway. Suppose that these rates are maintained in the future. While vehicle miles double (a 100% increase), by roughly what percentage will the amount of highway increase?

7.8 The law of entropy: *why you can't break even*

The second law has been stated in two different ways. There is also a third way, which puts it in terms of nature's microscopic behavior.

One statement of the second law says that thermal energy flows naturally from hot to cold. For instance, if a box of hot gas and a box of cold gas are put into contact, the hot one will heat the cold one. The hot gas will cool and the cold gas will warm until they reach the same temperature.

What happens microscopically? Molecules of both the hot and the cold gas move around randomly, colliding with other molecules and with the walls of their box. The molecular motions are **disorganized**. Thermal energy is just this disorganized molecular motion.

Microscopically, the difference between the two boxes is that the hot box's molecules move around faster than do the cold box's molecules. Despite the disorganization, there is a kind of organization among the pair of boxes, in the fact that one box contains faster molecules and the other box contains slower ones. When the boxes are put into contact, the molecules of the hot gas slow down, and the molecules of the cold gas speed up. If the two gases are of the same type (O_2, for instance), they will reach the same temperature when the average speed of the molecules in the first box equals the average speed of those in the second box.*

Once the two boxes have reached the same temperature, we no longer have the nice separation of fast and slow molecules with which we started. Since fast molecules are no longer separated from slow molecules, *disorganization has increased*. When thermal energy flowed from hot to cold, microscopic disorganization increased.

This turns out to be the general situation, no matter whether the materials are gases or anything else. When thermal energy flows from hot to cold, microscopic disorganization always increases. In fact, this idea—that microscopic disorganization must increase—is equivalent to the fact that thermal

---

10. $T = 70/P = 70/8 \approx 9$ years. At this rate, production in 1995 alone (65 years after 1930—about seven doubling times) would be 128 billion barrels, nearly the entire amount of recoverable U.S. oil in the ground as of 1992.

11. For vehicle miles, $T = 70/P = 70/4 = 18$-year doubling time. For miles of highway, $T = 70/0.1 = 700$ yrs. During 18 years, the number of miles of highway will increase by about $(0.1\%/\text{yr}) \times (18 \text{ yrs}) = 1.8\%$.

* In other words, their energies become equal. If the two gases are of different types, it is their average molecular kinetic energies that become equal, rather than their speeds.

For some reason, the universe at one time had a very low entropy for its energy content, and since then the entropy has increased. So that is the way toward the future. That is the origin of all irreversibility, that is what makes the processes of growth and decay, that makes us remember the past and not the future, remember the things which are closer to that moment in the history of the universe when the order was higher than now, and why we are not able to remember things where the disorder is higher than now, which we call the future.

Richard Feynman, in *The Feynman Lectures on Physics*

energy flows from a high to a low temperature. In other words, this idea is another way of stating the second law.

Physicists have found a quantitative measure of the disorganization, at the microscopic level, of any system. It is called **entropy**. There is no particular reason for us to delve into the precise definition of entropy here. Suffice it to say that it can be specified by measurements of temperature and thermal energy.

So we have yet a third way of stating the second law:

> ## THE SECOND LAW OF THERMODYNAMICS, ENTROPY FORM (LAW OF ENTROPY)
> The total entropy of all the participants in any physical process must either increase or remain unchanged; it cannot decrease.

The law of entropy is similar to the law of energy. Both laws place restrictions on natural processes: The total energy of all the participants in any process must remain unchanged, and the total entropy of all the participants must not decrease.

But unlike the law of energy, the law of entropy has an "irreversible" quality. Processes must go in the direction of increasing, not decreasing, entropy. We have noted this quality of the second law before. Except for a very subtle effect at the subatomic level, the second law is the only principle of physics that distinguishes between a "forward" and a "backward" direction of time.* If it weren't for the second law, everything could just as well run backward: Water could run uphill by converting its molecular thermal energy to macroscopic kinetic and gravitational energy. Thermal energy could flow from cold to hot. And people could remember the future and grow younger instead of older.

The law of entropy helps us understand the inefficiency of heat engines. When a heat engine converts thermal energy to work, it turns disorganized energy into organized energy. But the law of entropy says that disorganization must increase, not decrease. That is why there must be an exhaust. There must be increased disorganization somewhere to compensate for the conversion of thermal energy to work. This increased disorganization occurs when a certain fraction of the energy flows from hot to cold.

The law of entropy suggests a deeper reason behind the second law. Increased disorganization is common in everyday life. For example, suppose that you have a partially organized deck of cards, perhaps with all the spades collected together. If you now shuffle the deck, you will almost certainly disorganize the deck further. It is possible, but improbable, that the shuffling will further organize the cards, perhaps by leaving the spades together and, luckily, collecting the hearts together also. But the deck of cards is much

Humpty Dumpty sat on a wall.
Humpty Dumpty had a great fall.
All the king's horses,
And all the king's men,
Couldn't put Humpty together again.

Mother Goose Rhymes

---

* The effect involves the "decay" of a particle known as the K-meson. Physicists James Cronin and Val Fitch discovered in 1964 that K-mesons distinguish between forward and backward in time. It is not known whether this subatomic physics discovery is in any way related to the second law or to our sense of a forward direction in time. One member of the 1980 committee that awarded Cronin and Fitch their Nobel Prize remarked, "It would take a new Einstein to say what it means."

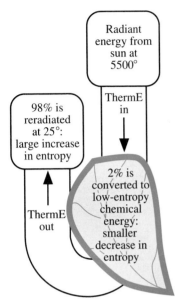

FIGURE 7.27
Energy flow through a leaf. The leaf is similar to a heat engine. A growing leaf illustrates how Earth has become more organized, despite the universe's trend toward increased entropy.

A living organism. . .feeds upon negative entropy. Thus the device by which an organism maintains itself at a fairly high level of orderliness (= fairly low level of entropy) really consists in continually sucking orderliness from its environment.
Erwin Schroedinger, physicist, in
What Is Life?

more likely to evolve toward a more disorganized state, simply because there are so many more ways to disorganize the deck than there are ways to organize it. As you know whenever you clean your house, it is easy to disorganize things, but it takes effort—or a lot of luck—to organize them.

There is an underlying "statistical" or "probabilistic" reason for the second law. Like cards, molecular systems are much more likely to evolve toward greater disorganization than toward greater organization, simply because there are many more ways for a molecular system to become disorganized than to become organized. It is highly likely (in fact, overwhelmingly likely for systems composed of very many molecules) that such systems will become more disorganized.

Biological systems obey the second law and all other laws of physics, and they provide interesting examples of the law of entropy. For example, a growing leaf manufactures complex glucose molecules out of simple CO_2 and H_2O molecules. Glucose is highly organized when compared with the randomly moving CO_2 and H_2O that goes into its formation. The leaf must create this organization. How does it manage to produce this decrease in entropy, in apparent violation of the second law?

The answer is that the leaf had help. The second law says that the total entropy of all the participants in any process cannot decrease. In the growth of a leaf, the other vital participant is the sun. Solar radiation has a temperature, the 5500°C surface temperature of the sun. When this radiation is absorbed by a leaf, only about 2% of the energy is converted to chemical energy. The remaining solar energy is reradiated out into space, at the 25°C temperature of the leaf. So most of the solar energy flows from 5500°C to 25°C, and the large entropy increase of this hot-to-cold thermal energy flow allows the remaining solar energy to be organized into low-entropy chemical energy (Figure 7.27). Solar radiation organizes the leaf despite (but not in violation of) the second law.

The sun works its powerful ways on Earth through both of the great laws, the law of energy and the law of entropy. Solar radiation is both the energizer and the organizer of all life on Earth.

Creationism, the belief that the various biological species (and especially humans) originated from specific acts of instantaneous creation rather than by the natural process of biological evolution, argues that biological evolution is impossible because it violates the second law. It is an idea that seems plausible at first glance. After all, it does seem paradoxical that life on Earth could have evolved on its own from simple single-celled structures into the much more organized plants and animals of today. It seems overly lucky, and it seems to violate the second law's demand that disorganization should always increase.

But like a leaf, biological evolution had help from the sun. The increase in entropy that occurs when sunlight is absorbed and reradiated by Earth is much more than adequate to compensate for the decrease that occurs when plants evolve into new organisms. And animals, which do not use solar energy directly, are organized by the highly organized food they eat. Biological evolution is fully consistent with the second law. Over billions of years, Earth has evolved toward greater organization at the cost of a much greater disorganization of the solar radiation that has passed through biological systems.

Until evolutionists can not only speculate, but demonstrate, that there does exist in nature some vast program to direct the growth toward higher complexity of the [biosphere]. . .the whole evolutionary idea is negated by the Second Law. . . .There seems to be no way of modifying the basic evolutionary model to accommodate this Second Law.

Henry M. Morris, director of the Institute for Creation Science, in *Scientific Creationism*, Creation-Life Publishers, San Diego, 1974

Your brain is one result of this long evolution toward greater organization. As an information-storage device, the human brain is the most highly organized form of matter on Earth.* It is surely the most organized form within some 4×10^{13} km from Earth—the distance to the nearest star beyond the sun—and perhaps within the entire Milky Way galaxy (see Chapter 12). In the human brain, nature has finally produced a self-aware organization of molecules, molecules so organized for information processing that they know they are a collection of molecules!

After billions of years of evolution, Earth has spent a long time getting to this point. So take good care of yourself, and of all of us.

---

* Carl Sagan makes this point in *The Dragons of Eden* (New York: Random House, 1977), in which he estimates the information content of the brain's neurons and compares it with computers, with other animal brains, and with the information content of genetic material.

Summary of Ideas and Terms

Second law of thermodynamics Thermal energy can flow spontaneously from a high to a low temperature. Also, any process that uses thermal energy to do work must have a thermal energy exhaust. And finally, the total entropy of all the participants in any physical process cannot decrease.

Temperature A quantitative measure of warmth. Unit: degrees Celsius.

Heat engine Any cyclic device that uses thermal energy to do work. Its **efficiency,** the fraction of its input thermal energy that is converted to work, will be higher if the input temperature is higher and the exhaust temperature is lower. The portion of the input energy that is not converted to work is the **exhaust**.

Irreversibility The second law says that physical systems proceed spontaneously toward states of higher entropy and do not proceed spontaneously in the reverse direction. This principle is responsible for the difference between forward and backward in time.

Internal combustion engine A heat engine in which burning occurs within the hot gases that push directly on mechanical parts, such as a **piston**, to provide useful work. Typically, its energy efficiency is low, largely because of the second law.

Transportation efficiency Useful output (such as distance traveled, passengers moved, freight moved, or mass moved) per unit of fuel input.

Steam-electric power plant Uses thermal energy from an external source, such as burning coal, to turn water into steam that pushes on a **steam turbine** that provides work to generate electricity. This is an example of an **external combustion engine**. Typically, its energy efficiency is less than 50%, mostly because of the second law.

Exponential growth Growth by a fixed percentage in each unit of time. It has a fixed **doubling time** related to its fixed percentage **growth rate** by $T \approx 70/P$.

Nonrenewable resource A natural resource that can be used up. Its use follows a curve that begins exponentially, levels off, and declines.

Renewable resource A natural resource that is continually replaced. Its use typically begins exponentially and then levels off at some sustainable level (provided that it is not overconsumed).

Entropy The amount of microscopic disorganization that a system possesses.

Microscopic interpretation of the second law An increase in entropy results from the fact that microscopic disorganization is overwhelmingly likely to increase.

Biology and the second law The entropy of a growing leaf decreases, at the expense of a great entropy increase of absorbed and reradiated solar energy. The same is true of an evolving species. Growth and biological evolution are organized by solar radiation and fully obey the second law.

Review Questions

HEATING

1. What is heating?
2. In your own words, state the heating form of the second law.
3. Give an example showing that thermal energy and temperature are really two different things.

HEAT ENGINES AND ENERGY QUALITY

4. Give an example of each of these energy transformations: kinetic to thermal, gravitational to thermal, thermal to kinetic.
5. In your own words, state the heat engine form of the second law.
6. What properties of the input and the exhaust are the most important to determining a heat engine's efficiency? Describe the manner in which the efficiency depends on these properties.
7. Is the actual overall energy efficiency of an automobile closest to 98%, 90%, 40%, 10%, or 2%? What about a steam-electric power plant?

8. If a heat engine operated entirely without friction, would it then be 100% efficient? Explain.
9. In terms of energy, what happens when the motion of a rock swinging from a string "dies down"? In what sense is this behavior irreversible?

THE AUTOMOBILE AND TRANSPORTATION

10. Name two general types of heat engines of major social importance.
11. Why do we call it an "internal combustion" engine? Describe in your own words how it works.
12. Name two alternative fuels (not gasoline or diesel fuel) for the automobile.
13. What is the source of the largest inefficiency in an automobile's operation?
14. Describe two different ways of measuring a transportation mode's efficiency, and give an appropriate measurement unit for each.

THE STEAM-ELECTRIC POWER PLANT

15. In a steam-electric power plant, what is the purpose of the turbine? Generator? Condenser? Cooling tower? Stack?
16. What is thermal pollution?
17. What is the source of the most important inefficiency in a steam-electric power plant's operation?
18. What part of a steam-electric power plant is analogous to the piston in an automobile?

EXPONENTIAL GROWTH

19. When is the difference between exponential and linear growth?
20. Your savings account grows at 7% per year. What is its doubling time?

21. Draw a typical life-history graph for a nonrenewable resource. Is any part approximately exponential?
22. Repeat the previous question, but for a renewable resource.

THE LAW OF ENTROPY

23. What is entropy?
24. In your own words, state the entropy form of the second law.
25. Which law or laws of physics distinguish between forward and backward in time?
26. When a growing leaf increases its organization, does it violate the second law? Explain.

Home Projects

1. Line up three containers filled, respectively, with hot (not scalding) water, warm water, and ice water. For a few moments, put one finger into the hot water and one finger into the cold water. Now plunge both fingers into the warm water. Can you always trust your sense of hot and cold?
2. Technology assessment: the automobile. How much of your neighborhood is devoted to the automobile? Choose a single city block and measure the area devoted to the automobile, using a tape measure or pacing. Include garages, parking lots, gas stations, car dealers, and the like. Include only half the total area of streets bordering your block; the other half belongs to the adjoining block. Find the fractional area devoted to the car; it is over 60% in most U.S. cities. Your class could divide into groups and measure several blocks and then pool its results for an overall figure.
3. Study your home electric bill. How many kW·h of electric energy were used during the month? Use this figure to calculate your home's average power consumption in kilowatts. Use Table 6.1 to estimate the electric energy, in kW·h, that your home appliances consume during one month, and compare your estimates with the total shown on the bill. Compare this with your classmates' results.

4. Is electricity a good buy? Look at your electric bill to see how much you pay for 1 kW·h. Calculate the height to which 1 kW·h of energy could lift you. *Hint*: Recall that GravE = weight × height; find your weight in newtons from the fact that 1 pound = 4.5 newtons and 1 kilogram weighs about 10 newtons.
5. The growth game. Put 100 pennies into a box. Bring a penny into "penny world" by taking one penny out of the box and putting it on a table. Flip it. "Heads" means it had a baby that year, so you should bring another penny into penny world, and "tails" means it didn't have a baby. Now the penny world's population is 1 or 2. Flip all the pennies in penny world, getting 0 or 1 or 2 new babies during the second "year." Continue in this fashion, and draw a graph showing the population of the penny world versus the number of years beginning at 0 years with a population of 1. If 100 represents overpopulation, how many years will it take until overpopulation is reached? Play the game several times. Is it the same number of years each time? This game models population growth. Is this growth approximately exponential? From your graph, measure the doubling time and yearly growth rate.

For Discussion

1. Technology assessment of transportation. (a) List several transportation needs of your region, such as campus to shopping, suburbs to downtown, campus to evening entertainment, travel to nearest large city. List all the possible ways to fill each need, assuming that the proper facilities (roads, bikeways, sidewalks, and so on) existed. Include city-planning options such as relocation of workplaces and reduction in the distance to outlying suburbs. Underline the modes that actually do fill each need.
2. Discuss the pros and cons of the different modes listed for each need. Include energy consumption as one consideration. Other considerations might include conve-

nience, pollution, safety, cost, traffic congestion, community quality of life, and so forth.

3. List ten significant inventions or technologies, including the automobile and electric power. Which two or three of these seem to have the greatest impact on our way of life?

4. List several quantities whose growth is exponential. Which ones are likely to continue growing exponentially, which are likely to follow a bell-shaped growth-and-decline curve, and which are likely to level off at a sustainable level?

Exercises

HEATING

1. What is your approximate body temperature, in °C (98.6°F is normal)?

2. How does the flow of thermal energy through a closed window illustrate the second law? Which direction is this flow when it is cold outside? Hot outside?

3. In the operation of a refrigerator, does thermal energy flow from hot to cold, or is it from cold to hot? Does this happen spontaneously, or is outside assistance required?

HEAT ENGINES AND ENERGY QUALITY

4. Which are not heat engines: hydroelectric generating plant, steam-electric power plant, gasoline-fueled automobile, steam locomotive, bicycle, solar-thermal electric power plant?

5. What does the second law tell us about the efficiency of heat engines?

6. When we say that the motion of a rock swinging on a string is irreversible, do we really mean that it is impossible to get the rock back to its starting condition? Explain.

7. Can you think of any way to drive a ship across the ocean by using the ocean's thermal energy, without violating the second law?

8. Farswell Slick approaches you with plans for a revolutionary transportation system. He has noticed that when he drives an automobile without accelerating, all the input energy eventually shows up as thermal energy. Slick proposes to use this thermal energy to drive the car at a constant speed. The car will still need fuel, but only for accelerating. It will be possible to travel cross-country on only a few gallons of gasoline. He describes his scheme as a "computerized advanced-technology exhaust feedback afterburner." Should you invest in Slick's scheme? Explain.

9.*In one cycle of its operation, a heat engine does 100 J of work while exhausting 400 J of thermal energy. What is its energy input? Its efficiency?

10.*In one cycle of its operation, a heat engine consumes 1500 J of thermal energy while performing 300 J of work. What is its efficiency? How much energy is exhausted in each cycle?

THE AUTOMOBILE AND TRANSPORTATION

11. Suppose an automobile's fuel could be made to burn hotter without harming the engine's operation (for instance, without cracking the engine). Would you still get the same amount of useful work from each gallon of gasoline?

12. Suppose an automobile could run on hard wheels that were not squeezed by the weight of the car on the road. Would this alter the car's efficiency? How might this affect the gas mileage? What kind of wheels and road might you suggest?

13. According to Figures 7.7, 7.8, and 7.9, which of the three main sectors of the U.S. economy (industry, residential–commercial, transportation) consumes the most oil?

14. One car has twice the overall energy efficiency (in km/l) of a second car. Compare the amounts of pollution they produce when they both travel the same distance.

15.*Out of every 100 barrels of gasoline, about how many actually go into driving a car down the road?

16.*A bus carries 30 people 200 kilometers using 300 liters of gasoline. Find its passenger-moving efficiency.

THE STEAM-ELECTRIC POWER PLANT

17. Which type of generating plant would you expect to be more energy efficient, steam-electric or hydroelectric? Defend your answer.

18. Would it be more energy efficient to heat your home electrically or to heat it directly using a natural gas heater, assuming that the electricity comes from a steam-electric plant?

19. Which method of fueling your car is likely to be more energy efficient, and why: gasoline used in a standard car engine or electricity taken from a coal-fueled generating plant and stored in lightweight car batteries? Assume that the batteries convert electricity to work at 100% efficiency.

20.*Out of every 100 tons of coal fed into an electric generating plant, roughly how many tons produce the electricity you can use at your home, and how many go into waste energy? Use the energy flows given in Figure 7.19.

21.*For every 100 kilograms of coal entering a generating plant (recall that this much enters every second), about 15 kilograms of sulfur oxides and ash are removed, producing a significant solid-waste disposal problem. For a typical 1000 MW plant, how much of this solid waste is produced every day? Express your answer in tonnes (1 tonne = 1000 kg).

22.*How would the pollution from two coal plants compare if the first plant were twice as energy efficient as the second? Assume that they both produce the same amount of electric power.

23.*Making estimates. In the United States, solar energy strikes a single square meter of ground at an average rate (averaged over day and night and over the different seasons) of 200 watts. At what average rate does solar energy strike a football field (about 100 m by 30 m)? A typical home in the United States consumes electricity at an average rate of 1 kW. How much surface area would be needed to provide this electric power, assuming a 10% conversion efficiency? What dimensions would a square-shaped photovoltaic collector need to cover this area?

EXPONENTIAL GROWTH

24. A lily pond doubles its number of lilies every month. One day, you notice that 2% of the pond is covered by lilies. About how long will it be before the pond is entirely covered?

25. Company X increases its profits every year by $50 million. Is its growth in profits exponential? Company Y increases its profits by 1% every year. Is its growth in profits exponential?

26. Which of the following are renewable energy resources: coal, firewood, solar radiation, wind, water behind a dam, oil? Which can be traced back to the sun?

27. According to Figure 7.23, did electric power production grow exponentially between 1910 and 1935? Estimate the number of kW·h produced in 1935, 1945, 1955, 1965, and 1975, and verify that production grew approximately exponentially between 1935 and 1975.

28.*How much electric energy would have been produced in 1985 if exponential growth had continued for another 10 years beyond 1975? If this growth had continued, roughly how many more power plants would have been needed in 1985, as compared with 1975?

29.*Between 1985 and 1990, annual U.S. population growth was 0.8% per year; for Mexico it was 2.2 %; and for Kenya (the highest) it was 4.2%. At these rates, how long will it take for the populations of each of these countries to double?

30.*World population is now about 6 billion. The growth rate has been roughly 2% per year since the end of World War II (1945). If a 2% per year growth rate continues, when will the world's population be 12 billion?

THE LAW OF ENTROPY

31. A pan of liquid water freezes when you place it outside on a cold day. Liquid water has greater molecular disorder than ice does. Is the freezing process an exception to the law of entropy? Explain.

32. When orange juice and grapefruit juice are mixed, does entropy increase?

8

LIGHT AND ELECTROMAGNETISM

Light is all around us, yet it is not easy to say what light is. How are you able to see this page? Are invisible rays emitted by your eyes that move from your eyes toward the page, as the ancient Greek thinkers Plato and Euclid thought? Or does the page send out or reflect a stream of particles that is received by your eyes, as the Pythagoreans (Section 1.3) and Isaac Newton thought? Or is light, as the Greek thinker Empedocles and Newton's contemporary Christian Huygens thought, a high-speed wave?

What is light? It is one of science's oldest questions, and it is the focus of this and the following chapter. Pursuit of this question led to both of the great post-Newtonian theories, relativity theory and quantum theory. Light and its near relatives are also important to understanding societal topics like solar energy, ozone depletion, and global warming.

This chapter presents two topics that are essential to understanding the nature of light: waves (Sections 8.1 through 8.3) and the electromagnetic force (Sections 8.4 and 8.5). Sections 8.6 and 8.7 discuss the planetary atom, a model that incorporates electromagnetic forces and helps explain light. Chapter 9 then presents the electromagnetic wave theory of light and explores the human impacts of these ideas.

8.1 Waves: *something else that travels*

You are probably familiar with some kinds of waves (Figure 8.1). Stretch a few meters of flexible rope along the floor, fix one end (perhaps under a friend's foot), and give the free end a single shake. Something travels down the rope. Figure 8.2 is a series of pictures taken with a movie camera, showing a similar

FIGURE 8.1
Water waves.

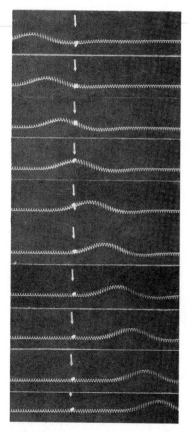

FIGURE 8.2
A series of pictures taken with a movie camera, showing a wave moving along a spring. A ribbon is tied to the spring at the point marked by the arrow. The ribbon moves up and down as the wave goes by but does not move in the direction of the wave.

"something" traveling down a long spring that has been given a single up-and-down shake at the right-hand end. As another example, imagine (better yet, try it!) stretching a Slinky toy between your two hands along a tabletop. Quickly move the left end a short distance toward the right and then back to the left, holding the right end fixed. Something travels along the Slinky from your left to your right hand (Figure 8.3).

This "something" that travels across the water, along the rope, and along the Slinky, is called a **wave**. As another example, the continued shaking of one end of a rope causes a long, continuous wave to travel along the rope (Figure 8.4). But *wave* is just a word that names the behavior without telling us what it really is. What actually happens here?

Observe the motion carefully. In Figure 8.2, how do the individual parts of the spring actually move? As you can see from the motion of the small ribbon tied to the spring, each loop just moves up and then back down.

How does a particular part of the water move in Figure 8.1? Fill a bowl with water, float a small cork in it, and drop a small pebble in, several centimeters from the cork, to create ripples. Observe the cork as ripples pass by. If the ripples are small, the cork will move up and down, not outward along with the ripples. Each portion of the water just moves up and down, shaking or **vibrating** up and down, but they do not travel along the water surface. The Slinky wave is similar, except that the vibrations are parallel instead of perpendicular to the Slinky.

One thing that is traveling along with each of these waves is energy. You can verify this for yourself by holding the fixed end of a rope while a friend shakes the other end. Your hand vibrates as the pulse arrives. It takes work to force your hand back and forth this way, and we know that work requires energy. So waves transfer energy.

On the other hand, no material substance is transferred by waves: No water is transferred outward in Figure 8.1; no part of the spring is transferred from left to right in Figure 8.2; and no part of the Slinky is transferred from the left to the right hand in Figure 8.3. This type of motion is unlike any motion we have examined before. Previously we studied balls, books, molecules, and other material objects actually moving from one place to another. We call such cases **projectile motion**, because a material object is actually projected (thrown or transferred) from one place to another.

In discussing waves, it is important to distinguish between the small vibrational motions of the substance through which the wave travels and the nonvibrational motion of the wave itself.

What do we see traveling down the spring in Figure 8.2? Very simply, we see a bump in the otherwise straight spring. In Figure 8.3, we see a compression, a squeezed region (followed by a stretched region), traveling down the otherwise evenly stretched Slinky. We could describe both as *disturbances* that travel down the otherwise undisturbed rope or Slinky. The situation is similar for water waves. The material through which the disturbance travels—the spring or Slinky or water—is called the **medium** for the wave.

So we have a useful definition: A **wave** is a disturbance that travels through a medium, and that transfers energy without transferring matter. Such **wave motion** is different from projectile motion.

FIGURE 8.3
With the right-hand end of the Slinky held fixed, a quick motion of the left-hand end to the right and back again to the left creates a pulse that travels down the Slinky.

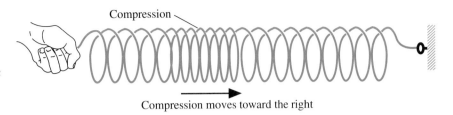

Compression

Compression moves toward the right

FIGURE 8.4
Continued shaking of the end of a rope creates a continuous wave that travels down the rope.

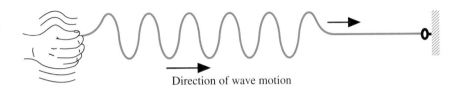

Direction of wave motion

FIGURE 8.5
The meaning of wavelength and amplitude. The wave speed is the speed at which a crest or a trough moves down the rope.

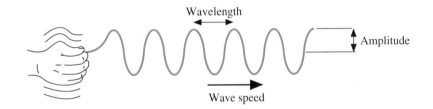

Wavelength

Amplitude

Wave speed

We need some quantitative terms. The **wavelength** of a continuous, repeated wave is the distance from any point along the wave to the next similar point, for example, from crest to crest or from trough to trough in Figure 8.5.

A wave's **frequency** is the number of vibrations that any particular part of the medium completes in each second. Waves are usually sent out by a vibrating source of some kind, in which case the wave's frequency must be the same as the source's frequency. The frequency could also be defined as the number of waves that the source sends out during each second. The unit for measuring frequency is the "vibration per second," also called a **hertz** (Hz).

A wave's **amplitude** refers to its width of vibration, the distance that each part moves back and forth during that part's vibrations. The amplitude is measured quantitively from the midpoint of vibration to the farthest point, so it is just half of the overall vibration width.

The **wavespeed** of a wave is the speed at which the disturbance moves through the medium, for example, the speed at which a compression moves along a Slinky or the speed at which a crest moves along the rope in Figure 8.5.

Disturbances are able to travel through a medium because of the connections between the parts of the medium. For example, when you shake one end of a rope, this disturbance is transferred down the rope because the different parts of the rope are connected, so that when one part is lifted, its neighbor soon is lifted also. So it is reasonable to suppose that the wavespeed is determined mainly by the medium and is roughly the same for different disturbances within the same medium. Experiments confirm this notion that all disturbances travel through any particular medium at roughly the same wavespeed.

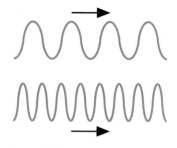

FIGURE 8.6
Which wave has the higher (larger) frequency, assuming that both have the same wavespeed?

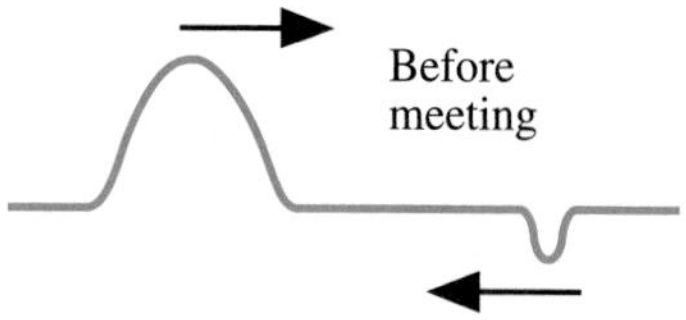

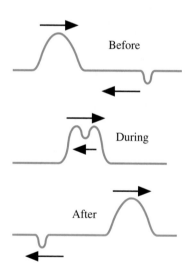

FIGURE 8.7
Two waves travel in opposite directions along a rope. What happens when they meet?

Before

During

After

FIGURE 8.8
Two waves meeting: interference.

DIALOGUE 1   Surfers are able to ride large water waves coming into a beach. Are these really waves as we defined them? What about a row of falling dominoes, in which each domino knocks over its neighbor as it falls—is this a wave? If either answer is yes, then what is the medium?

DIALOGUE 2   Which wave has the larger or "higher" frequency: a long-wavelength wave moving along a rope or a short-wavelength wave moving along the same rope (Figure 8.6), assuming that both have the same wavespeed? Which carries more energy, assuming their amplitudes are the same? Note these useful rules: Short wavelength means high frequency,* and high frequency means high energy.

8.2 Interference: *a behavior unique to waves*

How do different waves in the same medium interact with one another? For example, what happens when the large upward wave and the small downward wave shown in Figure 8.7 meet? If we do an experiment like this, we will find that the two waves just pass through each other without distortion (Figure 8.8). This is what we might have expected, because each wave just lifts or lowers the rope as it travels along the rope, so when the two waves meet, the rope is raised a lot by the large wave at the same time that it is lowered a little by the small wave. Experiments like this show that when different waves move through the same medium, they pass through each other without disturbing each other and continue on their way as though nothing had happened.

The effects that occur when two waves are present at the same time and place are called **wave interference** (or just interference) effects.

DIALOGUE 3   What would be the interference effect if the two waves of Figure 8.7 had identical heights, one upward and one downward? What if both were upward?

Two equal waves of opposite orientation interfere by canceling each other (Figure 8.9). Two equal waves of the same orientation interfere by adding up to a disturbance that is twice the size of either wave (Figure 8.10). These two cases, cancellation and reinforcement, are called **destructive interference** and **constructive interference**.

---

1. Breaking waves carry water into the beach (and then back out again as an undercurrent). This is projectile motion, not wave motion. A row of falling dominoes is a true wave, because the dominoes don't go anywhere. The medium is the row of dominoes.
2. The short-wavelength wave had to be sent out more frequently; the rope must vibrate more frequently as this wave passes through it; and the short-wavelength wave must carry the most energy. For instance, it could vibrate any "receiver" twice as often, which would take more work.
* Quantitatively, a wave's wavelength λ, frequency f, and wavespeed v are related by $v = f\lambda$. For example, if three waves are sent out by the wave source every second ($f = 3$ vib/s) and each wave has a length of 2 meters ($\lambda = 2$ m), then the speed at which the wave moves is $3 \times 2 = 6$ m/s.
3. See Figures 8.9 and 8.10.

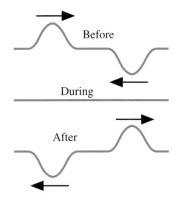

FIGURE 8.9
Two waves meet and interfere destructively.

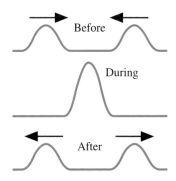

FIGURE 8.10
Two waves meet and interfere constructively.

Wave interference shows, once again, the stark difference between projectile motion and wave motion. Two moving material objects—say two freight trains moving toward each other on the same track—do not interfere in the way that waves do. Two trains certainly don't pass through each other undisturbed!

Interference effects become more interesting when they happen in two or three dimensions. An undisturbed rope or a Slinky has only one significant "dimension," length. The surface of a lake is "two dimensional" because it has length and width. And the space in a room is "three dimensional" because it has length, width, and height.

In a two- or three-dimensional medium, the waves created by a small source spread out into circles or spheres. If you send out a continuous water wave from a small vibrating source such as your finger tapping on the water surface, waves will spread out into circles, as in Figure 8.1, and eventually cover the entire surface.

Suppose you fill a rectangular pan with water and tap your fingers at the same steady rate on the surface at two points along one side of the pan. Continuous waves will soon cover the water surface, spreading out from each of the two sources. What will the interference effects look like? The wave crests spreading out from each individual source form circles (Figure 8.1), as do the wave troughs (the low points). The two waves from the two sources have the same frequencies because the tapping rates are identical, and so the two wavelengths also must be identical. In Figure 8.11, these waves are drawn looking down from above. The two sources are marked A and B. The colored circles represent the crests from source A acting alone, and the black circles represent the crests of the waves from source B. The troughs, not drawn, lie midway between the crests.

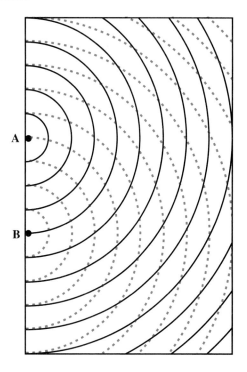

FIGURE 8.11
Continuous surface waves spreading out from two sources. What will the interference effects look like?

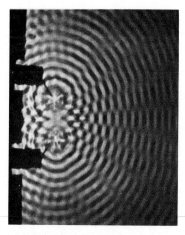

FIGURE 8.12

Interference between continuous surface waves spreading out from two sources: experimental results.

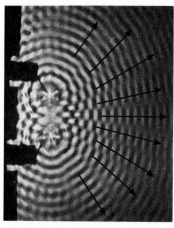

FIGURE 8.13

Lines of constructive and destructive interference remain fixed in place, and constructive (large) waves move outward within the constructive regions in the directions indicated by the arrows.

DIALOGUE 4   But the sources do not act alone. Predict the appearance of the water's surface as it would look in a snapshot. Draw an x (a color enhances the effect) at every point of constructive interference, in other words, at points where crest meet crest and at points where trough meets trough. Can you see a pattern? Next, draw a small o (in a different color) at every point of destructive interference. Now can you see a pattern?

Figure 8.12 is a photograph of this experiment, looking down onto the water's surface. The photographic technique causes crests to appear bright and troughs to appear dark. As you can see, the interference pattern has lines of undisturbed water radiating outward as though they came from a point somewhere between the two sources. The interference is destructive along these lines. Between these undisturbed lines are other lines of constructive interference, with large crests and troughs. This is just what Dialogue 4 predicted.

All of our analysis so far has been at one instant in time. Now "turn on time" by imagining a moving picture that begins with the snapshot in Figure 8.12. Since the individual circular waves move outward from A and B, the entire interference pattern must move outward also. The rays of destructive and constructive interference remain fixed in place, and the large waves within the constructive rays move outward, as shown by the arrows in Figure 8.13.

Finally, think about how all this would appear to an observer who could only see these waves roll in to the far right border of Figure 8.13. Imagine that the body of water is a rectangular swimming pool. How does water-wave interference appear to an observer who examines the waves arriving along the right-hand wall but who for some reason cannot see any appreciable portion of the pool's surface?

We can predict the answer using Figure 8.13. The observer should find some points where large waves pound against the wall and other points where no waves roll in. These are the points of constructive and destructive interference along the wall, as diagrammed in Figure 8.14. An observer examining waves arriving at the right-hand wall would find alternating points (marked **X**) where large waves roll in and points (**O**) where no waves roll in.

The difference here between the pattern from a single source of water waves and the pattern from two sources is especially striking. Waves from a single source, say source A, spread out in circles that cover the entire surface. These waves roll into all parts of the bordering wall (Figure 8.15). If we now also turn on the second source, B, the pattern along the wall will shift to an interference pattern. The most dramatic change is that now no waves come into the points marked **O**, even though they did come into these points when only one source was operating. It seems paradoxical: We added a second source and got a reduced (in fact, zero) effect at the points marked **O**. Interference effects such as these give us a good way to identify wave behavior.

Now let's look at light.

---

4. See the small x's and o's in Figure 8.14.

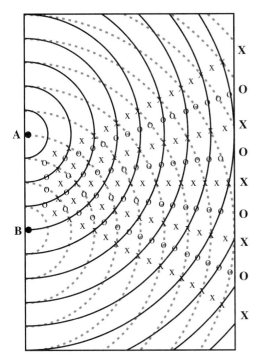

Observer stands here and looks down at pool, observing large waves rolling into points marked **X** at side of pool, and no waves rolling into points marked **O**. Small x's and o's are places on surface of pool where interference is constructive (x's) and destructive (o's).

FIGURE 8.14
An observer scanning a wall at the far border of the pool finds points where large waves come into the wall, interspersed with points where no waves come in.

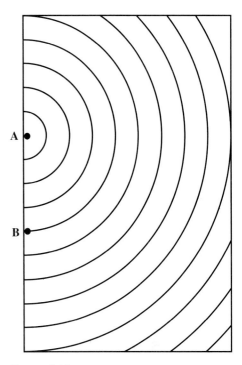

FIGURE 8.15
If one of the two wave sources shuts off, an observer scanning the wall will find that waves arrive at all points. Since waves now spread out from only a single source, there is no longer any interference.

8.3 Light: *particles or wave?*

When you turn off the light at night, it gets dark. So the light in your room must have come from the lightbulb. It couldn't, for example, have come from your eyes, because even though your eyes were still open after you turned off the lightbulb, you could not see. So light is something that enters your eyes from the outside. When you look at a luminous (light-emitting) object like a lightbulb, light goes from the bulb to your eyes. In order for you to see a nonluminous object such as the wall of your room, light from the light-bulb must bounce (*reflect* is the official word) off the wall and then into your eyes. The light reflecting from the wall does not give you a nice mirror reflection, however, as light does when it reflects from a glass mirror, because the rough surface of most walls scatters the incoming light in many different directions.

But what enters your eyes when you see light? The question has been debated for centuries. Most of the suggested answers fall into one of two different categories: particles and waves. Perhaps luminous objects send out tiny particles that enter our eyes. Or perhaps light is a wave emitted by some kind of vibrations within the luminous object. Experiment is the ultimate judge.

We need an experimental test that distinguishes between the projectile and wave models of light. The preceding section of this chapter suggests a good candidate: wave interference. When waves meet, they exhibit wave interference. Particles might interact in other ways, but they do not interfere in the way that waves do. What does light do? To answer this, we need an experiment like the water-wave interference experiment, but with light.

The first experiment you might think of is just to shine two flashlights on a movie screen. This gives no observable interference effects, as you can see by trying it yourself. So maybe light is a stream of particles.

But recall that in order to observe water-wave interference, we assumed that the two wave sources had identical and synchronized vibrations. For example, if both sources A and B in Figure 8.12 were changing their frequency all the time and in different ways, we would not expect to see any recognizable interference pattern. It is possible that light is a wave but that the waves from the two flashlights are not synchronized. A flashlight bulb's light is produced by heating up the bulb's thin wire, or "filament," until it glows. The microscopic thermal motions that make the filament hot enough to glow are mixed up and random, so we would not expect two different bulbs to have synchronized vibrations.

HOW DO WE KNOW? THE DOUBLE-SLIT EXPERIMENT WITH LIGHT.

In 1801, Thomas Young solved the problem of finding two synchronized light sources. Young's trick was to use a single light source but to split its light into two parts that should then have identical vibrations. He then recombined these parts to see whether they interfered.

FIGURE 8.16
The double-slit experiment with light. What will we see on the screen?

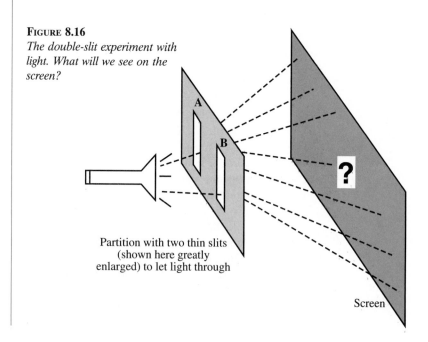

Partition with two thin slits (shown here greatly enlarged) to let light through

Screen

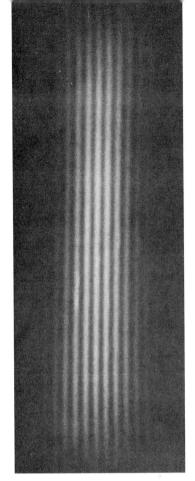

FIGURE 8.17
The double-slit experiment with light: experimental results.

Figure 8.16 shows one way to do this. A light source sends light to two very narrow parallel slits in a partition that blocks all the light except that going through the slits. The two slits then act as the two sources of light, like sources A and B in Figure 8.12. With this arrangement, if light is a wave, these two sources should have synchronized vibrations, because the light from each slit originated in the same filament.

Using this arrangement, Young found an experimental result like that shown in Figure 8.17. This photograph was made by placing photographic film at the position of the receiving screen in Figure 8.16. Figure 8.18 depicts the experimental arrangement and its outcome. As we will soon see, this outcome is excellent evidence that light is a wave.

In order to interpret Figure 8.17, let us return to water-wave interference. Figure 8.14 shows the interference pattern observed along a wall placed in the path of water waves from two synchronized sources. The receiving screen of Figure 8.16 is just like this wall: The screen is placed in the path of the light from the two sources. But water waves occur on the two-dimensional surface of water, whereas light fills up three-dimensional space. The two sources of light are not tiny points like A and B in Figure 8.14 but instead are slits that extend into the third dimension. If these sources send out light waves, we would expect the interference pattern seen on the receiving screen to be alternating lines of constructive and destructive interference running parallel to the slits, not small points like the points marked **X** and **O** along the right-hand border in Figure 8.14. In other words, we would expect alternating bright (lit) and dark lines. We would, in fact, expect precisely the outcome seen in Figure 8.17.

What would we expect if we closed one of the slits, leaving only one slit open? If light is a wave, we would expect that waves would spread out from

FIGURE 8.18
The double-slit experiment with light: the experimental setup and results.

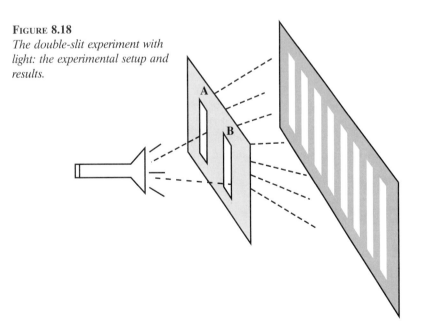

FIGURE 8.19

If one of the two slits is closed, light will arrive at all points along the receiving screen. Compare Figures 8.18 and 8.19 with Figures 8.14 and 8.15.

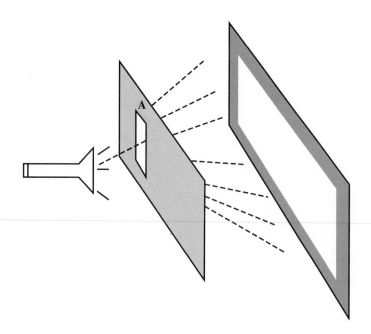

the open slit, without interference, just as water waves spread out from a single source without interference in Figure 8.15. A broad band of light should then cover a large area of the receiving screen (Figure 8.19). This is, in fact, what happens.

The clearest evidence that the flashlight is sending out waves and not particles can be found at the positions of the dark lines in the double-slit experiment, the points where no light arrives. With only slit A open, light spreads out over the entire receiving screen (Figure 8.19). How is it, then, that if we simply open slit B, no light will arrive at these particular positions? It is difficult to see how particles coming through the two slits could cancel one another out in this way, but it is just what we expect of waves. We conclude that light is a wave.

By measuring the distance from one bright constructive-interference line to the next such line in a pattern such as Figure 8.17 and using a little geometry, it is possible to calculate the wavelength of the light that created the pattern. Measurements of interference patterns like this are the usual method of measuring the wavelength of light. This wavelength turns out to be very small. Light sources have wavelengths ranging from about 0.4×10^{-6} m to 0.7×10^{-6} m (0.4 to 0.7 millionths of a meter).

HOW DO WE KNOW?

You can demonstrate light-wave interference yourself, using a single-slit wave-interference effect that occurs when the slit is much wider than the wavelength of the light. In order to get the spread-out, noninterfering result shown in Figure 8.19, the single slit must be very narrow—comparable in width to the wavelength of the light being used. In the following demonstration, we use a slit whose width is hundreds of times larger than the wavelength of the light. With such a wide slit, the light coming through the slit does not behave as though it came from a single tiny source. Instead, the slit acts like hundreds of tiny sources. If light is a wave, then all the individual waves from these hundreds of sources should interfere with one another to form an interference pattern.

Here's the experiment: Focus your eyes on a well-lit wall or other surface. Make a slit by holding your thumb and forefinger close together but not touching, about a millimeter apart and several centimeters in front of your eye. Focus on the source, and not on your fingers, because you want to learn about the light from this source. Your fingers should look blurred, and where the blurs overlap, you should be able to see narrow bright and dark lines running parallel to your fingers.

These light and dark lines are constructive and destructive interference regions, formed at the position of your eye, created by the light coming through the single slit. A half-millimeter-wide slit between your thumb and forefinger is about 1000 times larger than the wavelength of light, so as explained earlier, the observed pattern is just what we would expect if light is a wave.

Since light is a wave, what kind of wave is it? What is it a wave in? Water waves, rope waves, and Slinky waves are waves in water, ropes, and Slinkies. What medium vibrates when light waves travel? It's not an easy question. We don't see light beams directly the way we see water waves (Figure 8.20). It is as though we could see the impact of water waves against a bordering wall, but without being able to see the water. We can see light beams in dusty air (Figure 8.21), but only because the light is reflected off the dust particles. The medium for light waves is itself invisible.

Could the medium be air? This sounds plausible because you cannot see air, and air does fill up space, at least near Earth. But what about light traveling far from Earth? In any appreciable amounts, air extends only a few miles above Earth's surface, yet light arrives here from distant stars, across great reaches of nearly empty space. So air cannot be the medium for light waves.*

The odd thing about light is that it moves through empty space where there is essentially no matter at all. But something must be out there, in so-called empty space, because you can't have a wave without having a medium to do the waving. After all, we began our study of waves with the idea that a wave is a disturbance in a medium. You can't have a disturbance without having something to disturb. The medium for light, then, must be nonmaterial, not made of atoms or other forms of matter.

Nineteenth-century scientists devoted a lot of time and effort to learning what kind of wave light is. It turned out that the answer is bound up with a new phenomenon that was beginning to be understood during that century: electricity. The answer, briefly, to the question about light waves is that light is an "electromagnetic wave." In order to understand this important idea, we will devote the rest of this chapter to electricity. Then we will see, in the following chapter, what electricity has to do with light.

FIGURE 8.20
Light beams cannot be seen from the side. What invisible medium is carrying the light waves?

FIGURE 8.21
You can see a light beam by allowing it to reflect off dust particles in the air.

---

* Air is the medium for sound waves, not light waves. Since sound does not bear directly on the major purposes of this book, we won't discuss it further here.

DIALOGUE 5   If we compared two water-wave interference experiments, one using a short wavelength and the other using a long wavelength, how would the interference patterns observed along the right-hand wall in Figure 8.14 differ? How would the use of longer-wavelength light waves affect the interference pattern in Figure 8.17?

DIALOGUE 6 MAKING ESTIMATES   Roughly, how does a typical light wavelength compare with the thickness of a piece of paper?

8.4 Electric force: *part of the electromagnetic force*

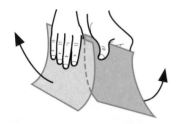

FIGURE 8.22
Two rubbed transparencies exert electric forces on each other.

FIGURE 8.23
An extremely bad hair day: electric hair. The source of this effect is the "charged" metal sphere that the woman is touching.

Find a couple of plastic transparencies, the kind that teachers use on overhead projectors. Rub them vigorously with tissue paper and hold them from one edge in separate hands (Figure 8.22). If you hold them parallel and just a few centimeters apart but not touching, they should repel each other. At short separations, they repel each other strongly enough to stand noticeably apart. The force weakens at larger separations. Now spread out the tissue on a level surface, and hold one transparency directly above it. The tissue is pulled upward and clings to the transparency. You may experience similar forces when you take a synthetic shirt from a clothes dryer or when you brush your hair. Figure 8.23 shows an extreme example.

What is the source of these forces? This force can act across a distance such as that between the transparencies. The only force acting at a distance that we have encountered so far in this book is gravity. This force has other similarities with gravity. For example, the force acting between the two transparencies weakens as the separation widens, just as gravity does.*

Can this force actually be gravity? There are several reasons that it cannot be. First, the transparencies repel each other, whereas gravity attracts. Second, this new force is far stronger than gravity could possibly be between relatively low-mass objects such as transparencies and tissues. The force by the transparency on the tissue is so strong that it easily lifts the tissue against the downward gravitational pull on the tissue by the entire Earth. Third, the existence and size of this force depend on whether the transparencies are rubbed, and it is hard to see how rubbing could affect the gravitational force.

---

5. Longer-wavelength water waves would create a more spread-out pattern, with more distance between the **X**'s and **O**'s in Figure 8.14. Longer-wavelength light would produce broader, more spread-out lines in Figure 8.17.
6. Choose a typical wavelength of light, say 5×10^{-7} m (in making estimates, choose simple but reasonable numbers). To estimate the thickness of a sheet of paper, estimate the thickness of, say, 500 sheets (about 5 cm) and divide by 500 (5 cm / 500 = .01 cm = 10^{-4} m). The number of wavelengths in this thickness is $10^{-4}/5 \times 10^{-7} = 10^{-4+7}/5 = 10^3/5 = 1000/5 = 200$.
* In fact, this force even turns out to be inversely proportional to the square of the separation distance between the two objects experiencing this force, precisely like gravity. This similarity between gravity and this new type of force has always fascinated physicists. Nature seems to favor "inverse square" force laws.

This force has properties that are different from those considered so far in this book. We call it the **electric force**. An "electrified" object such as the transparency or the tissue (after rubbing) is said to be **electrically charged**, or "charged," and any process that produces this state is called **charging**.

The experiments show that when we electrically charge two identical objects in identical ways, they repel each other. But the charged transparency and the charged tissue attract each other. So the transparency must be charged differently from the tissue. After experimenting with all sorts of other charged objects, one finds that every charged object falls into just two categories: those that repel the transparency but attract the tissue, and those that attract the transparency but repel the tissue.

We name these two categories of charged objects **positive** and **negative**. Do not attach much significance to the names—we could just as well call them red and blue, or charming and revolting.

Our experiments demonstrate:

> THE ELECTRIC FORCE LAW
> Electrically charged objects exert forces on each other at a distance. Objects may be charged in either of two ways, known as positive and negative. Two objects possessing like charges repel each other, and two objects possessing unlike charges attract each other.*

8.5 Magnetic force: *the other part*

Have you ever played with magnets? Everybody should have the opportunity to experience the intriguing forces of attraction and repulsion that magnets exert on each other. You can buy inexpensive magnets at a toy store.

If you bring two bar magnets near each other, you will discover that the ends of two magnets either attract or repel each other even when they are separated by a distance. The ends are called **magnetic poles**. The two ends are called the "north" and "south" pole of the magnet and might be indicated by "N" and "S" printed on the ends. Experiment shows that similar poles repel each other and dissimilar poles (north and south) attract each other. This reminds us of the forces between electric charges: Likes repel, and unlikes attract.

It seems plausible to hypothesize that the force acting between the magnets actually is the electric force. This hypothesis is easy to check, for it predicts that magnets should exert forces on electrically charged objects such as a rubbed transparency or tissue. If you try this, you will find that the magnets

---

* The electric force is quantitively proportional to the product of the "strengths" (or magnitudes) of the charge of each of the two objects and inversely proportional to the square of the separation distance: $F \propto q_1 q_2 / d^2$. Note the similarity with the gravitational force law: $F \propto m_1 m_2 / d^2$.

do not exert forces on a rubbed transparency or tissue.* So our hypothesis is wrong. Despite the similarity to electricity, the force between bar magnets is not the electric force, and the two ends of a bar magnet are not electrically charged.

There are other important differences between magnetism and electricity. First, a bar magnet's magnetism is permanent and has nothing to do with rubbing. Second, every magnet has both a north and a south pole—magnetic poles always come in pairs. Nobody has ever found an object that possessed either kind of pole without the other kind, although many experiments have looked for evidence of such "monopoles." On the other hand, it is easy to find objects that are entirely positively charged or entirely negatively charged.

This new kind of force is called the **magnetic force**. Despite their differences, the similarities between the electric force and the magnetic force suggest that they might be related. One of the great triumphs of nineteenth-century physics was the demonstration that electricity and magnetism are in fact related. The most concrete evidence was an experiment, first conducted in 1820, in which electrically charged particles that were in motion exerted a measurable force on a magnet.

Note that only charged objects that are *moving* exert forces on magnets. The experiment involving a magnet and *stationary* charged objects such as a transparency or tissue shows that charged objects *at rest* do not exert forces on magnets.

Further experiments during the nineteenth century showed that *all* magnetic forces can be traced to the motion of charged objects. Moving charged objects exert and feel magnetic forces over and above whatever purely electric forces they would feel if they were at rest. This additional force, due solely to the motion of charged objects, is the magnetic force.

For example, the forces observed with bar magnets are due to subatomic charged particles moving inside each magnet (Section 8.7). For another example, Earth's magnetic effects are due to electrically charged material flowing inside Earth.

We summarize this idea as:

> THE MAGNETIC FORCE LAW
> Charged objects that are moving exert and feel an additional force beyond the electric force that exists when they are at rest. This additional force is called the **magnetic force**. All magnetic forces are caused by the motion of charged objects.

----

* A small attractive force is sometimes obtained with the transparency, an electric effect called *electrical polarization*. This force, if it does occur, is attractive at both the north and south poles of the magnet, and it occurs equally strongly even if an unmagnetized piece of metal is used in place of the magnet, so it is not caused by the presence of the magnetic poles. Rather, it is caused by the redistribution of electric charge that occurs in a metal when a charged object like the transparency is brought near it. A similar electric polarization occurs quite dramatically when a charged transparency is brought near an empty aluminum can that is free to roll. Try this!

This means that the separate concept of magnetic poles is not needed, that we can drop the idea of magnetic poles and just think of moving charges instead.

A goal of science is to find connections between apparently different phenomena. For example, Newton united the heavens and Earth by finding similarities between a falling apple and the moon. Similarly, nineteenth-century scientists found that both electricity and magnetism are due to the existence of electrically charged objects and that these two forces can be united into a single **electromagnetic force** between charged objects.

DIALOGUE 7   Could the force between two bar magnets be due to gravity? Defend your answer.

DIALOGUE 8   If you charge two transparencies by rubbing and then hold them at rest several meters apart, will they exert an electrical force on each other? Will they also exert a magnetic force? What if you shake both of them back and forth?

DIALOGUE 9   What will you have if you saw off one end of a magnet?

8.6 The electric atom: *the planetary model*

One outgrowth of the nineteenth-century developments surrounding electricity and magnetism was a new model of the atom, an atom made of electrically charged subatomic parts. Because light is created by the motions of these subatomic parts, as part of our effort to understand light, we will devote the rest of this chapter to the electric atom.

Physicists have always sought microscopic explanations of macroscopic events. So far in this book, we have discussed, from a microscopic point of view, chemical reactions, thermal energy, and much more. All of these phenomena are comprehensible on the basis of the **Greek model of the atom** (Chapter 2). According to this theory, all matter is made of small unchangeable particles, *atoms* (Greek for "uncutable"), which come in many varieties or "elements" and move and combine in various ways. This model of the microscopic world, conceived by the ancient Greeks, formed the background for Newtonian physics. During the nineteenth century, this model was developed into the theory of chemical elements, compounds, and reactions.

It is difficult to fit electromagnetic phenomena into this indivisible-particle picture of the atom. Where does electric charge come from? How can

---

7. Although the magnetic force can be repulsive, gravity cannot. Also, the force between bar magnets is far too strong to be caused by gravity acting between the small bars.
8. With both at rest, you will get only an electric force. With both in motion, you will get not only an electric force but also a magnetic force. However, if you try an experiment like this, the magnetic part of the overall electromagnetic force will be immeasurably small.
9. Two magnets. Each piece will have its own north and south poles. It is impossible to isolate a north or south pole.

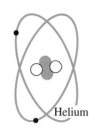

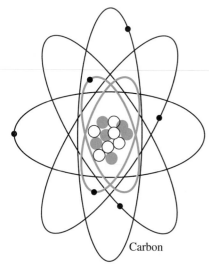

Helium

Carbon

FIGURE 8.24

Two examples of the planetary model of the atom. Color code: Protons are green, neutrons are white, and electrons are black. The diagrams are not drawn to scale! If it were drawn to scale on the page, the nuclei would be too small to be seen.

rubbing produce it? Why are there two kinds of charge? These and other electromagnetic phenomena led, early in the twentieth century, to the **planetary model of the atom**.* According to this theory, the atom is not an unchangeable, solid, indivisible particle. To the contrary: The planetary atom is continually changing, is almost entirely empty, is divisible, and is made of many parts.

As its name implies, the planetary model resembles a miniature solar system. Figure 8.24 portrays single atoms of two different elements, helium and carbon. The defining feature of the planetary atom is the tiny **nucleus** at the center, surrounded by a number of even tinier **electrons** ("electrified ones") that orbit the nucleus at a relatively great distance, much greater than the size of the nucleus itself. The overall size of an atom, the distance across its electron orbits, is typically about 10^{-10} m. This is the typical size of an atom as it is visualized in the single-particle model, and it is roughly the distance between neighboring atoms in solid materials. But a typical nucleus is 10,000 times smaller, on the order of 10^{-14} m. To put this into perspective, if we built a scaled-up model of an atom in which the nucleus were represented by a soccer ball, the orbiting electrons would be dust specks several kilometers away! Atoms are nearly totally empty. Despite the fact that the nucleus is far smaller than the atom, more than 99.9% of an atom's mass resides in its nucleus.

Electrons always repel negatively charged objects and attract positive ones. You cannot remove the charge from an electron—it is permanently negatively charged, and the electron has a mass that is about two thousand times smaller than the mass of even the least massive atom. Nobody has any inkling of why either fact is so.

The nucleus is itself made of several subatomic particles, of two different types: protons and neutrons. A **proton** ("positive one") is another permanently charged particle, like an electron. It is charged precisely as strongly as the electron, but positively instead of negatively. When we say that electrons and protons are charged "equally strongly," we mean that when they are placed near some other charged object, both exert the same amount of force (but in opposite directions) at the same distance away. Although the electron and proton have equally strong charges, they do not have equal masses. The proton is about two thousand times more massive. The **neutron** ("neutral one") is an uncharged, or neutral, particle whose mass is nearly the same as the proton's mass. Between one and a few hundred protons and neutrons form the nucleus of any atom.

The "glue" that holds electrons in their orbits around the nucleus is the electric attraction between the electrons and the protons in the nucleus. The glue that holds the nucleus together, however, must be some nonelectric force, because the electric force between the positively charged protons is repulsive and neutrons do not exert an electric force (Chapter 15).

---

* The term *planetary atom* is self-contradictory. *A-tom* means "indivisible," and *planetary* refers to the parts into which the atom can be divided! The old Greek name, *atom*, has stuck, but not its essence.

Since an atom's electrons are relatively distant (compared with the size of the nucleus) from the nucleus, it is not surprising to learn that the forces binding them to atoms are rather weak, and that it is not difficult to remove electrons from atoms. On Earth, unattached atoms (atoms that are not combined into molecules) usually have just as many electrons as protons. The reason is that any atom having fewer electrons than protons carries a net positive charge and so tends to attract electrons from its environment, while any atom having more electrons than protons carries a net negative charge and tends to lose its outermost electrons to its surroundings. This is why it took so long to discover many electrical phenomena: Most individual atoms are normally uncharged and exhibit no obvious electrical effects. Any atom having an excess or deficiency of electrons is called an **ion**.

FIGURE 8.25
J. J. Thomson.

HOW DO WE KNOW? THE DISCOVERY OF THE ELECTRON

One key experiment leading to the planetary atom was by the English physicist J. J. Thomson (Figure 8.25). In 1897, Thomson was investigating a type of invisible beam known as a *cathode ray*. Cathode rays were produced in a nearly evacuated (emptied of air and other gases) glass tube whose two ends were attached by metal wires to a source of electric power. When the power was switched on, rays of unknown composition streamed along the length of the tube, as could be observed by the flashes of light where they hit one end of the tube.

Suspecting that these rays were electrically charged, Thomson placed electric charges and magnets around them. The flashes of light shifted in position. Because the charges and magnets deflected the rays, the rays themselves had to be electrically charged. The only charged microscopic objects then known were ions, observed in certain chemical experiments. Thomson hypothesized that the electrically charged cathode rays were streams of such ions.

He then measured the deflections of the rays. Using the known electric and magnetic force laws, he deduced from these measurements that these rays were streams of charged particles whose charge was the same as the charge of typical ions but whose mass was some two thousand times smaller.* These, then, were not ions.

This was revolutionary. It established that atoms had parts. According to Thomson, "At first there were very few who believed in the existence of these bodies smaller than atoms. . . .It was only after I was convinced that the experiments left no escape from it that I published my belief in the existence of bodies smaller than atoms."

Thomson had discovered the electron. Cathode rays are now also called *electron beams*. Today, electron-beam tubes are in wide use as TV tubes, fluorescent bulbs, computer screens, and many other devices.

---

* More precisely, he found that the *ratio* of the mass to the charge was two thousand times smaller than it was for any known ion.

*The energy produced by the breaking
down of the atom is a very poor kind
of thing. Anyone who expects a
source of power from the
transformation of these atoms is
talking moonshine.*

Ernest Rutherford, discoverer of the
atomic nucleus, made this famous
wrong prediction in 1933, while
addressing the British Association for
the Advancement of Science in the
same hall where physicist Lord Kelvin
had asserted, in 1907, that the atom
was indestructible. The *New York
Herald Tribune* article, dated
September 12, 1933, carried the
headline "ATOM-POWERED
WORLD ABSURD, SCIENTISTS
TOLD. Lord Rutherford Scoffs at
Theory of Harnessing Energy in
Laboratories"

HOW DO WE KNOW? THE DISCOVERY OF THE NUCLEUS

New Zealander Ernest Rutherford's work was at least as revolutionary as Thomson's (Figure 8.26). Like others around 1910, Rutherford was trying to determine the atom's internal structure. He knew that atoms contained electrons and that a positive charge must be present too. It was known that atoms are pressed right up against one another in solid materials and that huge forces are required to compress solids into smaller volumes. This means that it is difficult to squeeze atoms into other atoms. So Rutherford and others hypothesized that atoms were filled with matter throughout most of their volume.

To probe the atom's structure and test this hypothesis, Rutherford and his coworkers used what has become a traditional physics technique: He threw tiny things at other tiny things in order to see what would happen. He threw a recently discovered ray known as an alpha ray at the atoms in pieces of thin metal foil, similar to aluminum foil. The alpha ray was a high-energy stream of positively charged and fairly massive (about four times the proton's mass) "alpha particles" that emerged from certain substances (Chapter 15).

The idea was to observe how far the foil deflected the fast-moving alpha particles from their original directions and, from this, to deduce how matter must be distributed within the foil's atoms (Figure 8.27). The deflection was measured by observing flashes of light where the alpha particles hit a screen placed partially around the foil. Similar experiments had been done before, and it had been found that the foil had surprisingly little effect on the motion of the alpha particles. Most deflections were less than one angular degree. Since even the thinnest foils were about 500 atoms thick, alpha particles apparently passed straight through most atoms without deflection. Apparently, atoms were fairly porous, open structures.

Then in 1911, Rutherford decided to see whether any alpha particles were deflected through very large angles, perhaps greater than 90 degrees. His coworkers studied this by surrounding the foil with the detection screen (with a gap to allow the alpha ray to enter). Rutherford expected to see no large deflections, because a fast-moving and massive alpha particle was thought to pass through an atom somewhat like a high-velocity cannonball through pudding. An alpha particle would have to experience an enormous force to be deflected by very much.

His coworkers came to Rutherford a few days later with the news that a few alpha particles had been deflected backward. In Rutherford's words, "It was quite the most incredible event that has ever happened to me in my life. It was almost as incredible as if you fired a 15-inch shell at a piece of tissue paper and it came back and hit you."

The reason was, apparently, that nearly all of the matter and all of the positive charge of the atom were concentrated in a tiny, dense region in the center. The cannonball had struck an even more massive cannonball and bounced back. Rutherford had discovered the atomic nucleus.

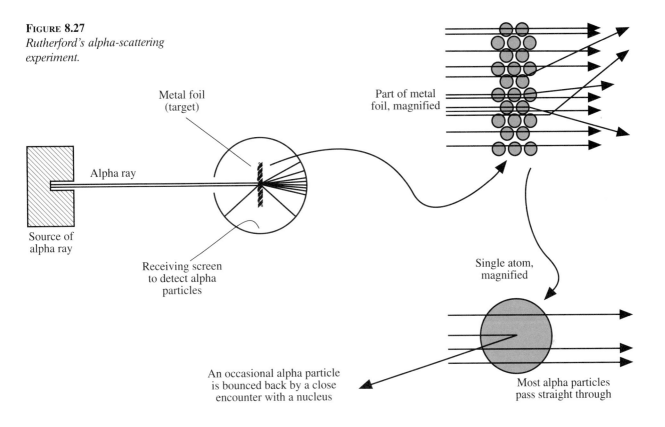

FIGURE 8.27
Rutherford's alpha-scattering experiment.

Metal foil (target)

Part of metal foil, magnified

Alpha ray

Source of alpha ray

Receiving screen to detect alpha particles

Single atom, magnified

An occasional alpha particle is bounced back by a close encounter with a nucleus

Most alpha particles pass straight through

Not exist—not exist! Why I can see the little beggars there in front of me as plainly as I can see that spoon!
   Rutherford, when asked over a dinner table whether he believed that atomic nuclei really existed

8.7 The planetary atom: *a useful theory*

Like all good theories, the planetary atom explains many things.

It explains the experiments with charged objects. When you rub a transparency with tissue, some of the loosely attached outermost electrons in the transparency's atoms are rubbed off and transferred to the tissue's atoms, charging the transparency positively and the tissue negatively. That is why the two attract each other after rubbing. The two rubbed transparencies repel each other because both are positively charged.

When the ends of a copper wire are attached to the positive and negative attachments (called *electrodes*) of a battery, electric forces are suddenly exerted on every charged particle in the wire (Figure 8.28). Chemical processes in the battery cause its electrodes to be permanently charged, positively and negatively, and these charged electrodes exert electric forces on every charged particle in the wire. These forces produce practically no disturbance of the positively charged copper nuclei, which are fixed in position. But in copper or any other metal, each atom's outermost electrons are nearly free of their parent atom. As soon as these electrons feel the electric forces established by a battery, they all begin to move along the wire. Such a flow of charged particles is called an **electric current**.

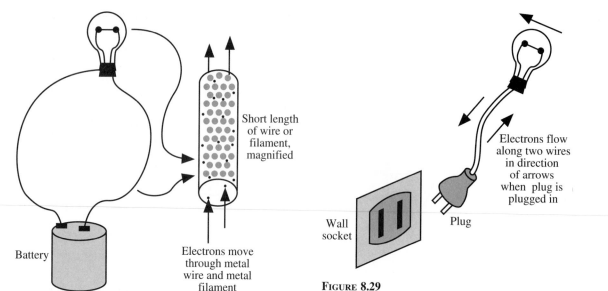

FIGURE 8.28

A battery produces electrical forces that cause electrons to move through the metal wire.

FIGURE 8.29

Electrical appliances are based on the motion of unseen, microscopic, charged particles called electrons that move through appliances. A wall socket (more precisely, a generating station connected by wires to a wall socket) is the electrical equivalent of a water pump. It "pumps" (energizes) the electrons so that they can flow around the circuit.

If this "electric circuit," from one electrode around to the other electrode, also includes a lightbulb filament (Figure 8.28), electrons will flow through the thin filament and heat it by simply bumping into lots of atoms. The filament becomes so hot that it creates light. All of the standard electrical appliances—toasters, lightbulbs, electric motors, and so forth—are based on the flow of the unseen, microscopic electrons that ordinarily orbit in atoms. Figure 8.28 shows how the process works when a battery is the source of the electric force. Electrical outlets in your home work in a similar way (Figure 8.29).

The magnetic force law says that all magnetic effects come from electric charges in motion, in other words, from electric currents. Where, then, are the electric currents responsible for permanent magnets? The answer is that they are found at the subatomic level, in the motions of electrons in atoms.* Each moving electron in a bar magnet causes its own tiny magnetic force on each moving electron in another bar magnet. In most materials, these tiny magnetic forces cancel one another because the electrons all are oriented differently. But the treatment of the iron when a magnet is made locks many electron orbits into similar orientations. This causes the many atomic-level magnetic forces to add up to a large macroscopic effect, creating the forces you see between permanent magnets.

---

* More precisely, in orbiting and spinning electrons. Electrons not only orbit around a nucleus, but they also spin on their own axis. This spin is a kind of electric current also, with magnetic effects. Electron spin, rather than orbital motion, causes the large effects seen in ordinary magnets.

Permanent magnets can temporarily magnetize objects such as nails by forcing many of the nail's electrons to orient themselves in similar directions. This is why nails are attracted to permanent magnets.

Earth's magnetism is thought to be caused by molten iron and electric currents flowing in Earth's hot interior. Its magnetism is, in turn, responsible for the behavior of compass needles, which are small permanent magnets, and for Earth's other magnetic effects.

The planetary model explains many chemical phenomena. Chemical reactions and other properties of the elements occur because of the behavior of the orbital electrons of the atoms of that element. Table salt, NaCl, makes a good example. Sodium (Na) combines readily with chlorine (Cl) because the Cl atom has a stronger attraction for electrons than does Na, so one electron from Na is attracted to Cl, leaving an overall positive charge on the Na and a negative charge on the Cl. The electric force between opposite charges then attracts and holds the Na and the Cl atoms together.

The property that really defines an "element" is the structure of the electron orbits in the atoms of that element. Two neutral atoms with the same number of electrons have the same chemical properties because their electron orbits have the same structure. The different elements are in fact numbered according to their number of electrons or, what is the same thing, their number of protons. This number is called the element's **atomic number**.

Since an atom is mostly empty space, what keeps solid matter solid? Why don't you fall through the floor, for instance? The planetary model's answer is that the repulsion of electrons by electrons keeps atoms from penetrating one another. Although an atom itself is mostly empty space, it occupies a considerable amount of space because of the forces that its electrons exert on surrounding objects. All contact forces can be interpreted at the microscopic level as forces by the orbital electrons in atoms on the orbital electrons in other atoms. Every time you touch or feel something, you are experiencing the electric force between orbiting electrons! In fact, all the forces in your daily environment come down to just two kinds of forces. The gravitational force explains weight, and the electromagnetic force explains all the contact forces and also the forces between charged or magnetized objects. Just two fundamental forces underlying all ordinary phenomena. This is quite a unification!

Despite the many successes of the planetary model, more recent experiments require it to be replaced with yet another model, the *quantum atom*. Today, the quantum atom is regarded as the correct theory, and the planetary model and the Greek model are regarded as approximations that are useful for many purposes.

DIALOGUE 10   The magnetic north pole of a compass needle points roughly toward Earth's geographic North Pole. What type of magnetic pole must be at Earth's northern end?

---

10. A magnetic *south* pole.

"THESE DAYS EVERYTHING IS HIGHER."

Summary of Ideas and Terms

Wave motion A transfer of energy without a transfer of matter. A disturbance that travels through a **medium**.

Projectile motion or **particle motion** An actual transfer of matter.

Wavelength The distance from any point to the next similar point along a continuous wave.

Wave frequency The number of vibrations that any part of the medium completes in each second. Also the number of complete wavelengths sent out in each second. Higher-frequency waves have shorter wavelengths and (assuming equal amplitudes) higher energies. Unit: vib/s = **hertz** (Hz).

Amplitude The half-width of vibration of each part of a wave's medium.

Wave speed The speed at which a disturbance (a wave) moves through a medium.

Wave interference The effects that occur when two waves of the same type are present at the same time and place. Interference can be either **constructive** or **destructive**, depending on whether the two waves reinforce or cancel each other.

Double-slit experiment with light The interference of light coming from two synchronized sources is observed on a screen, demonstrating that light is a wave.

Electrically charged object Any object that can exert or feel the electric force. There are two types of charge, **positive** and **negative.**

Electric force law Electrically charged objects exert forces on each other at a distance. Like charges repel each other, and unlike charges attract each other.

Electric current A flow or motion of charged particles. Electric currents in wires are due to electrons moving along the wire.

Magnetic poles The ends of a permanent magnet. There are two types, north and south. Like poles repel, and unlike poles attract.

Magnetic force law Charged objects that are moving exert and feel an additional force, called the **magnetic** force, beyond the electric force that exists when they are at rest.

Electromagnetic force The total (electric and magnetic) force between charges.

Planetary model of the atom Atoms are made of negative **electrons**, positive **protons**, and uncharged **neutrons**. From one to a few hundred protons and neutrons form a tiny central **nucleus**, which the electrons orbit.

Ion Any atom having an excess or deficiency of electrons.

Atomic number The number of protons in an atom. Also the number of electrons in a neutral atom. An atom's atomic number determines its chemical properties and the **element** to which it belongs.

Review Questions

WAVES

1. Describe the motion of the parts of the medium when a wave travels along a rope, along a Slinky, and across the water.

2. What is the difference between wave and projectile motions?

3. What does "hertz" mean?

4. Waves A and B have the same wave speed, but A's wavelength is longer. Which has the larger frequency, or are they the same?

5. Choose the correct answer(s): Two different continuous waves on the surface of a pond would be expected to have roughly the same (wave speeds, wavelengths, frequencies, energies, amplitudes, none of these).

INTERFERENCE

6. Two continuous water waves each have a 2 cm amplitude. What is the water surface's displacement (a) when a crest meets a trough? (b) When crest meets crest? (c) When trough meets trough?

7. Give an example of (a) a one-dimensional medium. (b) A two-dimensional medium. (c) A three-dimensional medium.

8. You tap one finger on the water's surface, on one side of a rectangular pan of water. Describe the waves rolling into the far side. How will the wave pattern at the far side change if you now begin tapping, at the same rate, at two different points, using one tapping finger at each point?

LIGHT

9. Describe the experimental evidence supporting the claim that light is a wave.

10. How would the pattern on the screen in Figure 8.18 change if one of the slits were closed?

11. How do we know that the medium for light waves is not air?

ELECTRIC FORCE, MAGNETIC FORCE

12. When you rub two transparencies with tissue and hold them close together, they stand apart. Give two reasons that the force causing this cannot be gravity.

13. Cite the evidence supporting the claim that there are two, and only two, types of electric charge.

14. Give two reasons that the force between bar magnets cannot be the electric force.

15. What is meant by an "electric current"?

16. Magnetic forces are always caused by what types of objects?

THE PLANETARY ATOM

17. List several phenomena that require the planetary atom, rather than the Greek atom, for their explanation.

18. Name and briefly describe the three kinds of subatomic particles found in atoms.

19. Give evidence supporting the claim that most of an atom's mass is concentrated in a tiny nucleus at the center.

20. Explain what happens at the microscopic level when a battery creates an electric current in a wire.

21. What is an "atomic number," and how is it related to the chemical elements?

Home Projects

1. Float a cork at one end of a tub of water. Drop a pebble into the other end to make a wave that reaches the cork. Observe the cork's motion. Does it vibrate back and forth? Up and down? Try it with larger pebbles. Is there some point at which the water's motion is no longer a true wave?
2. In a tub of still water, drop two small pebbles in at different places. Observe the interference effects as the two spreading circular waves cross. Do they appear to pass through each other without distortion? Try a tiny pebble and a larger pebble. Try tapping one finger at the water's edge to produce a continuous wave.
3. Use a razor blade to cut a thin slit in an index card, a millimeter or less in width and a few centimeters long. Look at a well-lit wall or other surface through the slit, focusing on the wall. Hold the card at an angle to narrow the width through which light can come. You should see rather striking light and dark lines running parallel to the long dimension of the slit. What causes these lines?
4. Ask a friend to stand at the far end of a carpeted room. Scuff your way across the rug until your noses are close together. What happens? Do you get a stronger effect when either of you are barefoot or when wearing shoes? Would the weather make any difference? Try touching wood, metal, and plaster. Explain any differences.

Exercises

WAVES

1. Is a mountain stream rushing downhill an example of wave motion? Defend your answer.
2. A gust of wind moves across a wheat field, causing a noticeable ripple that crosses the field. Is this ripple a wave? If not, why not? If so, then what is the medium?
3. When you send a brief wave down a rope or Slinky, it eventually dies out. What has become of the energy?
4. A cork floats on the water as a water wave passes by. What happens to the cork? Will the cork's vibrational frequency be related to the water wave's frequency, and if so, how?
5.*If you doubled the wavelength of a wave without changing its wave speed, what would happen to the frequency? What if you halved the wavelength?

INTERFERENCE

6. The two waves shown in Figure 8.30 are moving toward each other along a string. Sketch the string during the interference of the two waves.

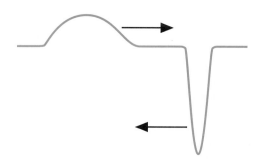

FIGURE 8.30
How will the two waves interfere?

7. Suppose the frequency of the waves in Figure 8.12 is 1 Hz. Describe the appearance of the water surface one-half second later (one-half second after the snapshot was taken). One second later.
8. What happens to the energy of the two waves in Figure 8.9 when they interfere destructively, as shown in the second of the three sketches? Did the energy vanish? (*Hint*: What are the parts of the rope doing as the wave moves? Which type of energy is the wave carrying?)
9. Two small sources of water waves send out circular waves. Each source produces waves whose amplitudes are 1.5 cm. What is the water level (relative to the undisturbed level) at the point where crest meets crest? Where trough meets trough? Where trough meets crest?

LIGHT

10. Do waves of any sort travel to Earth from the moon? Explain.
11. How do you know that light is not something that comes out of your eyes?
12. You shine two flashlights on a wall. Why don't you see an interference pattern?
13. Which of these objects are luminous (light sources), during their normal operation: camera, polished chrome, firefly, electric stove heating element, camera flashbulb, mirror, diamond, sun, moon?
14.*Making estimates. Estimate the number of wavelengths of light in 1 millimeter.

ELECTRIC FORCE, MAGNETIC FORCE

15. Since matter is made of electrically charged particles, why don't we and the objects around us feel electric forces all the time?
16. When you remove a wool dress from a garment bag, the sides of the bag might stick together. Explain.

17. Figure 8.31 shows an "electroscope." The leaves (made of metal foil) normally hang down, but they spread apart when the metal sphere on top touches a charged object. Explain.

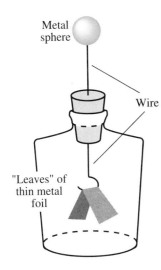

FIGURE 8.31
Why do the leaves stand apart?

18. How does the operation of the electroscope (previous exercise) demonstrate electrical conduction?

19. You have three iron bars, only two of which are permanent magnets. Because of temporary magnetization (Section 8.7), all three bars at first appear to be magnetized. How can you determine which one is not magnetized, without using any other objects?

THE PLANETARY ATOM

20. While brushing your hair, you find that the hairs tend to stand apart from one another and that they are attracted toward the brush. Explain this in microscopic terms.

21. A covered mystery shoebox is placed on a table. What are a few ways that you could learn something about its contents without directly touching it or having it lifted?

22. After you walk across a rug and scuff electrons off the rug, are you positively or negatively charged?

23. According to Figure 8.24, what are the atomic numbers of carbon and helium? Roughly how much more massive is the carbon atom than the helium atom?

24. Some science fiction stories portray atoms as true miniature solar systems populated by tiny creatures. What are some differences, other than size, between our solar system and the planetary model of an atom?

25. An atom loses its two outermost electrons. How does the resulting ion behave when it is near a positively charged transparency? A negatively charged tissue? Would anything be different if it lost only one electron?

26.*Making estimates. About how many atoms thick is a sheet of paper?

27.*Making estimates. Which is bigger, an atom or a wavelength of light? Roughly how much bigger?

9

ELECTROMAGNETIC WAVES

This chapter continues our quest for an understanding of light. We found in the preceding chapter that light must be a wave because it shows interference. Then we looked at a new force: electromagnetism. Now we will see what electromagnetism has to do with light.

In this chapter we meet a radically new physical concept: fields. "Radically" is not too strong a term, for our studies so far have centered on the Newtonian worldview of a universe made of particles moving mechanically in empty space, and fields are in many ways the opposite of particles. Particles are discrete and isolated in particular places, whereas fields are continuous and spread out over a region of space. Fields were first introduced into physics during the nineteenth century, when physicists discovered that light is an electromagnetic phenomenon. Fields are characteristic of post-Newtonian physics, cropping up as gravitational fields, electromagnetic fields (this chapter), psi fields (Chapters 13 and 14), and quantized force fields (Chapter 18).

Section 9.1 explains the field concept, and Section 9.2 explains light in terms of electromagnetic fields. Section 9.3 asks whether fields are actually real physical entities or simply imaginary figments of our theories and looks at the implications for the Newtonian worldview. The electromagnetic field theory of light leads to an understanding of a wide assortment of other light-like "radiations," explained in Section 9.4. Section 9.5 studies the radiations coming to us from the sun.

Finally, we discuss a new kind of technology-related problem that has only recently emerged: environmental problems that cannot be contained locally and so are truly global. Science is often part of the cause and part of the solution of such problems. We all must think seriously about such matters, including their scientific dimensions, if Earth is to pull through its current ex-

periment with powerful technologies. Two of these problems, ozone depletion (Section 9.6) and global warming (Section 9.7), are closely related to
the electromagnetic field theory of light.

9.1 Force fields: *the possibility of a force*

In order to visualize electric and magnetic forces, nineteenth-century scientists invented a new idea called a **force field**, the effect that the source of a
force has on the surrounding space: not on the things in the space, but on
the space itself. Think of a field as a stress in space. When this idea was invented, force fields were generally considered to be only mental images—
imaginary concepts that are made up because they are useful. At that time,
forces were thought to affect only material objects and not space itself. It
seemed impossible, for example, that a charged object could have real effects on distant regions of empty space. Nevertheless, scientists invented the
notion that force fields could exist even in empty space.

The force field idea applies to every type of force. Let's begin with the
gravitational force. A **gravitational field** is said to exist throughout any region of space where an object could feel a gravitational force if it were placed
there. A gravitational field is "the possibility of a gravitational force." For
example, Earth's gravitational field exists wherever another object could feel
a gravitational force from Earth if another object were placed there. This
means that Earth's gravitational field exists even outside Earth in empty
space. Since a gravitational field can exist even in empty space, we must think
of it as a property of space itself—as a stress or alteration of space. We imagine that Earth alters the space around Earth.

Fields help us think about forces by one object on another object some
distance away. Consider the force by Earth on the moon. Rather than thinking of this force as something that happens directly between Earth and the
moon, we can imagine that Earth creates a gravitational field that fills the
surrounding space in the way that smoke from a fireplace can fill a room.
The moon then feels the force exerted by this field, in the same way that a
person can smell the smoke.

Every material object can exert gravitational forces and so is surrounded
by its own gravitational field. We can speak of the gravitational field of (or
created by) the sun, the moon, a rock, your body, and so forth. Strictly speaking, each object's gravitational field fills all space, but from a practical point
of view, most fields can be neglected at large distances from the objects that
create them.

Just as gravitational fields surround massive objects, an **electromagnetic
field** surrounds every charged object; this field exists wherever any other
charged object would feel electromagnetic forces exerted by the first charged
object. An electromagnetic field is the possibility of an electromagnetic (electric or magnetic or both) force. An electromagnetic field with no magnetic
component is referred to as an *electric field*, and the term *magnetic field* is
used in a similar way.

For instance, an electrically charged transparency creates an electric field
around itself, and this field exists even in the absence of other material

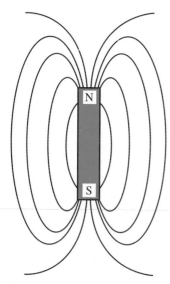

FIGURE 9.1

A representation of the magnetic field of a bar magnet.

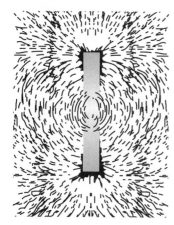

FIGURE 9.2

An experimental demonstration, using iron filings, that the field of a bar magnet actually does have the shape drawn in Figure 9.1.

objects around the transparency. Even if the transparency is isolated in outer space, it still creates an electric field throughout the space around it. Magnetic fields surround electric currents and permanent magnets. Earth's magnetic field is created by electric currents within Earth and can be detected by a compass needle on Earth or in space.

We often draw "field lines" to represent fields and to help make the field concept more concrete. For example, Figure 9.1 shows the magnetic field lines in the vicinity of a bar magnet. The lines are drawn parallel to the direction of the magnetic force that would be exerted on any other magnetic pole that happened to be placed in the vicinity of this magnet. Figure 9.2 is an experimental demonstration that a bar magnet's field actually does have the shape drawn in Figure 9.1. Figure 9.2 is made by sprinkling small iron filings in the vicinity of a bar magnet. The long slender filings are temporarily magnetized by the magnetic field (in the way that an iron nail is magnetized), and the magnetic forces on their north and south poles then cause the filings to line up parallel to the bar magnet's field.

All of these ideas can be made quantitative and mathematically precise, although there is no need to do that here. For example, Earth's gravitational field has been precisely measured and tabulated out to large distances. Such a mapping can be useful, for example, in planning satellite missions.

DIALOGUE 1   A proton is placed at rest in the middle of a vacuum chamber, an enclosure that has been emptied of all matter. Consider some point X near a particular corner of the chamber. Neglecting all influences other than the proton, is there an electric field at X? Is there an electric force at X? Is there a magnetic field at X?

9.2 The electromagnetic wave theory of light

Suppose you pull a rubber comb through your hair, scuffing electrons from your hair onto the comb. The charged comb then creates an electric field in the surrounding space, a field that can be detected by a rubbed transparency held near the comb.

If you quickly shake the comb once, up and back down, the electric field in the surrounding space will shake too. This can be detected by the transparency, which will shake in response to the comb's motion. The comb also creates a magnetic field during the brief time that it is moving. This temporary magnetic field can be detected as a brief shake of a magnet placed near the comb. In summary, the motion of the comb causes an alteration in the electromagnetic field around the comb, an alteration that can be detected by other charged objects and magnets (Figure 9.3).

There is an interesting question about this experiment, a question that many scientists asked during the nineteenth century: *When* will a distant detector feel the altered force? Is the effect instantaneous? Suppose you shake the comb precisely at noon. Does the transparency shake precisely at noon, too, or a little later? Is the electromagnetic force transmitted instanta-

---

1. Yes, no, no (because the proton is not moving).

FIGURE 9.3
If you shake a charged object such as a charged comb, other charged objects will shake in response.

FIGURE 9.4
James Clerk Maxwell, the "Isaac Newton of electromagnetism."

neously? At that time, it was not easy to perform an experiment to answer this question.

During the 1860s, British physicist James Clerk Maxwell (Figure 9.4) did some hard thinking about electromagnetism. He gathered together all that was then known about the subject and put it together in a single consistent quantitative theory. Maxwell's theory emphasized the field concept and described how electrically charged objects create electromagnetic fields.

Three basic principles of the electromagnetic force were known at that time. The first was the electric force law, and the second was the magnetic force law. Stated in terms of fields, these two laws say that charged objects create electric fields and that electric currents create magnetic fields.

Maxwell's third principle was one we have not yet mentioned. It is called *Faraday's law*. This idea can be expressed entirely in terms of fields, without direct reference to electric charge or current. It states that any change in a magnetic field must create an electric field. This means that electric fields can be created in two ways: by electric charge (this is the electric force law) and by a change in a magnetic field (this is Faraday's law).

To these three principles, Maxwell added a fourth idea that he invented for reasons that were as much aesthetic as physical. Maxwell was impressed with the similarities between electric and magnetic effects. Because of these similarities, it seemed to Maxwell that since any change in a magnetic field creates an electric field, then any change in an electric field ought to create a magnetic field. To Maxwell, it seemed that the relationships between the fields should be symmetric in this manner. It turned out that Maxwell's invention was experimentally correct. Once again (compare Chapter 1) we see that beauty is important to science.

All of this worked remarkably well. Maxwell's theory, which he formulated in precise mathematical language, described all electromagnetic phenomena correctly and quantitatively and, as we will see, predicted important new phenomena.

Maxwell's theory predicted an answer to our question about the time delay for electromagnetic forces. The key ingredient was the symmetry between electric and magnetic effects. Since a changing magnetic field creates an electric field and also since a changing electric field creates a magnetic field, these two fields can create and recreate each other. Once the fields at one point in space are altered—for instance, by giving a charged comb a single shake—Maxwell's theory implied that the alteration is transmitted outward as an alteration in the nearby fields a short time later, and these changing fields in turn transmit the change farther outward, and so forth. It follows from this that electric and magnetic forces are not transmitted instantaneously. There must be a time delay between the motion of a charged comb and the resulting force felt by a charged transparency.

Maxwell's theory showed that an electromagnetic field behaves like a continuous elastic substance such as foam rubber. The different places in the field are connected in the way the different places in a piece of foam rubber are connected. Just as with elastic foam, if you disturb the field at one point, these connections will cause the disturbance to move outward through

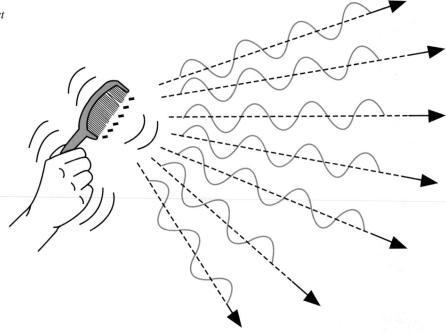

the field. Any alteration of an electromagnetic field, as might be caused by shaking a charged comb, is transmitted as a pulse, a wave pulse, that spreads outward through the field. But the medium through which such a pulse travels is not foam rubber and in fact is not a material (made of matter) medium at all. Rather, the medium is the electromagnetic field itself. Any such disturbance that moves through an electromagnetic field is called an **electromagnetic wave.**

You can't directly see electromagnetic waves in the way that you can see water waves, because the medium for electromagnetic waves is an invisible electromagnetic field instead of a material substance such as water. Nevertheless, an electromagnetic wave can be detected by other charged or magnetized objects at some distance from the source of the wave and at some later time after the wave was sent out. Figure 9.5 pictures these invisible waves.

This all was worked out quantitatively in Maxwell's theory. The theory predicted not only that there must be a delay in the transmission of electromagnetic forces but also how long that delay must be for a given separation between the source and the recipient of any such force. In other words, the theory predicted the speed of transmission of electromagnetic waves and forces. The predicted speed was 300,000 km/s.

This particular speed had come up before, in experiments performed nearly two centuries before Maxwell invented his theory. But these previous experiments seemed entirely unrelated to the electromagnetic effects that Maxwell was studying. This speed, 300,000 km/s, was the known speed at which light travels!

People once thought that light requires no travel time—that its speed is infinite. Light certainly travels much faster than sound, as you can verify whenever you see lightning before you hear thunder.

Galileo was one of the first to try to measure lightspeed, by measuring the total round-trip time for light to travel to a distant mountain and back again. His experiment didn't work because the time turned out to be far too short to measure using Galileo's timing methods. Either greater timing accuracy or a greater travel distance was needed.

The first evidence for a finite, and not infinite, speed of light came from astronomical observations several decades later. Pointing telescopes at a moon of Jupiter, astronomers found that the time they measured for this moon to orbit Jupiter did not remain constant, as they had expected. Why should a moon take longer for some orbits than for others? Earth's moon takes 27.3 days, every time. The astronomers found that these variations were not caused by Jupiter's moon at all but were instead related to Earth's motion around the sun. They found that the variations were just what would be expected if the light from Jupiter's moon traveled at a finite, and not infinite, speed. The light from Jupiter's moon would then arrive at Earth sooner or later, depending on whether Earth was closer to or farther from Jupiter in Earth's orbit around the sun (Figure 9.6). From the measured variations in orbital time, lightspeed could be estimated.

So electromagnetic waves were predicted to spread through electromagnetic fields at the speed of light. Was this because light actually was an example of Maxwell's predicted electromagnetic waves, or was it just a coincidence? Perhaps light had nothing to do with electromagnetism. Scientists knew of no way to verify the tantalizing suggestion that light was an electromagnetic wave. So Maxwell's work didn't attract much attention until two decades later, when it became possible to check Maxwell's theory directly, by causing charged objects to oscillate and observing the effects some distance away.

FIGURE 9.6

If Earth moves from point A to point B while Jupiter's moon is orbiting Jupiter one time, the orbital time as measured on Earth will be longer than the true orbital time, because it takes longer for light to get from the moon to point B than it takes for light to get from the moon to point A. This effect creates a variation in the measured orbital time and can be explained by assuming that light has a finite, not infinite, speed.

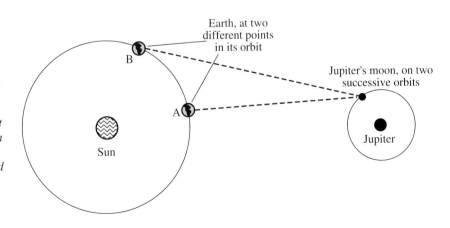

One way to verify Maxwell's theory would be to shake a charged object
at the frequency of visible light, about 10^{15} hertz, and to see whether the
shaking created light. But such high frequencies of oscillation are hard to
achieve, even today.

German physicist Heinrich Hertz (the hertz is named for him) figured
out how to do an experiment of this sort, but at a frequency of only about
10^9 hertz, far below that of visible light. He constructed an electric circuit
that contained a small open space or air gap. Ordinarily, such a gap stops
the flow of electric charge, but Hertz built the endpoints of the gap in
such a way that large amounts of charge could build up on them. After
enough buildup, the charge was forced to jump across the gap. We observe such charge jumping as a spark. Lightning is a spark of this sort.

In Hertz's circuit, the first jump of charge triggered a brief series of
such jumps back and forth across the gap, at a rate of about a billion per
second. If Maxwell was correct, these oscillations should create electromagnetic waves with a frequency of a billion hertz.

At some distance from this predicted source of electromagnetic waves,
Hertz placed a second circuit. This circuit was entirely passive, with no
battery or other internal source of electric current. If Maxwell's theory
was correct, electromagnetic waves from the first circuit should cause
an electric current to oscillate in the second circuit, also at a billion
hertz. The transmission from one circuit to the other should occur at
lightspeed.

Hertz performed several experiments during 1887/1888. The results entirely confirmed Maxwell's predictions. Hertz found oscillations in the second circuit at the expected frequency. He showed that the transmission
occurred at a speed that was at least the same order of magnitude as lightspeed (as you might expect, this was a difficult measurement). And he
showed that these waves had many other properties that light waves have,
such as reflection.

Although Hertz's waves were not light waves, his work convinced scientists that electromagnetic waves really existed and that light is actually
an electromagnetic wave. As a by-product, Hertz's basic scientific work
came to the attention of an ingenious Italian inventor named Guglielmo
Marconi, launching the radio and television revolution. Today, we know
Hertz's waves as radio waves.

Maxwell and Hertz had uncovered a vast new range of physical phenomena, invisible waves similar to light. In Hertz's words, "The connection between light and electricity, of which there were hints and suspicions and even
predictions in the theory, is now established. . . . Optics [the study of visible
light] appears merely as a small appendage of the great domain of electricity. We see that this latter has become a mighty kingdom."

It is a stunning unification: Maxwell's theory correctly describes electricity, magnetism, light, and radio. All of these are different aspects of one underlying reality: electric charge.

We summarize these ideas as follows:

> **THE ELECTROMAGNETIC WAVE THEORY OF LIGHT**
> Every vibrating charged object creates a disturbance in its own electromagnetic field, which spreads outward through the field at 300,000 km/s.* Light is just such an electromagnetic wave.

DIALOGUE 2   If you electrically charge a comb by running it through your hair and then shake it back and forth at a frequency of 1 Hz, will this produce an electromagnetic wave? What will be the wave speed? The frequency? How long is the wavelength?

DIALOGUE 3   Hertz's waves had a frequency of 10^9 hertz. How many megahertz is this? How many kilohertz? Is this on your radio dial?

DIALOGUE 4 MAKING ESTIMATES   About how long does it take light to get to your eyes from a lightbulb in your room?

9.3 The decline of the Newtonian universe

Since ancient Greek times, scientists have generally thought that the universe was made of tiny material particles, atoms. As Democritus put it, there are only atoms and empty space (Chapter 2). It is a view that is remarkably compatible with Newton's physics, which can be thought of as the rules according to which atoms move. This general worldview—atomic materialism coupled with Newtonian physics—dominated science at least during the eighteenth and nineteenth centuries (Section 5.5).

When the electromagnetic field was first proposed around 1850, it was thought of as only a useful way to picture electromagnetic forces and not as a real physical entity. Then Maxwell and Hertz showed that waves can travel through electromagnetic fields and that light is one example of these electromagnetic waves. This meant that electromagnetic fields were not just a useful fiction. They were physically real, as real as light.

The most convincing argument for the reality of electromagnetic fields comes from the conservation of energy. Suppose that a radio transmitter sends a message (an electromagnetic wave) to a receiver on the planet Mars and that the message's travel time is 20 minutes. Energy must travel from the sender to the receiver because it takes energy to cause the receiver to respond. Where is this energy during the 20 minutes between sending and

---

* More precisely, the speed of light, and all electromagnetic waves, is 299,792.458 km/s in a vacuum. When traveling through matter, however, light is slower than this because it is continually absorbed and reemitted by atoms.

2. Yes, you produce an electromagnetic wave whose wave speed is 300,000 km/s and whose frequency is 1 Hz. Since you produce one complete wave every second and the wave travels outward at 300,000 km/s, the wavelength must be 300,000 km.

3. One thousand megahertz; 10^6 kilohertz; this is on neither your AM nor your FM radio dial.

4. The travel time for light is the distance to the lightbulb divided by lightspeed, 3×10^8 m/s. To make the arithmetic easy (remember that in estimates, you want to choose approximate numbers that make the arithmetic easy), suppose the distance is 3 m: $3 \text{ m}/(3 \times 10^8 \text{ m/s}) = 10^{-8}$ s, or 0.00000001 seconds, a hundredth of a millionth of a second.

receiving the message? Not in the sender. Not in the receiver. And energy never just vanishes. So it must be in the space between sender and receiver, in the electromagnetic field.

So *electromagnetic fields contain energy*. And to a physicist, nothing is more real than energy.

"Empty space" is a contradiction in terms. Even when devoid of matter, space cannot be empty. The universe is filled with a sea of nonmaterial and mostly invisible fields.

Physicists resisted giving up the Newtonian clockwork universe. So ingrained was the Newtonian worldview that scientists could not contemplate that energy might exist apart from tiny material particles. They developed the idea that a gaslike material substance called **ether** filled all space. Ether was assumed to be a form of matter but made of some unknown substance rather than of the atoms that are familiar to us. Electromagnetic forces and other forces that act at a distance were assumed to be transmitted by ether. Light and other electromagnetic waves could then be explained mechanically, in terms of the motions of material ether particles and in keeping with the Newtonian tradition. Maxwell, Hertz, and others accepted this ether theory of the electromagnetic force. After two centuries of the Newtonian worldview, it was difficult to think in any other way.

Early in the twentieth century, Albert Einstein was one of the first to break out of this mold. His work showed that the ether theory had to be rejected. But surprisingly, this had no effect on Maxwell's theory or on the interpretation of light and radio as electromagnetic waves. It only affected the underlying mechanistic interpretation of the electromagnetic field. After Einstein's work, electromagnetic fields could no longer be interpreted as properties of a material substance.

So the electromagnetic field turned out to be philosophically revolutionary, the first of many post-Newtonian physical ideas. Although not made of particles and not made of any material substance at all, electromagnetic fields are real nonetheless. Apparently the universe does not act quite like a mechanical clock.

Although the Newtonian worldview still dominates much popular culture and even lies behind many scientists' intuitive view of nature, the mechanical universe is seriously out of tune with twentieth-century physics. The two major post-Newtonian theories, relativity theory and quantum theory, contradict both the specific predictions and the conceptual underpinnings of Newtonian physics. Physics is still in the middle of the post-Newtonian revolution, and it is not clear what new overall scientific worldview will emerge from it. But it is clear that the Newtonian assumptions have broken down.

9.4 The complete spectrum

Light is only a small portion of the "mighty kingdom" (as Hertz puts it) of electromagnetic waves. The entire kingdom is called the **electromagnetic spectrum** (Figure 9.7). *Spectrum* simply means range, or types.

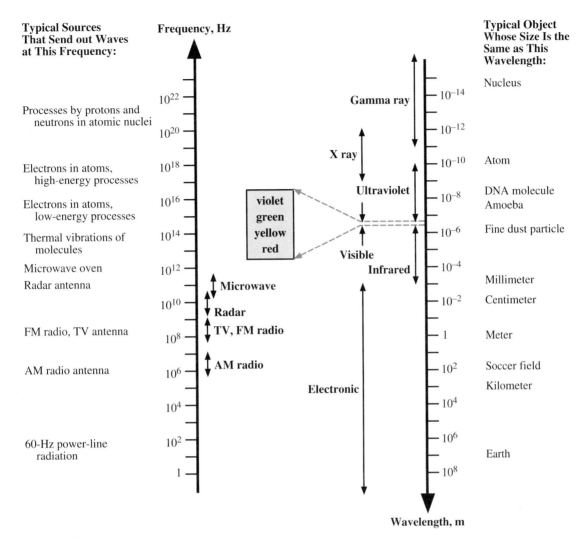

FIGURE 9.7
The electromagnetic spectrum. There are no definite ends to the spectrum and no sharp boundaries between the regions.

As you can see from the figure, scientists have observed electromagnetic waves over an enormous range of wavelengths. Wavelengths are arranged from the longest at the bottom to the shortest at the top, with typical objects having the size of these wavelengths listed for comparison. Frequencies are shown from the lowest at the bottom to the highest at the top, with typical sources of these frequencies listed.

In order to display a large range, Figure 9.7 shows the wavelengths and frequencies on a "logarithmic scale" in which each increment is a factor of 10: 1, 10, 100, 1000, and so forth. A "linear scale"—for example, 10, 20, and 30—could not do justice to the entire range. Logarithmic scales are useful to graph quantities that span a large range.

All these waves are basically the same phenomenon, namely, an electromagnetic field disturbance that is created by a vibrating charged object and that travels outward through the field in the form of an electromagnetic wave moving at lightspeed. All of these waves can be received by other charged objects that are caused to move by the wave as it passes by. All can travel through empty space.

All these waves carry energy. The energy of an electromagnetic wave is called **radiant energy**. For example, the energy in transit from the sun to Earth is radiant energy. Electromagnetic waves are often called **electromagnetic radiation** because they "radiate" out from charged objects. Since higher frequency means higher energy (provided that other things such as amplitude are not changed), energies increase as we move up the scale from bottom to top.

It is useful to arrange the electromagnetic spectrum into the six regions shown in Figure 9.7. These regions correspond to different ways of either sending or receiving electromagnetic radiation. The longest wavelengths, down to about a millimeter, form the **electronic waves**, or radio waves. Humans can create and control these waves electronically by causing electrons to vibrate in human-made electric circuits. Hertz's waves fall into this category, and so does a lot of modern technology. AM radio waves at around 1000 kilohertz (10^6 Hz) and FM radio and TV waves at around 100 megahertz (10^8 Hz) are created by electrons vibrating along a metal antenna that is part of an electric circuit. Radar and microwaves, with frequencies up to a trillion (10^{12}) hertz, also are created electronically.

Many natural processes also create radio waves. Radio astronomers use radio receivers pointed at stars or other space objects to learn about the universe. In fact, stars and other space objects produce electromagnetic radiation in all parts of the spectrum. During the past few decades, many new sorts of receivers or "telescopes," stationed on or above Earth, have produced an explosion of astronomical knowledge.

Infrared radiation has wavelengths ranging from 1 millimeter to below one-thousandth of a millimeter, which is the size of a fine particle of baby powder. Infrared is typically created by the chaotic thermal motion of molecules due to their thermal energy. Since all objects have thermal energy, all objects produce infrared. Hotter objects produce more than cooler objects do. Infrared detectors, such as certain infrared-sensitive chemicals, can detect warmer objects against a cooler background, which is the basis for night-vision devices and infrared photography.

Although you cannot see infrared radiation, you can feel it. Since it is created by thermal motion, it is not surprising that it has the proper frequency to shake molecules into thermal motion. So it can warm the objects it hits. When you feel the warmth of a fire or a hot plate at some distance away, you are using your skin as an infrared detector, "seeing" with your skin. Through competitive evolution, many animals have highly developed infrared sensors for nocturnal vision.

Most higher animals have sensors that detect a narrow range of frequencies just above infrared. The wavelengths of **light** or **visible radiation**

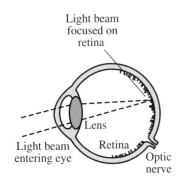

FIGURE 9.8
The human eye.

span less than a single factor of 10, centering on 5×10^{-7} m. This is smaller than the finest dust particles and some five thousand times larger than an atom. Light is created typically by electrons moving within individual atoms.

This region's defining characteristic is simply that the human eye is sensitive to it. Light waves entering the pupil of the eye strike the retina at the back (Figure 9.8). The retina is covered with light-sensitive cells that act like tiny antennae to receive electromagnetic waves in the visible range. Some cells respond differently to different wavelengths, and the brain interprets these as different colors. All of these cells contain light-sensitive chemicals, within which the light wave's vibrating field causes electrons to move, which in turn initiates chemical reactions in the cell. The signal received by the retina is transmitted to the optic nerve, which transmits the signal along the optic nerve to the brain.*

Suppose that you have a variable-frequency source of electromagnetic waves and that you set it to produce green light with a frequency around 6×10^{14} Hz. If you gradually decrease the frequency, the green light will change from green to yellow, orange, and finally red. As you continue decreasing the frequency, the red becomes deeper and darker until, at about 4×10^{14} Hz, the frequency is so low that your retina can no longer respond to it. The source no longer emits visible light. The waves have crossed the boundary into *infra-* (below) red. The source still radiates, but your eye cannot detect it.

Now go in the other direction. Beginning with green, increase the frequency. The color changes from green to blue to violet. The violet light darkens until, around 8×10^{14} Hz, your eye can no longer detect it. The waves have crossed into the *ultra-* (above) violet region. **Ultraviolet radiation** is created in the same way that light is created, by electrons moving within individual atoms. Although similar to light, ultraviolet's higher energy has important consequences.

Ultraviolet radiation has the proper frequency to shake many biological molecules, so it is readily absorbed by living things. It has enough energy to split molecules, which can disrupt or kill living cells. When absorbed by DNA, ultraviolet radiation can cause genetically significant changes, or **mutations**. This can produce a cancerous growth if the mutated cell reproduces many copies of itself, or it can lead to a mutant offspring if the cellular mutation occurs in a sperm or egg cell.

X-ray radiation also comes from electrons in individual atoms, but only from the highest-energy electron activities within atoms. X-ray wavelengths span a range around 10^{-10} m, which is about the size of an individual atom. Humans make X rays in high-energy X-ray tubes (Figure 9.9). Electrons are boiled off a thin heated wire filament and are accelerated toward a metal plate at the other end of the vacuum tube. When the electrons smash into this plate, they cause the plate's atoms to create X rays.

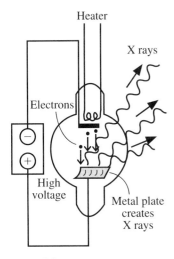

FIGURE 9.9
The operation of an X-ray tube.

---

* Nerve cells transmit signals not as electromagnetic waves but, rather, as waves of chemical reactions moving at only about 30 m/s along the cell.

X rays have important interactions with biological matter. They have enough energy to ionize molecules within biological cells, that is, to knock the electrons right out of some molecules. The evicted electrons can then move on to cause still more damage. Like ultraviolet, this radiation can cause mutations and cancers. Radiation energetic enough to ionize biological matter is called **ionizing radiation**. X rays and gamma rays are ionizing radiations, but ultraviolet has insufficient energy to be an ionizing radiation. X rays are able to penetrate deeply into biological matter, and they are put to the useful cause of examining the interior of the human body without surgery.

There is a certain logic to our tour through the regions of the electromagnetic spectrum. As we move to smaller wavelengths, we are moving toward higher frequencies, which imply higher-energy radiation, which in turn implies higher-energy processes to create the radiation. We also have progressed toward processes that occur in smaller and smaller regions of space: Electronic waves are created in electric circuits; infrared is created in molecules; and the next three (visible, ultraviolet, and X ray) are created in atoms.

It should be no surprise that the shortest-wavelength radiation, **gamma radiation**, carries the highest frequency and highest energy and comes from the highest-energy processes and the smallest regions of space. Gamma rays are created within atomic nuclei by high-energy nuclear processes caused by the strong forces that hold the nucleus together. Gamma rays are created in radioactive materials and in the nuclear reactions known as *fission* and *fusion*, topics that we will discuss in Chapters 15 and 16.

Like X rays, gamma rays are a form of ionizing radiation and can damage biological matter. But this very feature is often put to use to destroy diseased cells and so cure some cancers. Since gamma-ray wavelengths are much smaller than individual atoms, atoms cannot readily respond to them, and so they can penetrate rather deeply into matter.

Consider the electromagnetic waves all around you. The room you are in is full of all these waves. Hundreds of television and radio broadcasts, radio pulsations from neutron stars, radio noise from millions of normal stars, the faint cosmic background radiation, possibly communications from extraterrestrial life, radiations from the sun and from the center of our galaxy, and much more are passing through your room right now. Your body is equipped to receive directly only the tiny visible portion of the complete spectrum of these waves. With the proper receiver, you could tune in to any of the other frequencies. The universe would appear far different in other wavelength ranges and would appear complex indeed if we could directly receive the entire spectrum. The reality that meets the eye is only a tiny fraction of nature's mighty kingdom.

DIALOGUE 5   Your radio is tuned to 98 on the FM dial. What electromagnetic wave frequency is your radio receiving? How fast is this wave traveling? Roughly how big is the wavelength? How about the sound that comes from this radio—is it moving at the same speed as the radio wave? Is sound an electromagnetic wave?

DIALOGUE 6 MAKING ESTIMATES   In the following list, which have wavelengths much bigger than your room (a few meters), which have wavelengths between a millimeter and a few meters, and which have wavelengths of less than a millimeter? AM radio, light, electromagnetic waves from the alternating current that oscillates 60 times each second in your house circuits, warming rays from a fire, rays from a microwave oven, radar, electromagnetic radiation from shaking an electrically charged blouse that you remove from the dryer.

9.5 Solar radiation: *the light from our star*

The sun, "Sol," transmits electromagnetic waves in every region of the spectrum (Figure 9.10, see color insert). Most solar radiation is visible, infrared, and ultraviolet and is created at the sun's visible surface. Other solar radiation is created in a rarefied, very hot gas that surrounds the sun in the same way that Earth's atmosphere surrounds Earth. Processes within the sun's atmosphere create high-energy X rays and some gamma rays, along with radio waves. The intense radiation created by high-energy processes deep within the sun is absorbed and altered within the sun, and little of it escapes directly.

Figure 9.11 graphs the relative amounts of radiant energy created by the sun at different wavelengths. When you sit in the sunlight, your eyes detect the sun's visible radiation, and your skin detects its infrared as warmth. Your skin also detects ultraviolet, but you don't notice it until a little later, as the cellular damage that we call *sunburning*.

The solar radiation graph (Figure 9.11) reaches its highest point around a yellow-green wavelength of 5.5×10^{-7} m and remains high throughout the

FIGURE 9.11

The relative amounts of energy at different wavelengths in the solar spectrum.

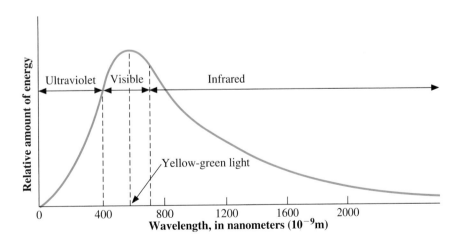

---

5. 98 megahertz = 98×10^6 hertz = 98,000,000 hertz. 3×10^8 m/s. Roughly 5 m (Figure 9.7). Sound moves much more slowly; sound is not an electromagnetic wave.

6. Use Figure 9.7. Much bigger than a few meters: AM radio, waves from alternating current, waves from the blouse. One millimeter up to a few meters: microwaves, radar. Less than 1 mm: warming rays, light.

visible region. Solar energy is most concentrated throughout the region that our eyes can detect.* Even more strikingly, our eyes have evolved in such a way that their maximum sensitivity is to yellow-green light, precisely where the solar spectrum peaks. Some municipalities take advantage of this by painting their fire trucks yellow-green. What a strange coincidence that all this radiation happens to occur precisely in the region that our main sensors, our eyes, can detect. How lucky for us! But it was not just luck. Through biological evolution, animals developed detectors of the dominant radiation reaching Earth. An animal whose sensitivity peaked in, say, the microwave region would be at a biological disadvantage, because there isn't much naturally occurring microwave energy by which such an animal could see.

The amount of solar energy reaching Earth is different at different locations, in different seasons, in different weather conditions, and at different times of the 24-hour day. In the United States, an average 200 watts (200 joules every second) strikes every square meter. This figure, 200 watts per square meter, is an average that takes into account all of the mentioned variations.

DIALOGUE 7 MAKING ESTIMATES   Photovoltaic cells are devices that transform solar energy into electric current. If such devices were 100% efficient, about how much area would need to be covered by these cells in order to provide the average of 1 kilowatt of electric power that a typical family home uses? Actual photovoltaic cell are only about 15% (one-seventh) efficient. At this efficiency, how much area must be covered? Could you put this on your roof?

DIALOGUE 8   Once radiant energy from the sun is absorbed by your skin, is it still radiant energy? What has it become?

9.6 Global ozone depletion _____

The concept is not obvious: a perfume spray in Paris helps to destroy an invisible gas in the stratosphere and thereby contributes to skin cancer deaths and species extinction half a world away and several generations in the future. Neither traditional environmental law nor traditional diplomacy offers guidelines for confronting such situations.

Richard Benedick, chief U.S. negotiator of the Montreal Ozone Treaty

Ozone depletion is a classic science and society problem, involving technology, politics, business, scientific understanding, public understanding, and the daily decisions you and I must make. It is perhaps the first global environmental problem, because local actions have worldwide effects on ozone. Public understanding of science is essential if democracies are to cope with such problems. Ozone is a good historical case study because its societal history is mostly completed: Humankind has damaged Earth's ozone, belatedly recognized that damage, and done nearly all that can be done to rectify it. The verdict, as to how bad the damage is and what its effects will be, is now largely back in nature's hands. That verdict will not be in until the mid-twenty-first century.

---

* More precisely, the solar energy per unit wavelength range is largest in this region.
7. One kilowatt (1000 watts) of solar energy falls on 5 square meters (1000/200) of surface. At an efficiency of about one-seventh, it would take seven times this much area: 35 square meters. If square shaped, this would be about 6 m on a side. You could probably fit it on your roof.
8. No. It has been transformed into thermal energy, plus some chemical energy from chemical reactions caused by ultraviolet.

FIGURE **5.12**
Star birth. This cloud of dust and gas in
space is associated with a group of
newly forming stars.

FIGURE **5.14**
An imaginary view of the newborn sun
during formation of small rocky bodies
that will soon coalesce to form planets.
Dust clouds partially obscure the sun.

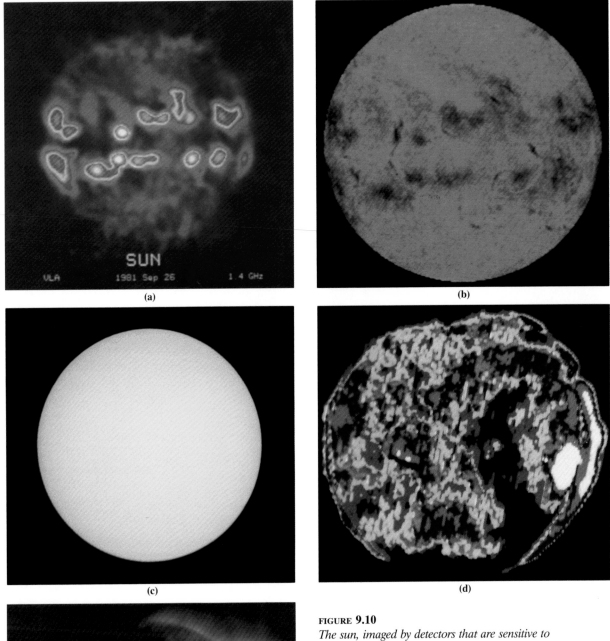

(a)   (b)

(c)   (d)

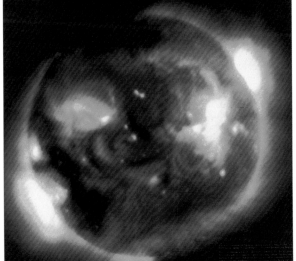

FIGURE **9.10**

The sun, imaged by detectors that are sensitive to each of five different regions of the electromagnetic spectrum: (a) electronic or radio waves, (b) infrared radiation, (c) visible light, (d) ultraviolet radiation, (e) X rays. Since ultraviolet, X-ray, and gamma radiation are strongly absorbed by Earth's atmosphere, detectors in these spectral regions must operate above the atmosphere, in satellites or balloons. No gamma-ray image is included here, because gamma-ray detectors today are unable to "resolve" the sun's image into a recognizable disk-shaped object.

FIGURE 12.7a

Artist's concept of a space habitat for 10,000 people. To provide artificial gravity, the colony would spin, like a bicycle wheel, around an axis through its center. The wheel might be two kilometers in diameter.

FIGURE 12.7b

Interior view of the habitat. It could contain air, clouds, lakes, mountains, and towns. Gravity would feel normal for inhabitants on the inside of the outer surface.

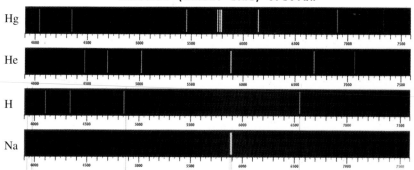

EMISSION (BRIGHT LINE) SPECTRA

Hg

He

H

Na

FIGURE **14.2**

The continuous spectrum created by an incandescent bulb, and the line spectra produced by several different kinds of gases: Mercury (Hg), helium (He), hydrogen (H), and sodium (Na). Wavelengths are shown in angstrom units. One angstrom is 10^{-10} m, or a tenth of a billionth of a meter.

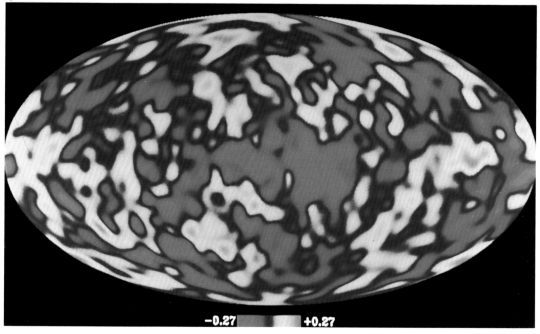

−0.27   +0.27

FIGURE **16.6**

A fifteen-billion-year-old "fossil": map of the ripples in the background radiation emitted by the big bang. The radiation responsible for this map was emitted just 100,000 years after the initial creation event. Before 100,000 years, the universe was so hot that protons and electrons could not unite to form hydrogen, and the resulting mix of charged particles absorbed any radiation present. At 100,000 years the universe had cooled enough for electrons to combine with protons to form neutral hydrogen atoms and radiation was able to propagate through the universe for the first time. Although it is now cooled into the microwave region of the spectrum, you are looking at the pattern of the universe's first light.

Chlorofluorocarbons (CFCs), molecules made from atoms of chlorine, fluorine, and carbon, were first synthesized in 1930. CFC molecules are chemically "inert" or "stable," meaning that they do not readily react with other substances. They normally form a gas but become liquid when put under high pressure. Being inert, they are nontoxic in the human body, noncorrosive in mechanical devices, and nonflammable in the atmosphere. And they are inexpensive to make.

Such a chemical can have many uses, one of which is as a coolant. Refrigerators and air conditioners operate in the same way that heat engines do, only in reverse (Figures 9.12 and 9.13). Just as heat engines use hot gases as the "working fluid" that does the work, so refrigerators and air conditioners use a "coolant" that has work done on it while extracting thermal energy from a refrigerator or from a house.

Because CFCs had the right properties, it soon became a universal coolant, marketed as "Freon." Production soared after 1931. In the 1940s, CFCs were found to be useful as pressurized gases to propel aerosol sprays. Then they began to be used as blowing agents to form lightweight, closed bubbles in plastics that could be used as "thermal insulators" against heat and cold and to make soft foam rubber. They caused the air-conditioning revolution that cooled shopping malls and automobiles and facilitated population shifts to the U.S. Southwest. They played a role in the computer revolution, in which their nonreactive properties made them useful solvents to clean small electronic parts.

In 1973, two university scientists were studying chemical emissions from rocket launches into the upper atmosphere or **stratosphere**, 10 to 50 kilometers

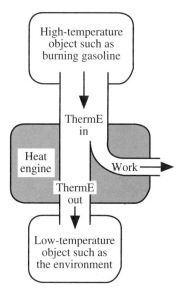

Figure 9.12

Energy-flow diagram for a heat engine.

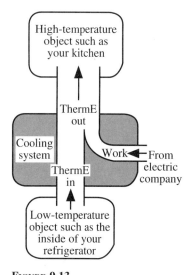

Figure 9.13

Refrigerators operate in the same way that heat engines do, but in reverse. An outside energy source does work to push thermal energy "uphill," from a low to a high temperature.

TABLE 9.1

Composition of the Atmosphere

Molecule	Concentration
Major constituents	
N_2	78.084 %
O_2	20.946 %
Ar	00.934 %
total	99.964 %
Trace gases (in ppm[a])	
CO_2	350
H_2O	20–20,000
Ne	12
He	5
NO_2	2
CH_4	2
Xe	2
Kr	1
O_3	<1
H_2	<1
NO	<1
CO	<1
N_2O	<1
SO_2	<1
NH_3	<1

[a] Parts per million.

overhead. They were concerned about how the emissions reacted with O_3, or **ozone**, one of the **trace gases** in the atmosphere (Table 9.1). Ozone is produced naturally in the stratosphere from O_2 with the help of ultraviolet radiation from the sun at that altitude. The radiation breaks up the O_2 molecule, and the resulting oxygen atoms then combine with O_2 to create O_3.

They discovered that the chlorine released by rockets could destroy ozone. The ozone-destroying reaction is

$$Cl + O_3 \rightarrow ClO + O_2$$

This produces normal molecular oxygen, O_2. Under the bombardment by ultraviolet radiation at that altitude, a second reaction then occurs:

$$ClO + ClO + sunlight \rightarrow Cl + Cl + O_2$$

The second reaction frees the chlorine, which is then able to destroy more ozone. Scientists found that a single Cl atom broke down about 100,000 ozone molecules.

This didn't seem too important, because not much chlorine is released from space rockets. But the next year, in 1974, other university chemists suggested a more alarming possibility. Mario Molina and Sherwood Rowland discovered that because CFC molecules are inert and gaseous, they are not chemically broken down or rained out in the lower atmosphere. Instead, they migrate slowly up to the stratosphere, where they may remain intact for decades to even centuries. Molina and Rowland theorized that solar ultraviolet radiation should eventually break down CFCs, releasing large quantities of ozone-destroying chlorine into the stratosphere.

Nobody had dreamed that CFCs, seemingly safe and thoroughly tested, could pose dangers many kilometers above Earth's surface. With annual CFC production at 800,000 tonnes in 1974 and because of CFCs' long atmospheric lifetimes, people became alarmed. The reason for the alarm is that stratospheric ozone protects us from the sun's ultraviolet radiation and so is important to life on Earth.* As you know, ultraviolet radiation is present in the solar spectrum and is biologically harmful. The ozone molecule vibrates naturally at ultraviolet frequencies and so is a strong absorber at these frequencies.

Ozone is extremely dilute in the atmosphere, forming only about 0.00003 percent, or 0.3 parts per million (ppm) (less than 1 molecule in 1 million) of the atmosphere. If compressed to atmospheric pressure, atmospheric ozone would form a layer only about 2 millimeters thick. And yet this dilute gas absorbs a significant fraction of the sun's ultraviolet radiation. Ozone illustrates the possible importance of each of the many atmospheric trace gases (Table 9.1). Being dilute, they are easily altered by human activities (see Dialogue 9).

A debate, of a type that has now become familiar, ensued, continuing from 1974 to 1978. The chemical industry argued that the theory was speculative and that there should be no economically damaging CFC restrictions until there was clear evidence of harm. Environmentalists countered that the the-

---

* *Ground-level* ozone, on the other hand, is a toxic pollutant. It is a part of automobile exhaust and is the primary component of urban smog.

ory was plausible enough that we should not risk further delays. Consumer pressures for non-CFC goods played a role. In 1978, the United States and a few other governments initiated a ban on the least-essential CFC technology, spray propellants. It was the first time that a substance suspected of causing global harm had been regulated before the effects had been fully demonstrated.

This ended the debate. By 1980, the problem had faded from public view. But CFCs continued building up in the atmosphere, because U.S. spray propellants formed only 25% of the world's CFCs and other production was on the increase.

It was fortunate that a British meteorological survey team had been performing routine operations in Antarctica since 1956. The team observed a new trend beginning in 1977. Ozone concentrations dropped every spring and then returned to normal in a few months. The decline was greater each year, plummeting 40% by 1984. The area involved was larger than the United States. So unbelievable at first were these measurements that the scientists delayed publication for several years while rechecking their data and instruments. Their report, in 1985, was greeted skeptically. Perhaps the effect was coincidental or part of a natural cycle and would soon vanish.

In early 1986, U.S. atmospheric chemist Susan Solomon (Figure 9.14) organized an expedition to Antarctica to study the possible causes of ozone depletion there. With the arrival of the Antarctic spring in October, she found not only that ozone was now 50% below normal but also that high levels of chlorine were present. There was indeed an "ozone hole," and it was getting bigger. It was entirely unpredicted. Molina and Rowland and others had thought that the effects would occur only gradually and worldwide. Was this sudden localized depletion at the South Pole actually caused by chlorine from CFCs?

The data amassed by the Solomon expedition provided the key to understanding the ozone hole. Solomon offered a hypothesis involving the ice clouds that form high above Antarctica during the long dark polar winter. When the sun reappeared in the Antarctic springtime, solar energy caused chemical reactions on the surfaces of the tiny ice particles forming these clouds. The reactions released ClO into the stratosphere from chlorine compounds that had been mixed in the ice particles, and sunlight then split the ClO molecule to free the chlorine. It is a complex mechanism, not the sort of thing that scientists were likely to have predicted in advance. The stratosphere turned out to be surprisingly subtle. A second ozone-observing expedition, in 1987, confirmed this hypothesis.

FIGURE 9.14
Atmospheric chemist Susan Solomon organized the first National Ozone Expedition in 1986.

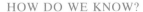

From time to time we are all a bit taken aback by the fact that this thing that fascinates us from an intellectual point of view has serious consequences. I know a number of others feel as I do, that it would be nice to be doing medical research. . .and say "Wow, I have a cure for some terrible disease." But we don't have that kind of message. Still, intellectually it is incredibly exciting. It's so complicated and so beautiful.
　　Susan Solomon, atmospheric ozone scientist

HOW DO WE KNOW?

Figures 9.15 and 9.16 show some of the evidence that implicated chlorine as the culprit. The two graphs show data recorded by aircraft flights into the Antarctic stratosphere on August 23, 1987, as the sun was reappearing, and three weeks later on September 16. The ClO and ozone concentrations are graphed in parts per billion (ppb) for ClO and parts per million (ppm) for ozone, at different southern latitudes from 62 degrees (just south of South America) to 73 degrees (inland over Antarctica). On August 23, ClO appeared in the stratosphere over Antarctica,

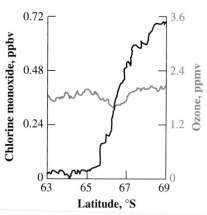

FIGURE 9.15
ClO concentrations (black) and ozone concentrations (green) measured on August 23, 1987, before the ozone had begun vanishing near the South Pole for that year.

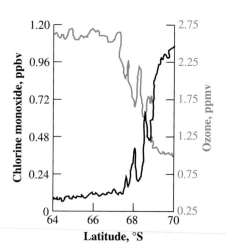

FIGURE 9.16
Concentrations measured on September 16, 1987, after the ozone was depleted.

south of 65 degrees, but had not yet begun to destroy ozone. By September 16, ozone had been destroyed in the polar regions, and at precisely the latitudes where ClO was most concentrated. The correlation between the presence of ClO and the depletion of ozone in Figure 9.16 is exquisite: Ozone zigs precisely where ClO zags. This detailed correlation leaves little doubt of a causal relationship between ClO buildup and ozone depletion.

The Antarctic ozone hole has continued deepening, with springtime depletions reaching 70% and spreading across 10% of the Southern Hemisphere. Australian monitoring stations have noted sharp yearly depletions since 1987, and shifting patches of ozone-depleted atmosphere have occasionally pushed ultraviolet radiation in Australia 20% above normal. Due to the longevity of CFCs in the atmosphere, such deep Antarctic ozone holes may be the norm for the twenty-first century, no matter what we do now to stop further CFC production.

Ozone depletion is now showing up around the globe, and a smaller hole has opened over the Arctic. The worldwide trend of declining ozone levels began in the 1980s; by 1992, ozone levels over the United States had dropped by 8%.

Science is demonstrating that this planet is more vulnerable than had previously been thought.
Richard Benedick

In 1987, in Montreal, Canada, forty-seven nations forged a treaty that was unique to world history.* The **Ozone Treaty**, developed independently of and mostly before the Solomon Antarctic expedition, was not based on clear experimental evidence of ozone depletion and was not a response to a disaster that had already occurred. Instead, the treaty was preventive action on

---

* I recommend U.S. chief negotiator Richard Benedick's account, *Ozone Diplomacy* (Cambridge, MA: Harvard University Press, 1991).

TABLE 9.2

Ozone-Destroying Chemicals in the Atmosphere

Chemical	Uses	Average lifetime in atmosphere (yr)	Share of problem (%)	U.S. share of use (%)
CFC-11	coolant, aerosol, foam	60	28	22
CFC-12	coolant, aerosol, foam	130	47	30
CFC-113	solvent	90	5	45
Carbon tetrachloride	solvent	50	15	27
Methyl chloroform	solvent	7	2	50
Halon 1211	coolant, foam	25	1	25
Halon 1301	fire extinguisher	110	2	50

SOURCE: U.S. Office of Technology Assessment, 1991

a global scale, based mostly on scientific theories about what might occur. The wisdom of this treaty, in which the United States played a leading role, has now been amply confirmed.

The Ozone Treaty calls for CFCs and other ozone-destroying chemicals (Table 9.2) to be essentially phased out by the year 2000. The restrictions are extraordinarily broad and rapid. The entire treaty is unprecedented. The treaty will definitely make a difference (Figure 9.17). Since the 1987 treaty, the chemical industry, sensing an expanding new market for ozone-friendly products, has moved rapidly to develop replacements. In fact, industries have been adopting these so quickly that the phaseout has moved faster than expected.

The verdict is now mainly up to nature. Because of the long atmospheric lifetimes of ozone-destroying chemicals (Table 9.2), that verdict will take a while. As the lower graph of Figure 9.17 shows, concentrations of CFCs are expected to continue rising until 2000 and then slowly fall to below 2 ppb by 2050. Because the effects in the Antarctic were first observed when global chlorine concentrations had reached 2 ppb, this level might be safe. For

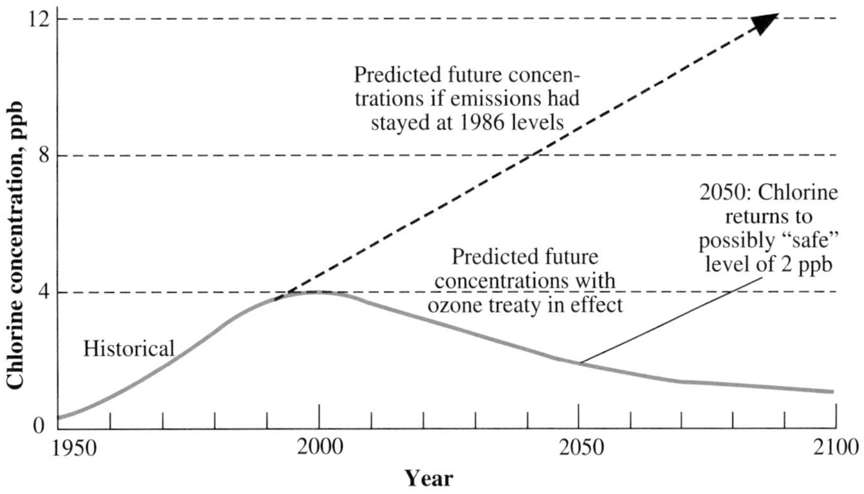

One should never make the mistake of underestimating the power of consumers when they are educated and motivated. . . .What consumers were reading about and seeing on television persuaded them to buy something different. And this provided a very powerful signal to industry that they should come up with substitutes, which they soon did.
Richard Benedick, on CFC-powered sprays

FIGURE 9.17
The effect of the Ozone Treaty on atmospheric chlorine concentrations, which are in parts per billion. (Source: U.S. Office of Technology Assessment)

comparison, Figure 9.17 also shows the predicted chlorine concentrations if there had been no restrictions (upper graph).

What effects are likely? The U.S. Environmental Protection Agency estimates 12 million additional U.S. skin cancer cases, including 200,000 additional deaths, during the next fifty years. This is a 50% increase above current U.S. skin cancer rates. Other expected human effects include eye cataracts, suppression of immune system responses, and premature aging of the skin. Some plant species are sensitive to ultraviolet, and crop production is expected to decrease. Microscopic plants and animals living near the ocean's surface may be depleted, along with shellfish and other fish. Scientists are divided on the threat to life in the most affected region, Antarctica. Most of the organisms there live underwater, where they are somewhat protected. But microscopic plants living near the surface are fundamental to the Antarctic food chain and may be vulnerable. By 1993, there was evidence of a 5 to 10% decline in the number of plankton in the ocean around Antarctica. Possible scenarios run the gamut from a minor drop in these microorganisms to a collapse of the entire Antarctic ecosystem.

DIALOGUE 9 MAKING ESTIMATES: THE MEANING OF PPM
Although Earth's atmosphere extends upward for about 250 km, if it were squeezed down in such a way as to bring the entire atmosphere to the density and pressure that it normally has at sea level, it would extend only 8 km. From this information, estimate the thickness of atmospheric ozone (concentration of 0.3 ppm) if it all were compressed to atmospheric density and pressure.

DIALOGUE 10   Refrigerators (Figure 9.13) cool the objects inside. How do they affect your kitchen?

9.7 Global warming

If you took Earth's temperature from space, you would find it to be a chilly $-19°C$ ($-2°F$). The way you could take Earth's temperature, if you were in space, would be by studying its electromagnetic radiation output. This would be like taking the approximate temperature of an electric hot plate by looking at it: If it glowed white, it would be very hot, bright red means hot, and a dimmer red means less hot. A refinement of this method, using devices that measure the precise spectrum of the radiation, would tell you the precise temperature. Earth is not hot enough to glow visibly in the way a hot plate or the much hotter sun glows. Instead, Earth "glows infrared."

---

9. $(0.3 \times 10^{-6}) \times 8$ km $= 2.4 \times 10^{-6}$ km $= 2.4 \times 10^{-3}$ m $= 2.4$ millimeters.
10. The thermal energy removed from the refrigerator, and the energy from the electric company, must be exhausted from the refrigerator into your kitchen. This warms your kitchen.

Fortunately for us, Earth's average surface temperature is $+14°C$, far warmer than the $-19°C$ detected from space. The extra 33 degrees of warming is due to Earth's surrounding "blanket" of atmospheric gases. This warming is called the **greenhouse effect**.

Surprisingly, the gases that create the greenhouse effect are not the nitrogen and oxygen that form the bulk of Earth's atmosphere. Rather, the effect is due to certain trace gases, mainly water vapor and carbon dioxide (Table 9.1). The reason that these gases, called **greenhouse gases**, are so important to causing the greenhouse effect is that they are strong absorbers of infrared radiation.* We might expect any atmospheric gas that absorbs infrared to be important to Earth's energy balance, since Earth radiates in the infrared. Again, we see the importance of trace gases.

Like many of nature's ways, the greenhouse effect is best understood in terms of energy flows. For starters, let's study the energy flow near the surface of an imaginary Earth that has no greenhouse gases (Figure 9.18). We further imagine that the sun has suddenly turned on for the first time and begun warming a cold Earth.† Some of the solar radiation hitting Earth is reflected back into space, and the rest is absorbed by the atmosphere and the surface. The absorbed energy warms the atmosphere and the surface. As it warms, Earth itself begins radiating energy out to space, just as any object radiates energy because of the thermal motion of its molecules. As Earth warms, its radiated energy increases. Earth's radiated energy is infrared because Earth is not hot enough to radiate at higher frequencies. Eventually, Earth reaches a temperature at which its radiated energy just balances the energy it absorbs from the sun. This balance temperature is $-19°C$. Earth then ceases getting warmer and maintains this $-19°C$ temperature, because the radiated energy just balances the absorbed energy. The energy flow is then as shown in Figure 9.18.

When we add the greenhouse gases, something new happens (Figure 9.19): Greenhouse gases absorb and reradiate the infrared energy that Earth radiates. When these gases reradiate, they do so in all directions, downward as well as upward. This makes Earth's surface warmer than it would have been because there is now additional, "recycled," energy striking the surface. Notice that the top of the atmosphere is still at $-19°C$, even though the temperature of Earth's surface has been raised to $+14°C$. The greenhouse gases act like a thermally insulating blanket that warms Earth underneath, but without warming the outside of the blanket.

The glass on a greenhouse operates in a roughly analogous way. Sunlight passing inward through the glass is absorbed inside the greenhouse and is reemitted as infrared, which is absorbed and reradiated both outward and

---

* Three-atom molecules such as H_2O and CO_2 are efficient infrared absorbers because they have many different ways of vibrating in response to radiation. The dominant atmospheric molecules, N_2 and O_2, have only two atoms, and so each molecule has only one way it can vibrate (atoms moving closer together and then farther apart).

† Something like this did happen once, about 5 billion years ago.

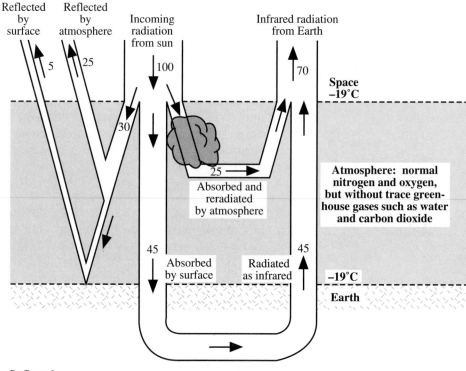

FIGURE 9.18

Energy flow near the surface of an imaginary Earth that has no greenhouse gases but otherwise has a normal atmosphere. The numbers represent percentages relative to the radiant energy received from the sun.

Reflected by surface

Reflected by atmosphere

Incoming radiation from sun

5

25

100

Infrared radiation from Earth

70

Space −19°C

30

25

Absorbed and reradiated by atmosphere

Atmosphere: normal nitrogen and oxygen, but without trace greenhouse gases such as water and carbon dioxide

45

45

Absorbed by surface

Radiated as infrared

−19°C

Earth

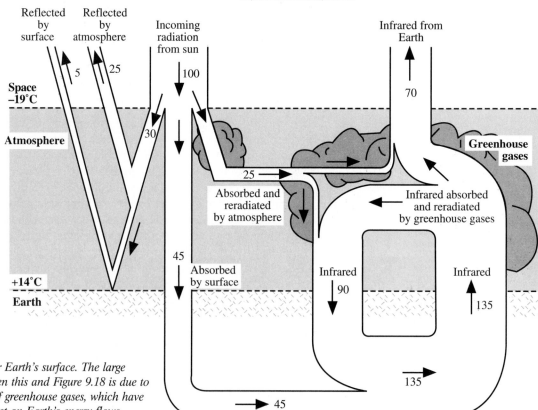

Reflected by surface

Reflected by atmosphere

Incoming radiation from sun

5

25

100

Infrared from Earth

70

Space −19°C

Atmosphere

30

25

Absorbed and reradiated by atmosphere

Greenhouse gases

Infrared absorbed and reradiated by greenhouse gases

45

Absorbed by surface

Infrared

90

Infrared

135

+14°C

Earth

135

45

FIGURE 9.19

Energy flow near Earth's surface. The large difference between this and Figure 9.18 is due to trace amounts of greenhouse gases, which have a significant effect on Earth's energy flows.

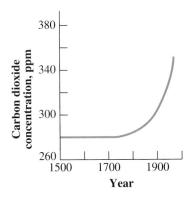

FIGURE 9.20
Measured or estimated past atmospheric carbon-dioxide concentrations in parts per million. (Source: U.S. Office of Technology Assessment)

Of course, the overarching issue now is global warming, which is admittedly much more complex than the ozone problem.
Richard Benedick

First, no evidence for the existence of the greenhouse effect can be found in the temperature records of the last 100 years; and second, current forecasts of global warming for the 21st century are so inaccurate and fraught with uncertainty as to be useless to policy-makers. . . .It is difficult to argue with the contention that "the only question concerning the warming is how much, and by when." However. . .the answer to the question "how much" may turn out to be very little.
Our second recommendation [is] that major policy actions not be undertaken until the implications are better understood.
Richard Lindzen, meteorologist, and William Nierenberg, oceanographer

inward by the glass.* The inside of your car also warms this way when you park it in sunlight.

Combustion, the source of most of the energy for our industrial society, produces carbon dioxide, a greenhouse gas. This is beginning to create a problem. For hundreds of years before the Industrial Revolution, which began around 1750, carbon dioxide maintained a natural atmospheric concentration of about 280 ppm (Figure 9.20). The history of the fossil-fueled industrial age is reflected in the CO_2 accumulation since that time. The concentration is now over 350 ppm, 25% higher than the preindustrial concentration. This increase in greenhouse gases might warm Earth above its current temperature, a possibility known as **global warming**. It is a problem that has been predictable ever since the beginning of the industrial age and that actually was predicted during the nineteenth century but that was seldom discussed as long as the problem was still thought to be far in the future. Today, most scientists consider it to be a problem for the present generation.

It isn't easy to predict the weather. There is a lot of uncertainty about when global warming will become a problem and how serious it will be. Scientists approach this as they approach other questions, with theories and observations.

One important approach is computerized models of the climate. Modelers predict atmospheric quantities throughout the atmosphere, at each point of a three-dimensional grid of points separated by several hundred kilometers. Sophisticated climate models, with closely spaced grid points and using lots of data, are complex and involve many uncertainties.

One uncertainty in all the predictions is future CO_2 levels, which will be determined mainly by fossil-fuel use. To separate this societal part of the problem from the science part, modelers simply assume some CO_2 level and ask, "What will happen if we reach this level?" A common level to choose is an effective CO_2 doubling, an increase of all human-created greenhouse gases to concentrations having the same effect as 560 ppm of CO_2, twice the preindustrial CO_2 concentration.

Putting together the uncertainties, the models predict that an effective CO_2 doubling will cause the global average surface temperature to increase by 1 to 5°C (2 to 9°F).

HOW DO WE KNOW?

How valid are the computer models? One way to check is to see what they predicted about past events. Most models predict (retroactively) a 1°C increase during the past century. It is not easy to check this because 1°C is smaller than typical local variations and because it is not easy to measure Earth's overall surface temperature. The most comprehensive assessment reveals a warming of 0.3 to 0.6°C during the past century (Figure 9.21), which is roughly consistent with, but less than, the predicted increase. It is hard to tell whether any or all of the observed increase is actually due to greenhouse warming. For example, it could be

---

* But the analogy isn't perfect because a greenhouse also traps the air inside so that it cannot cool by mixing with outside air. Without this trapping, a greenhouse wouldn't get very warm.

FIGURE 9.21

Global average surface temperatures measured relative to the average temperature between 1951 and 1980. The bar graph shows the temperature measured each year, and the smooth curve is a "running average" obtained by averaging over the five years around each year. (Source: National Academy of Sciences)

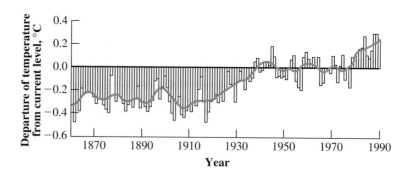

caused instead by increased solar activity. It is also possible that other effects have cooled the globe and partially masked an even greater global warming. In another decade or two, when the predicted warming will be greater, temperature data should give a clearer check on the models.

Long-term geological records also provide useful checks. Figure 9.22 shows temperatures and CO_2 concentrations during the past 160,000 years. Air bubbles in ancient Antarctic ice sheets were analyzed to obtain the changing atmospheric CO_2 content over this period. The air temperature at the time the ice was deposited can be inferred by studying the oxygen in the snow.* The resulting temperatures and CO_2 concentrations are correlated (they track each other), which may indicate a causal relationship between the two.

The U.S. Environmental Protection Agency predicts that unless emissions change significantly, the amount of CO_2 will double during the 2030s. Simply to hold the amount of CO_2 at its present level, there would have to be a net 50 to 80% reduction in emissions during the next few decades. Figure 9.23 shows the history of CO_2 emissions since 1950. Emissions increased at

Tapping the limited resources of our world—particularly those of the developing world—simply to fuel consumerism, is disastrous. If it continues unchecked, eventually we all will suffer. We must respect the delicate matrix of life and allow it to replenish itself.
     The Dalai Lama, speaking at the
          1992 Earth Summit in
               Rio de Janeiro

FIGURE 9.22

Carbon-dioxide levels (left-hand scale, upper graph) and temperatures (right-hand scale, lower graph) during the past 160,000 years. Temperatures are departures from the current level. The period of low temperatures, from 120,000 to 15,000 years ago, was the last ice age. Carbon-dioxide levels have been about 280 ppm for the past few thousand years. During the past 250 years, an interval too short to plot on this graph, the industrial age has pushed the level to over 350 ppm.

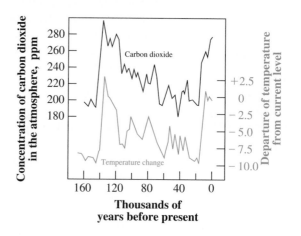

* The ratio of the oxygen isotope $^{18}_{8}O$ to ordinary oxygen $^{16}_{8}O$ is dependent on the temperature.

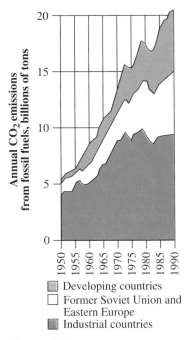

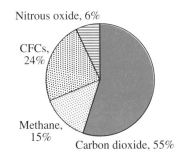

FIGURE 9.23
The history of carbon-dioxide emissions from fossil fuels. Annual emissions in billions of tons, from three groups of countries and total. (Source: Oak Ridge National Laboratory)

FIGURE 9.24
The main human-related greenhouse gases and the contribution of each to global warming in the 1980s. (Source: U.S. Office of Technology Assessment)

4% per year between 1950 and 1973, dropped following the 1973 and 1979 world oil price increases, and rose at 2 to 3% per year between 1983 and 1990.

Figure 9.22 gives a historical perspective on the importance of all this. The highest CO_2 concentration of the past 160,000 years was about 300 ppm and occurred 130,000 years ago at the warmest part of Earth's last interglacial (between the ice ages) period. Today's 350 ppm is already considerably higher than that previous maximum and occurred just during the past 250 years, a time interval too short to show on the figure. The last ice age began 120,000 years ago and ended 15,000 years ago. Earth warmed by 5 degrees as it emerged from that ice age.* Because enormous changes in world climate accompanied that 5-degree increase, we should expect a significant climate change if anything like a 5°C warming occurs.

But CO_2 is not the only problem. As Figure 9.24 shows, three other gases contribute, including the CFCs that are also causing ozone depletion. Tree removal also warms Earth, because trees store CO_2. Destroyed trees release their CO_2 to the atmosphere either rapidly (when burned) or gradually (when used for wood products). Permanent deforestation in the tropics causes between 7 and 30% of worldwide CO_2 emissions. However, the long-term effects are reduced, perhaps to zero, if the destroyed trees are replaced.

A temperature increase is only one possible consequence. Another is a rise in ocean levels, due mainly to the expansion of the oceans as they are heated and also due to partial melting of polar ice caps. A doubled CO_2 concentration is expected to cause a 10- to 30-centimeter rise in the sea level. Sea levels have risen between 10 and 20 centimeters during the past century, but it is not known whether this is related to global warming.

Many consequences will have effects that further influence their original cause. Such **feedback effects** can be unpredictable and dramatic. As one example, warming should increase the evaporation of water, which might increase the cloud cover, cooling Earth and reducing global warming. Such a reduction would be a negative feedback effect. On the other hand, increased evaporation might create additional atmospheric water vapor, which would enhance the warming, because water vapor itself is a greenhouse gas. This would be a positive feedback or amplifying effect. Positive feedbacks can "run away" unpredictably. For example, warming could cause increased evaporation, causing increased water vapor, causing even further warming, causing even further evaporation, and so forth. Scientists do not know which of these effects will actually dominate.

What can be done? To hold greenhouse gases to current levels, there would have to be a net 50 to 80% reduction in the annual emissions of all

---

* Figure 9.22 exaggerates this warming because temperatures in the Antarctic, where the data for the figure were gathered, vary more than do global temperatures.

TABLE **9.3**

Carbon Dioxide Emissions from Fossil Fuels, per person, in Selected Countries in 1989.

Country	CO_2 per person (tonnes)
United States	17.9
Soviet Union	12.0
West Germany	9.5
Japan	7.7
China	2.0
Brazil	1.3
India	0.7
Zaire	0.1

SOURCE: Oak Ridge National Laboratory

Analysis of the CO_2 budget illuminates the diversity in processes and feedbacks that control and respond to atmospheric CO_2. . . .[B]oth land and sea will respond dynamically, not passively, to anthropogenic [human-caused] increases in atmospheric CO_2. Several important feedbacks will extend to millennia and longer. . . .Thus, fundamental uncertainties. . .extend to time scales well beyond the period of foreseeable management policies.

Eric Sundquist,
U.S. Geological Survey

Teach your children what we have taught our children—that the earth is our mother. What ever befalls the earth, befalls the sons of the earth.

Chief Seattle

greenhouse gases during the next few decades. This could be accomplished only with large reductions in fossil-fuel use.

Are large reductions in fossil-fuel use possible? Economic growth went hand in hand with fossil-fuel growth up until the 1973 Mideast oil crisis. Since then, oil price increases have given industrialized countries incentives to achieve growth without more fossil fuel (Figure 9.23), primarily by using energy more efficiently and also by switching to nuclear and solar energy. Emissions from developing countries have increased recently (Figure 9.23), for two reasons: Their fossil-fuel use per person is increasing rapidly as they begin to industrialize, and their population is increasing rapidly as well. Their per capita fossil-fuel use is still quite low today (Table 9.3) but could triple or quadruple. If this happens, world emissions will rise greatly no matter what happens elsewhere (Figure 9.25).

Action on global warming has been much weaker than on ozone depletion, partly because the costs of making a big dent in the CO_2 problem are much greater than they are for the ozone problem. Western Europe has agreed to maintain its greenhouse emissions at a constant level. Some analysts have suggested that future large reductions of greenhouse gases, of up to 80%, are both feasible and necessary for all industrialized nations.

Recommendations typically include installing better building insulation, manufacturing more efficient automobiles, using nuclear or solar energy instead of coal to generate electricity, reducing deforestation, and reforesting. Figure 9.26 graphs the effects of several energy-efficiency measures recommended by the U.S. Office of Technology Assessment. We will discuss some of these options in Chapter 17. All these measures taken together would reduce the total annual U.S. CO_2 emissions by over 50%. Studies indicate that the United States could make significant (25%) reductions in total greenhouse gas emissions if it were willing to invest about $10 per ton of CO_2 emissions. In more familiar terms, this is equivalent to an 11¢ rise in the price of a gallon of gasoline. Such a rise in gasoline would also slightly inhibit gasoline use and further reduce CO_2 emissions.

In fact, many studies suggest rebates and taxes as incentives for energy conservation. A tax on fossil fuels would encourage switching to noncarbon fuels. Revenue-neutral taxes and rebates on, respectively, gas-guzzling and gas-sipping cars would promote fuel efficiency. An underlying view is that environmental costs that are not charged to the user amount to hidden subsidies that amount to incentives to harm the environment. In order to prevent a free market from harming the environment, environmental costs such as global warming must be incorporated into the economic system.

DIALOGUE 11   How do you suppose we know that the visible surface of the sun has a temperature of about 5400°C?

---

11. By measuring the radiation emitted at different frequencies.

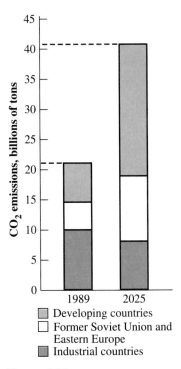

Carbon-dioxide emissions from fossil fuels in 1989 and projections for 2025 from three groups of countries. The estimate for 2025 assumes a 20% reduction by the industrial world from the 1989 emissions. (Source: Worldwatch Institute)

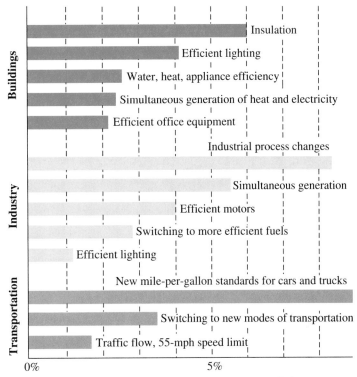

Percentage reduction in total annual CO₂ emissions

FIGURE 9.26
The effects on CO_2 emissions of several possible energy-efficiency measures. Each measure could, if vigorously pursued, reduce the total U.S. emissions by the percentage indicated. (Source: U.S. Office of Technology Assessment, 1991)

Summary of Ideas and Terms

Gravitational or electromagnetic field Exists wherever an object can feel a gravitational or electromagnetic force. Every material object creates a gravitational field throughout the space around it, and every charged object creates an electromagnetic field throughout the space around it.

Electromagnetic wave theory of light Every vibrating charged object creates a disturbance in its own electromagnetic field, which spreads outward through the field at 300,000 km/s. Light is just such an **electromagnetic wave**.

Lightspeed 300,000 km/s, or 3×10^8 m/s.

Ether The gaslike material substance that nineteenth-century scientists believed filled the universe. Ether's mechanical motions were thought to be responsible for light waves and for the transmission of forces, such as the electromagnetic force, that act at a distance.

Electromagnetic spectrum The complete range of electromagnetic waves. Divided into the following regions, listed from longest to shortest wavelength (lowest to highest frequency):

- **Electronic**, or **radio waves** Created by humans in such forms as AM and FM radio, TV, radar, and microwave.
- **Infrared** Created by the thermal motion of molecules.
- **Visible** Detectable by the human eye. Created by lower-energy motions of electrons in atoms. Colors are due to different wavelengths, ranging from red (longest) to violet (shortest).
- **Ultraviolet** Created by higher-energy motions of electrons in atoms. Has sufficient energy to damage biological molecules but not enough to ionize them. Can cause mutations and cancers.

- **X rays** Created by the highest-energy motions of electrons in atoms. X rays are one type of **ionizing radiation** because they can ionize biological molecules.
- **Gamma rays** Created in atomic nuclei. Very high energy and highly penetrating.

Radiant energy The energy of an electromagnetic wave.

Solar radiation Electromagnetic radiation from the sun. Concentrated mainly in the infrared, visible, and ultraviolet.

Trace gases Gases that form only a minute fraction of the atmosphere. Examples: ozone, carbon dioxide, and water vapor.

Ozone The O_3 molecule. A dilute layer of ozone fills the upper atmosphere or **stratosphere**. Ozone absorbs and shields biological life from most ultraviolet radiation.

Chlorofluorocarbons or **CFCs** Chemicals whose molecules are made of chlorine, fluorine, and carbon. Manufactured as coolants, spray propellants, foaming agents, and solvents. They destroy stratospheric ozone.

Ozone Treaty Calls for a nearly total phaseout by A.D. 2000 of CFCs and most other ozone-destroying chemicals.

Greenhouse effect The warming created by Earth's surrounding blanket of atmospheric gases.

Greenhouse gas The atmospheric gases, mostly water vapor and carbon dioxide, that cause the greenhouse effect.

Global warming Additional greenhouse-effect warming of Earth, caused by fossil-fuel use, deforestation, and other human activities. Although the concentration of CO_2 has increased significantly, evidence that this has caused warming is inconclusive.

Feedback effect Any effect that further influences the phenomenon that caused it. **Negative feedback** diminishes its cause, and **positive feedback** enhances its cause.

Review Questions

FORCE FIELDS

1. Name two kinds of force fields.
2. Which force fields would be felt by a stationary charged object: gravitational, electric, or magnetic? By a moving uncharged object? By a moving charged object?
3. What does it mean to say that there is an electric field throughout this room?

ELECTROMAGNETIC WAVES

4. Make a list of phenomena that are caused by electric charge.

5. What is an electromagnetic wave, and how fast does it go?
6. Are electromagnetic fields physically real? Defend your answer.
7. Describe the ether theory of light.
8. In what way do electromagnetic waves violate the philosophical underpinnings of Newtonian physics?

THE COMPLETE SPECTRUM

9. List the six main regions of the spectrum, from longest to shortest wavelength.

10. Describe a typical source of each: X rays, infrared, gamma rays, electronic waves, and ultraviolet.
11. Describe physically how humans see.
12. List several practical applications of electronic waves.
13. Which parts of the spectrum are ionizing radiations?
14. Which three regions of the spectrum dominate solar radiation? Which single region dominates?

GLOBAL OZONE DEPLETION

15. Where does ozone occur naturally on Earth, and what is its human significance?
16. What are CFCs? How have they have been useful to humans?
17. Outline the process by which CFCs destroy ozone.
18. List three atmospheric trace gases.
19. What is the Ozone Treaty?

20. List some expected effects of ozone depletion.

GLOBAL WARMING

21. What are the two major greenhouse gases, and what property makes them greenhouse gases?
22. What would happen if we removed all the CO_2 from the atmosphere? All the ozone?
23. Explain the greenhouse effect.
24. Which would increase global warming, forestation or deforestation? Why? Would this also increase ozone depletion?
25. What is a feedback effect? Describe one feedback effect related to global warming.
26. List three recommendations that have been made to combat global warming.

Home Projects

1. Count seconds during the next thunderstorm, to estimate the distance to several lightning strokes. Sound travels at 340 m/s. If a lightning stroke is 1 km away, about how many seconds will elapse before you hear it? Verify that the travel time for light is negligible, by calculating light's travel time for a few of the lightning strokes.
2. Turn on the heating coil of an electric stove, heater, or toaster. Before it begins to glow, hold your hand at some distance from the coil and feel its warmth. You are feeling infrared. When you sense a warm object near you, you are using your body as an infrared detector.
3. Find a black sheet of paper (you could paint it black with a marker pen) and a white sheet of paper of similar composition. When illuminated with white light, which one absorbs the visible radiation? Where does this energy go? Try putting the two sheets out in strong sunlight, sheltered from wind. Which one warms up more?

For Discussion

1. What scientific topics are important to understanding ozone depletion and global warming? Should we be educating people about such problems? How should we go about it? Are we doing this now? At what age (if any) should students learn about these problems?
2. At what points in the history of ozone depletion were good decisions or poor decisions made, and what were the consequences? What would have been the consequences of different decisions? In dealing with global warming, what (if any) lessons can we learn from the history of the ozone problem?
3. Dealing with uncertainty. There often are large uncertainties about complex systems such as the atmosphere. How should society deal with these uncertainties when they involve possible human or environmental harm?

What principles should be applied when making decisions about products such as CFCs? "Presumed innocence" is an accepted principle of criminal law: Defendants are considered innocent until legally proven guilty. Should new products and technologies be considered environmentally harmless until proven harmful? Some people support the "precautionary principle" that says that in case of uncertainty, society should act as though new technologies are harmful, in order to ensure that harm does not occur. Yet the precautionary principle can be taken too far. There are economic and other costs in not permitting a new technology. What is the proper balance between caution and innovation?

Exercises

FORCE FIELDS

1. Do the electric circuits in your home produce magnetic fields? Suggest a measurement that might check your answer.
2. Is an electric field a form of matter? Explain. What about a gravitational field?
3. If you place a proton at some point in an electric field and then release it, what will happen? How would the proton's motion differ from the motion of an electron placed at the same point in the same electric field? What if the field were gravitational instead of electric?

ELECTROMAGNETIC WAVES

4. What evidence is there that the medium that carries light waves is not air?
5. Imagine a region of space that had no atoms and no subatomic particles. Would it contain matter? Might it contain energy? If so, then would it really be empty?
6. Which travels faster, light or radio waves? Which has a longer wavelength? Which has a higher frequency?
7. It takes light 20 minutes to travel from Mars to Earth (depending on where Mars and Earth are in their orbits). How might this affect a conversation between an astronaut on Mars and people on Earth? Suppose an ultrapowerful telescope were used for visual communication. Would this speed things up?
8. Are radio waves filling your room right now? Describe a simple experiment that could demonstrate your answer.
9. If your ears actually detected radio waves instead of sound waves, what would your ears be "hearing" right now?
10.*How long does it take a radio signal to travel from New York to San Francisco, about 5000 km?
11.*A radar transmitter pointed at the moon receives a reflection 3 seconds after the signal is sent. How far is it to the moon?
12.*Light from the sun takes about 8 minutes to reach Earth, whereas light from the nearest star (other than the sun) takes about 4 years. How many times farther is it to this star than it is to the sun?

THE COMPLETE SPECTRUM

13. Is there any physical difference between light and a radio wave? Explain.
14. You can get a sunburn even on a cloudy day. Why?
15. You do not get a sunburn, even on a sunny day, if you are behind glass. Why?
16. What kinds of animals might evolve on a planet orbiting a star whose radiation is mostly in the infrared?

17. Suppose you viewed Earth's dark side from space. Could you "photograph" Earth using ultraviolet-sensitive film? Light-sensitive film? Infrared sensitive film? If so, what would you expect to see in the photo?
18. What energy transformation occurs (from what type to what type?) when solar radiation warms your skin? In a flashlight? When microwaves warm your food?
19.*Your radio is tuned to 92 on the FM dial. What frequency of electromagnetic wave is your radio receiving? How fast is this wave traveling? About how long is one wavelength? Does an electromagnetic wave actually reach your ear? What does reach your ear?

GLOBAL OZONE DEPLETION

20. Which of these worsens the ozone problem: coal-fired power plants, gasoline engines, CFC coolants, nuclear power plants, hydroelectric power plants, solar heating of homes, deforestation? Which contributes to global warming?
21. Why is it important that spray-can aerosols be propelled by inert gases? Why does the inertness of CFCs lead to problems with stratospheric ozone?
22. According to Table 9.2, which will be the first ozone-destroying chemical to decrease to harmless levels, and how long will it take? Which will be the last, and how long will it take?
23. What was the chlorine concentration in 1950? Under the Ozone Treaty, what is the highest predicted level? When is the level predicted to return to the 1975 level? What would the level be at that time if there were no Ozone Treaty?
24. Which of the ozone-destroying chemicals has the longest atmospheric lifetime? Today, which one forms the largest share of the problem?

GLOBAL WARMING

25. Which of these automobile fuels would not contribute to global warming: natural gas, synthetic gasoline made from coal, electricity from a solar power plant, electricity from a coal power plant, electricity from a nuclear power plant, ethanol (grain alcohol) made from corn?
26. What are some alternatives to burning fossil fuels? Consider electric power generation, transportation, home heating.
27. Will the Ozone Treaty help in solving global warming? How?
28.*According to Figure 9.22, how far back before the present century must one go to find a time when the CO_2 concentration was above 280 ppm? How large was the

CO$_2$ concentration at that time? How does the temperature at that time compare with current temperatures?

29.*According to Figure 9.26, by what total percentage would U.S. CO$_2$ emissions be reduced if the simultaneous generation of heat and electricity were vigorously pursued? What if all three listed transportation options were vigorously pursued?

30.*Making estimates. The mass of 1 liter of gasoline is about 0.7 kg, and the mass of 1 gallon is about 2.5 kg. A CO$_2$ molecule is about three times as massive as a carbon atom. Assuming that gasoline is pure carbon (actually only about 85% of its mass is carbon), estimate the mass of CO$_2$ that a typical automobile puts into the atmosphere each year. Compare this with Table 9.3.

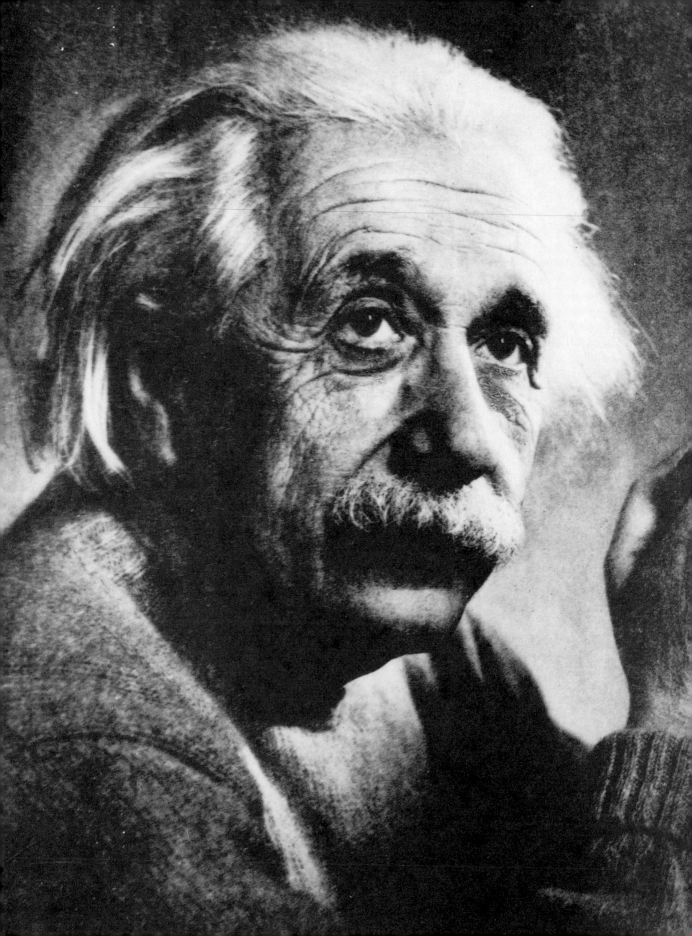

PART **IV**

THE POST-
NEWTONIAN
UNIVERSE
the observer intrudes

10

THE THEORY OF RELATIVITY

Nature and Nature's laws lay hid in night:
God said, "Let Newton be" and all was light.

Alexander Pope

It did not last: the Devil, shouting "Ho
Let Einstein be" restored the status quo.

Sir John Collings Squire

This chapter enters the domain still referred to as "modern" physics, even though it is now nearly a century old. "Post-Newtonian" physics is a more accurate term. The introduction of fields during the nineteenth century represented a break from the materialistic, mechanical tradition of Newtonian physics. The theory of relativity steps even further away from the Newtonian tradition. As will be explained in this and the next chapter, time, space, and mass now become to some extent "observer dependent" or "relative" rather than "absolute." Philosophically, an important innovation of the new theory is the significance attached to the observer's state of motion, or "reference frame." This emphasis on the observer is even more characteristic of the other great twentieth-century physical theory, quantum theory.

The transition from Newtonian to post-Newtonian physics is scientifically nothing like the transition from Aristotelian to Newtonian physics. Whereas Newtonian physics rejected and replaced Aristotelian physics, post-Newtonian physics does not reject Newtonian physics. Instead of being found wrong, Newtonian physics was found to be of limited validity. As discussed in Section 5.6, the predictions of Newtonian physics agree with observation in "normal" human situations, defined as those for which speeds are very small compared with lightspeed; distances are far larger than molecules but far smaller than galaxies; and gravitational forces are not many times stronger than on Earth. But outside this "normal" or "Newtonian" range, Newtonian physics differs significantly from observation—in other words, Newtonian physics is wrong outside this range. The major post-Newtonian theories, namely, relativity theory and quantum theory, were invented to deal with phenomena outside the Newtonian range. Within the Newtonian range, these theories make the same predictions (to within experimental accuracy) as does Newtonian physics.

But from the philosophical point of view, the transition from Newtonian to post-Newtonian physics represents a more complete revolution, a revolution that is still in progress today. The field concept, which contradicts the atomistic materialism that is at the heart of the Newtonian worldview, has already given us a taste of this revolution (Chapter 9). Here we will see that relativity theory and quantum theory contradict the Newtonian worldview in other fundamental ways.

Although it has the reputation of being difficult, Einstein's theory of relativity is actually fairly simple. It is based on two very simple ideas, and all of its major principles are offshoots of these two. Rather, the theory's reputation for difficulty comes from its strangeness. Its conclusions seem to violate common sense. To understand this theory, mental flexibility is required. One must be willing to follow some simple but unusual ideas wherever they may lead, even when they violate one's cherished beliefs about the way one thinks the universe is.

This and the next chapter consider the theory of relativity. This chapter focuses on the two key ideas behind the theory and the theory's most basic new prediction. Chapter 11 looks at several other predictions and presents Einstein's extension of these ideas to form a new theory of gravity.

Section 10.1 provides historical context, and Section 10.2 discusses the older "Galilean" (Newtonian) way of viewing the phenomena with which Einstein was concerned. Sections 10.3 and 10.4 cover the principle of relativity and the principle of the constancy of lightspeed, the two key ideas behind the new theory. In line with two of the themes of this book, Section 10.5 discusses why we believe Einstein's theory and compares the new view with Newtonian views. Sections 10.6 and 10.7 explain one of Einstein's most fascinating predictions: the relativity of time.

10.1 Einstein: *rebel with a cause* _____

Albert Michelson, whose highly accurate measurements of lightspeed earned him America's first Nobel Prize in physics, was one of the most highly esteemed physicists of his time. In 1894, in a speech dedicating a new physics lab at the University of Chicago, Michelson proclaimed that the era of new discovery in fundamental physics was probably finished:

> While it is never safe to affirm that the future of Physical Science has no marvels in store even more astonishing than those of the past, it seems probable that most of the grand underlying principles have been firmly established and that further advances are to be sought chiefly in the rigorous application of these principles to all the phenomena which come under our notice. . . . The future truths of Physical Science are to be looked for in the sixth place of decimals [that is, in improvements in the accuracy of known results].

Common sense is nothing more than a deposit of prejudices laid down by the mind before you reach eighteen.
Einstein

FIGURE 10.1

Never one to take himself too seriously, Einstein stuck his tongue out when asked to smile on his seventy-second birthday.

To punish me for my contempt for authority, Fate made me an authority myself.

Einstein

FIGURE 10.2

Albert Einstein during his student days in Zurich, a few years before he created his special theory of relativity.

My intellectual development was retarded, as a result of which I began to wonder about space and time (things which a normal adult has thought of as a child) only when I had grown up.

Einstein

It was a powerful acknowledgment of the triumph of Newtonian physics. Many scientists of that day shared Michelson's confidence that the known "grand unifying principles"—Newton's laws and the laws of thermodynamics and electromagnetism—were complete and permanent.

But the world changed only six years later. In 1900, the physicist Max Planck introduced a revolutionary idea, the quantum of energy (Chapters 13 and 14). The new idea was little noticed. The post-Newtonian revolution began quietly.

It is remarkable that in a span of only five years at the outset of the twentieth century, the seeds of both of the great post-Newtonian physical theories were planted. In 1905, five years after Planck's introduction of the quantum, a quite different but equally revolutionary idea was hatched in the brain of an obscure patent clerk in Bern, Switzerland. His name was Albert Einstein (Figures 10.1 and 10.2).

Einstein was a rebel in more ways than one. In his midteens, he got fed up with high school and dropped out. This surprised no one, for he had been a mediocre student and a daydreamer since beginning elementary school. Before that he had been a slow child, learning to speak only at three years of age. His high school teachers were glad to see him go, one of them informing Einstein that he would "never amount to anything" and another suggesting that he leave school because his presence destroyed student discipline. Einstein was only too happy to comply. He spent the next few months as a model dropout, hiking and loafing around the Italian Alps.

After deciding to study engineering, he applied for admission to the Swiss Federal Polytechnic University in Zurich, but unfortunately he failed his entrance exams. It seems he had problems with biology and French. To prepare for another try, he spent a year at a Swiss high school where he flourished in this particular school's progressive and democratic atmosphere. He recalled later that it was here that he had his first ideas leading to his theory of relativity.

The university now admitted Einstein on the basis of his high school diploma. He was known as a charming but indifferent university student who attended cafes regularly and lectures sporadically. He managed to pass the necessary exams and eventually graduate with the help of friends who shared their systematic class notes with the nonconforming Einstein.

Einstein recalled that his college experience

> had such a deterring effect upon me that, after I had passed the final examination, I found the consideration of any scientific problems distasteful to me for an entire year. . . . It is little short of a miracle that modern methods of instruction have not already completely strangled the holy curiosity of inquiry, because what this delicate little plant needs most, apart from initial stimulation, is freedom; without that it is surely destroyed. . . . I believe that one could even deprive a healthy beast of prey of its voraciousness, if one could force it with a whip to eat continuously whether it were hungry or not.

Following his graduation in 1900, Einstein applied for an assistantship to do graduate study, but it went to someone else. After looking unsuccessfully for a teaching position, in 1902 a friend helped him land a job as a patent examiner. Einstein often referred to his seven years at this job as "a kind of salvation" that paid the rent and occupied only eight hours a day, leaving him the rest of the day to ponder nature.

The two great theories of twentieth-century physics are the theory of relativity and quantum theory. Quantum theory is the creation of many physicists, including Einstein and several others whom we will meet in Chapters 13 and 14. One remarkable aspect of the theory of relativity is that it was invented nearly single-handedly, by Einstein.

10.2 Galilean relativity: *relativity according to Newtonian physics*

Here is a typical relativity question: Suppose that a train passenger, call her Velma, throws a baseball toward the front of the train. Both she and Mortimer, who is standing on the ground watching the passing train, measure the speed of the baseball (Figure 10.3). Will they get the same answer? If not, how will their answers differ?

This question concerns two observers who are moving differently. We say that Velma and Mort are in **relative motion** whenever they are moving differently. A **theory of relativity** is any theory that works out answers to questions concerning observers who are in relative motion. We can think of the train as being Velma's laboratory, or her **reference frame**, within which Velma measures things like the speed of the ball. We can think of the ground beside the tracks as a second reference frame, Mort's reference frame, for his measurements. The standard question that any theory of relativity asks is how measurements made in one reference frame compare with those made in another. Scientists have thought about questions like this since at least the time of Galileo.

To be more specific, suppose that the train moves at 70 meters per second (150 miles per hour, a typical modern train speed). Suppose that Velma throws the baseball toward the front of the train at 20 m/s **relative to Velma** (as measured on the train, using meter sticks and clocks that are on the train). How fast does the baseball move **relative to Mort** (as measured on the ground)?

FIGURE 10.3
Velma throwing a ball, observed by Mort.

The answer: Mort observes the baseball to move at 90 m/s. The reason is that during each second, the baseball moves 20 meters toward the front of the train as measured by Velma within the train and (as by observed by Mort) the baseball moves an additional 70 meters, because the train itself moves 70 meters during that second.

Centuries ago, Galileo would have given the same answer, 90 m/s. This straightforward, fairly intuitive form of relativity is called **Galilean relativity**. Here are a few more questions along this line:

DIALOGUE 1   Alphonse moves eastward in a car at 30 m/s. Velma, moving eastward in a train at 70 m/s, passes him. Are Velma and Al in relative motion? What is Velma's velocity (including direction) relative to Al? What is Al's velocity relative to Velma?

DIALOGUE 2   While the train moves eastward at 70 m/s, Velma throws a ball toward the rear of the train, throwing it just as hard as she did earlier when she threw it at 20 m/s toward the front of the train.
(a) Find the velocity (including direction) of the ball relative to Velma.
(b) Find the velocity of the ball relative to Mort, who is standing beside the tracks. (c) Since the two answers differ, whose answer is correct, Velma's or Mort's?

DIALOGUE 3   The train moves eastward at only 20 m/s, and Velma throws a ball at 20 m/s toward the rear of the train. Find the ball's velocity relative to Mort.

DIALOGUE 4   Let's give the ball to Mort. (a) He throws it at 20 m/s eastward as measured by him. What is the ball's velocity relative to Velma, who is in the train moving eastward at 70 m/s? (b) What if Mort had instead thrown the ball westward at 20 m/s?

Let us turn our attention to a similar example that involves light beams instead of baseballs. We learned in Chapter 9 that light is an electromagnetic wave moving at 300,000 km/s, a speed that we will symbolize by the letter c. It is difficult to imagine such a high speed. A light beam travels from New York to Los Angeles in a hundredth of a second and travels a distance equal to 7.5 times around Earth in 1 second. Trains, jet planes, and even Earth satellites moving at 8 km/s are slowpokes compared with light.

Now we are going to imagine that Velma is piloting a futuristic rocket ship that is much faster than today's fastest vehicles. Suppose Velma is moving past Earth at 75,000 km/s, or $0.25c$ (25% of lightspeed). Mort is standing on Earth holding a source of light such as a flashlight or a laser. Suppose that Velma is flying eastward past Mort and that Mort directs the light beam

---

1. Yes; 40 m/s eastward; 40 m/s westward because Velma observes Al moving 40 meters farther away on her west side every second.
2. (a) 20 m/s westward. (b) 50 m/s eastward. (c) Both answers are correct—they differ because they refer to different observers.
3. Zero velocity relative to Mort—the ball falls straight down.
4. (a) 50 m/s westward. (b) The ball's velocity relative to Velma would be 90 m/s westward.

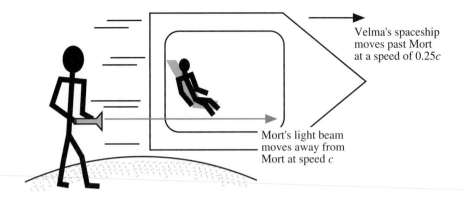

Velma's spaceship moves past Mort at a speed of 0.25*c*

Mort's light beam moves away from Mort at speed *c*

eastward. What would be the speed of the light beam—the moving tip of the beam—relative to Mort and relative to Velma? Experiments with light beams from sources on Earth have repeatedly shown that a light beam moves away from its source at speed *c*. So Mort's light beam moves eastward at speed *c* relative to Mort. What speed would Velma measure for the same light beam (Figure 10.4)? The intuitively sensible answer is 0.75*c*.

This is the answer Galileo would have given, the answer given by Galilean relativity. It is the answer that all scientists would have given up through the end of the nineteenth century. It is certainly a sensible answer. Nevertheless, it is experimentally wrong. Nature does not comply with our intuitions. To see why there might be something wrong with this answer and to see what nature's answer is, we turn in the following section to some thoughts that Albert Einstein had around the beginning of the twentieth century.

10.3 The principle of relativity: *relativity according to Einstein*

Suppose you are riding in a smoothly moving unaccelerated vehicle such as a jet airplane in level flight. Suppose that the flight attendant pours you a cup of coffee. Where should you hold your cup: a little behind the spout, to take account of the fact that your airplane is moving forward, or under the spout? In other words, does the coffee pour straight downward relative to the airplane? Try it sometime and see. Or try dropping a coin from one hand to the other in a moving vehicle (but not when you are driving): Is the catching hand directly beneath the dropping hand?

The result is that the coffee pours straight downward relative to the plane.

You could carry out other, more sophisticated experiments along this line, all on a smoothly moving reference frame such as a jet airplane. You might take multiple-flash photos of a falling ball (Figure 3.13) or perform experiments with frictionless air gliders (Figure 3.6) or experiments designed to check the validity of Newton's laws. You could perform electromagnetic experiments involving electric charges, electric currents, or magnets. Just as for the poured coffee, you would find in all these cases that the experimental result is exactly the same as it is when the experiment is performed in a reference frame that is at rest on the ground: A falling ball falls straight downward with an acceleration of 9.8 m/s^2; an object's acceleration equals the force exerted on it divided by its mass; and so forth— all as measured relative to the moving reference frame (the interior of the jet airplane). In other words, all physics experiments performed entirely within the jet airplane come out precisely as they would if performed at rest on Earth.

Suppose that you were the only passenger on an airplane with no windows in the passenger compartment and that you went to sleep and woke up sometime later to find yourself alone in the compartment. Could you tell, without receiving information from the outside world,* whether your airplane was flying at a constant velocity or parked on the ground? The answer is no. You could throw a ball, do handstands, pick up nails with magnets, and the like, and everything would be the same, regardless of whether your plane was in flight or parked.

As long as you are in a nonaccelerating reference frame (no bumps, no turns, no speed changes), everything is normal. All physics experiments, and consequently all the laws of physics, are normal. We summarize this idea as:

> THE PRINCIPLE OF RELATIVITY
> Every nonaccelerated observer observes the same laws of nature. In other words, no experiment performed within a sealed room moving at an unchanging velocity can tell you whether you are standing still or moving.

Unless you look outside, you can't tell how fast you're going. This is a fairly plausible idea and was the key to Einstein's thinking about relativity. It is called the *principle of relativity* because it says that as far as the laws of physics are concerned, all motion is just relative motion. When we say, for example, "the car is moving at 25 km/hr westward," we really mean that "the car is moving at 25 km/hr westward relative to the ground" or that "the car and the ground are in relative motion at a speed of 25 km/hr." In regard to the basics (the laws of physics), we could just as well say that the car is standing still and that the ground is moving eastward at 25 km/hr. We could even say that the ground is moving eastward at 1600 km/hr (which it is, relative to Earth's center, due to Earth's spin) and that the car is moving eastward at only 1575 km/hr. It is only the relative velocity, the 25 km/hr, that really counts.

DIALOGUE 5 If you drop a coin inside a car that is speeding up, where will the coin land? What if the car is slowing down? Turning a corner to the right? Try it sometime (when you are not driving).

---

* Information from the pilot would be from the outside world, because the pilot's information enters through the cockpit window and through radio receivers.

DIALOGUE 6   The principle of relativity states that unless you look outside, you cannot detect the velocity of your own reference frame. What about acceleration of your own reference frame—can this be detected without looking outside?

10.4 The constancy of lightspeed: *strange but true*

One of the many questions that Einstein enjoyed pondering in high school was "What if I could catch up with a light beam? How would a light beam appear if I were moving along with it?"

The possibility of moving along with a light beam seemed paradoxical, contradictory, to Einstein. The reason is that to an observer moving along with a light beam, the light beam itself would be at rest. To this observer, the light beam would appear as an electromagnetic "wave" that was standing still! That seemed to Einstein to be impossible.

Our understanding of electromagnetic waves, such as light, is based on Maxwell's theory of electromagnetic fields (Chapter 9). Recall that Maxwell's theory predicts that any disturbance in an electromagnetic field, such as a disturbance caused by the motion of an electrically charged object, must propagate as a wave moving outward through the field *at speed c*. This particular speed, 300,000 km/s, is built into Maxwell's theory.

Einstein believed that Maxwell's theory should, like all other laws of nature, obey the principle of relativity. In other words, Maxwell's predictions should be correct even within any moving reference frame. Since speed *c* is built into the laws of nature, Einstein concluded that *every observer ought to observe every light beam to move at speed c*, regardless of the observer's motion. No matter how fast you move, a light beam always passes you at speed *c*, relative to you. This is why the idea of catching up with a light beam seemed contradictory to Einstein. If every observer sees every light beam move at speed *c*, then nobody can even begin to catch up with a light beam, much less catch all the way up with one and observe it at rest.

This idea is simple but odd, which is why it took an original genius like Einstein to think of it. We will discuss a few of its oddities later. This idea is so odd that other turn-of-the-century physicists who might have discovered it did not. It is the second important principle underlying Einstein's theory. We summarize it as:

> THE PRINCIPLE OF THE CONSTANCY OF LIGHTSPEED
> The speed of light (and of other electromagnetic radiation) in empty space is the same for all nonaccelerated observers, regardless of the motion of the light source or of the observer.

---

5. To the rear of the drop point. Toward the front. To the left of the drop point.
6. Yes, you can detect accelerations by doing experiments like those of Dialogue 5.

Like the principle of relativity, this principle is again valid only for nonaccelerated observers. The reason is that Maxwell's theory, like most laws of physics, is valid only for nonaccelerated observers.

To get a feel for Einstein's idea, we will apply the principle of the constancy of lightspeed to several "thought experiments," experiments that could in principle be performed but that are impractical for one reason or another. Each experiment involves a light beam, which we take to be a laser beam but which could just as well be a flashlight beam or any other electromagnetic radiation such as a gamma ray or radio wave.

If both our observers, Velma and Mort, are on Earth and Mort switches on the laser beam, they will observe the beam moving at speed c. No surprises here.

If Velma and Mort are together in any other reference frame, such as a spaceship moving rapidly past Earth or on a distant galaxy that is moving away from Earth at 90% of lightspeed, the result will be the same: When Mort switches on the laser beam, Einstein's principle predicts that both observers will see the laser beam move at speed c. This is not very surprising, perhaps, since both observers are in the same reference frame.

Things get more interesting when Velma and Mort are in relative motion. Let's give the laser to Mort and imagine that Velma is in a spaceship approaching Mort at 75,000 km/s, or $0.25c$ (Figure 10.5). You can imagine Mort to be on Earth, but Mort could as well be on a distant galaxy moving at a high speed relative to Earth or on a spaceship moving rapidly past Earth. The important point is that Velma moves at $0.25c$ relative to Mort. Mort switches on the laser and observes the beam to move away from him at speed c. What speed does Velma observe for the laser beam as it passes her spaceship? Galilean relativity and our intuitions answer $1.25c$, or 375,000 km/s. But Einstein's relativity predicts that the answer is c, or 300,000 km/s.

At this point, many people are inclined to ask, "But why? Why should it be c rather than $1.25c$?" We'll save that question for the next section.

Now let's give a laser beam to Velma, who is still approaching Mort at a speed of $0.25c$ (Figure 10.6). She switches on her laser and observes that the beam moves away from her at speed c relative to her reference frame. What speed does Mort measure for the same beam, relative to his reference frame? Galileo and common sense again predict $1.25c$, but Einstein predicts c.

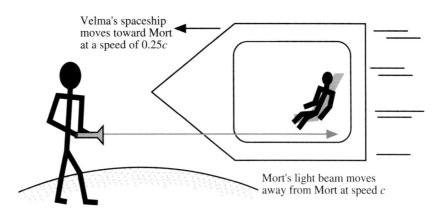

Velma's spaceship moves toward Mort at a speed of $0.25c$

Mort's light beam moves away from Mort at speed c

FIGURE 10.5
What is the speed of Mort's light beam relative to Velma?

FIGURE 10.6

*What is the speed of Velma's light
beam relative to Mort?*

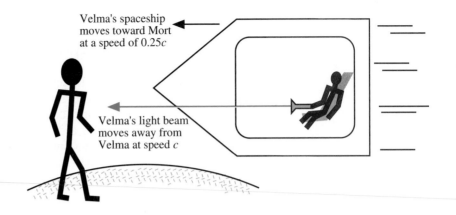

Velma's spaceship
moves toward Mort
at a speed of 0.25*c*

Velma's light beam
moves away from
Velma at speed *c*

Let's return to the example posed toward the end of Section 10.2 (Figure 10.4). Mort has the laser, and Velma moves away from Mort at a speed of 0.25*c*. Mort switches on the laser, observing the beam to move away from him at speed *c*. Galilean relativity predicts that this beam will pass Velma at a speed of 0.75*c* as measured by her, but Einstein predicts that she will measure this beam to move at speed *c*.

To dramatize the oddness of this, let's assume that Velma is moving away from Mort at a speed of 0.999,999*c*, just a little slower than lightspeed (Figure 10.7). Mort switches on his laser and sees the light beam move away from him at speed *c*. As observed by Mort, Velma is moving only 0.000,001*c* slower than this light beam—he says that she is nearly keeping up with the light beam. Galilean relativity predicts that Velma will observe the light beam passing her very slowly, at only 0.000,001*c*. This is one-millionth (10^{-6}) of lightspeed, or 0.3 km/s, or 300 m/s—about the speed of sound and of fast jet airplanes. If the Galilean prediction is correct, she should be able to watch the tip of the beam as it moves past her at about the speed of a jet airplane! The Galilean prediction says that from Velma's point of view, she is nearly keeping up with the light beam.

But Einstein's relativity says that from her point of view Velma isn't even beginning to catch up with the light beam. She sees it move past her at precisely *c*, 300,000 km/s, despite the fact that she is moving away from the light source at nearly lightspeed.

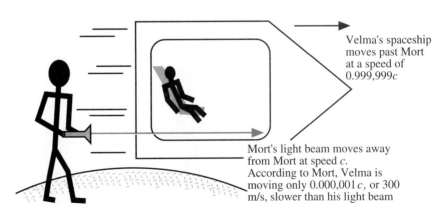

Velma's spaceship
moves past Mort
at a speed of
0.999,999*c*

Mort's light beam moves away
from Mort at speed *c*.
According to Mort, Velma is
moving only 0.000,001 *c*, or 300
m/s, slower than his light beam

FIGURE 10.7

*How fast is Mort's light beam
moving, as observed by Velma?*

You might have noticed that we didn't allow Velma to have precisely speed c in these examples. We did allow her to move at nearly, but not quite, speed c relative to Mort. If we imagine that she moves at precisely speed c, then we get into the difficulty that Einstein noted about moving along with a light beam: An observer moving at speed c would observe a light beam to be at rest. So, according to Einstein's relativity, an observer can move at nearly, but not precisely, speed c relative to another observer. We will find in the next chapter that there are additional reasons that no observer can move at lightspeed relative to any other observer.

HOW DO WE KNOW?

As strange as these ideas may seem, they are experimentally verified every day in laboratories around the world. However, the experiments do not usually involve macroscopic objects like spaceships moving at speeds near lightspeed, because it isn't easy to get large objects moving so fast. Instead, most experiments involve fast-moving microscopic particles.

For example, in 1964 an experiment was performed in which a subatomic particle moving at nearly lightspeed emitted electromagnetic radiation in both the forward and backward directions. Galilean relativity predicts that the forward-moving radiation should move at a speed much faster than c and that the backward-moving radiation should move much slower than c, as measured in the laboratory. But the actual measurement shows that both radiation beams move at speed c relative to the laboratory, in agreement with the principle of the constancy of lightspeed.

10.5 Einstein's logic, materialism, and the logic of science

Nobody thought of the principle of the constancy of lightspeed before Einstein did, because it is so odd. Maxwell and other nineteenth-century scientists had a more conventional view of light beams. As we explained in Section 9.3, they believed that light waves were caused by vibrations of a material medium, just as, for example, water waves were caused by vibrations of water. They called this medium **ether**. Nobody had ever observed ether. It was known not to be made of ordinary chemical atoms, because light waves travel through regions of space where there are essentially no atoms. Instead, ether was thought to be a gaslike substance filling the entire universe and made of some unknown, nonatomic form of matter.

The ether theory predicts that the "natural" speed of light, 300,000 km/s, is the light's speed relative to the ether. Observers moving through the ether should then measure other speeds for light beams, speeds that should depend on the observer's speed through the ether and on the light beam's direction (Figure 10.8).

The **Michelson–Morley experiment**, performed by Albert Michelson (whom we met at the outset of this chapter) and Edward Morley in 1887, eighteen years before Einstein's theory, provided the earliest evidence that light has the same speed for all observers, in disagreement with the ether

FIGURE 10.8

The ether theory of light predicts that an observer moving through the ether will measure different light beams moving at different speeds.

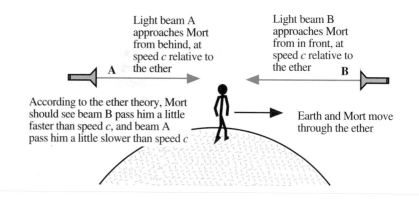

Light beam A approaches Mort from behind, at speed *c* relative to the ether

Light beam B approaches Mort from in front, at speed *c* relative to the ether

According to the ether theory, Mort should see beam B pass him a little faster than speed *c*, and beam A pass him a little slower than speed *c*

Earth and Mort move through the ether

theory. Michelson and Morley made accurate comparisons of the speeds of light beams that were moving in different directions. Since their laboratory (like all of Earth) was presumed to be moving through the ether, they predicted that different light beams would have slightly different speeds relative to the lab. But they found that all light beams had the same speed relative to their lab. The physicists of that time, including Michelson and Morley, failed to appreciate this experiment's fundamental implications, preferring instead to assume that light beams merely *appeared* to have the same speed in that particular experiment and to seek explanations for that appearance. Michelson and Morley themselves were disappointed at their "failure" to detect the expected difference in the speeds of different light beams. Actually, their experiment was a stunning breakthrough to an exciting new experimental fact. It's just that nobody could see it that way.

Einstein, in 1905, was the first to state that light has the same speed for all observers. Furthermore, and surprisingly, Einstein paid little attention to the Michelson–Morley experiment in developing his theory. To him, it seemed obvious that light should have the same speed relative to all observers, for the reasons given at the beginning of Section 10.4. Here was a truly independent and original mind: Other physicists rejected the constancy of lightspeed even despite the Michelson-Morley experiment's evidence that lightspeed is constant, but Einstein accepted the constancy of lightspeed even without paying attention to the evidence of the Michelson–Morley experiment!

The Lord is subtle, but He is not malicious.

Einstein

Because it contradicts a key prediction of the ether theory, Einstein's theory undermines the notion that light waves are caused by vibrations of a material medium. So light waves and other electromagnetic waves must be nonmaterial. Since Einstein, electromagnetic waves have been viewed as caused by the vibrations of an electromagnetic field, which itself is not made of any material substance. As we discussed in Section 9.3, this represents an important deviation from the materialist worldview associated with Newtonian physics.

The principle of the constancy of lightspeed is the key new idea in the theory of relativity. This idea gives the theory its odd, nonintuitive quality. It is natural that we should question this principle. Why is it true? How can it be true?

The answer is very simple and goes to the heart of science. The principle of the constancy of lightspeed is true because nature says so. Observation of nature, not our beliefs, determines truth in science. And numerous experiments show that every light beam in nature moves at speed *c*, regardless of the motion of the observer. It is a simple answer, but profound.

The two key ideas of Einstein's theory are the principle of relativity and the principle of the constancy of lightspeed. These should be regarded as the "first and second laws of Einstein's relativity." They play a role in the theory of relativity that is identical to the role of Newton's laws in Newton's theory of force and motion: They form the logical basis of the theory, from which everything else in the theory is derived, and which are themselves justified directly by experience.

Physicists call the theory founded on these two principles the **special theory of relativity**. The word *special* distinguishes this theory from the Galilean theory of relativity and also from another, related theory of Einstein's called the **general theory of relativity**. The distinguishing feature of the general theory of relativity is that it does not assume an unaccelerated observer. That is, the general theory of relativity applies to all observers, both accelerated and unaccelerated, and so it is more general, broader, than the special theory. We will look at Einstein's general theory of relativity in the latter part of Chapter 11.

Strictly speaking, Earth itself is an accelerated reference frame, both because it spins on its axis and because it rotates around the sun. But these accelerations are small compared with the accelerations that are significant in the general theory of relativity. So we can assume, as a good approximation, that the predictions of the special theory of relativity are correct for any Earth-based observer.

Beginning in the next section and continuing through most of the next chapter, we explore five of special relativity's most important predictions: the relativity of time, the relativity of space, the relativity of mass, c as the limiting speed, and $E = mc^2$. These results are predictions that follow from the two basic principles of the theory; in other words, if you believe the two basic principles, then you must logically believe these predictions. Not surprisingly, these predictions seem fairly odd to our intuitions. The source of this oddness is the principle of the constancy of lightspeed.

DIALOGUE 7   Velma is moving away from Mort at a speed of 0.75 c. She turns on two lasers, one pointed forward and the other pointed backward. (a) What speed does Velma observe for each of the two laser beams? (b) According to Galilean relativity, what speed does Mort observe for each beam? (c) According to Einstein's relativity, what speed does Mort observe for each beam? (d) What speed does Mort actually observe for each beam (according to nature)?

If you do not ask me what is time, I know it; when you ask me, I cannot tell it.

St. Augustine, A.D. 400

10.6 The relativity of time: *time is what humans measure on clocks*

What is time? The question is so fundamental that it is hard even to understand what the question means. The problem is that we always exist in time. Time is so much with us that we cannot get away from it to study it from a distance. So we have deeply embedded intuitions about time, unconscious intuitions that might turn out not to be true in nature.

Einstein's basic insight into time was that it was physical, part of the physical universe, rather than apart from or beyond the physical universe. Just as one can measure the properties of a stone or of a light beam, one can

7. (a) c. (b) 1.75c for the forward beam, 0.25c for the backward beam. (c) c. (d) c.

measure the properties of time itself. And how should we measure the properties of time? With clocks, obviously. This trivial-sounding reply is more profound than it appears. The only way we can measure time is with real, physical clocks (including all time-keeping phenomena such as a heartbeat and Earth's rotation around the sun). Physically, the concept of a clock really defines time. In other words, time is what humans measure on clocks.

Absolute, true, and mathematical time, of itself, and from its own nature, flows equably without relation to anything external, and by another name is called duration. Relative, apparent, and common time, is some sensible and external measure of duration by means of motion, which is commonly used instead of true time; such as an hour, a day, a month, a year.
Newton, justifying the idea that time is absolute

So if we want to investigate the properties of time, we must investigate the properties of clocks. How do clocks really behave? Einstein's goal in studying time was to predict the properties of clocks using as his starting point only the two principles of the special theory of relativity.

An ordinary spring-wound or battery-driven clock would be hard to study using only Einstein's two simple principles because these clocks are too complex. They involve springs, electric current, vibrational motions, gears, and so forth. So Einstein invented a simple kind of clock, a simple thought experiment, really. His clock involves no mechanically moving parts; its only motion is the motion of a light beam.

Figure 10.9 shows this **light clock**. Two parallel mirrors face each other, one above the other, and a light beam bounces up and down between them. It is convenient for our discussion (but not too practical for the clock maker) to imagine that the two mirrors are separated by 150,000 kilometers, because then the time for one complete round-trip of the light beam is just 1 second. We know that this time is 1 second because the principle of the constancy of lightspeed tells us that all light beams travel 300,000 kilometers in 1 second.* In fact, this device defines what we mean by "1 second": It is the interval during which the tip of the light clock's light beam makes one round-trip. We will assume that this light clock ticks at the end of every round-trip, in other words, that it ticks once every second.

Now that we have a simple physical definition of time, we can investigate the properties of time. We begin by installing one light clock in Velma's spaceship and another in Mort's laboratory. We imagine that Mort is on Earth and Velma is moving eastward past Earth in a spaceship. Let's begin by thinking about Velma's light clock. As seen by her, her light beam bounces straight up and down, covering 150,000 km upward and then 150,000 km downward, ticking when the light beam gets back to the bottom mirror (Figure 10.10(a)). Simple enough.

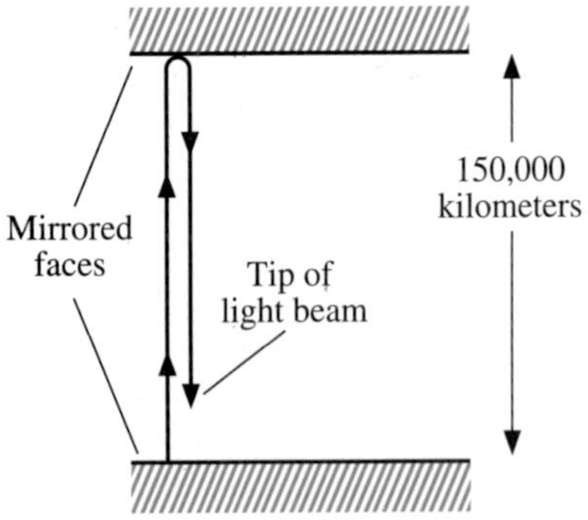

FIGURE 10.9
A light clock. A light beam bounces up and down between two mirrors. If the distance between mirrors is 150,000 km, then 1 second will elapse during one complete round-trip up and back down.

How does Velma's light clock appear to Mort? From his point of view, the tip of Velma's light beam is not only moving up and down, it is also moving eastward because of the eastward motion of Velma's spaceship. This means that the tip of Velma's light beam, as seen by Mort, moves along a diagonal path as it moves up to the top mirror and then back to the bottom mirror. Figure 10.10(b) shows this effect by depicting Mort's observations of Velma's spaceship at three different instants: when the tip of her light beam is at the bottom mirror, when her light beam has moved up to the top mirror, and when her light beam is back at the bottom mirror.

Recall that the top-to-bottom distance between the two mirrors is 150,000 km. As you can see from the figure, the distance along one of the two diagonals must then be greater than 150,000 km. This means that the total round-trip distance traveled by Velma's light beam, as measured by Mort, must be greater than 300,000 km. There is nothing surprising or subtle about this; Galileo would have said the same thing.

* We are assuming that the clock is carried by an unaccelerated observer.

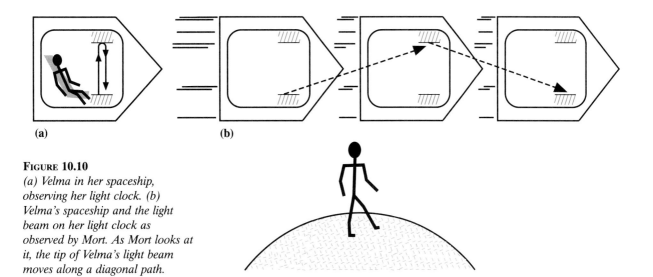

(a)   **(b)**

FIGURE 10.10
(a) Velma in her spaceship, observing her light clock. (b) Velma's spaceship and the light beam on her light clock as observed by Mort. As Mort looks at it, the tip of Velma's light beam moves along a diagonal path.

Now comes the part that Galileo (and our intuitions) wouldn't agree with: The constancy of lightspeed predicts that Mort observes Velma's light beam to move at just 300,000 km/s. Since Mort observes the round-trip distance to be greater than 300,000 km, it follows that according to Mort it takes more than 1 second for Velma's light beam to make the round-trip! So, as measured by Mort using his clock, more than 1 second elapses between Velma's ticks. According to Mort, Velma's clock runs slow.

Velma's second is different from Mort's second. The two observers measure different time intervals for the same event (one round-trip of Velma's light beam). That is, *time is relative to the observer.*

Let's turn things around. How does Mort's clock appear to the two observers? To Mort, his own clock's light beam travels 300,000 km in one round-trip and requires 1 second to do so. But from Velma's viewpoint, Mort's clock is moving westward, so the tip of Mort's light beam is moving along a diagonal, so the total round-trip distance traveled by Mort's light beam as observed by Velma is greater than 300,000 km. But because Velma observes Mort's light beam to move at 300,000 km/s, she must observe that more than 1 second elapses between Mort's ticks. According to Velma, Mort's clock runs slow.

Mort observes that his own clock is accurate and that Velma's runs slow. Velma says that her clock is accurate and that Mort's runs slow. This is not your normal situation caused by an inaccurate clock, in which if my clock runs slow according to your clock, then your clock must run fast according to my clock. Instead, if you and I are in relative motion, then my clock runs slow according to you, and your clock runs slow according to me.

This raises an interesting question: Who is right? Whose clock is really running slow, and whose clock is really accurate?

The answer is that both Velma and Mort are right. Velma observes that Mort's clock is slow, and Mort observes that Velma's clock is slow, and both observations are accurate. This situation is not caused by inaccurate observations or inaccurate clocks; it is instead a property of time itself. There is no single "real" time in the universe, no "universal time"; there is only Mort's time and Velma's time and all the other possible observed times.

The general rule here is that moving clocks run slow.

This effect—the slowness of moving clocks—occurs whenever the two observers are in relative motion. The effect is small if the relative speed is small compared with lightspeed, because for such small speeds the diagonal path length (Figure 10.10) is only slightly longer than 300,000 km. At familiar human speeds, the effect is so small that it wasn't noticed before Einstein's work. Even at a speed as high as 3000 km/s (1% of lightspeed, which would carry you across the United States in about a second), the effect is very small.

As you might expect, there is a formula that quantitatively describes the relativity of time.* Table 10.1 gives a few of the numerical results that can be calculated from this formula, and Figure 10.11 is a graph based on the same formula. As you can see from the table, the effect is negligible at jet airplane speeds (0.3 km/s) and at orbiting satellite speeds (10 to 20 km/s). It is not until speeds of 0.1c are reached that the effect amounts to even a half of 1%. But at large fractions of lightspeed, the effect becomes quite large: If Velma passes Mort at 99.9% of lightspeed (0.999c), their seconds will be more than 22 seconds long as measured by the other observer. The relativity of time is also called **time dilation**, because as you can see from Table 10.1, a time interval of 1 second on a moving clock is expanded or "dilated" to more than 1 second as measured by an observer past whom the clock is moving.

Although we investigated the relativity of time by studying light clocks, the conclusion holds for every type of clock. Remember that a clock is a specific physical definition of time. We are investigating time, not just light

I do not define time, space, place and motion, as being well known to all.

Newton

It requires a very unusual mind to undertake the analysis of the obvious.

Alfred North Whitehead, twentieth-century philosopher

TABLE **10.1**

The Relativity of Time: Some Quantitative Predictions

Relative speed (km/s)	Relative speed as a fraction of light speed (c)	Duration of one "tick" on a moving clock, as measured by an observer past whom the clock is moving (s)
0.3	10^{-6}	1.000,000,000,000,5
3	10^{-5}	1.000,000,000,05
30	10^{-4}	1.000,000,005
300	0.001	1.000,000,5
3000	0.01	1.000,05
30,000	0.1	1.005
75,000	0.25	1.03
150,000	0.5	1.15
225,000	0.75	1.5
270,000	0.9	2.3
297,000	0.99	7.1
299,700	0.999	22.4

* This formula can be derived by studying Figure 10.10 and using the "Pythagorean theorem," which states that a right triangle's short side lengths a and b are related to its diagonal length c by $c^2 = a^2 + b^2$. The formula that results from this is $T = T_0/\sqrt{(1 - v^2/c^2)}$, where v is the relative speed, T_0 is the time between two of Velma's ticks as observed by Velma ($T_0 = 1$ second for a normal clock), and T is the time between two of Velma's ticks as observed by Mort.

FIGURE 10.11
Relativistic time dilation. The graph shows the duration of one clock tick (representing 1 second in the clock's reference frame) on a moving clock, for various speeds of the clock relative to the observer.

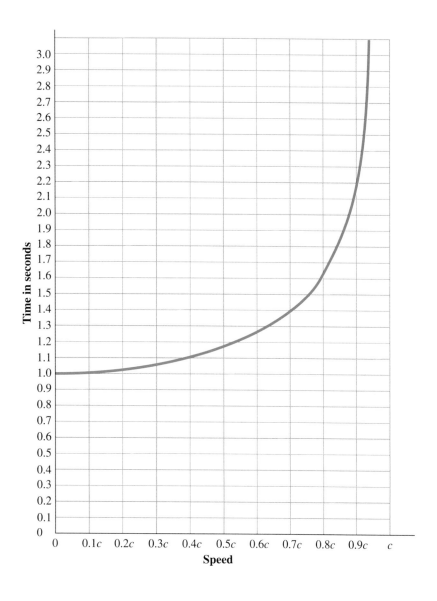

clocks. All clocks must behave the way that light clocks behave: spring-driven clocks, pendulum clocks, and so forth, because time behaves this way.

Furthermore, everything that happens in time, every phenomenon that occurs during an interval of time, must also behave in this way. Think, for example, of an ice-cream cone melting. Suppose that an ice-cream parlor makes ice-cream cones that melt in exactly 10 minutes. Both Velma and Mort have one of these cones. These cones are a kind of clock, a clock that ticks in 10 minutes.

The testimony of our common sense is suspect at high velocities.
Carl Sagan, astronomer and author

DIALOGUE 8   Suppose that Velma passes Mort at 75% of lightspeed. Use Table 10.1 to find (a) how long Velma's 10-minute cone will take to melt as measured by Mort; (b) how long Velma's cone will take to melt as measured by Velma; (c) how long Mort's cone will take to melt as measured by Velma; and (d) how long Mort's cone will take to melt as measured by Mort.

DIALOGUE 9 Answer the questions of Dialogue 8, assuming that Velma moves past Mort at 99% of lightspeed.

DIALOGUE 10 How fast must Velma move, relative to Mort, in order for Velma's ice-cream cone to melt in 20 minutes as measured by Mort (use Figure 10.11)? How fast must she move in order for her cone to melt in 30 minutes as measured by Mort?

Instead of ice-cream cones, it could be frogs. Suppose your local biology department hatches guaranteed 10-day frogs, having a 10-day lifetime. Biological life occurs in time, too, just as clocks do, so these frogs can be thought of as a kind of clock. So if Velma passes Mort at 75% of lightspeed, he says that her frog lives 15 days but that his frog lives only 10 days (see Dialogue 8). And she says that his frog lives 15 days but that her frog lives only 10 days.

This raises the interesting question of which frog dies first. According to Mort, his frog dies first, and Velma's frog "ages slowly." According to Velma, her frog dies first, and it is Mort's frog that hangs on 5 days longer. Whose frog really dies first? Again, the answer is that both observations are right. Mort observes his frog to die first, Velma observes her frog to die first, and both observations are accurate.

Again, the unspoken belief in the back of our mind when we ask "whose frog really dies first?" is that there is one universal real time. Without that belief, we wouldn't have asked the question in the first place.

HOW DO WE KNOW?

The relativity of time has been verified again and again in laboratories, by observing fast-moving subatomic particles. One experiment, similar to the frog example, involved a type of subatomic particle known as a *muon*. Muons, unlike most ordinary matter, are not permanent objects. Instead, they have a "lifetime" after which they disintegrate spontaneously into other particles. The lifetime of a muon is only 2.2 microseconds (a microsecond is a millionth of a second). If you observe a muon that is at rest relative to you, its lifetime is 2.2 microseconds as measured by you. But a muon moving rapidly past you lives much longer as measured by you, because of time dilation. For example, if the muon is moving past you at 99% of lightspeed (muons often move this fast in high-energy physics labs), then Table 10.1 says that its lifetime will be lengthened by a factor of 7.1 and that it will not disintegrate until $7.1 \times 2.2 = 15.6$ microseconds have passed. This experiment has been done, and the muons were observed to have lifetimes that were lengthened by just the predicted amount.

The relativity of time is one of the deep and surprising predictions of the theory of relativity. It is partly a consequence of the way that time is measured,

You could see that Einstein was motivated not by logic in the narrow sense of the word but by a sense of beauty. He was always looking for beauty in his work. Equally he was moved by a profound religious sense fulfilled in finding wonderful laws, simple laws in the universe. It was really a religious experience for him, of the most profound sort, even though he did not believe in a personal god.
    Banesh Hoffmann, mathematician and author, in *Some Strangeness in the Proportion*

---

8. (a) Since a 1-second tick on Velma's clock is observed by Mort to take 1.5 s (1.5 times as long), Velma's 10-minute ice-cream cone must be observed by Mort to take 15 minutes to melt; (b) 10 minutes; (c) 15 minutes; (d) 10 minutes.
9. (a) 71 minutes; (b) 10 minutes; (c) 71 minutes; (d) 10 minutes.
10. About 0.87c. About 0.94c.

a process that really defines time. When this measuring process is analyzed carefully, incorporating the principle of the constancy of lightspeed, it turns out that time is relative. This emphasis on the measuring process and on the observer is a new element in physics, one that appears in both relativity theory and quantum theory. Although these two theories are quite different in many respects, both are founded on the idea that reality is what humans observe and measure it to be. Reality cannot be separated from the observation process.

10.7 Time travel: *you can't go home again* _____

As you might have suspected, the next step is to investigate the life span of Velma and Mort themselves. Suppose they are born at the same time* and that they have lifetimes of 80 years. In other words, Velma measures her lifetime to be 80 years, and Mort measures his lifetime to be 80 years. If Velma and Mort spend their lives moving at 75% of lightspeed relative to each other, then Table 10.1 informs us that Mort's descendants (his grandson perhaps) will observe that Velma lives for 120 years, as measured by Mort's clocks. And Velma's granddaughter will observe that Mort lives for 120 years, as measured by Velma's clocks. According to Mort and his grandson, Velma ages slowly, by just a year during each of Mort's 1.5 years; he dies after just 80 of his years, and she dies after 120 of Mort's years, but having the physical appearance of a person who is only 80. According to Velma, all of this is reversed: He ages slowly, and she dies first. And both of them are correct.

DIALOGUE 11   (a) When Mort observes himself to be 60 years old, how old will he observe Velma to be? (b) When Velma observes herself to be 45 years old, how old will she observe Mort to be?

This suggests a perplexing question. Suppose that Velma and Mort are born at the same time on Earth, as twins perhaps, and that then Velma boards a spaceship, takes a fast trip to a far star, and returns to Earth. Will they disagree, after her return to Earth, about who has aged less—about whose biological clock ran slow? Surely, once they both are back in the same reference frame (Earth), they must agree on who is older, because disagreements about time occur only between observers who are in differently moving reference frames. In any single reference frame, there is only a single time.

What will the two twins observe once they are back in the same reference frame? This question, known as the "twin paradox," is an old one in the history of the theory of relativity. Here is the answer to it:

Recall that the special theory of relativity applies only to nonaccelerated observers. But in the scenario given for the two twins, Velma had to leave Earth, speed up enormously, reverse directions at least once in order to get

The idea that time can vary from place to place is a difficult one, but it is the idea Einstein used, and it is correct—believe it or not.
    Richard Feynman, physicist

*   You might wonder what "at the same time" means, since we are assuming that Mort and Velma are in different reference frames. To simplify matters, suppose that Mort and Velma are just passing each other. Then "at the same time" means that as either one comes into the world, he or she observes that the other is coming into the world too.

11. (a) Forty, because he observes that 1.5 of his years pass for every 1 of hers. (b) 30.

back to Earth, and then slow down and come to rest on Earth. Since this trip necessarily involves enormous accelerations, the special theory does not apply to Velma's observations.

The special theory does, however, apply to Mort's observations, since he did not accelerate. As we have seen, the theory predicts that he will observe Velma to age slowly during her entire trip, because she is moving relative to him. For example, if she moves at an average speed of 75% of lightspeed, then he should observe that 1.5 of his years elapse for every 1 of her years (Table 10.1). Suppose that Velma's trip takes 60 years as measured by Mort. He observes that during these 60 of his years, only 40 of her years elapse. So he observes that when they get back together on Earth, he is 60 and she is 40! Her observations must agree with this, since the two are now in the same reference frame. This is how you can get to be 20 years younger than your twin brother.

I asked [Einstein] once about a theory and he said, "When I am evaluating a theory, I ask myself, if I were God, would I have made the universe in that way?" If the theory did not have the sort of simple beauty that would be demanded of a God, then the theory was at best only provisional.

Banesh Hoffmann

HOW DO WE KNOW?

This conclusion has been experimentally verified, but in a less dramatic way. Atomic clocks were flown around the world on commercial jet plane flights. Although the predicted time difference was measured in only billionths of a second, it was measurable using high-accuracy clocks based on beams of atoms. As predicted, the clock that went on the trip came back younger (it hadn't ticked as many times) than the clock that stayed home. And the quantitative differences in elapsed time agreed with the predictions of the special theory of relativity.

This suggests some strange and wonderful possibilities. Suppose your mother leaves Earth for the star Vega, a sunlike star lying relatively close to our sun and a possible candidate for a planetary system. The distance to Vega is 26 light-years, by which we mean that it takes light 26 years to reach Vega from here. A **light-year** is the distance light travels in one year.

Suppose that your mother's spaceship averages a colossal 99.9% of lightspeed. She spends 3 years on a planet that is orbiting Vega and returns home. Since she travels at nearly lightspeed, each one-way trip takes slightly more than 26 years, as measured on Earth. So she is gone for about $26 + 3 + 26 = 55$ years, as measured on Earth. If you were 5 and your mother was 30 when she departed, you would now be 60 years old. But your mother would no longer be 25 years older than you! Table 10.1 informs us that during the 52 "Earth-years" of space travel at 99.9% of lightspeed, she aged by only 1 year for every 22.4 years of "Earth time." So she aged by only $52/22.4 = 2.3$ years during the 52 Earth-years. Including the 3 years spent on Vega, she aged by only 5.3 years during the entire trip. So she is 35.3 years old when she returns, and you are 60! This is how you can get to be older than your mother.

Your mother took a trip to the future. She could travel much further into the future, hundreds or thousands of years into the future, by moving much faster during a longer Earth time. But she couldn't go home again, in time. You can go as far forward in time as you like, if you only move fast enough, but you can never go backward in time.

Time dilation holds out the physical possibility of traveling to distant stars within a single human lifetime. Suppose you travel to a star 200 light-years away, at an average speed of 0.999 c relative to Earth. Even though the trip takes a little over 200 years as measured on Earth clocks, it takes you only $200/22.4 = 9$ years as measured in your spaceship. You can get there within your lifetime. But you can't go home again in time, to your own time on Earth. By the time you arrive at the star, two centuries will have elapsed on Earth. Even if you immediately hurry back to Earth, you will time-travel four Earth centuries into the future during the round-trip but will age by only 18 years. On Earth, you will be a relic from four centuries earlier. Although there are many practical difficulties in accelerating humans to speeds like $0.999c$, it might be possible someday (Section 12.5).

DIALOGUE 12   Could you leave Earth and travel fast enough to return before you were born? Could your mother leave Earth after you were born and return before you were born? Explain.

_____

12. No to both questions. The relativity of time allows time travel into the future but not into the past.

Summary of Ideas and Terms

Relative motion Two objects are in relative motion whenever they are moving differently.

Reference frame The laboratory or other surroundings within which an observer makes measurements. Measurements made in a particular reference frame are said to be "relative to" that frame. The phrase **relative to** means "as compared with."

Theory of relativity Any theory that provides answers to questions about observers in relative motion.

Galilean relativity The intuitive theory of relativity, in which time and space are absolute (in other words, different observers measure the same time intervals and the same distances) and light has different speeds relative to different reference frames.

Einstein's special theory of relativity The theory based on the principle of relativity and the principle of the constancy of lightspeed. In this theory, time and space are not absolute, and light has the same speed in all nonaccelerated reference frames.

The principle of relativity Every nonaccelerated observer observes the same laws of nature. "Unless you look outside, you can't tell how fast you're moving."

The principle of the constancy of lightspeed Light (and other electromagnetic radiation) has the same speed for all nonaccelerated observers, regardless of the motion of the light source or of the observer.

Ether theory The idea that a gaslike, nonatomic, material medium, the ether, fills the entire universe and that light waves are waves traveling through this medium. This theory was rejected after 1905 because it contradicted Einstein's theory. Philosophically, this amounts to a rejection of the idea that every physical thing is made of a material substance. Light waves are physical, but they are not made of matter.

Michelson–Morley experiment An experiment in 1887 to compare the speeds of light beams moving in different directions. The ether theory predicted that the light beams should have slightly different speeds, but the experiment detected no such difference.

Time Time is defined by clocks, in other words, by the operations we perform to measure time. The light clock, based on the motion of light beams, is a simple instrument to define time.

The relativity of time or **time dilation** The elapsed time (the number of seconds) between two particular events, such as two ticks on a particular clock or the birth and death of a person, is different for two observers who are in relative motion. The duration of one clock tick is longer for observers who are moving relative to the clock than it is for observers for whom the clock is at rest: Moving clocks run slowly.

Light-year The distance that light travels in one year.

Time travel An observer who accelerates to a high speed and then returns to the initial reference frame experiences a shorter elapsed time than does an observer who remains in the initial frame. So objects in the initial reference frame have aged more than the traveler has. This effect makes it possible to travel to stars that are many light-years distant in only a few years' travel time. It also makes travel to the future possible, by going on a fast trip and returning.

Review Questions

INTRODUCTION

1. Name the two revolutionary physical theories initiated at the beginning of the twentieth century.
2. Which inaccuracies of Newtonian physics are corrected by relativity theory? By quantum theory?
3. Describe the new element found in both relativity theory and quantum theory that is not found in Newtonian physics.

GALILEAN RELATIVITY

4. What is meant by relative motion, reference frame, a theory of relativity?
5. A train moves at 70 m/s. A ball is thrown toward the front of the train at 20 m/s relative to the train. How fast does the ball move relative to the tracks? What if the ball had instead been thrown toward the rear of the train?

6. A spaceship moves at 0.25c relative to Earth. A light beam passes the spaceship, in the forward direction, at speed c relative to Earth. According to Galilean relativity, how fast does the light beam move relative to the spaceship? Is this answer experimentally correct? If not, then what answer is correct?

THE PRINCIPLES OF RELATIVITY AND CONSTANCY OF LIGHTSPEED

7. How does travel in a jet airplane illustrate the principle of relativity? How must the airplane be moving in order to illustrate this principle?
8. State the principle of relativity in your own words. Does it apply to every observer? Explain.
9. State the principle of the constancy of lightspeed in your own words. Does it apply to every observer? Explain.

10. Why can't you keep up with a light beam—what law of physics prevents it?
11. Use the principle of the constancy of lightspeed to explain why no observer can move at precisely speed c relative to any other observer.

EINSTEIN'S LOGIC

12. Describe the ether theory. What prediction does this theory make regarding the speeds of light beams? Is this prediction correct?
13. What did the Michelson–Morely experiment measure, and what was the result?
14. Describe the philosophical implications of the rejection of the ether theory.
15. In Galilean relativity, space and time are absolute, and lightspeed is relative. How are these ideas different in Einstein's relativity?
16. What distinguishes the special from the general theory of relativity?
17. List the basic "laws" of the special theory of relativity.

THE RELATIVITY OF TIME

18. How is time defined in physics?
19. Describe the light clock.

20. Use the principle of the constancy of lightspeed to show that moving clocks must run slowly.
21. Velma passes Mort at a high speed. Both observers have clocks. What does each observer say about Velma's clock? What do they each say about Mort's clock?
22. Velma passes Mort at a high speed. Mort and Velma have identical ice-cream cones, frogs, and life spans. What does each observer say about the melting of both ice-cream cones, the aging of both frogs, and the aging of Velma and Mort?
23. One twin goes on a fast trip and returns. Does the special theory of relativity apply to the observations of both twins? Why, or why not?
24. One twin goes on a fast trip and returns. Have the two twins aged differently during the trip? If so, how do their ages differ?
25. Could you be older than your mother? If so, how?
26. A certain star is 200 light-years distant from Earth. If you traveled there at nearly lightspeed, how many years would the trip take as measured by observers on Earth? Would the trip take this many of your years? Explain.
27. Explain how you can travel to the future.

Home Project

Do some experiments while traveling on a smoothly moving, unaccelerated jet plane, train, or car. Also try to do the same experiments while your vehicle is speeding up, slowing down, or turning. Suggested experiments: Drop a ball with one hand held directly beneath the drop point; drop a piece of tissue paper (close all windows and vents to avoid air currents); roll a ball on a smooth tabletop. Don't try this while you are driving.

For Discussion

1. Does it seem to you that Einstein's personality, or character, might have had much to do with his creation of the theory of relativity? Discuss.
2. Does Einstein's comment on his college experience (Section 10.1, including the quotations in the margins) seem to you to be a valid criticism of education? Is this criticism valid today, at the grade school, high school, or college level?

Exercises

GALILEAN RELATIVITY

1.*Velma bicycles northward at 4 m/s, and Mort jogs southward at 4 m/s. Are Velma and Mort in relative motion? What is Mort's velocity relative to Velma?
2.*Velma bicycles northward at 4 m/s, and Mort jogs northward at 4 m/s. Are Velma and Mort in relative motion? What is Mort's velocity relative to Velma?
3.*Velma bicycles northward at 4 m/s. Mort, standing by the side of the road, throws a ball northward at 10 m/s.

What is the ball's speed and direction of motion relative to Velma? What if Mort had instead thrown the ball southward at 10 m/s?
4.*Velma bicycles northward at 4 m/s. She throws a ball northward at 10 m/s relative to herself. What is the ball's speed and direction of motion relative to Mort, who is standing beside the road? What if Velma had instead thrown the ball southward at 10 m/s? What if she had thrown the ball southward at 4 m/s?

5. A desperado riding on top of a freight-train car fires a gun pointed forward. Compare the speed of the bullet as observed by a passenger on the train and as observed by the sheriff standing beside the tracks. What if the desperado instead points his gun backward toward the rear of the train?

6.*In the preceding question, suppose that the gun's muzzle velocity (speed of the bullet relative to the gun) is 500 m/s and that the freight-train's speed is 40 m/s. What is the bullet's speed and direction of motion as observed by the sheriff if the desperado fires the gun forward? What if he instead fires toward the rear of the train?

7.*Referring to the preceding two exercises, suppose that the sheriff fires his own gun at the desperado. The muzzle velocity of the sheriff's gun is 600 m/s. How fast does the bullet move relative to the desperado if the sheriff is standing on the ground behind the receding train? What if the sheriff is instead standing in front of the train as it approaches him?

THE PRINCIPLES OF RELATIVITY AND CONSTANCY OF LIGHTSPEED

8. Although everything is normal inside a smoothly moving jet airplane, things are not normal when the ride is bumpy; you spill your coffee, you experience moments of partial weightlessness, and so on. Does this conflict with Einstein's principle of relativity? Explain.

9. Think of several ways that you could determine from inside an airplane whether the plane was flying smoothly or parked on the runway. Do each of these ways involve some direct or indirect contact with the world outside the airplane?

10. How can you tell from inside a jet airplane that the plane is speeding up? Slowing down? Turning a corner? Does your ability to determine these things conflict with Einstein's principle of relativity?

11. How fast are you moving right now? What meaning does this question have?

12. Is it physically possible (that is, would it be consistent with the principles of physics) for a person to move past Earth at 99% of lightspeed? If an Earth-based light beam were turned on, how fast would this person see it moving?

13. Is it physically possible for a person to move past Earth at exactly lightspeed? Explain.

14. Velma's spaceship approaches Earth at $0.75c$. She turns on a laser and beams it toward Earth. How fast does she see the beam move away from her? How fast does an Earth-based observer see the beam approach Earth?

15.*A desperado riding on top of a freight-train car fires a laser gun pointed forward (compare this with Exercise 5).

What is this gun's "muzzle velocity"? Suppose the train is moving at 40 m/s (0.04 km/s). How fast does the tip of the laser beam move relative to the sheriff, who is standing on the ground beside the train? What answer would the Galilean theory of relativity give to this question?

16.*Earth orbits the sun at a speed of about 30 km/s. If we assume that ether fills all space and that the sun is at rest in the ether, then it follows that Earth is moving at about 30 km/s through the ether. If these assumptions are true, then what speed will the observer in Figure 10.8 observe for light beam A? For light beam B?

17.*Michelson and Morley measured the *difference* between the speeds of two light beams moving in different directions. Under the assumptions stated in the preceding exercise, what would be the difference between the speeds of the two light beams in Figure 10.8? Express your answer as a fraction of c.

THE RELATIVITY OF TIME

18. Because of their relative motion, Mort says that Velma's clock runs slow, and Velma says that Mort's clock runs slow. Whose clock is right?

19. Your pulse could be considered to be a kind of clock. Think of several other natural (but not necessarily biological) clocks. Would an observer moving rapidly past these clocks observe all of them to run fast, slow, or on time?

20. Velma passes you at a high speed. According to you, she ages slowly. How does she age according to her own observations? How do you age according to her?

21. If you were moving away from Earth at nearly lightspeed (but with no acceleration), would you notice any change in your pulse? Any change in your breathing rate? Anything unusual about your clocks?

22. Suppose you have a twin brother. Is there anything you two could do that would make you older than him? Is there anything you two could do that would make you younger than him? Explain.

23. The center of our galaxy is about 30,000 light-years away. How long does light take to get here from there? Could a person possibly travel there in less than 30,000 years as measured on Earth? Could a person possibly travel there in less than 30,000 years of his or her own time? Could a person get there within his or her own lifetime? Explain.

24.*Velma passes Earth moving at 50% of lightspeed. She watches a TV program that runs for 1 hour as measured by her. How long does her TV program run as measured by an Earth-based observer?

25.* Your fantastic rocketship moves at 30,000 km/s. If you took off, moved at this speed for 24 hours as measured by you, and returned to Earth, by how much time would your clock differ from Earth-based clocks? Would you have aged more or less than people on Earth? By how much?

26.* Your extraordinarily fantastic rocketship moves at 99% of lightspeed. If you took off, moved at this speed for 24 hours as measured by you, and returned to Earth, by how much would your clock differ from Earth-based clocks? Would you have aged more or less than people on Earth? By how much?

11

THE UNIVERSE
ACCORDING TO EINSTEIN

Einstein's Pegasus

There's Einstein riding on a ray of light,
Holding a mirror up, at arm's length in his hand,
In which he cannot see his face in flight
Because his jesting image, I now understand,
Won't ever reach the mirror since its speed,
Too, is the speed of light. He rides, this fleeting day,
As if on Pegasus, immortal steed
Of bridled meditation, past the Milky Way,
Out to my mind's Andromeda, where I,
Also transported, staring at a windless pool,
Watch his repaired reflection whizzing by.
Though he can't see himself, this self-effacing fool
Who holds all motion steady in his head,
I won't forget his facing what he cannot see
In thought that binds the living and the dead,
And ride with him, outfacing fixed eternity.

Robert Pack, Middlebury College, 1991

Humans stand about 1 or 2 meters tall and can run at a few meters per second. Since our brains have evolved to deal with these normal sizes and speeds, it is not surprising that phenomena at speeds approaching lightspeed, at intergalactic distances, and at atomic dimensions are hard to imagine. This chapter continues our study of Einstein's special theory of relativity, which is essentially the theory of high-speed phenomena. Also in this chapter we look at phenomena involving large distances or intense gravitational fields, ranges that are dealt with by Einstein's general theory of relativity.

Chapter 10 presented the foundations of the special theory of relativity: the theory's two basic laws and its most fundamental prediction, namely, time dilation. The first four sections of this chapter present four more of the extraordinary phenomena predicted by this theory: the relativity of space (Section 11.1), the non-Galilean relativity of velocity (Section 11.2), the relativity of mass (Section 11.3), and the equivalence of energy and mass (Section 11.4). The most fundamental of these, and the aspect of special relativity that Einstein himself thought was the most significant, is the equivalence of energy and mass. We discuss the philosophical significance of this idea in Section 11.5.

The special theory of relativity describes the observations of nonaccelerated observers. What about accelerated observers? Einstein found a significant connection between acceleration and the force of gravity. His general theory of relativity, which describes the observations of accelerated observers, is essentially a new theory of gravity. Sections 11.6 and 11.7 present this theory's key ideas and discuss its predictions concerning nature's largest object, the universe.

11.1 The relativity of space: *space is what humans measure with rulers* _____

What is space? Just as time means "what is measured by clocks" (Section 10.6), space means "what is measured by rulers." In other words, the process of using meter sticks to measure distance defines what we mean by space.

What operations should Mort (remember Mort? We met him in Chapter 10, along with Velma) perform to measure the length of a particular object such as a table? For a table at rest relative to Mort, the prescription is to place a meter stick along the table (or several meter sticks if it is a long table) and compare the ends of the table with the marks on the meter stick.

But what if the table is moving past Mort? As we saw in Chapter 10, we must be careful because it is easy for unstated assumptions to slip in unnoticed. Mort should continue to use meter sticks that are fixed in his own reference frame, because he wants to know the length of the moving table as measured in his own reference frame. If the length being measured lies along the direction of motion, it is important for Mort to measure the positions of the two ends of the table simultaneously. The reason is that if, for example, the position of the table's front end is measured earlier than is the position of its back end, the back end will have moved forward during the time between the two measurements, and the measured length will be shorter than it would have been if the two ends had been measured at the same time.

In order to ensure that the two measurements, front end and back end, are simultaneous, Mort must use a clock (two clocks, actually—one at each end). This means that the question of measuring the length of a moving object is mixed up with the question of measuring time; time and space are tangled up with each other! Considering, then, that time is relative, it comes as no surprise to learn that space is relative too. We won't go through the argument that proves this result; it is similar to the argument in Section 10.6 by which we proved that moving clocks go slow.

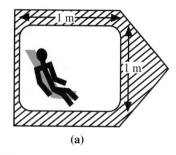

(a)

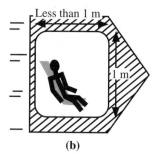

(b)

FIGURE 11.1

The window in Velma's spaceship as measured by (a) Velma and (b) Mort.

More specifically, Einstein's theory predicts that the table's length along its direction of motion, as measured by Mort, will be shorter than its length as measured by Velma, who is traveling along with the table. This effect is called **length contraction**. There is no length contraction along directions perpendicular to the table's direction of motion. Figure 11.1 shows how this affects Mort's measurements of a window in Velma's spaceship that is 1 meter square as measured by Velma.

As with time dilation, length contraction works both ways: Velma finds that objects in Mort's reference frame are contracted along the direction of motion, just as Mort finds that Velma's objects are contracted. The general rule is that moving objects are contracted along their direction of motion.

A quantitative analysis leads to a formula that predicts an object's measured length for various speeds of the object.* Figure 11.2 graphs this formula. It shows the predicted length of a 1-meter-long object such as a meter stick, held parallel to its motion, for various speeds of the object. This graph should be compared with Table 10.1 and Figure 10.11 for the predicted time between ticks on a moving clock. Both length contraction and time dilation are barely detectable for speeds below about 0.1c but become large above about 0.9c.

Recall that time dilation is not simply something that happens to clocks. Rather, it is time itself that is dilated. Similarly, length contraction is not just something that happens to spaceships, windows, and meter sticks. Since space is defined by objects like meter sticks, it is space itself that is contracted. Velma's space is different from Mort's space, just as Velma's time flow is different from Mort's time flow. Instead of a single, universal space, we must speak of "Velma's space" and "Mort's space."

Einstein's basic insight into space and time is that space and time are not abstractions that exist independently of observers and clocks and meter sticks. Space and time are *physical*; they exist only because physical objects such as observers, ice-cream cones, tables, clocks, and meter sticks exist.

FIGURE 11.2

The relativity of space. The predicted length of a meter stick for various speeds of the meter stick relative to the observer.

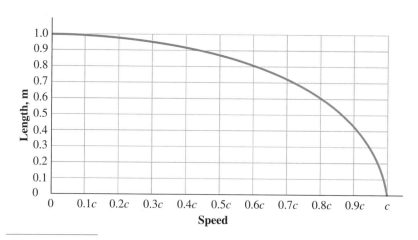

---

* The formula is $L = L_0\sqrt{(1 - s^2/c^2)}$, where s is the object's speed, L_0 is the length of the object as measured by an observer moving along with the object ($L_0 = 1$ meter if the object is a meter stick), and L is the length of the object as measured by an observer who is moving at speed s relative to the object.

DIALOGUE 1   Velma's spaceship is 100 meters long and 10 meters high as measured by Velma. (a) She passes Mort at 60% of lightspeed. How long and how high is her spaceship as measured by Mort? (b) What if she passes him at 90% of lightspeed?

DIALOGUE 2   (a) Velma observes Mort's automobile as she passes Mort at 60% of lightspeed. Mort's automobile is 4 meters long as measured by Mort. How long is it as measured by Velma? (b) What if she passes him at 90% of lightspeed?

DIALOGUE 3   Velma's spaceship is 100 meters long and 10 meters high as measured by Velma. (a) Roughly how fast must she be moving past Mort in order for Mort to measure the spaceship's length to be 50 meters? (b) Roughly how fast must she be moving past Mort in order for Mort to measure her spaceship to have a square shape, 10 meters long by 10 meters high?

11.2 The relativity of velocity according to Einstein

We have seen (Section 10.4) that when light beams are involved, speeds don't add up in the normal Galilean way. How do velocities add up in situations that don't involve light beams?

To be specific, consider Figure 11.3. Velma's spaceship passes Mort, and Velma fires a baseball launcher capable of launching baseballs at a considerable fraction of lightspeed. What is the baseball's speed relative to Mort? According to Galilean relativity, the answer is found by simply adding the spaceship's speed relative to Mort and the baseball's speed relative to Velma. For example, if Velma passes Mort at $0.75c$ and if the launcher's muzzle velocity is also $0.75c$, Galilean relativity predicts that the ball will pass Mort at $0.75c + 0.75c = 1.50c$.

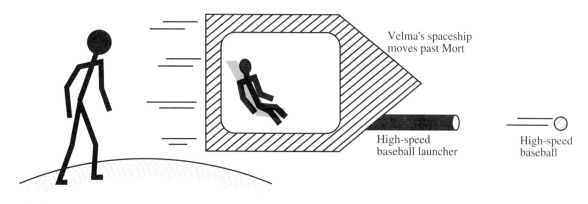

Velma's spaceship moves past Mort

High-speed baseball launcher

High-speed baseball

FIGURE 11.3

If Velma moves past Mort and she launches a baseball, how fast will it move relative to Mort?

1. (a) 80 meters long and 10 meters high. (b) About 45 meters long and 10 meters high.
2. (a) 0.8×4 meters = 3.2 meters. (b) 0.45×4 meters = 1.8 meters.
3. (a) About $0.87c$. (b) About $0.99c$.

TABLE **11.1**

The Relativity of Velocities: Some Quantitative Predictions, Based on Einstein's Theory of Relativity. The table gives the speed of the baseball relative to Mort (Figure 11.3) for various speeds of Velma relative to Mort and various speeds of the baseball relative to Velma.

		Velma's speed relative to Mort					
		0.10c	*0.25c*	*0.50c*	*0.75c*	*0.90c*	*0.99c*
Baseball's speed relative		Baseball's speed relative to Mort					
to Velma	0.10c	0.198c	0.341c	0.571c	0.791c	0.917c	0.992c
	0.25c	0.341c	0.470c	0.667c	0.842c	0.939c	0.994c
	0.50c	0.571c	0.667c	0.800c	0.909c	0.966c	0.997c
	0.75c	0.791c	0.842c	0.909c	0.960c	0.985c	0.9986c
	0.90c	0.917c	0.939c	0.966c	0.985c	0.994c	0.9995c
	0.99c	0.992c	0.994c	0.997c	0.9986c	0.9995c	0.99995c
(light beam):	c	c	c	c	c	c	c

Einstein's relativity gives a different prediction because speed measurements involve distance and time, which are subject to length contraction and time dilation. A quantitative analysis leads to a formula that gives the baseball's speed relative to Mort in terms of Velma's speed and the baseball's speed relative to Velma.* Table 11.1 gives some of the results predicted by this formula.

For example, if Velma passes Mort at 0.75c and if the baseball's launching speed is also 0.75c, the ball's speed relative to Mort will be 0.960c rather than the 1.5c that Galilean relativity predicts. As we expected, the ball passes Mort at less than speed c.

Even if Velma is moving at nearly lightspeed and launches the ball at nearly lightspeed, Mort will observe the baseball pass him at less than lightspeed. For example, if Velma is moving at 0.99c and launches the ball at 0.99c, Table 11.1 shows that it will pass Mort at a little slower than lightspeed. This is similar to the behavior of a light beam "launched" by Velma, shown in the bottom line of the table. It is not only light beams that have the strange behavior stated in the principle of the constancy of lightspeed; material objects such as baseballs also approximate this same strange behavior at high speeds. Einstein's theory describes space and time themselves, and not just light beams.

For slower speeds, Einstein's predictions are nearly the same as the "normal" Galilean predictions. For example, if Velma passes Mort at 0.1c and launches her baseball at 0.1c, Einstein's relativity predicts that the ball will pass Mort at 0.198c, nearly as large as the 0.2c predicted by Galilean relativity. And as expected, Einstein's predictions are experimentally indistinguishable from the Galilean predictions at normal speeds. Returning to the example we started with in Chapter 10 (Figure 10.3): If Velma's train passes Mort at 70 m/s and Velma throws a baseball forward at 20 m/s, Einstein's formula for relative velocities predicts that the ball will pass Mort at 89.999,999,999,999,982 m/s. This is so close to the

---

* The formula is $S' = (s + S)/(1 + sS/c^2)$, where s is Velma's speed relative to Mort, S is the baseball's speed relative to Velma, and S' is the baseball's speed relative to Mort.

Galilean prediction of 90 m/s that no known experiment could tell the difference.

DIALOGUE 4   Suppose that Velma passes Mort at 0.99c and that she launches a baseball at 0.99c. According to Einstein's theory, how fast will the ball move relative to Mort? What does Galilean relativity predict? Which prediction agrees with nature?

11.3 The relativity of mass: *inertia is relative*

Einstein's new idea, the constancy of lightspeed, affects nearly everything in physics: time, space, relative velocities, and much more. Relativity also alters Newton's law of motion (Chapter 4). This law states that an object's acceleration is equal to the net force exerted on the object divided by the object's mass, or in symbols

$$a = F/m$$

Newton's law of motion implies that if you exert an unchanging force on an object, the object will maintain an unchanging acceleration. Eventually the object will be going at lightspeed and still accelerating. An observer riding on this object could keep up with a light beam.

So Newton's law of motion, the centerpiece of Newtonian physics, is not consistent with the theory of relativity! Apparently, relativity alters Newton's law in such a way as to prevent objects from being accelerated up to lightspeed. To describe this alteration of Newton's law of motion, let's imagine that Velma and Mort have identical 1-kilogram objects, 1-kilogram melons perhaps. If Mort pushes on his melon with a force of, say, 1 newton, he will find that his melon accelerates at 1 m/s^2, just as Newton's law of motion predicts. If he now pushes on Velma's melon (which is moving past him) with an identical 1-newton force, Newton's law of motion predicts that Velma's melon will accelerate at 1 m/s^2, but *relativity theory predicts that Velma's melon will accelerate at less than* 1 m/s^2. As was the case for other relativistic effects, this effect is negligibly small at normal speeds. But if Velma's melon is moving at a significant fraction of lightspeed, Mort will find that its acceleration is noticeably smaller than 1 m/s^2. Relativity theory correctly predicts this effect.

From Mort's point of view, a 1-newton force applied to each melon produces a smaller acceleration in Velma's melon than in his own melon. From Mort's point of view, Velma's melon has more **inertia** than does his own melon (recall that a body's inertia is its resistance to acceleration). But this is the same as saying that Velma's melon has more mass, because the fundamental meaning of **mass** is "amount of inertia" (see Chapter 4). In other words, Mort measures Velma's melon as having a larger mass than does his own melon, even though they are identical melons. As usual, the effect works the other way around: Relative to Velma, her own melon has a mass of 1 kg, but Mort's melon has a mass of more than 1 kg.

---

4. 0.99995c. 1.98c. Einstein's prediction agrees with experiment.

Mass is relative: An object's mass increases with its speed. This effect is known as **relativistic mass increase**.

A quantitative analysis leads to a formula that predicts an object's mass for various speeds.* Figure 11.4 is a graph based on this formula, along with the graphs for time dilation (Figure 10.11) and length contraction (Figure 11.2). The formulas for mass increase and time dilation have identical forms, so these two graphs have identical shapes.

When we first met the idea of mass (Chapter 4), it was introduced in its most familiar and intuitive meaning as quantity of matter or amount of material. We then found it necessary to redefine the word *mass* as "amount of inertia," in order to be able to compare different masses. This was not a ma-

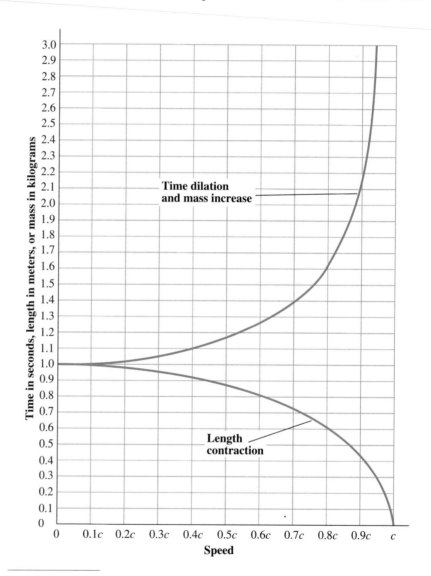

FIGURE 11.4

Relativistic mass increase, length contraction, and time dilation. The graph shows the duration of one clock tick (representing 1 second in the clock's reference frame) on a moving clock, the length of a moving meter stick, and the mass of a moving standard kilogram, for various speeds of the clock, meter stick, and kilogram relative to the observer.

---

* The formula is $m = m_0/\sqrt{(1 - s^2/c^2)}$, where m_0 is the object's "rest mass" (the mass as measured by an observer for whom the object is at rest), and m is the mass of the object when it is moving at speed s.

jor shift in meaning at the time, because in Newtonian physics an object's inertia is precisely the same as its quantity of matter, so the two definitions of mass yield the same quantitative value for the mass of any object. But relativity implies that these two ways of looking at mass are different, because an object's inertia increases with its speed, but its quantity of matter doesn't change, because it is still the same object, containing the same atoms and hence the same amount of matter that it had when it was at rest.

Our words need to distinguish between these two different concepts of "mass." An object's mass as measured by an observer for whom the object is at rest is called its **rest mass**. For example, Velma's and Mort's melons both have rest masses of 1 kg, regardless of who observes them. An object's rest mass measures the quantity of matter the object contains; it is a property of the object itself, regardless of who observes the object. The amount of inertia possessed by an object is called its **inertial mass**. Inertial mass is relative, but rest mass is not relative. The inertial mass and rest mass of a slow-moving object are essentially the same, but the inertial mass of a high-speed object is significantly greater than its rest mass. Because the defining meaning of mass is "amount of inertia," the unadorned word **mass** will henceforth always mean inertial mass.

HOW DO WE KNOW?

Relativistic mass increase is an everyday fact of life in high-energy physics labs. A subatomic particle can be accelerated to speeds in excess of 0.99 c, speeds so close to lightspeed that relativity predicts its mass to be thousands of times greater than its rest mass. One way to check this prediction is by applying electric or magnetic forces to a fast-moving particle, in order to bend its path, and measuring the curvature of the resulting path. If fast-moving particles really do have larger masses, their paths should bend less than they otherwise would, because their larger inertia tends to keep them moving straight ahead. Measurements show that the curvatures of the paths of fast-moving particles are less bent than they would be in the absence of a relativistic mass increase and that the amount of bending agrees with the predictions of the theory of relativity.

Time, space, and mass are relative. Is everything relative? Is there anything that is the same for every observer? The answer is that many things are not relative. It is a fundamental principle of the theory of relativity that the speed of any light beam is not relative. And it is a fundamental principle that every law of physics is the same for every observer.

Relativistic mass increase explains why you cannot accelerate objects up to lightspeed. At high speeds, an object's mass becomes very large, increasing without limit as the speed approaches c (Figure 11.4). To accelerate a fast-moving object requires an enormous force, and eventually the required force becomes so large that no combination of forces in the universe can provide it.

But there is something that moves as fast as lightspeed: light itself. In fact, light never moves slower than 300,000 km/s.* When you turn on a lightbulb,

---

* However, light travels through a material substance such as water or glass at an average speed that is sometimes far less than lightspeed. When moving through matter, light momentarily vanishes when it is absorbed by an atom and is recreated when it is emitted by the atom. Whenever the light actually exists as light, it moves at 300,000 km/s.

the light that is created does not accelerate from zero up to lightspeed; instead, it is moving at precisely lightspeed the instant it is created. Light is quite different from any material object. When you put a material object down in front of you, it has rest mass. Light beams must have no rest mass, because if they did, then relativistic mass increase would make their mass infinite while moving at lightspeed.

Anything, such as light or other forms of electromagnetic radiation, that moves at speed c and has no rest mass is classified as "radiation." The distinction between **matter**, which has rest mass and always moves slower than lightspeed, and **radiation**, which has no rest mass and always moves at lightspeed, is a useful one.

DIALOGUE 5 Mort finds that Velma ages slowly, that her body is shortened along its direction of motion, and that she is more massive than her normal mass. What does Velma say about herself? What does Velma say about Mort?

DIALOGUE 6 Use Figure 11.4 to estimate how fast Velma must move relative to Mort, for Mort to observe that her body's mass is 50% larger than normal. How fast must she move for Mort to observe that her spaceship is 50% shorter than normal?

DIALOGUE 7 Which of these is matter, and which is radiation: (a) red light, (b) the invisible waves drawn in Figure 9.5, (c) the invisible carbon-dioxide gas emitted by automobiles, (d) a radio wave, (e) chlorofluorocarbons drifting into the upper atmosphere from air conditioners, (f) X rays, (g) the electron beam that creates the picture on a TV tube (Section 8.6), (h) the so-called solar wind, which is a stream of protons emitted outward from the sun, (i) a gamma ray, and (j) an electric current flowing in a copper wire (Section 8.7)?

11.4 $E = mc^2$: energy has mass, and mass has energy

The connection that Einstein found between mass and energy starts from the relativity of mass: If you accelerate a rock from rest, the work you do on the rock will increase the rock's kinetic energy (Chapter 6), and at the same time the rock's mass will increase because of relativistic mass increase. So energy increase and mass increase go hand in hand.

---

5. To Velma, her aging process, body size, and mass all are normal. But she says that Mort ages slowly, that his body is shortened, and that he is more massive.
6. About $0.75c$. About $0.87c$.
7. (a) Radiation, (b) radiation, (c) matter, (d) radiation, (e) matter, (f) radiation, (g) matter, (h) matter, (i) radiation, and (j) matter.

Working from the principles of the theory of relativity and the law of conservation of energy, Einstein found that mass must be connected to energy in this fashion, regardless of the form of the energy that is given to an object. Instead of increasing the object's kinetic energy, you could give it thermal energy by warming it up, gravitational energy by lifting it, electromagnetic energy by charging it, or any other form of energy, and its mass would still increase.

This is surprising. If you lift a rock from the floor to the table, you don't expect its mass to increase. It is still the same rock, after all. Furthermore, since the rock on the table is at rest, Einstein is saying that its *rest* mass increases. Similarly, the theory of relativity predicts that the rest mass of a heated object, of an electrically charged object, and so forth are increased because of their energy increase. This is a new result: An object's mass increases not only because of its motion but also because of any form of energy increase, even those such as gravitational energy that don't involve motion.

For another simple example, find a rubber band and stretch it. In stretching it, you give it elastic energy. The theory of relativity states that you also increase its mass!

Einstein's analysis yields a simple formula that quantitively relates the increase in mass to the increase in energy. The formula states that the amount of mass given to an object equals the amount of energy given to it divided by the square of lightspeed:

$$\text{change in mass} = \frac{\text{change in energy}}{\text{square of lightspeed}}$$

In symbols:

$$\Delta m = \frac{\Delta E}{c^2}$$

where Δm and ΔE mean "the change in (or additional amount of) mass" and "the change in energy." If the standard metric units, kilograms and joules (J), are used for Δm and ΔE, then meters/second must be used for c. So $c = 3 \times 10^8$ m/s, and $c^2 = 9 \times 10^{16}$ m²/s².

To develop a feel for this result, let's look at examples. Suppose you accelerate a standard kilogram mass up to 100 m/s (360 km/hr, a fast train's speed). The kilogram's kinetic energy (Chapter 6) is then

$$\Delta E = \left(\tfrac{1}{2}\right) ms^2 = \left(\tfrac{1}{2}\right) \times (1 \text{ kg}) \times (100 \text{ m/s})^2 = 5000 \text{ J}$$

Einstein's formula tells us that the kilogram's relativistic mass increase is

$$\Delta m = \frac{\Delta E}{c^2} = \frac{5000}{9 \times 10^{16}} = 5.6 \times 10^{-14} \text{ kg} = 0.000,000,000,000,056 \text{ kg}$$

Not much of an increase. The kilogram's mass is now 1.000,000,000,000,056 kg. The increase is small because c^2 is so large. This is why relativistic mass increase wasn't noticed before Einstein: At ordinary speeds, it is too small to notice.

Suppose you lift a kilogram. One kilogram on Earth weighs 10 newtons. So if you lift it by, say, 400 meters (about the height of New York's World Trade Center), the kilogram's gravitational energy will increase by

$$\Delta E = \text{weight} \times \text{height} = 10 \text{ N} \times 400 \text{ m} = 4000 \text{ J}$$

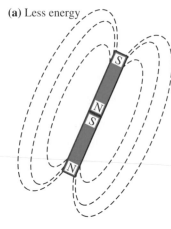

(a) Less energy

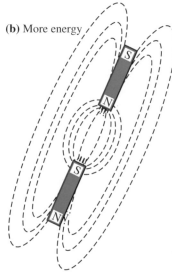

(b) More energy

FIGURE 11.5

The separated magnets of (b) have more energy, and hence more mass, than do the two joined magnets of (a). The excess energy and mass in (b) reside in the invisible and nonmaterial magnetic field, indicated by dashed lines. Compare this with Figures 9.1 and 9.2.

which is a little smaller than the 5000 J found in the previous example, and the mass increase is again too small to measure.

Heating an object typically involves larger amounts of energy. For example, in heating a pot of soup, you might give it 100,000 J of thermal energy. This is twenty times larger than the 5000 J in the first example, so the soup's mass gain is twenty times larger than the mass gain found in the first example, or about a trillionth of a kilogram. Still undetectable.

As another example, suppose that you have two bar magnets and that the north pole of one is joined to the south pole of the other so that they cling together (Figure 11.5(a)). Since you must do work on the two combined magnets to pull them apart, we see that the separated magnets have more energy than do the combined magnets (Figure 11.5(b)). But more energy means more mass. So if Einstein's idea is correct, the total mass of the two combined magnets is increased simply by pulling them apart! As the magnetic field lines shown in Figures 11.5(a) and (b) indicate, the work done in pulling the magnets apart creates a magnetic field that extends from one magnet to the other. The excess energy in the separated magnets resides in this invisible and nonmaterial magnetic field. We have encountered such "field energy" before, in the energy of electromagnetic radiation. But now we see that the energy of nonmaterial fields such as magnetic fields means that these fields also have *mass*.

The work involved quantitively in separating the two magnets is only a few joules, so the mass difference is again negligibly small compared with the original mass of the magnets.

Conceptually, it seems extraordinary that a system's mass should change simply by lifting it or heating it or separating it into two parts. But quantitatively, the predicted mass changes turn out to be negligibly small in all of these commonly encountered situations.

Things are different when we turn to uncommon examples, involving nature's more powerful forces. **Nuclear reactions**, for example, involve nature's strongest forces, the forces acting within the atomic nucleus (Chapters 15 and 16). For now, all we need to know about nuclear reactions is that they are similar to chemical reactions but that they involve changes in the structure of the nuclei of one or more atoms instead of changes in the way that different atoms combine to form molecules.

For example, in nuclear power reactors and nuclear weapons, the element uranium undergoes a nuclear reaction known as **nuclear fission**, in which the nucleus of each uranium atom is altered.* The fission reaction is a little like combustion, but the forces involved in fission are so strong that the thermal energy created is far larger than it is in any chemical reaction. So the mass loss, after removing the thermal energy, is far larger. If a kilogram of uranium is fissioned, the mass loss is about 0.01 kg (10 grams), which is a 1% mass decrease and easily detected. Measurements show that the predicted amount of mass actually is lost in fission experiments.

Nineteenth-century scientists believed matter to be indestructible, in other words, that rest mass is conserved in every physical process. This is certainly plausible. Since the days of the early Greek materialists (Section 2.1), sci-

---

* Each uranium nucleus is split to form two nuclei of various lighter-weight elements.

entists and others have felt that matter is indestructible—that although its form might change, its total amount cannot change. Nineteenth-century chemists performing high-precision mass measurements concluded that rest mass is conserved even in highly energetic chemical reactions.

But Einstein's relativity contradicts the idea of the conservation of matter. Matter, that is, rest mass, is not conserved in chemical reactions, in the stretching of a rubber band, and so forth. But these changes in rest mass are so small that they are experimentally undetectable. In high-energy processes such as nuclear fission, however, the changes are easily detected, and the results show that matter is not conserved.

MAKING ESTIMATES: THE ENERGY AVAILABLE IN A FEW GRAMS

We have seen that it takes a lot of energy to make even a little mass. It follows that even a small amount of mass can make a lot of energy. Suppose, for example, that 1 kilogram of uranium is entirely fissioned* and that all the energy "released" (converted to thermal and other forms of energy) during this process is used to do work. If this work were used to do some large task such as lifting the entire U.S. population, about how high would it be lifted?

The rest mass loss during the fissioning of 1 kg of uranium is 0.01 kg, or 10 grams—about 1/3 ounce. The amount of energy converted during this process is found from the formula $\Delta m = \Delta E/c^2$, only "turned around" to read $\Delta E = (\Delta m)c^2$:

$$\Delta E = (\Delta m)c^2 = (0.01 \text{ kg}) \times (9 \times 10^{16}) = 9 \times 10^{14} \text{ J}$$

How high could you lift the U.S. population with this much energy? The population is about 250×10^6, and a typical person's mass is perhaps 600 newtons (135 pounds). So

$$\text{weight of U.S. population} \approx 250 \times 10^6 \times 600 = 15 \times 10^{10} \text{ N}$$

Recall from Chapter 6 that gravitational energy equals weight times height. If the 9×10^{14} J of energy goes into lifting the population, then

$$9 \times 10^{14} \text{ J} = (15 \times 10^{10} \text{ N}) \times (\text{height})$$

This means that the height to which you could lift the U.S. population is

$$\text{height} = \frac{9 \times 10^{14} \text{ J}}{15 \times 10^{10} \text{ N}} = 6000 \text{ m}$$

This is 6 kilometers. That's a lot of work, out of just 10 grams.

Einstein's result, $\Delta m = \Delta E/c^2$, says that any energy given to a body increases that body's mass. In other words, "energy has mass." The preceding estimate shows, for example, that 9×10^{14} joules of energy has a mass of 10 grams: When you add this much energy to any body, you add 10 grams to it.

Now we take this reasoning one step further: Einstein believed that this result extended to any body's *entire* mass, so that a body's mass was entirely created by various forms of energy, such as kinetic energy, gravitational

---

* We assume, in this example, that our 1 kg of uranium is made entirely of the "fissionable isotope" of uranium, ^{235}U (see Chapter 16).

energy, and nuclear energy (energy due to the strong nuclear force). According to Einstein, there was no such thing as "pure mass" apart from energy—not only did all energy have mass, but also all mass had energy.

This means that $\Delta m = \Delta E/c^2$ extends to the total mass and energy:

$$\text{total mass of any object} = \frac{\text{object's total energy}}{c^2}$$

or, in symbols,

$$m = \frac{E}{c^2}$$

This implies Einstein's famous formula,

$$\text{total energy of any object} = (\text{object's total mass}) \times (c^2)$$
$$E = mc^2$$

Since energy means the ability to do work, the practical, experimental meaning of $E = mc^2$ is that any object of mass m should be able to do mc^2 units of work.

HOW DO WE KNOW?

If Einstein's idea is correct, there should be some physical process by which mc^2 units of work can be obtained from any object of mass m. Such processes, known as **matter–antimatter annihilation**, have been discovered.

In addition to the protons, neutrons, and electrons that form ordinary matter, physicists have discovered three other material particles, known as **antiprotons**, **antineutrons**, and **antielectrons**. If one of these "antiparticles" is brought close to its corresponding particle, the two particles will vanish entirely, and high-energy radiation will be created. It's an extreme example of the nonconservation of matter: Matter entirely vanishes, to be replaced by radiation.

Since any proton, neutron, or electron can be annihilated in this manner, this experiment shows that in principle, any material object can be turned into radiation—although it would be difficult in practice to collect enough antiparticles to annihilate a macroscopic object. Since this radiation could then be used to do work, we have here a demonstration of Einstein's idea. Furthermore, when the radiation's energy is measured, it is found to equal the total mass of the pair of particles times c^2.

It is easy to misinterpret the mass–energy idea. For example, it is sometimes said, incorrectly, that "mass can be converted to energy." One problem with this statement is that $E = mc^2$ implies that mass (meaning *inertial* mass) is always conserved because energy is always conserved, so mass is never converted to anything else. In an annihilation event, for example, the mass of the particle–antiparticle pair is precisely equal to the mass of the created radiation because the energies are equal. Rather, it is *rest* mass, or matter, that is destroyed here. So it is correct to say that "matter, or rest mass, can be converted to radiation." Concerning mass (inertial mass), it is correct to say that "all mass has energy and all energy has mass" or even that "all mass is energy and all energy is mass" or, as Einstein put it, "mass and energy are equivalent." One must be careful with the word *mass*.

We will summarize Einstein's idea as follows:

> **THE PRINCIPLE OF MASS–ENERGY EQUIVALENCE**
> Energy has mass; that is, energy has inertia. And mass has energy; in other words, mass is capable of doing work. The quantitative relation between the amount of energy and the amount of mass is $E = mc^2$.

DIALOGUE 8   In each of the following processes, does the system's mass change? Does its rest mass change? (a) An automobile speeds up from rest to 50 km/hr. (b) A rubber ball is squeezed. (c) Two positively charged objects at rest, such as two charged combs, are moved closer to each other and placed at rest. (d) An electron and an antielectron, at rest, spontaneously annihilate each other.

11.5 Relativity and the Newtonian worldview

Mass–energy equivalence represents a sharp break with the Newtonian worldview. Newtonian physics follows the Greek materialists in the belief that the predictable interactions between indestructible atoms moving in empty space determine everything that happens in the physical universe. Toward the end of the nineteenth century, electromagnetic phenomena intruded on this worldview by implying the physical existence of nonmaterial, nonatomic, electromagnetic fields. Apparently, the universe was made of atoms and fields, interacting and moving in predictable ways.

If all mass has energy, it seems reasonable to ask what kinds of energy an atom's mass might have. Part of an atom's mass is due simply to its motion, because all atoms are in motion and because of relativistic mass increase. But what about the remaining mass, the atom's rest mass? Since an atom is made of subatomic particles and these particles orbit the nucleus and vibrate within the nucleus even when the atom as a whole is at rest, some of an atom's rest mass must be due to the relativistic mass increase that arises from this subatomic kinetic energy. In addition, all these subatomic particles exert electromagnetic forces and nuclear forces on one another, so electromagnetic fields and nuclear force fields surround these subatomic particles. All of these fields possess energy, just as the magnetic fields in Figure 11.5 possess energy, and this **field energy** must contribute to an atom's mass.

Some portion of an atom's mass is due to kinetic energy and field energy. Apparently, atoms themselves are partly made of fields and "made of motion." This suggests an interesting question: Is that all there is? Are atoms made *only* of fields and motion? Is there ultimately no such thing as "pure rest mass," apart from internal motions and nonmaterial force

---

8. (a) The mass increases, but the rest mass doesn't change. (b) The mass increases, and the rest mass increases (because the ball is at rest). (c) The mass increases because work must be put into the system to bring the two objects closer together against the repulsive electrical force. The rest mass increases also, because the system is at rest. (d) The system's mass is unchanged because no work is done by external agents, and so energy is conserved. But the entire rest mass vanishes, since radiation has no rest mass.

fields? If so, atoms are not only "mostly empty space," they are *entirely* empty space and are made only of fields, similar to the magnetic fields in Figure 11.5.

The answer to this question is not yet known experimentally. This question, concerning the fundamental origin of mass, is one that might be answered by high-energy physics experiments. Einstein appears to have believed the "pure field" answer, that all mass arises from fields and the motion of those fields. The fundamental theories of contemporary physics, known as *relativistic quantum field theories*, also lean in this direction while still leaving an opening for the possible existence of pure rest mass. For example, Steven Weinberg, a leading high-energy theorist, stated:

> [According to the physical theories developed during the 1920s] there was supposed to be one field for each type of elementary particle. The inhabitants of the universe were conceived to be a set of fields—an electron field, a proton field, an electromagnetic field—and particles were reduced to mere epiphenomena. In its essentials, this point of view has survived to the present day, and forms the central dogma of quantum field theory: *the essential reality is a set of fields* [Weinberg's emphasis] subject to the rules of special relativity and quantum mechanics; all else is derived as a consequence of the quantum dynamics of these fields.

In this **field view of reality**, there is no "there" there (to quote the poet Gertrude Stein). There are no "things" at all. Electrons and other material particles are only force fields in space, just like the magnetic fields in Figure 11.5, and the particle is simply the place where the field is strongest. In this view, all mass is due only to the energy of force fields. Since fields are "possible forces" (Section 9.1), and forces are interactions, this view implies that every "thing," everything, is interactions and motion. It is the interactions and motion that are fundamental rather than the material particles that we had always supposed were doing the interacting and the moving. It's the interactions and the motion that make the particles, and not the other way around.

The theory of relativity tilts the philosophical balance further from the Newtonian view that indestructible atoms moving in empty space are the cause of everything that happens. Indeed, the field view of reality would stand materialism on its head and argue that it is motion and interactions, "everything that happens," that cause atoms to exist. According to this view, reality is made of fields filling all space and moving in predictable ways. The only notion still standing within the Newtonian worldview is the view that nature is predictable. As we will see (Chapters 13 and 14), quantum theory sweeps even this away.

There is no "there" there.
Gertrude Stein, poet

Perhaps particles are essentially "boxes" that hold energy. Indeed, modern views of the origin of the mass of elementary particles have very much this character. Perhaps there is no meaning to the concept of mass separate from energy!
Robert Adair, physicist

We are such stuff As dreams are made on
Shakespeare, *The Tempest*

Perhaps Shakespeare understood this universe of ours better than we do ourselves!
John A. Wheeler, physicist, *The Frontiers of Time*, referring to the preceding lines from *The Tempest*

11.6 Einstein's gravity

The special theory of relativity begins with the idea that the laws of physics are the same for all unaccelerated observers. What about accelerated observers? This question is the starting point for the general theory of relativity.

"Elevator," inside a rocket ship in outer space, accelerates at 9.8 m/s² in this direction

Because of acceleration, observer feels force by floor against shoes

(a)

Because of gravity, observer, at rest on the ground, feels force by floor against shoes

(b)

FIGURE 11.6
The effects of acceleration are just like the effects of gravity. The observer inside the "elevator" cannot tell the difference.

Imagine that you are in an elevator accelerating upward from a building's ground floor. As you might have noticed in your own experience, the acceleration makes you feel squashed down, heavy, as though there were more gravitational pull on you than usual. This connection between acceleration and "apparent gravity" runs fairly deep. For example, imagine that you are inside an accelerating elevator in outer space, far from all planets and stars so that there are no gravitational forces (Figure 11.6). If the elevator's acceleration is 9.8 m/s² or "1g" (pronounced "one gee"), you will feel the same as you do when you are stationary on Earth. The reason is that according to Newton's law of motion, the elevator's floor must push upward on the bottoms of your shoes in order to provide the 1g acceleration. This push quantitively equals your "Earth weight" (your weight on Earth). So you feel the same accelerating through space at 1g as you do at rest on Earth.

If you are in an elevator accelerating at 1g through space, how can you tell, without communicating with the world outside the elevator, that you are actually in space and not at rest on Earth's surface? You might try dropping a stone, to see how it falls (Figure 11.7). But your elevator is accelerating at 9.8 m/s², so the floor accelerates upward to meet the stone. From your point of view in the elevator, the ball "falls" down to the floor with an acceleration of 9.8 m/s². Furthermore, from your point of view inside the elevator, Galileo's law of falling (Chapter 3) is valid: Both a high-mass stone and a low-mass stone, released together, will contact the floor at the same time.

You could even throw a stone horizontally to see whether it curves down to the elevator floor (Figure 11.8) the way it does when you throw a stone horizontally on Earth. Because of the elevator's upward acceleration, the stone gets closer to the floor as it moves across the elevator. As you view it from inside the elevator, the stone "falls" to the floor exactly as though it were thrown horizontally on Earth.

It seems that it is not easy to find an experiment that can tell you whether you are at rest on Earth's surface or moving through space with a 1g acceleration.

Einstein made this reasoning into a fundamental new principle. The new principle is similar to the principle of relativity, which says you cannot determine your own speed from inside your own laboratory. The new principle states that the laws of physics are unaffected by acceleration, so that there is no experiment that you can perform inside an elevator that can tell you whether you are accelerating or, on the other hand, sitting still and feeling gravitational effects.

We summarize this as follows:

THE PRINCIPLE OF GENERAL RELATIVITY*
Every accelerated observer experiences the same laws of nature. In other words, no experiment performed inside a sealed room can tell you whether you are accelerating in the absence of gravity or are at rest in the presence of gravity.

---

* Often called the principle of equivalence.

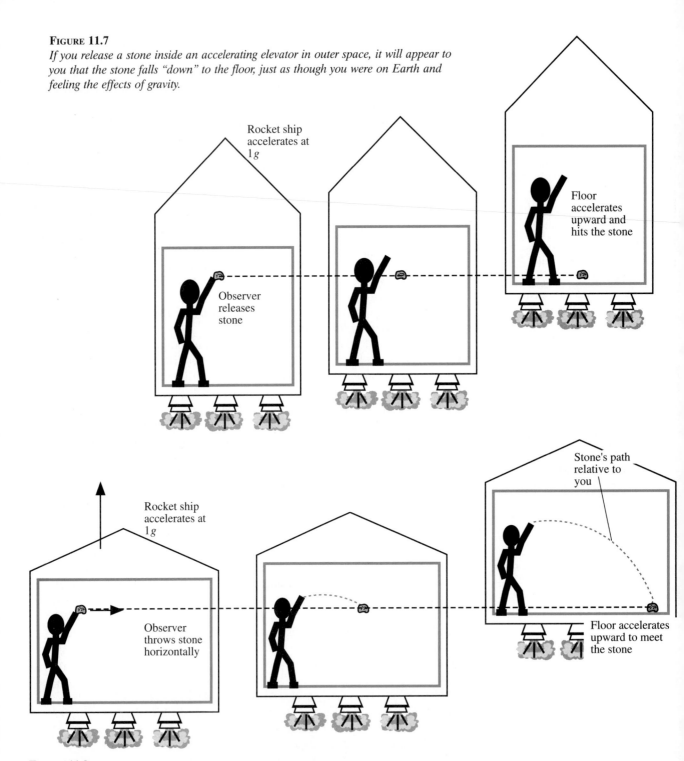

FIGURE 11.7
If you release a stone inside an accelerating elevator in outer space, it will appear to you that the stone falls "down" to the floor, just as though you were on Earth and feeling the effects of gravity.

Rocket ship accelerates at 1 *g*

Observer releases stone

Floor accelerates upward and hits the stone

Rocket ship accelerates at 1 *g*

Observer throws stone horizontally

Stone's path relative to you

Floor accelerates upward to meet the stone

FIGURE 11.8
If you throw a stone inside an accelerating elevator in outer space, it will appear to you that the stone falls to the floor as though you were on Earth and feeling the effects of gravity.

FIGURE 11.9
If you turn on a flashlight inside an accelerating elevator in outer space, the light beam will bend relative to you.

Light beams play a central role in the general theory of relativity, just as they do in the special theory. If you are accelerating through outer space and you turn on a flashlight horizontally, what will happen to the light beam? Just like the horizontally thrown stone (Figure 11.8), the light beam bends downward relative to you, because of your acceleration (Figure 11.9). The light beam curves far less than the stone does, but only because of the light beam's far greater speed.

The principle of general relativity implies that this experiment must come out the same way if performed in a stationary room in the presence of gravity. So *gravity must bend light beams*. This is a new, non-Newtonian, prediction of Einstein's theory.

HOW DO WE KNOW?

Earth's gravity is too weak to bend light beams very much. But the sun is massive enough to measurably bend light from the distant stars as the light beam passes close to the sun. The first measurement of this effect was made during a total eclipse of the sun in 1919, when astronomers could photograph the stars that appear near the edge of the sun (Figure 11.10). Measurements of these stars' positions showed that the starlight bends while passing the sun and that the amount of curvature agrees with the quantitative predictions of the general theory of relativity. Since 1919 there have been several more experiments of this type, often using radio telescopes to observe radio waves from stars observed near the edge of the sun. The advantage of using radio waves instead of light beams is that because the sun is a fairly "dim" radio star, stars near the edge of the sun can be detected by radio waves even when the sun is visible, so astronomers needn't wait for an eclipse. The results agree with Einstein's predictions.

FIGURE 11.10
Because the sun bends light beams, we can see stars that are behind the sun.

Einstein's special theory and general theory are laid out in similar ways. The two starting points, the principle of relativity and the principle of general relativity, state, respectively, that the laws of physics are unaffected by the reference frame's velocity and acceleration. In the special theory the next step is the principle of the constancy of lightspeed. Similarly, in the general theory the next step is the prediction that gravity must bend light.

Now recall that the constancy of lightspeed led Einstein to the surprising new conclusion that time is relative. Similarly, we will see that the gravitational bending of light implies a new and surprising property of space, related to the concept of "straightness." Just as "time" can be defined by light clocks, "straightness" can be defined as the path followed by a light beam. In fact, surveyors often use laser light beams to determine straightness, and you use light beams to determine straightness when you aim a gun by sighting along its barrel.

But what can it mean to say that gravity bends light beams, when light beams themselves are the definition of straightness? Einstein saw the answer. Just as the slowing down of moving light clocks implies that time itself slows down, the bending of light beams means that *space itself is bent by gravity*. Just as Einstein saw that time is physical and is defined by physical objects such as light clocks, he saw that straightness is physical and is defined by physical objects such as light beams.

It turns out that gravity affects not only space but also time, in a way that is best described as the **bending of space–time**. The idea of space–time arises because space and time are intertwined in both the special and general theories. It turns out that time is intertwined with space in a way that is mathematically analogous to the way that each of the three spatial dimensions (length, width, height) are intertwined with one another. Because of this, time is sometimes thought of as a "fourth dimension," even though neither Einstein nor anybody else can visualize four physical dimensions. As Stephen Hawking (Figure 11.11), one of today's most profound relativity and quantum theorists, remarked, "It is hard enough to visualize three-dimensional space," let alone four. Einstein's general theory mathematically describes this imagined **four-dimensional space–time**.

What does a curved three-dimensional space mean? Just as it is impossible to visualize four dimensions, it also is impossible to visualize a three-dimensional space that is curved. The reason is that space has only three dimensions (length, width, and height), and we exist within these three dimensions. There is no "other" dimension, no higher dimensionality, from which we can visualize the curvature of our three-dimensional space. Furthermore, the phenomenon of curved space is entirely outside our intuitions, because the bending of space is too small to be noticeable over the relatively short distances on Earth with which we are familiar.

The best we can do is visualize an analogy to this important idea: Instead of visualizing a curved three-dimensional space, we visualize a curved *two*-dimensional space, and this will give us insight into a curved three-dimensional space.

FIGURE 11.12
If it were extended forever in both length and width, a flat tabletop would be a flat two-dimensional space.

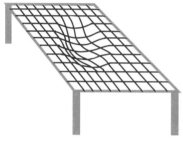

FIGURE 11.13
If you put a warp in a flat two-dimensional space, it becomes a curved two-dimensional space.

FIGURE 11.14
The surface of a sphere is a curved two-dimensional space of finite total extent.

What do we mean by a "space"? We call our three-dimensional space a "space" because we can continue through it forever and never come to a boundary or edge beyond which there is no more space. This is the essential idea of a **space**: an extended region that has no edges or boundaries. A two-dimensional space, then, is a two-dimensional region that has no edges. For example, a flat tabletoplike area that is extended forever in both directions is a two-dimensional space (Figure 11.12).

Now imagine that we begin with this infinite flat tabletop, this *flat* two-dimensional space, and we put a warp in it, a depression perhaps (Figure 11.13). This makes it into a *curved* two-dimensional space.

A flat two-dimensional space must have infinite extent, infinite total area, in order not to have edges. A *finite* (meaning "not infinite") tabletop is not a space because it has edges. But if you imagine squeezing a flat tabletop and bending it around on itself to form the surface of a sphere, the edges will vanish (Figure 11.14). This spherical surface is a curved two-dimensional space of finite extent. Note that this "spherical two-dimensional space" consists of only the sphere's surface; the sphere's interior and exterior are not part of the space.

Suppose you are a two-dimensional creature inhabiting a two-dimensional spherical space, something like a flat ant walking around on the surface of a beach ball. How can you tell that your space is curved? You couldn't stand outside or inside the sphere's surface, in the third dimension, and see that your space is a spherical surface, because you are confined within your two dimensions. One way you can learn that your space is curved is by performing geometry experiments. For instance, two lines, beginning as parallel and extending as straight (or straightest) lines, should eventually meet (Figure 11.15).

Although we cannot visualize the curvature of three-dimensional space, we can visualize experiments to determine whether or not our space is curved. The 1919 experiment that measured the curvature of light near the sun was just such an experiment. It found that even the straightest path, the path of a light beam, curves near the sun. We conclude that three-dimensional space is curved. Like two-dimensional creatures living in a curved two-dimensional space, we three-dimensional creatures live in a curved three-dimensional space, but we cannot visualize the curvature because we cannot stand outside our three dimensions. Nevertheless, the curvature is real because we can observe (but not visualize) and measure it.

HOW DO WE KNOW?

Despite the bending of light beams, one might still imagine that this bending doesn't really show space to be curved but instead shows only that light beams curve in a "flat" space. This possibility is ruled out by an experiment in 1972 in which a spacecraft orbiting Mars beamed back radar signals sent from Earth (Figure 11.16). When the line of sight from Earth to Mars passed near the sun, the radar beam's travel time was measured. This travel-time measurement can tell whether the curved light beam is traveling through a flat space or through a curved space: It is easy to use the curved path to predict the travel time in a flat space, by making a

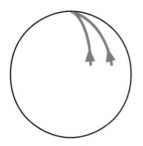

FIGURE 11.15

In a two-dimensional spherical space, two lines that start out parallel and extend as "straight" (or straightest) lines will eventually meet.

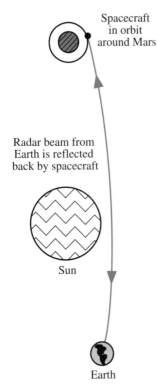

Spacecraft in orbit around Mars

Radar beam from Earth is reflected back by spacecraft

Sun

Earth

FIGURE 11.16

An experiment to measure the total travel time for a radar beam to get to Mars and back.

scaled-down drawing of the curved path of the radar beam on a flat sheet of paper and seeing how much longer it is than a straight line. In the experiment, the answer was about 10 m, so in a flat space the signal should have been delayed by about 30 billionths of a second, the time that it takes light to travel 10 m.

But you cannot use a flat sheet of paper to measure distances in a curved space, for the same reason that you cannot accurately determine the distance from Los Angeles to London by making measurements on a flat map: The "scale" keeps changing because of the curvature. Einstein's formulas for curved space–time predict a delay of 200 millionths of a second, about 7000 times longer than the predicted delay through a flat space–time. The experiment confirmed the longer delay.

There still are scientific doubts about the validity of Einstein's general theory of relativity, although there no serious doubts about the special theory. One reason for the doubts is that unlike the special theory, it is difficult to find practical experiments that can test the general theory. Despite these doubts, all the other possible theories that have been proposed as competitors to the general theory are curved-space theories. The consensus is that space is curved.

DIALOGUE 9   The surface of a sphere is a two-dimensional space. (a) Is the interior of a sphere a three-dimensional space? (b) Is it a two-dimensional space? (c) How many dimensions does the interior of a sphere have? (d) Is the interior finite or infinite in extent?

DIALOGUE 10   (a) How many dimensions does a circle (meaning the perimeter, or edge, of a circular area) have? (b) How many dimensions does the interior of a circle have? (c) Is a circle a space? (d) Is the interior of a circle a space?

11.7 The shape of the universe _____

The general theory of relativity is primarily a new theory of gravity. Conceptually, it is radically different from Newton's theory of gravity, although its predictions for "normal" situations such as the fall of a stone to Earth or the orbits of the planets are nearly identical to Newton's predictions. As we have seen, general relativity predicts that gravity bends space–time; in other words, masses bend space–time (Figure 11.17). According to the theory, gravity is this bending of space–time. In Einstein's theory, gravitational effects such as Earth's circular motion around the sun are not

---

9. (a) No, because it has a boundary. (b) No, because it's not even two dimensional. (c) Three. (d) Finite.
10. (a) One. (b) Two. (c) Yes. (d) No.

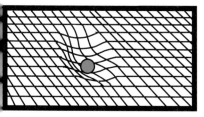

FIGURE 11.17
Masses such as the sun cause bends in space-time.

caused by forces at all but are instead entirely due to the curvature of space–time.

When Einstein's theory is applied to space–time as a whole, it describes the possible ways that our curved three-dimensional universe could evolve throughout past and future time. Einstein's theory and several significant pieces of experimental information about the origin and evolution of the universe (Section 16.3) lead to the following picture of the evolution of the universe:

The universe began in a single event some 15 billion years ago, a violent explosion called the **big bang** that created the different forms of energy and matter and that hurled matter and energy outward. Although the universe is expanding, it is not expanding into anything because there is no "outside" of the universe. And just as there is no space outside the universe, there was no time before the big bang. That is, the big bang created space and time. Questions such as What is outside our universe? and What happened before the big bang? may be meaningless. The big bang was not an event, like the explosion of a firecracker, that happened in time and space. Rather, the big bang *created* time and space.

The general theory of relativity predicts that the universe's three-dimensional space must either expand or contract. It is remarkable that space and time themselves are in a state of continual change. Everything, it seems, changes: Time and space expands, the stars are born and die, life on Earth evolves, you and I are born and will die, the atoms perpetually change (Chapters 14 and 18), and even so-called empty space or vacuum is undergoing perpetual change (Chapter 18). It seems that the only unchanging aspect of the universe is change itself.

The earliest evidence for the **expansion of the universe** and for the big bang came from astronomical observations of other galaxies outside our own Milky Way galaxy. Distant galaxies are moving away from us, and the more distant galaxies move away faster. But the galaxies are not just moving away from our particular galaxy; they all are moving away from one another. Regardless of which galaxy you live in, you will observe the other galaxies moving away from you.

According to general relativity, there are three possible overall shapes or "geometries" for the three-dimensional universe: spherical, flat, and "hyperbolic." Two factors determine the shape: the universe's expansion rate and the large-scale average mass density (the average amount of mass there is per unit of volume throughout the universe). Neither of these two factors is very well known. Current observations can pin down the expansion rate to only within a factor of about 2, and the mass density is even more uncertain. So today we cannot determine which shape the universe actually has, out of the three possibilities.

Let's look at just one of these possible geometries: a **spherical universe.** If the expansion is slow enough and the mass density is large enough, the universe's mass will cause space to bend in on itself and form a three-dimensional spherical space. This spherical universe has a finite total volume but no edge, just as the two-dimensional surface of a sphere has a finite total area but no edge. The expanding universe is then analogous to the two-dimensional curved surface (not the inside) of a balloon that is being inflated (Figure 11.18), with the galaxies represented by small dots on the balloon's surface. As the balloon expands, the distance between the dots increases. In agreement with science's Copernican attitude, no galaxy can be at the center of a spherical universe because such a universe has no center, just as the surface of a balloon has no center.

If the universe's expansion is fast enough or its mass density is small enough, the universe will bend outward away from itself instead of inward on itself, in which case its spatial extent is infinite and it has a "hyperbolic" shape. At the dividing line between the spherical and hyperbolic geometries, if the universe's expansion rate and mass density are balanced in just such a way that the universe bends neither inward nor outward, the universe is said to be "flat."

The universe's future evolution is, like its shape, also determined by the balance between its expansion rate and its mass density. Like a stone thrown upward and slowed by Earth's mass, if the expansion is slow enough or the mass density is large enough, the universe will eventually stop expanding and fall back inward on itself. But if it has a rapid expansion or a low mass density, it will continue expanding forever.

What do observations indicate about the shape and fate of the universe? The mass in the universe that we can directly observe with today's telescopes is only about 10% of the amount needed to stop the expansion. But observation and theory both suggest that still-unobserved "dark" matter—such as black holes or other "dark stars" or yet-unknown forms of matter—may be sufficient to stop the expansion.

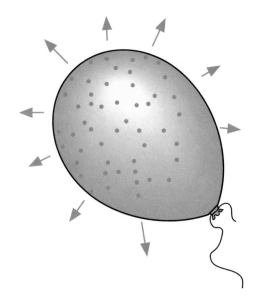

FIGURE 11.18
A two-dimensional representation of the expansion of the three-dimensional universe. As space, represented by the balloon's surface, expands, the galaxies, represented by the dots, move farther apart. Although this two-dimensional analogy shows the universe expanding into the empty space outside the balloon, there is no space outside the real three-dimensional universe. Our universe is all there is.

The general theory of relativity holds all sorts of fascinating possibilities, which are, to quote philosopher-scientist John Haldane, "not only queerer than we suppose, but queerer than we can suppose." Perhaps our universe is one among many universes, each having its own space and time, perhaps different space–time dimensionality, and its own physical laws. Perhaps, over many different "times," an infinity of different universes passes into and out of existence, forming collectively a reality that occurs not in space and time at all but that is in some sense beyond space and time.

In one such universe, in one galaxy called Milky Way, on one planet called Earth, you who read these words and I who write them are privileged to hold such ideas in our minds.

Summary of Ideas and Terms

The relativity of space or **length contraction** Moving objects are contracted along their direction of motion, so an object's length is different for different observers.

The relativity of velocities Both Galileo's and Einstein's theories predict that velocities are relative, but velocities "add" differently in Einstein's theory.

The relativity of mass An object's inertia (in other words, its mass) increases with its speed, so its mass is different for different observers.

Rest mass The mass of an object at rest. Rest mass represents quantity of matter.

Inertial mass or **mass** The amount of inertia possessed by an object.

Matter has rest mass, but **radiation** has no rest mass. Matter moves at less than lightspeed, but radiation always moves at lightspeed.

The principle of mass–energy equivalence, or $E = mc^2$ All mass has energy, and all energy has mass. A system with m units of mass has mc^2 units of energy. A system with E units of energy has E/c^2 units of mass.

Nuclear reaction A change in nuclear structure. **Nuclear fission** is one type of nuclear reaction.

Matter–antimatter annihilation The transformation of matter into high-energy radiation that occurs when a subatomic particle such as an electron is brought close to its antiparticle (antielectron).

Field view of reality The view that reality is a set of fields in motion and that material objects arise from the energy of these fields.

The principle of general relativity No experiment performed inside a sealed room can tell you whether your room is accelerating in the absence of gravity or is at rest in the presence of gravity. In other words, every accelerated observer experiences the same laws of nature.

Einstein's general theory of relativity describes the observations of accelerated observers, and his **special theory of relativity** applies to nonaccelerated observers.

Four-dimensional space–time The intertwined combination of space and time described by Einstein's relativity. Gravity affects space and time in a manner best described as the **bending of space–time**.

Gravity, light beams, and curved space Gravity bends light beams, and so gravity must bend space. In the general theory of relativity, gravity is the curvature of space–time caused by masses.

Space A space is a region without edges. A space may have any number of dimensions and may be curved or flat, infinite or finite.

Big bang An event some 15 billion years ago that created time, space, energy, and matter and sent them hurtling outward. The universe has been expanding ever since.

A spherical universe One possible geometry allowed by general relativity. The three-dimensional analogue of the two-dimensional surface of a sphere.

Review Questions

1. Name the two basic laws of the special theory of relativity, and list four phenomena predicted by this theory.
2. What distinguishes the special theory from the general theory of relativity?

THE RELATIVITY OF SPACE

3. What do we mean by "space" or "distance"?
4. What does "space is relative" mean?
5. Velma passes Mort at a high speed. Each of them holds a meter stick parallel to the direction of motion. What does each observer say about Velma's meter stick? What does each say about Mort's meter stick?
6. In the preceding question, suppose that Mort and Velma hold their meter sticks perpendicular to the motion. Now what does each observer say?

THE RELATIVITY OF VELOCITY

7. Velma passes Mort at a speed of $0.75c$, and she launches a high-speed projectile in the forward direction at $0.75c$ relative to herself. According to Galilean relativity, what is the projectile's speed relative to Mort?

8. In the preceding question, how does Einstein's prediction compare with Galileo's: larger than Galileo's prediction, equal, or smaller? How does Einstein's prediction compare with lightspeed: larger, equal, or smaller?

THE RELATIVITY OF MASS

9. According to Galilean relativity, which of the following items are relative (different for different observers): time, space (distance), velocity, rest mass, inertial mass, lightspeed?
10. In the preceding question, which items are relative according to Einstein's theory?
11. Velma passes Mort at a high speed. Both observers carry a standard kilogram (whose rest mass is 1 kg). What does Mort say about each of the standard kilograms? What does Velma say?
12. Mort exerts a 1-newton force on his standard kilogram. What acceleration does this give to the kilogram?

What will he find if he exerts the same force on Velma's standard kilogram if Velma is passing him at a high speed?

13. What is the distinction, if any, between rest mass, inertial mass, and mass? Which ones increase with speed?

14. What is the distinction between matter and radiation?

15. Why can't material objects be sped up to lightspeed? Does anything move at lightspeed?

16. According to the theory of relativity, is there anything that is not relative? If so, what?

$E = MC^2$ AND THE NEWTONIAN WORLDVIEW

17. What does $E = mc^2$ mean? Does it mean that mass can be converted to energy? Explain.

18. Is matter always conserved? Is mass always conserved? Is rest mass always conserved? Is energy always conserved?

19. Name the process in which about 1% of the rest mass vanishes.

20. Is rest mass precisely conserved in chemical reactions? Can this effect be measured?

21. Describe an experiment in which a system's entire rest mass vanishes. Is matter conserved here? Mass? Energy?

22. Special relativity suggests that the physical world is made of fields in motion. List some similarities and differences between this view and the Newtonian view.

EINSTEIN'S GRAVITY AND THE SHAPE OF THE UNIVERSE

23. List two experiments you could do in a spaceship accelerating at 1g through outer space, which might make you think you are at rest on Earth.

24. What is acceleration equivalent to?

25. As observed in an accelerating reference frame, does a light beam bend? What does this tell us about the effect of gravity on light beams?

26. Give an example of a flat two-dimensional space, a curved two-dimensional space, and a two-dimensional space of finite extent.

27. How can we tell from inside our actual three-dimensional space whether space is curved?

28. According to Newton, gravity is a force exerted by material objects on other material objects. What is gravity according to Einstein?

29. What are the three possible long-term fates of the universe, and what physical properties of the universe determine which way it will go?

Home Project

Try geometry on a spherical surface. Use a sturdy balloon or a smooth ball and a marker pen. Draw a short "straight" (straightest possible) line. Draw a second short line parallel and next to the first. Extend both line segments as straight lines. Try extending them all the way around the sphere. Draw a large triangle, with "straight" sides. What can you say about the sum of the triangle's angles (*Hint*: What is the sum of the angles of a triangle in a flat space)? Can you draw a triangle whose angles are each 90 degrees? What if the triangle is far smaller than the size of the sphere—What is the sum of the angles?

For Discussion

In considering the birth, shape, and fate of our universe, is science trying to answer a basically religious question? Can science answer such questions? Should it try?

Exercises

THE RELATIVITY OF SPACE

1. How fast must Velma move past Mort if Mort is to observe her spaceship's length to be reduced by 50%? If Velma is flying over the United States (about 5000 km wide) at this speed, how wide will she observe the United States to be?

2. Mort's swimming pool is 20 m long and 10 m wide. If Velma flies lengthwise over the pool at 60% of lightspeed, how long and how wide will she observe it to be?

3. How fast would Velma have to fly across Chicago in order to observe the city's width to be reduced 10% below the width measured on the ground?

THE RELATIVITY OF VELOCITY

4. Velma, passing Mort at 0.1c, launches an apple forward at a launch speed of 0.1c. How fast does the apple move relative to Mort?

5. Velma, passing Mort at 0.25c, launches a rocket forward at 0.5c. According to Galileo's relativity, how fast does the rocket move relative to Mort? How fast according to Einstein? Which answer is correct?

THE RELATIVITY OF MASS

6. If Velma passes Mort at a high speed, Mort will find her mass to be larger than normal. Will he also find her to be larger in size?

7. Velma's spaceship has a rest mass of 10,000 kg, and she measures its length to be 100 m. She moves past Mort at 0.8c. According to Mort's measurements, what are the mass and the length of her spaceship?

8. How fast must Velma move past Mort if Mort is to observe her spaceship's mass to be increased by 50%? How fast must she move if Mort is to observe her spaceship's length to be reduced by 50%?

9. A meter stick with a rest mass of 1 kg moves past you. Your measurements show it to have a mass of 2 kg and a length of 1 m. What is the orientation of the stick, and how fast is it moving?

10. A meter stick with a rest mass of 1 kg moves lengthwise past you. Your measurements show it to have a mass of 2 kg. How long is this meter stick relative to you?

11. You are in a spaceship moving past Earth at nearly light-speed. You measure your own mass, pulse rate, and size. How have they changed?

12. You are in a spaceship moving past Earth at nearly light-speed and you observe Mort, who is on Earth. You measure his mass, pulse rate, and size. How have they changed?

$E = MC^2$

13. When you throw a stone, does its mass increase, decrease, or neither? Can this effect be detected?

14.*You impart 90 J of kinetic energy to a 1 kg stone when you throw it. By how much do you increase its mass?

15.*If you had two shoes, an ordinary shoe and an "anti-shoe" made of antiparticles, and you annihilated them together, by how far could you lift the United States' population? Assume that each shoe's rest mass is 0.5 kg and that all the energy goes into lifting.

16.*Making estimates. Show that if all the energy "released" (transformed) when fissioning 1 kg of uranium were used to heat water, about 2 billion kg of water could be heated from freezing up to boiling. (Assume that the uranium's rest mass is reduced by about 1%. Roughly 4 J of thermal energy are needed to raise the temperature of 1 gram of water by 1°C.) How many tonnes of water is this (a tonne is 1000 kg). How many large highway trucks, each loaded to about 30 tonnes, would be needed to carry this much water?

EINSTEIN'S GRAVITY

17. If you were in a rocket ship in space (far from all planets and stars) accelerating at 2g, how heavy would you feel? What if your acceleration were, instead, 0.5g? What if you were not accelerating at all?

18. If you were in a rocket ship in space accelerating at 2g and you dropped a ball, how would it move as observed by you? What if your acceleration were instead 0.5g? What if you were not accelerating at all?

19. Longitudinal (north–south) lines on Earth are "straightest" lines that eventually meet. What about latitudinal (east–west) lines? Are they circles? Are they "straightest" lines? Do they eventually meet?

20. Is a circle (meaning the perimeter of a flat circular area) a space? How many dimensions does it have? Is it a curved space, or is it flat? Give an example of a flat one-dimensional space.

THE SHAPE OF THE UNIVERSE

21. Since the universe is about 15 billion years old, what would we expect to see if we could see a distance of 15 billion light-years (recall that a light-year is the distance that light travels in a year)?

22. If you could magically travel much faster than light and traveled in a straight line, would you get to the edge of the universe? Where would you get?

12

ARE WE ALONE?
the search for
extraterrestrial intelligence

For this world was created by Nature after atoms had collided spontaneously and at random in a thousand ways, driven together blindly, uselessly, without any results, when at last suddenly the particular ones combined which could become the perpetual starting points of things we know—earth, sea, sky, and the various kinds of living things. Therefore, we must acknowledge that such combinations of other atoms happen elsewhere in the universe to make worlds such as this one. . . .So we must realize that there are other worlds in other parts of the universe, with races of different men and different animals.
Lucretius, Roman poet, first century B.C.

Sometimes I think we're alone. Sometimes I think we're not. In either case, the thought is staggering.
Buckminster Fuller, architect and futurist

One of science's most intriguing questions is whether the universe harbors other instances of what we are pleased to call "intelligent life." Either answer to this question has immense consequences. If intelligent aliens exist, our culture is likely to eventually be influenced by them. Some anthropologists believe that the discovery of alien intelligence would irreversibly change humankind's self-image. The possibility of such a discovery has inspired many scientists to devote much of their professional lives to the search for extraterrestrial intelligence, or **SETI**.

On the other hand, the conclusion that intelligence evolved only on Earth would challenge the general Copernican view that nature is basically the same everywhere, even though the details differ.

A survey of the universe certainly indicates nothing strikingly unique about our own location. Even within our own planetary system, there are four other bodies (Mars, Jupiter, Saturn, and a moon of Saturn named Titan) that may have life-supporting potential or may have had it in the past. Furthermore, in our Milky Way galaxy alone, there are hundreds of billions of stars, many of them similar to our sun. And there are billions of other galaxies. It seems plausible that among so many stars in so many galaxies, there are some with planets similar enough to Earth that life and intelligence evolved there.

How can we bring scientific reasoning to bear on a question like this? One feature that makes this topic both difficult and fascinating is its interdisciplinary character. It involves lots of physics, along with astronomy, geology, chemistry, biology, and even anthropology and sociology. This is a good place to discuss this topic because we now have most of the physics tools we need for it.

Any discussion of extraterrestrial life must be based on incomplete evidence. There are many science-related issues for which the evidence is incomplete. Examples include global warming (Section 9.7), health risks from low-level nuclear radiation (Sections 15.6 and 15.7), and future energy technologies (Chapter 17). We must draw conclusions about such pressing issues despite the incomplete information. This chapter provides an example of reasoning from incomplete evidence. We will find that the evidence allows us to draw conclusions about some questions but not about others. In dealing with speculative topics, we must be careful to stay grounded in the observed evidence and to take note of the uncertainties in our conclusions.

This chapter follows a general method of thinking that has become standard in discussions about the likelihood of extraterrestrial life* and that is a good model for thinking about other speculative topics. We break the question into subquestions, each interesting in its own right: What is the likelihood that other Earth-like planets exist (Section 12.1)? Given an Earth-like planet, what is the likelihood that life would actually have started there (Section 12.2)? Given life, what is the likelihood that intelligence would have evolved (Section 12.3)? Given intelligence, what is the likelihood that a life-form would develop technology (Section 12.4)? And given technology, what is the likelihood that a life-form would make contact with us, or we with them (Section 12.5)?

Section 12.6 looks at a question that might be disturbingly relevant to *Homo sapiens*: If technology-based civilizations do develop on other planets, do they endure for long? Finally, Section 12.7 asks whether we have actually ever been visited by alien life and puts this question into the context of a topic that goes to the heart of human culture in our scientific age: pseudoscience.

A sad spectacle. If they [the stars] be inhabited, what a scope for misery and folly. If they be not inhabited, what a waste of space.

Thomas Carlyle, nineteenth-century Scottish essayist and historian

DIALOGUE 1   Based on the history presented in Chapter 1, at what time in history would you guess that people began actively speculating about the possibility of extraterrestrial beings?

12.1 Are there other "good" places for life?

We begin from an obvious piece of evidence: If "intelligence" means the ability to solve new problems and use abstract reasoning, then intelligent life did develop on Earth. This, along with the notion that the principles of physics are the same throughout the universe, suggests that Earth-like conditions elsewhere should lead to intelligent life elsewhere.

We adopt the conservative attitude of inquiring only about the possibility of intelligence that developed under Earth-like conditions. We will not speculate on such possibilities as, for example, life developing in interstellar space. Such possibilities seem implausible, and in any case it is hard to see how to even begin thinking about such possibilities.

---

* Although few physics texts discuss this topic, most astronomy texts do.
1. People began such speculation around 1600, as a result of the publication of Copernicus's theory in 1543.

Science's successes are due partly to the method known as *analysis*: breaking up a question into parts and examining them separately. Extraterrestrial intelligence is an especially good topic for analysis because it breaks up naturally into several distinct subquestions, as outlined in the introduction. Each subquestion corresponds to one or two particular scientific disciplines. Our first subquestion is an astrophysical one: How many "Earth-like" or "good" places are there, places that are sufficiently similar to Earth that life could have started there? Here and in the rest of the chapter we restrict our analysis to our Milky Way galaxy alone, because our galaxy is thought to be typical of galaxies throughout the universe. Our conclusions can then be extended to the universe by multiplying by the number of galaxies—at least 1 billion (the number within reach of present telescopes).

Astronomers believe that there are roughly 4×10^{11} (400 billion or 400,000 million!) stars in our galaxy alone. How many "good" planets or moons might orbit these stars? The theory and observations of star formation (Section 5.3) indicate that other stars developed the way that our sun developed, by the gravitational collapse of interstellar gas and dust. Planets should form from these clouds, just as planets formed around the sun. But many developing stars break up into double-star or triple-star systems rather than collapsing to become a single star. Observations of the stars in the sun's neighborhood shows that about 50% are not the single points of light that they appear to the unaided eye to be but are systems of two or more stars orbiting around each other. Because most planetary orbits around multiple stars are unstable, multiple stars are less likely to support life. To simplify our argument, we will neglect the possibility of life around multiple stars.

Many single stars have either too much or too little mass to have planets with conditions conducive to life. Stars much more massive than our sun "burn"* so brightly that they use up their fuel in only a few hundred million years. Life on Earth needed longer than that even to get started, and once it did get started, several billion years elapsed before it evolved beyond primitive single-cell forms. So very massive stars do not have "good" planets.

Stars much less massive than the sun burn so faintly that a planet would have to orbit close to the star in order to be warm enough for life. But any such close planet would experience gravitational forces that would lock it into a situation in which the same side of the planet would always face the star (in the way that the moon is locked into Earth). This would force any atmosphere that the planet might have around to the planet's dark side, where it would freeze out. Because Earth's atmosphere is essential to life here, we can rule out low-mass stars as candidates.

Astronomers estimate that about 10% of the single stars in our galaxy have masses close enough to the sun's mass to be candidates for having good planets. Of these sunlike stars, how many actually have planets, and how many of these planets are actually Earth-like?

HOW DO WE KNOW?

It is difficult to see planets around other stars because their faint light is overwhelmed by the light from their central sun. But both the theory of star formation and the available indirect evidence indicate that many single stars have planetary systems.

---

* Stars don't really burn; they "fuse" (Chapter 16).

Ten years ago, it was possible to argue that the solar system is unique. Today the evidence strongly suggests that planetary systems are abundant in the Galaxy. From statistical arguments alone, the likelihood is that many of them will have conditions favorable for life.
Anneila Sargen and Steven Beckwith, astronomers, in an April 1993 article in *Physics Today*

When a planet orbits a star, the star also orbits slightly because of the gravitational pull of the planet. A sufficiently large, Jupiter-sized planet causes its star to make an orbit large enough to be detectable by today's telescopes, at least if the star is close enough to Earth. This method has detected a few nearby stars that probably have at least one Jupiter-sized planet. More accurate telescopes will extend this method to more distant stars.

Most single sunlike stars are observed to rotate slowly rather than rapidly. A single star forming by the collapse of a gas cloud would be expected to spin faster and faster as the cloud collapses, just as a spinning ice-skater spins faster as she pulls in her arms. The only plausible mechanism for preventing a single star from rotating rapidly is the formation of a rapidly rotating flattened disk that is left behind as the central star collapses (Section 5.3), and such a rotating disk should eventually coalesce into planets.

Further evidence comes from thick gas clouds such as the one shown in Figure 5.12, where astronomers have found thousands of newly minted stars. Nearly all of these new stars are wrapped in flattened disks of dust grains and gas, just the sort of formation that the theory predicts will eventually coalesce into planets.

Since there is room for reasonable disagreement about the number of stars that have planetary systems, the most honest way to proceed is to establish a range of uncertainty. We will make two estimates, a high estimate that is plausible but optimistic about the chances for extraterrestrial intelligence, and a low estimate that is at the pessimistic end of the plausible range. A reasonable optimistic estimate is that nearly all of the sunlike stars have planets, and a reasonable but pessimistic estimate is that only about 10% of such stars have planets. A much smaller estimate, such as 1%, seems unreasonable, given the evidence.

Given a sunlike star with a planetary system, how many planets (or moons) are likely to have conditions conducive to the emergence of life? In addition to Earth, such conditions might also exist in our own solar system on Mars, in the gaseous atmospheres of Jupiter and Saturn, and on Titan, a moon of Saturn that has an atmosphere resembling the early atmosphere of Earth.

Mars probably does not contain life. The United States launched two spacecraft that landed there in 1975 and, among other things, tried to detect signs of life. It was thought that primitive microscopic single-celled life-forms, similar to those that existed for billions of years on Earth before the emergence of multicelled plants and animals, might exist in the Martian soil. But a search for organic molecules in the Martian soil was negative.

Optimists argue that there are probably one or more Earth-like planets or moons orbiting each sunlike star, because there is at least one orbiting our sun and there are four others that are at least close to being Earth-like. Pessimists argue that the other four places in our solar system aren't really conducive to life and that perhaps as few as 10% of the planetary systems have even a single Earth-like place.

To summarize, an optimistic estimate of the number of Earth-like planets in our galaxy is

(number of stars in galaxy)
× (fraction that are single stars)
× (fraction of single stars that are sunlike)
× (fraction of sunlike stars that have planetary systems)
× (number of Earth-like planets per planetary system)
$$= (4 \times 10^{11}) \times 0.5 \times 0.1 \times 1 \times 1 = 2 \times 10^{10}$$

A pessimistic estimate is
$$(4 \times 10^{11}) \times 0.5 \times 0.1 \times 0.1 \times 0.1 = 2 \times 10^{8}$$

We estimate that there are some 200 million to 20 billion Earth-like planets in our galaxy. This result reflects the consensus among scientists that there are many places where life might have arisen.

DIALOGUE 2 Suppose that some other galaxy contains 8×10^{12} stars, that only 25% of these are single stars, that 10% of the single stars are similar to the sun, that 50% of these sunlike stars have planetary systems, and that 1% of these planetary systems contain one Earth-like planet. How many Earth-like planets does this galaxy contain?

12.2 Does life develop on "good" planets? *how did life develop on Earth?*

Scientists have studied the origin of life on Earth for more than a century and now know enough to form a plausible hypothesis. We can explore the likelihood of life elsewhere by studying the origin of life on Earth.

Earth developed from the material left behind as gas and dust collapsed gravitationally to form the sun (Section 5.3). When Earth emerged from this process 4.6 billion years ago,* it was a solid, rocky planet having essentially no gaseous atmosphere.

Some of the materials that formed Earth were "radioactive" (Chapter 15), and their radioactivity heated Earth's interior, causing gases that had been trapped within Earth to rise to the surface during volcanic and other activity. These gases, mostly steam, CO_2, and N_2, formed Earth's earliest atmosphere. Once in the atmosphere, most of the steam cooled, condensed, and formed the oceans, with some of it remaining in the atmosphere as water vapor. Chemical reactions involving such surface materials as iron caused some H_2O and CO_2 to decompose, releasing carbon and hydrogen. Hydrogen gas is so light that most of it eventually escaped Earth's gravitational grip, but some of it participated in further chemical reactions to create gaseous methane (CH_4) and ammonia (NH_3).

So before the emergence of life, the primitive Earth was partly covered by oceans of liquid water and had an atmosphere composed of water, carbon

2. $(8 \times 10^{12}) \times 0.25 \times 0.1 \times 0.5 \times 0.01 = 10^{9}$ (1 billion)
* Section 15.5 presents evidence for this date.

dioxide, nitrogen, methane, ammonia, and hydrogen. There is substantial agreement among astronomers and geologists about this description, although there is uncertainty about the precise chemical composition.

The primitive Earth was subjected to a variety of energy sources, such as intense ultraviolet radiation from the sun (at that time, there was no ozone to shield the surface from ultraviolet), lightning, heating by volcanic activity, radioactivity, and radiations from outer space. Each of these energy sources can make and break chemical bonds and so promote new chemical reactions.

HOW DO WE KNOW?

Scientists have performed a variety of simulation experiments to determine what happens when a primitive Earth-like environment is subjected to these energy inputs. One of the best-known experiments was performed in 1953 by Stanley Miller and Harold Urey. Their experiment was simple: They reproduced a plausible early Earth atmosphere of methane, ammonia, water vapor, and hydrogen in a jar, placed it in contact with liquid water, and subjected it to electric-spark discharges to simulate lightning. The results were dramatic. Within a few days the liquid water turned brown, and chemical analysis revealed the presence of amino acids, nucleic acid bases, fatty acids, and a rich array of other molecules. These rather complex molecules are the building blocks of life on Earth. Biological proteins are formed from amino acids linked together into long chains, and DNA molecules are formed from nucleic acid bases strung together into a long "double helix."

Since the Miller–Urey experiment, numerous workers have performed many similar experiments using a variety of gas mixtures and energy sources. As long as these gases contain water, hydrogen, nitrogen, and a gaseous form of carbon, the outcome is consistent: Many biological building blocks are formed. It thus seems likely that energy interacted with the early Earth atmosphere to form the chemical building blocks of life.

Once amino acids and other organic molecules formed, there would have been many opportunities for them to aggregate to form a dense organic "soup." For example, if the water in a tidewater pool on the fringe of an ocean contained organic molecules and if the water then partially or completely evaporated, the organic molecules would be highly concentrated in the remaining water or on the dried-out land (Figure 12.1).

The remaining steps are less certain. Concentrated organic material may have dried out entirely and undergone dry reactions, or it may have remained suspended in liquid water and undergone wet reactions. Laboratory experiments of both types have been performed. In the dry experiments, amino acids react by linking up to form long protein molecules! When placed back into water, these proteins assume the shape of round objects similar to simple biological cells (Figure 12.2). Similar results are obtained in the wet experiments (Figure 12.3). These objects, called **protocells**, absorb material from the surrounding liquid,

FIGURE 12.1

On the primitive Earth, tidewater pools such as this might have contained a rich soup of organic molecules that became more highly concentrated as the water evaporated, allowing complex reactions whose products could then have "fertilized" the oceans with protocells or living organisms.

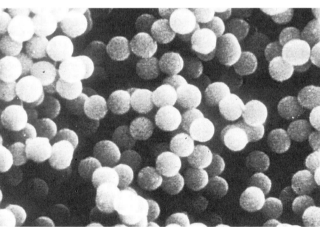

FIGURE 12.2

Electron microscope image of "protein microspheres." These are formed by heating a dry mixture of amino acids, which link up to form long protein molecules that, when placed back into water, assume the spherical shape shown. These "protocells" are similar to biological cells; they can even grow at the expense of the protein dissolved in their surroundings and can bud.

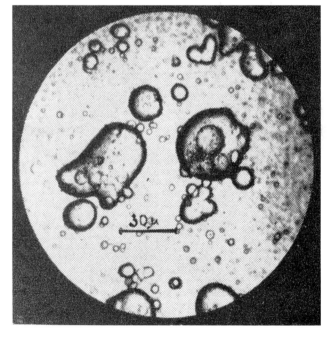

FIGURE 12.3

"Coacervate droplets," another type of protocell. When long chainlike molecules are mixed in water, they concentrate into cell-like "droplets" surrounded by a tough membranelike surface.

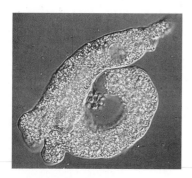

FIGURE 12.4
An amoeba, a living single-celled organism, flows to surround a nearby food particle.

grow by attaching to one another, and divide to produce additional such objects.

Biochemical theories indicate that if protocells are placed in water containing proteins and other nutrients, they can grow by absorbing the nutrients and can multiply by simple division. Today, single-celled organisms, such as the amoeba, perform essential life functions by just such mechanisms (Figure 12.4). Protocells could have undergone a process of "chemical evolution" in which those that were more efficient at these processes prospered at the expense of the less efficient protocells. Chemical evolution favors the development of ever more efficient and more elaborately organized protocells that could eventually have evolved into primitive single-celled life-forms.

The earliest fossil evidence for life is dated at 3.1 billion years ago (Figure 12.5), relatively early in Earth's 4.6-billion-year history. For about the next 2 billion years, the only life-forms were simple one-celled organisms with no central "cell nucleus." It was only after the biological evolution of more complex cells containing cell nuclei that multicelled organisms, and higher forms of life, became possible.

Biologists believe that life on Earth developed through a sequence of steps similar to those just outlined. We summarize this hypothesis as follows:

FIGURE 12.5
The wavy rock depositions result from the actions of blue-green algae and bacteria. Such biologically induced rock formations are a clue to early biological activities. This sample is from the earliest known rock formation of this type, in Rhodesia, and is dated at between 3.1 billion and 2.7 billion years old. A 5-millimeter-long ruler is shown for scale.

THE HYPOTHESIS OF A CHEMICAL ORIGIN OF LIFE
Life originated on Earth by physical and chemical processes, beginning with the action of various energy sources in the prelife atmosphere to create the organic chemicals that are the building blocks of life.

There is nothing that is necessarily unique to Earth in this scenario. What is needed is a good planet with water. Most biochemists agree that given conditions similar to those of the primitive Earth and given something like a billion years, life would develop elsewhere.

An optimistic estimate is that given a "good" planet and time, life is nearly certain to appear. A pessimistic view might assume that because of periodic catastrophes (such as collisions with another astronomical object) or because of a lack water or other necessary substances, life might develop on only some fraction, such as 10%, of the "good" planets. But because of the ease with which the precursors of life can be created in the laboratory, any much smaller estimate seems unreasonable.

Putting this together with our previous estimates, our optimistic estimate is that life developed at 2×10^{10} (20 billion) places in our galaxy alone, and our pessimistic estimate is a "mere" 2×10^7 (20 million) places. It appears that life is not a stranger in the universe.

DIALOGUE 3 Earth's prelife atmosphere contained very little oxygen (O_2), so where did Earth's oxygen come from?

12.3 Is intelligence a characteristic feature of life?

Are we the universe's only thinking creatures? If we are alone in being able to understand how nature operates, then it is only here on Earth that nature has evolved the ability to understand itself. This appears to violate the Copernican viewpoint.

Although there is little data to help us evaluate the likelihood of intelligence emerging from life, it appears that the concept of "biological convergence" is central to this question. On Earth, biological features sometimes emerge independently in two or more widely different organisms. For example, wings have appeared in insects, in the extinct reptile called the *pterodactyl*, in birds, and in bats. Each of these must have evolved its wings independently of the other three, and so wings evolved at least four times on Earth. The reason is that flying has survival value because flying animals can escape their enemies and can find food more easily. Such features that tend to evolve again and again are called **biologically convergent properties.**

There are many examples of biological convergence. For instance, North American hummingbirds are similar to Hawaii's hummingbird moth, although one is a bird and the other is an insect. Fishlike bodies developed

---

3. Green plants created O_2 from CO_2 and H_2O during photosynthesis (Section 2.8).

independently in fish, in the water mammals (dolphins, porpoises, whales), and in an extinct reptile called the *ichthyosaur*. Catlike and doglike shapes developed many times in unrelated creatures.

Is intelligence a biologically convergent property on Earth? If so, it seems likely that intelligence would evolve on other Earth-like planets having life. Let us examine the arguments on both sides.

We are the children of the eighth day.
Thornton Wilder

Those who argue that intelligence is not a biologically convergent property point out that intelligence evolved only once on Earth and that it emerged only a few million years ago, a brief moment ago in Earth's 4600-million-year history. If intelligence is a convergent property, then why didn't it emerge much sooner? A second argument is that the emergence of *Homo sapiens* depended on a series of unlikely events. One such event occurred 65 million years ago when a large astronomical body slammed into Earth. This impact helped make the dinosaurs extinct and provided an opportunity for the mammals to evolve in many new directions.

Those who regard intelligence as a convergent property reply that it is not surprising that 4.6 billion years were needed for intelligence to develop, because after all it took some 3 billion years just to get to multicelled organisms. Concerning the second argument, any property that actually is convergent will eventually occur despite unpredictable details such as catastrophic collisions. That is what we mean by a convergent property. For example, if the dinosaurs had not become extinct and if intelligence actually is a convergent property, there should eventually have appeared an intelligent reptile descended from the dinosaurs. Furthermore, intelligence does seem to have considerable survival value. How could a slow-moving, weak, soft-skinned, ground-dwelling creature that requires nearly two decades to rear its offspring have possibly survived and prospered for the past few million years if not by its wits? Since intelligence confers such a biological advantage, it should be a convergent property for the same reason that wings are a convergent property.

It appears that near intelligence has evolved independently in another animal: the dolphin. Dolphins can respond accurately to complex four- or five-element "sentences." They apparently communicate with one another via sound waves. They seem to be as intelligent as apes and more like human beings in their curiosity and eagerness to communicate. The apes are also nearly intelligent, but they are not an example of the independent evolution of intelligence, because apes and humans share recent common ancestors. Dolphins, on the other hand, are as far removed from humans as are dogs, bats, rabbits, and horses. How could dolphins have developed near intelligence if intelligence is not a convergent property?

So optimists contend that intelligence is a convergent property and should eventually appear on nearly all Earth-like planets that support life. Pessimists, reasoning that intelligence can arise only after a particular sequence of highly improbable events, estimate that life has a very low probability of evolving into intelligent life, perhaps only one chance in a million (10^{-6}). There is little consensus regarding this question.

Combining these figures with our previous estimates, our optimistic estimate is that intelligent life has arisen (or will eventually arise) at some 2×10^{10}

(20 billion) places in our galaxy, and our pessimistic estimate is that it has arisen or will arise in only 20 places.

DIALOGUE 4   Suppose that the optimistic estimate is correct. Then how many stars would we need to survey, on the average, before finding one having a planet on which intelligence has arisen or will arise?

12.4 Does intelligent life develop technology?

According to our optimistic estimate, one-twentieth of the stars have a planet where intelligent life has arisen or will arise (see Dialogue 4). We could experimentally check this estimate by surveying a few tens of other stars for signs of intelligent life. But what signs would we look for? We know only one way of detecting intelligent life on planets at other stars: detecting electromagnetic radiation sent out from those planets by radio transmitters or other devices.

So from an observational point of view, the relevant question is not how many places harbor intelligent life but, rather, how many places harbor life that is capable of sending out radio signals or some other form of communication. We will refer to any such form of life, having sufficient technology to be capable of sending out radio signals, as **technological.**

It is difficult to form any credible estimate of the likelihood of technological life. Humans began using tools some 2 million years ago, developed agriculture some 10,000 years ago, formed the first cities 5000 years ago, began industrialization 250 years ago, and invented radio only during the twentieth century. We have been technological for less than a century. Would other forms of intelligent life go through a similar development? The example of the dolphins shows that life can become intelligent without becoming technological—dolphins cannot develop technology because they don't have hands.

Optimistically, one could suppose that technology eventually develops wherever intelligence develops. But pessimistically, one could suppose that technology developed only here. There is wide disagreement on this question because there is no evidence as to what form of life might develop elsewhere. There is certainly no reason to believe that multicelled life-forms on other planets would resemble plants or animals on Earth. So life on Earth gives us little basis for any hypothesis about the development of intelligence or technology.

Although we cannot draw any conclusion about the likelihood of technological life elsewhere, our discussion has led to the interesting idea that life probably exists all over the universe, and this has provided a framework of questions for further thinking about this issue. This is the best we can hope for when treading such uncertain ground: Some of the questions are tentatively answered, and a framework is established for further study. Table 12.1 summarizes our conclusions.

4. Since there are some 400 billion stars in our galaxy, we would need to survey about (400 billion)/(20 billion), or about 20 stars.

TABLE 12.1
Extraterrestrial Life in Our Galaxy, Pessimistic and Optimistic Estimates

	Plausible upper estimate	*Plausible lower estimate*
Number of stars in our galaxy	4×10^{11}	4×10^{11}
Fraction of those that are single stars	0.5	0.5
Fraction of those that are sun-like	0.1	0.1
Fraction of those that have planetary systems	1.0	0.1
Number of Earth-like planets per planetary system	1.0	0.1
Number of Earth-like planets in our galaxy	2×10^{10}	2×10^{8}
Fraction of those on which life has arisen or will arise	1.0	0.1
Number of planets where life has arisen or will arise	2×10^{10}	2×10^{7}
Fraction of those where intelligence develops	1.0	small, perhaps 10^{-6}
Number of planets where intelligence arises	2×10^{10}	few, perhaps 20
Fraction of those that become technological	????	????

12.5 Interstellar communication and travel: *might we make contact?*

We have unintentionally sent out electromagnetic signals to space for most of this century, as radio and television communication. This evidence of our own technological civilization has now traveled some 100 light-years (the distance that light travels in 100 years) from Earth and has already arrived at hundreds of other stars. If technological life exists around these stars, they might be tuning in our old TV shows right now!

We have also sent out intentional signals. The first such radio message was sent from the large radio telescope at Arecibo, Puerto Rico, in 1974 and was beamed toward a cluster of stars 27,000 light-years away. Since the message will be 27,000 years in transit, nobody is holding his or her breath for an immediate reply.

Other communications have been sent out on space vehicles. The first was a plaque carried by *Pioneer 10* (Figure 12.6), launched toward Jupiter in 1972. The spacecraft was accelerated by the gravitational effects of Jupiter and its moons so that it became, in 1983, the first human artifact to leave the solar system. Since then, *Pioneer 11* and two *Voyager* spacecraft have exited the solar system. Because they are not moving toward any particular star, there is little expectation that their messages will ever be received. In any case, it would be hundreds of thousands of years before they could reach any distant star.

A new project got under way in 1992 to analyze radio-frequency radiation from many stars over a large part of the sky. The signals are detected by several radio telescopes that are used mostly for other non-SETI purposes and fed into an advanced radio receiver that simultaneously analyzes millions of frequency "channels." Although this project is far more sensitive and comprehensive than any previous SETI project, it is expected to succeed only

[The SETI project is] a legitimate part of astronomy. It is hard to imagine a discovery that would have greater impact on human perceptions than the detection of extraterrestrial intelligence.
National Academy of Sciences

Like Columbus, the only thing we can be certain of is that there is something beyond our shores.
Edward Stone, physicist

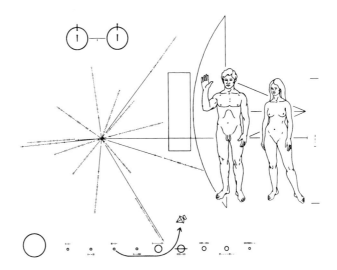

if there is an alien civilization in our part of the galaxy and then only if that civilization is purposely beaming a powerful signal at us.

Could extraterrestrial civilizations visit Earth, or could we someday visit them? It would be a long trip. It is 4 light-years to even the nearest other star, Alpha Centauri. This is a hundred million times farther then the distance to the moon, the farthest that humans have yet traveled. Travel to Alpha Centauri compares with travel to the moon in the same ratio that a trip to the moon compares with a trip across your living room.

Despite the huge distances, interstellar travel may be feasible. It is interesting to see what might be done using only current technologies. One approach to interstellar travel involves spaceships that move much slower than light and that would take hundreds or thousands of years to reach their destination. This approach could be used for robotic probes that would radio back information about a particular star's planets, including signs of life. A colony of humans could make a long-duration trip in several generations aboard a large "worldship" designed as a complete living environment (Figure 12.7, see color insert). Refrigeration techniques might allow humans to be in suspended animation during the trip.

It is not difficult to conceive of technologies that could accomplish such slow trips. The *Pioneer* spaceships, launched from Earth with conventional chemical rockets, used Jupiter and its moons as a slingshot to give them enough kinetic energy to exit the solar system. This produces rather slow speeds, however. It will take 80,000 years for *Pioneer* to reach a distance as far as the nearest star.

Controlled nuclear fission, which powers today's nuclear power plants (Chapters 16 and 17), could accelerate a rocket for several years until it attained a speed high enough to exit the solar system and to reach the nearest star in about 10,000 years.

Nuclear *fusion* bombs or hydrogen bombs (Chapter 16) could provide still more energy. According to physicist Freeman Dyson, a spaceship supporting many hundred crew members, with a fuel supply of 300,000 fusion bombs(!), could explode 1 bomb every 3 seconds to maintain an acceleration of 1*g* (9.8 m/s^2) for 10 days. This would accelerate the spaceship to one-thirtieth of lightspeed, enabling it to travel to Alpha Centauri in 130 years.

Much faster trips, well within a human lifetime, are possible using more futuristic technologies. Although these technologies are very demanding, they are based on technology currently being developed for other purposes and do not appear to violate any scientific principles. One possibility would use antimatter (Section 11.4), a material that is not normally found on Earth but that has been already created and stored for days at a time in high-energy physics facilities. The coming decades will probably see the production and storage of significant quantities of antimatter. With enough time and enough energy (from the sun, perhaps), it should be possible in the future to create and store as much as a few kilograms of antimatter. When combined with a few tonnes of liquid hydrogen, this would be enough to fuel a spaceship that could reach 10 to 50% of lightspeed and reach Alpha Centauri in 8 to 40 years.

Another concept for accelerating a spaceship to relativistic speeds is the "interstellar ramjet" first described by the astronomer R. W. Bussard in 1960. Its fuel is the dilute hydrogen gas that fills our galaxy. The spaceship consists of a payload, a nuclear fusion reactor (see Chapters 16 and 17), and a scoop some 100 km in diameter to collect hydrogen atoms in space. This hydrogen fuels the reactor, and the reactor's output (probably helium atoms) provides the vehicle's thrust. The beauty of this scheme is that the spaceship needn't carry its fuel, and so it can accelerate for as long as desired. An unchanging 1*g* acceleration creates apparent Earth-like gravity for the inhabitants and takes the spaceship to Alpha Centauri (4 light-years away) in about 7 years as measured on Earth. But relativistic time dilation (Chapter 10) would cause the travelers to experience a time duration of only about 4 years.

If the 1*g* acceleration could be maintained, much longer trips would be feasible, because the spaceship would now be moving at nearly lightspeed for nearly the entire trip and time dilation would become significant. For example, it would take only 21 "ship-time years" to travel to the center of our galaxy, a distance of about 30,000 light-years, whereas the elapsed time on Earth during this trip would be a little more than 30,000 years. Such travelers could, truly, not go home again.

Spaceships could be pushed with particle or radiation beams sent out from our solar system. For instance, a spaceship carrying light-reflecting sails could be pushed by a powerful light beam sent out from a laser array in our solar system. The lasers would be in orbit around the planet Mercury and would use the abundant sunlight there to produce a single combined beam that could travel as far as 40 light-years before spreading significantly. Within this distance, the spaceship could accelerate to nearly lightspeed in a few years and could reach any of the 20 nearest stars within about 17 years ship time and 20 years Earth time.

So travel to the stars appears difficult but is perhaps possible. On the other hand, interstellar travel might run into unforeseen difficulties. For example, collisions of spaceships with pebbles, dust, or gas particles might be catastrophic.

DIALOGUE 5   One of the closest stars that is a candidate for having life around it is Tau Ceti, 11 light-years away. At one-thirtieth of lightspeed, how long would a one-way trip to Tau Ceti take? How long would this trip take (as measured on Earth) if the spaceship moved at one-half of lightspeed? Would the crew members experience this same trip time?

DIALOGUE 6   Have our old radio or TV shows passed over Tau Ceti (Dialogue 5) yet? If we received a message from Tau Ceti today, when would it have been sent? How long might it take them to get here, assuming that the technologies discussed in this Section were feasible?

12.6 Where is everybody? *do technological civilizations endure?*

Enrico Fermi was a frequent visitor to Los Alamos. . . .A lunchtime conversation took place in 1950 when someone brought up the question of flying saucers. . . .They all agreed that flying saucers were not real alien spacecraft. . . .Fermi said "Don't you wonder where everybody is?". . .He followed up with a series of calculations. . . .He concluded that we ought to have been visited long ago and many times over. . . .He went on to conclude that the reason we hadn't been visited might be that interstellar travel is impossible, or if possible, it is always judged not to be worth the effort, or technological civilization doesn't last long enough for it to happen.
"Fermi's Question," in *Extraterrestrial Civilization*, American Association of Physics Teachers

If interstellar travel is possible, then any civilization having the drive to develop technology in the first place might be expected eventually to travel to other stars or at least send robotic probes. Many SETI specialists believe that such a civilization would, in only a few million years, explore the entire galaxy and colonize large parts of it. If even one such "spacefaring" civilization arose during the past history of our galaxy, it should have reached Earth by now. But as is pointed out in Section 12.7, there is no credible evidence that we have been visited; indeed, the overwhelming scientific consensus is that we have never been visited.

Where is everybody? This question was first asked by the great physicist Enrico Fermi in 1950. Fermi's reasoning was similar to ours: It is plausible that technological civilizations exist, and if they do exist, they should be here by now. But they are not here. So where is everybody? Before reading on, you might want to stop and speculate about that.

One suggestion is that the pessimistic estimates are correct and that technology has arisen only here. This is certainly possible. But it does seem wondrously strange that we should be the proud possessor of the only technology in our galaxy or in the universe. It certainly seems to violate the Copernican attitude underlying modern science.

Perhaps civilizations tire of technology, in the way that a child outgrows its toys, and enter a less adventurous phase that is not concerned with interstellar communication.

Perhaps aliens have long known about developments on Earth and are refraining from communicating because they do not want to disturb this experiment in intelligence.

Perhaps interstellar travel is far less feasible than it seems. If so and if many technological civilizations exist, the radio searches now under way should be successful before long.

[Civilization] is a highly complicated invention which has probably been made only once. If it perished it might never be made again. . . .But it is a poor thing. And if it is to be improved there is no hope save in science.
J. B. S. Haldane, biologist and philosopher

5. $30 \times 11 = 330$ years; $2 \times 11 = 22$ years. The crew members would experience a shorter trip time, owing to relativistic time dilation.
6. Yes. 11 years ago. If they could travel at close to lightspeed, it would take them just a little over 11 years (Earth time) to get here.

Many astronomers and SETI specialists feel that the most reasonable explanation is the **short-lifetime hypothesis**: *Perhaps technological civilizations do not survive for long because overindulgence in their own technology destroys their capacity to continue functioning as a technological civilization.* This explanation is more plausible than many of the others because it is suggested by our own experience on Earth and because it is consistent with the Copernican attitude that we are not unique. Unfortunately, it leads to the pessimistic conclusion that we, too, could have a short lifetime.

To make the short-lifetime hypothesis more concrete, let's return to our estimate of the number of places in which life arose: between 20 million and 20 billion. Let's also assume, somewhat optimistically, that technological civilizations developed at between a thousand and a million of these places. Now we take note of an additional point: Our galaxy is about 15 billion years old.

HOW DO WE KNOW?

The age of our galaxy is determined from the age of its oldest stars. To determine the age of a star, we must first determine its mass. For binary-star systems (two stars orbiting each other), this can be done by observing the gravitational effect that each star has on the motion of the other. Since 50% of all stars are binary, this method can determine the mass of about half the stars. Once the mass is known, the established theory of stellar evolution (Section 5.3) can predict the star's total lifetime. As explained in Section 12.1, lower-mass stars have longer lifetimes. Astronomers have observed stars in our galaxy whose mass is so low that they should have lifetimes as long as 15 billion years. As such a star begins to exhaust its fuel at the end of its life, it expands and cools and glows redder than its normal white or blue. Observation of the color of these low-mass stars shows that some of them, in our galaxy, are near the end of their life. These stars must be 15 billion years old. Since no older stars have been observed, we conclude that this is the age of our galaxy.

So our 5-billion-year-old sun is a relative newcomer. Stars formed, burned, and died for 10 billion years before our sun was created. If there have been thousands of technological civilizations, most of them should have become technological long ago. It is probable that ours, a mere century old, would be the galaxy's most recent newly born technological civilization. Calculation* shows that if a thousand to a million such civilizations have cropped up at various times during the entire 15-billion-year history of our galaxy, on the average one such civilization would start up every 15,000 to 15 million years. This means that the most recent technological civilization preceding ours would have started some 15,000 to 15 million years ago!

These numbers, plus our own experiences with technology on Earth, give concrete meaning to the short-lifetime hypothesis. Unless the typical technological civilization survives through some 15,000 to 15 million years of technology, we would expect no other such civilizations still surviving in our galaxy, even assuming some rather optimistic numbers about the number of

---

* $(15 \times 10^9 \text{ years})/(10^3 \text{ to } 10^6 \text{ tech civs})$ = 15 thousand to 15 million years.

technological civilizations that have arisen. Do technological civilizations generally last this long? Will ours? It's a good question.*

Consider the only evidence we have, our own experience with technology. Our most vigorous technological efforts have gone into weaponry. For decades humans have threatened themselves with intercontinental nuclear weapons that could have bombed the world back into a stone age. We have emerged, at least for now, from this threat, but smaller nuclear threats are even more possible than they were. Using conventional high-tech and low-tech weapons, humans continue the organized killing and brutalization of one another all over the world, both within and between nations.

But today the threat of war pales beside more insidious technology-related environmental problems. Driving all the other problems is the exponential growth of our species (Section 7.7). We are now nearly 6 billion strong, and it is by no means clear that Earth can support us. Yet our numbers will reach 12 billion during the twenty-first century, and that figure assumes all-out birth control efforts.

Other known problems include poverty, illiteracy, deforestation, animal extinctions, large-scale plant extinctions for the first time in Earth's history, global warming (Section 9.7), ozone depletion (Section 9.6), abuse of legal and illegal drugs, resource depletion, urban decay, air pollution, water pollution, solid-waste disposal, famine, desertification, subtle new epidemics, the gap between rich and poor nations, groundwater depletion, other water shortages, loss of soil fertility, overfishing, toxic algal blooms, weakened immune systems due to pollution, rising sea levels, new drug-resistant bacterial strains from legal drug overuse, and the problem-multiplier effect of an expected tenfold increase in global economic activity to meet the demands and aspirations of our exploding population. The unknown problems are anybody's guess.

Any one of these problems could prevent our development as a mature technological civilization. The combined long-term effect of all of them is difficult to predict. These problems are subtle and pervasive. They are caused by the small and apparently inconsequential daily actions of each one of us. Few of us pause to consider the global effects of, say, our own driving or air conditioning.

One root of the problem is our continued reliance on age-old instincts in an age of powerful technologies. For example, because we have enthusiastically accepted technology's ability to control disease while doing little to control our instinctive desire to have children, we have the population explosion. This is but one example of our willingness to accept the easy fruits of technology without accepting the hard choices and new thinking that must go with it if we are to survive. Intelligence-based technology demands that we apply our intelligence, rather than our instincts, to the uses of technology. It is difficult for our entire species to make such a cultural change during the short, explosive birth of the technological age. It

---

* Astrophysicist Richard J. Gott combined the general Copernican idea that "our time and our place are not special in the universe" with the fact that humans have been on Earth for only a brief portion of Earth's history, to show that the mathematical probability of an intelligent civilization achieving a long lifetime is low. See *Nature*, May 27, 1993, pp. 315–319.

is the fundamental contradiction of the technological age. A similar conflict between traditional biological behavior and the sudden new requirements of technology could have caused the demise of most technological civilizations.

If the short-lifetime hypothesis is correct, then humankind is faced with a challenging project, one of universal proportions: the challenge of becoming one of the breakthrough civilizations, a wise civilization that knows better than to use technology for self-indulgence and power.

Human beings have been described as natural systems having the curious property of self-awareness. We are perhaps among nature's most advanced attempts to become aware of itself. Will we meet the challenge of extending that awareness for millions of years following the explosive recent birth of technology on our planet? This question is being decided as you read these words.

12.7 Have we been visited? *UFOs and pseudoscience*

Pseudoscience, the dogmatic and irrational belief in an appealing idea that purports to be scientific but that has little or no observational support, comes in many guises: parapsychology, spoon bending, telepathy, the Bermuda Triangle, dowsing, quantified etherics, bioactochronics, levitation, occult chemistry, psychokinesis, pyramid power, crystal power, creationism, fortune-telling, ancient astronauts, flying saucers, dianetics, psionics, astrology, Velikovsky's colliding worlds, ancient astronauts, poltergeists, orgone boxes, and so forth.

It is a significant issue. For example, two-thirds of American adults read astrology reports periodically; 26 million read them often; 66 million believe that astrology is scientific; and 12 million say they sometimes change their plans after reading their horoscope. One and a half centuries after Darwin's *Origin of the Species*, 46% believe that human beings did not develop from earlier species of animals. And 43% believe it likely that some of the reported unidentified flying objects are really space vehicles from other civilizations, while tabloid newspapers bearing such headlines as "I bore an alien's child" continue to prosper. And 42% believe they have been in contact with someone who has died.

Most important, because rational thought is a threat to pseudoscientific beliefs, such beliefs create an anti-intellectual climate that inhibits the free expression of rational thought. Furthermore, habitually dwelling on pseudoscientific ideas surely weakens one's own ability to think rationally.

Let's look at three typical pseudoscientific beliefs: extraterrestrial visitations, astrology, and creationism.

UFOs are unidentified objects in the sky. Two **UFO beliefs** have gained a following in the popular media: The first is that some UFOs are visitations by aliens today; the second is that aliens visited Earth within the past few thousand years.

I shall tell you a great secret, my friend. Do not wait for the last judgment. It takes place every day.
Albert Camus, writer

By failing to confront superstition and pseudoscience, schools leave gaps in knowledge that are filled by the market place. And in the marketplace, there are few distinctions between science fiction and science facts. . . .Americans are evenly divided on the existence of extraterrestrial visitors, and half don't believe in the theory of evolution.
From a 1985 poll of 2000 Americans, Northern Illinois University Public Opinion Laboratory

He [the magician Houdini] threw down a challenge—offering any medium five thousand dollars if he could not duplicate any phenomenon of alleged spirits himself. Early in 1926 Houdini made a pilgrimage to Washington to enlist the aid of President Coolidge in his campaign "to abolish the criminal practice of spirit mediums and other charlatans who rob and cheat griefstricken people with alleged messages."
From *Houdini*, by B. R. Sugar

Every science that is a science has hundreds of hard results; but search fails to turn up a single one in "parapsychology."
John A. Wheeler, physicist

The problem with these ideas is not that UFO beliefs themselves are inherently antiscientific. In fact, our preceding discussion shows that visits to Earth by extraterrestrial beings are plausible. The problem, instead, is in the way in which these beliefs are supported. Let's examine the evidence.

There have been thousands of recent reports of sightings of strange lights, strange aircraft, and people being captured by aliens. Upon investigation, these reports fall into three categories. Most have normal explanations: automobile headlights reflected off high-altitude clouds, a flight of luminescent insects, unconventional atmospheric effects, unconventional aircraft, aircraft using searchlights for meteorological observations, aerial refueling operations, the setting planet Venus distorted by the atmosphere. These are honest reports, but nonetheless, self-deception, or "seeing what you want to believe," is common. For example, in 1968, the U.S. Air Force collected thirty UFO reports when a satellite reentered the atmosphere and broke into burning pieces in the night sky. Of these, 57% reported that the objects were flying in formation, implying intelligent control, and 17% claimed that the glowing objects were attached to a black "cigar-shaped" or "rocket-shaped" object, sometimes with glowing windows.

Other reports are hoaxes, often for profit. For example, a 1968 University of Colorado study, headed by physicist Edward Condon, established that many of the classic UFO photos are either fakes or photos of known natural phenomena. Nevertheless, these photos continue to reappear in new UFO publications. Great Britain's widely publicized "crop circle" phenomenon reported around 1990 was caused by pranksters.

Finally, a few UFO reports cannot be explained. In any investigation of unusual phenomena, there will always be cases that remain unexplained because of lack of data, false reporting, self-deception, and so forth. The unexplained UFO reports offer no positive evidence, such as unambiguous photographs or unambiguous sightings by many observers or an artifact (a tool or piece of material) left behind by aliens. Such a residue of unexplained cases, with no positive evidence, is not surprising and offers no support for UFO beliefs.

In fact, the evidence points the other way. The only real evidence we have is negative: Extraterrestrials have not come right out and revealed themselves to us. So if they exist, they prefer to conceal themselves. We have seen that any extraterrestrial civilization that visits Earth is overwhelmingly likely to be millions of years in advance of us. Any such civilization would surely be able to conceal themselves from us if they wanted to. They would not make simple mistakes like flying around in visually observable vehicles. So reports of UFO sightings are inherently implausible.

Furthermore, it is surprising that such beings would want to conceal themselves. It seems more reasonable that they would want to contact and investigate us. At least, this is what human explorers have done when they discovered new cultures.

A common fallacy of many UFO reports is that far from being overly fantastic, they are not nearly fantastic enough to be believable. The reported technologies are always just a little in advance of, or even behind, the current technology on Earth. The aliens are reported to have curiously humanlike features. But there is little reason to expect that alien technologies, or alien bodily features, would resemble ours.

These are some of the reasons that scientists who have thought about this matter overwhelmingly reject the hypothesis that we are being visited.

The second UFO belief, that we have been visited in the past, has even less supporting evidence. Ancient legends of superior beings mean little: Most cultures have had such legends, based on either real humans or stories promoted by the priesthood. Only an ancient legend containing "futuristic" information, such as instructions for an electronic circuit, would be convincing. Also convincing would be an ancient artifact that could not have been made by the ancient civilization, like an electronic microchip or an advanced metallic alloy.

There is a set of enormous geometrical figures in the high plains of Peru, not easily discernible from the ground but discernible from the air. Several books have reaped large profits by portraying these and other similar figures as a mystery that can be explained only by assuming that the figures were constructed by supertechnologists. But scientists visiting these markings found that they were made by simply pushing aside an inch-thick covering of dark stones. The simplest explanation is that they were intended for gods in the sky. Many cultures have believed in gods in the sky. The markings may be a graphical prayer; there is no reason to assume that they represent messages to ancient astronauts.

The idea that ancient astronauts taught early civilizations much of what they knew not only borders on deliberate fraud for profit; it also insults the accomplishments of past civilizations. Humans living many thousands of years ago were quite capable of building the sophisticated structures that archaeologists have discovered. To attribute these accomplishments to mythical visitations from space is dishonorable to both the living and the dead.

Popular UFO mythology illustrates several common features of pseudoscience: mistaken observations attributed to exotic causes, deliberate fraud, the belief in unexplained cases as proof of an exotic hypothesis, and self-deception caused by a desire to believe. A belief in extraterrestrial visitations can offer people excitement and psychological comfort. Believers are easily convinced by dubious sightings and outlandish claims. It is precisely for such reasons that scientists always ask, How do we know? What is the evidence?

Astrology, the belief that events on Earth are influenced by the positions and motions of the planets, began in ancient Babylonia. It seemed reasonable in an era that believed the planets existed for human purposes. Its central belief is that the configuration of the sun, moon, and planets at the moment of a person's birth affects his or her personality or fortune. A simplified form of astrology, based only on the position of the sun, is the mainstay of newspaper astrology columns.

Today, astrology is scientifically implausible, to say the least. The only known physical influences exerted on Earth by the planets are gravitational effects and electromagnetic radiation. It is hard to imagine how these effects at birth could influence our lives. For example, the gravitational effects*

---

* The effects referred to here are "tidal effects," by which the moon and sun cause tides in bodies of water on Earth. Tidal effects cause similar, but smaller, distortions in all objects on Earth.

exerted by the doctor and nurse and furniture in the delivery room far outweigh the effects of the planets; the walls of the delivery room shield us from many radiations; and the variations in the sun's radiation output (variations that are unrelated to a person's astrological sign) are far larger than the total radiation received from the moon and all the planets added together.

Despite its scientific implausibility, can we find any observational evidence that astrological predictions are correct? A number of researchers have studied this question, some of them using astrological predictions based on horoscopes (charts showing the orientation of the planets at the moment of a person's birth) calculated for thousands of people and found no evidence that astrology has any predictive power. There is no correlation, even in an average statistical sense, between astrological predictions and people's personalities or lives. If there is any validity to astrology, such correlations should be easy to find.

Even though astrology is incredible theoretically and disproved observationally, more than half of America's teenagers say they "believe in astrology"; newspapers continue their daily astrological predictions; and in the United States there are 20,000 professional astrologers, compared with only 2000 astronomers. Will humankind outgrow its most harmful instincts and develop into a mature culture that is able to control its own technology? Figures such as these give little cause for optimism.

Creationism is the belief that the Bible's Old Testament can be read "literally" as scientific and historical truth and that Earth and the main types of biological organisms, including humans, all were created separately and at roughly the same time, just a few thousand years ago. In the United States, creationism is perhaps the most important of the pseudoscience issues because it is believed by so many people, it has religious implications that are important to many people, and it affects the way that science is taught in the public schools. Like astrology, creationism was credible until a few centuries ago, and many scientists believed it. But today it conflicts with the observations and principles of astronomy, physics, chemistry, geology, biology, paleontology, and archaeology. There is a broad scientific consensus, supported by evidence from many sciences, that Earth is billions of years old, that humankind is millions of years old, and that humans are related through biological evolution to the other animals. Scientific ideas are never certain, however, and honest doubts about established theories should not be arbitrarily dismissed. But so far the creationist arguments have found essentially no scientific support.

In physics, the strongest evidence against creationism comes from radioactive dating and other methods of determining the ages of things. We will discuss some of this evidence in Chapter 15.

As with other pseudoscientific beliefs, the main danger lies not in creationist beliefs themselves but in the way that they are defended. For example, a leading creationist, Henry M. Morris, stated:

It should be emphasized that this order is followed not because the scientific data are considered more relevent than Biblical doctrine. To the contrary, it is precisely because Biblical revelation is absolutely authoritative and perspicuous that the scientific facts, rightly interpreted, will give the same testimony as that of Scripture. There is not the slightest possibility that the facts of science can contradict the Bible and, therefore, there is no need to fear that a truly scientific comparison of any aspect of the two models of origins [evolution and creationism] can ever yield a verdict in favor of evolution.*

The assumption here is that the answers are already known from the Bible and that science merely studies nature to verify those answers. This attitude precludes any true dialogue with nature. To scientists, it is a sin of pride to presume to tell nature the answer.

DIALOGUE 7   One common creationist argument is that all the fossils and other objects that appear old were created 5000 years ago in order to make Earth appear old. Is it possible to disprove this argument? Can this statement be investigated scientifically, or does it fall outside science? Consider the proposition that Earth was created just one hour ago and that all the fossils and other objects, including your memory, were created at that time, just to make everything seem old. Is it possible to disprove this proposition?

EINSTEIN SIMPLIFIED

s.harris

---

* Henry M. Morris, *Scientific Creationism* (San Diego: Creation-Life Publishers, 1974), pp. 15–16.

7. No, it is not possible to disprove this statement. Since observations cannot disprove it, it is not a scientific statement at all; it falls outside science.

Summary of Ideas and Terms

SETI The search for extraterrestrial intelligence. Usually conducted with **radio telescopes**, receivers that detect radio waves from space.

The Copernican viewpoint The view that Earth is not a unique place in the universe, that the same principles of nature apply everywhere. Since intelligent life occurred here, this view argues that it also should have occurred elsewhere.

Analysis Breaking a question into parts.

Protocells Nonliving precursors of single-celled organisms.

The hypothesis of a chemical origin of life Life originated on Earth by means of chemical processes, beginning with the action of various energy sources on the prelife atmosphere to create the organic chemicals that are the building blocks of life.

Biologically convergent property A biological feature that, because of its survival value, tends to occur again and again during biological evolution.

Technological life Any form of life having a sufficiently well developed technology to be capable of sending out radio signals.

Short-lifetime hypothesis Technological civilizations do not survive for long because overindulgence destroys their capacity to continue being technological. This is a possible answer to the question, Why haven't aliens visited Earth by now?

Pseudoscience The dogmatic and irrational belief in an appealing idea that purports to be scientific but that has little or no observational support.

UFO Unidentified object in the sky. Two popular **UFO beliefs** are that (1) UFOs are visitations by aliens today and (2) aliens visited Earth within the past few thousand years. There is no evidence to support either belief. Scientists overwhelmingly reject them as pseudoscientific and false.

Astrology The belief that events on Earth are influenced by the positions and motions of the planets. Scientists overwhelmingly reject astrology as pseudoscientific and false.

Creationism The belief that the Bible's Old Testament can be read "literally" as scientific and historical truth and that Earth and the biological organisms, including humans, were created separately just a few thousand years ago. Scientists overwhelmingly reject creationism as pseudoscientific and false.

Review Questions

ARE THERE OTHER GOOD PLACES?

1. This chapter analyzes the question of alien technological life by breaking it into four questions. What are these questions?

2. Is a multiple-star system likely to support life? Explain.

3. Is it likely that there are planets around other stars? What is the evidence?

4. Are there likely to be Earth-like planets orbiting very massive stars? Explain. What about very low-mass stars?

5. Which one or two of the following are the most plausible estimates of the number of Earth-like planets in our galaxy: billions, millions, thousands, less than one thousand, one? Or is there too little evidence to decide?

DOES LIFE DEVELOP ON GOOD PLANETS?

6. Describe an experiment, and its results, to determine what happens when a typical primitive Earth environment is subjected to typical early Earth energy inputs.

7. What are protocells?

8. Describe the most widely accepted (by scientists) hypothesis concerning the origin of life on Earth.

9. Which one or two of the following are the most reasonable estimates of the number of planets in our galaxy where life has emerged: billions, millions, thousands, less than one thousand, one? Or is there is too little evidence to decide?

10. How did Earth's molecular oxygen (O_2) get here?

DOES LIFE DEVELOP INTELLIGENCE OR TECHNOLOGY?

11. What is a "biologically convergent property"? Give an example.

12. Give one argument for and one against the proposition that intelligence is a biologically convergent property.

13. Which one or two of the following are the most credible estimates of the number of planets in our galaxy where intelligence has emerged: billions, millions, thousands, less than one thousand, one, too little evidence to decide?

14. How did we define "technological life" in this chapter?

15. Which one or two of the following are the most plausible estimates of the number of planets in our galaxy where technology has emerged: billions, millions, thousands, less than one thousand, one, too little evidence to decide?

INTERSTELLAR COMMUNICATION AND TRAVEL

16. What is SETI?

17. Describe one slow way and one fast way for humans to travel to the stars.

18. Give an argument supporting the notion that aliens should have visited here by now. Give at least two possible explanations of why they have not visited here.
19. What is the short-lifetime hypothesis?

UFOS AND PSEUDOSCIENCE

20. What is pseudoscience? List several examples.

21. What are the two popular UFO beliefs? What is the scientific consensus about them, and why?
22. Is the notion that aliens have visited Earth inherently implausible?
23. What is astrology? What is creationism?
24. Why do scientists consider UFO beliefs to be pseudoscientific? Answer the same question for astrology and for creationism.

Home Project

Former President Ronald Reagan stated in a 1980 newspaper interview that he consulted astrologer Carroll Righter's horoscope column every day, and he stated, "I believe you'll find that 80% of the people in New York's Hall of Fame are Aquarians" (Reagan, born on February 6, 1911, is an Aquarian). Perform an experimental check of sun-sign predictions of this sort by finding the birth dates of fifty persons who are outstanding in a single way. For instance, you might choose fifty musicians, athletes, or hall-of-famers from New York. How many of them were born under each of the twelve astrological signs? Do they fall preferentially into specific signs (for example, are 80% of them Aquarians?), or do your results appear to reflect a random distribution?

For Discussion

1. What would be the short- and long-term consequences of (a) aliens arriving at the UN building? (b) Radio signals from a star 10 light-years away, asking for two-way communication? (c) Communication from a civilization claiming to be one of the universe's few long-term technological civilizations, stating that there have been millions of technological civilizations but that only a few survived their own technology for more than a few centuries?
2. In view of the devastation wrought on many of Earth's cultures by contact with more advanced cultures, is it safe for us to (a) deliberately send out signals to possible aliens, (b) allow TV and radio signals to escape from our planet, and (c) search for extraterrestrial life with radio telescopes? How would the results of our listening program be affected if alien civilizations reached the conclusion that either (a) or (b) is not safe?
3. The sun will eventually become a red giant star (Section 5.3). When will this happen? How does this time compare with the roughly 10 million years that it took for apelike creatures to evolve into modern humans? Would

you expect the human species to be recognizable by the time the sun turns into a red giant?
4. Suppose that an earthquake occurs and that the following day five people report dreaming about an earthquake just the night before. Is this evidence that dreams can foretell catastrophic events? About how many dreams does the U.S. population have every night? Is it likely that five of them would be about earthquakes?
5. It has been said that "humankind cannot stand too much reality." Historically, science has caused some mental anguish by confronting cherished beliefs, such as an Earth-centered universe or creationism, with scientific findings. Are scientific ideas ultimately difficult to accept?
6. Beginning in the 1980s, lawsuits have attempted to require that every science course that teaches evolution also teach creationism, on the grounds that students should learn all of the competing theories. What is your opinion, and why? If you disagree, is there any other place in the public school system where it might be appropriate to teach creationism?

Exercises

ARE THERE OTHER "GOOD" PLACES?

1. Would life be likely on a planet associated with a white dwarf, neutron star, or black hole (Section 5.4)? Why?

DOES LIFE EVOLVE ON GOOD PLANETS?

2. Suppose that the soil on Mars contained organisms that undergo photosynthesis. What type of organic molecule would you expect to find in the soil?

3.*Suppose we make the following extremely pessimistic (conservative) estimates: Of the 400 billion stars in our galaxy, only 0.1% are sunlike with planetary systems; only 1% of these planetary systems have one Earth-like planet; and life emerged on only 1% of these Earth-like planets. In this case, on how many planets in our galaxy would life have emerged?

DOES LIFE DEVELOP INTELLIGENCE OR TECHNOLOGY?

4. Redo this chapter's calculation of the number of places supporting life, supporting intelligent life, and supporting technological life, substituting numbers that you think are more reasonable.

5. On which of the following five questions is there a rough consensus: How many "good" planets are there for life in our galaxy? On how many of these did life emerge? On how many of these did intelligence emerge? On how many of these did technology emerge? On how many of these does technology still survive?

INTERSTELLAR COMMUNICATION AND TRAVEL

6. If we listen at many frequencies and pick up no artificial signals, does that prove that life has not evolved elsewhere? Does it prove that technological life has not developed elsewhere? Explain.

7.*There are about 20 stars within 12 light-years from Earth. Of these 20, one of the stronger candidates for having a planet with life is Tau Ceti, 11 light-years away. How does this distance compare with the distance to our sun, 8 light-minutes away?

8.*There are about 400 billion stars in our galaxy. Based on the assumption that there are 10,000 technological civilizations in our galaxy today, roughly how many stars would astronomers have to scan to have a reasonable chance of finding one with technological life?

PSEUDOSCIENCE

9. Jeane Dixon, who also claims to have forecast John F. Kennedy's assassination, once claimed that an incredible vision informed her that aliens from another planet in our solar system would visit Earth the following August and announce their arrival to the entire Earth. She claimed that this other planet lies directly on the other side of the sun, which is why we have never seen it. Give one good scientific argument against the existence of any such planet.

10. Continuing Exercise 9: One answer is that any such planet should have a gravitational effect on the other planets and that this effect has not been observed. Suppose that Jeane Dixon then replied, "But these aliens are so advanced that they have been able to completely mask the effects of their planet's gravity, as well as all other observable effects of their planet." What is your response to this explanation? Does this supposed planet fall within the realm of science?

11. Continuing Exercise 10: Jeane Dixon's forecast was published on the front page of the *National Enquirer* on September 14, 1976. Have you heard of any reports, the following August, that her forecast was correct? Do you suppose that *National Enquirer* then printed a front-page story reporting that her forecast was wrong? Can you recall any instance when such forecasts were later reported as false when they turned out to be false?

12. Some supporters of ESP (extrasensory perception, or mind reading, causing objects such as spoons to move by means of mental concentration and the like) claim that ESP really exists but that it cannot be checked scientifically because scientific experiments always cause the ESP effect to vanish. What is your response to this argument? As far as science is concerned, then, would ESP exist or not?

13

THE QUANTUM THEORY

O amazement of things—even the least particle!

Walt Whitman

This and the next chapter discuss science's most accurate and complete description of physical reality: quantum theory. In this chapter we set the stage in Section 13.1 with a broad description of the general nature, aims, and cultural role of the theory. Section 13.2 uses a specific experiment, the photoelectric effect, to introduce quantization, at least as it applies to radiation. Quantization leads to some paradoxical questions about just what radiation really is, questions that are first raised in Section 13.3 and that will follow us all the way through these two chapters. Section 13.4 uses another specific experiment, the electron interference effect, to extend quantization to the study of matter. We find that matter at the microscopic level behaves a lot like radiation, that the concepts of quantum theory apply in much the same way to both matter and radiation. So questions arise about what matter really is, similar to the questions raised about radiation. We discuss these questions in Section 13.5. Section 13.6 introduces the theory's central concept, the psi-field. As its name indicates, this is a field, similar in some ways to the electromagnetic or gravitational field. Section 13.7 summarizes what we have found out so far, as well as some perplexing questions about radiation, matter, and the psi-field.

13.1 The post-Newtonian revolution _____

In 1900 the quantum idea slipped nearly unnoticed into physics. Although nobody realized it at the time, it was the dawn of the post-Newtonian era.

It is remarkable that the other big post-Newtonian idea, Einstein's relativity, was announced only five years later, in 1905. As we saw in Chapter 10, relativity altered Newtonian physics in some fundamental ways: It implied different spaces and times for differently moving observers. Matter was not an indestructible substance but was instead made of energy, motion, and fields. Following three centuries of clockwork determinism and impersonal objectivity, the universe now looked far less like a clock, and reality was somewhat dependent on the observer.

Einstein's special theory of relativity was fairly complete when first announced in 1905, and its revolutionary nature was already clear. Today, the special theory is accepted by all physicists and is not being actively evaluated.

Quantum theory is the set of ideas that physicists have used, since about 1925, to study the microscopic world. It started quietly in 1900 and developed slowly, but its impact ultimately went far beyond special relativity's impact, and today it is far from being a closed book. Although the theory's main principles had appeared by 1930 and despite the theory's wide testing and application, it is still not clear what the theory really means.

Because it predicts such a wide variety of phenomena so accurately, quantum theory is probably the most successful scientific theory ever invented. Its practical impact extends to every device or idea that depends on the details of the microscopic world: electronic devices such as transistors, silicon chips, and integrated circuits, and so all the information and communication technologies such as television and computers; most of modern chemistry and biology; lasers; our understanding of different types of matter ranging from superconductors to neutron stars; and nuclear physics, nuclear power, and nuclear weapons. Central to the entire high-tech world is an elusive and highly "non-Newtonian" particle: the electron.

Anyone who has not been shocked by quantum physics has not understood it.

Niels Bohr

Perhaps more significant but certainly less appreciated is quantum theory's philosophical impact. Quantum physics represents a more radical undoing of the Newtonian worldview than does relativity theory. We have emphasized throughout this book that a scientific worldview is by no means a trivial academic matter. Newtonianism is woven subtly into the fabric of Western civilization. The mechanical worldview has dominated Western culture for centuries, has been assimilated so deeply that it is accepted without even noticing it and without realizing that it is a particular worldview.

We will discover that contrary to the Newtonian worldview, quantum physics implies that randomness, or chance, is built into nature at the microscopic level: Nature doesn't know what she will do next! No longer can the universe be a predictable machine.

Also contrary to the Newtonian worldview, quantum physics implies that nature is deeply connected, that such parts of nature as electrons, protons, and light waves cannot be separated from their surroundings without fundamentally altering their character. No longer can the universe be viewed as a machine at all, even an unpredictable one, for the most basic feature of the machine metaphor has always been its separable parts.

The quantum revolution challenges not only the Newtonian assumptions but even more general scientific assumptions. An attitude underlying science itself has been the notion that there is an independent reality "out there" in nature, a reality that would be essentially unchanged even if we did not observe it. Scientists have always assumed that they were learning about this "objective reality." Quantum physics describes a microscopic world so delicate that it is altered significantly by the mere act of observation, even by observations that could not be said to physically disturb the measured object, and even by the mere possibility of observation. This challenges the notion of an independent and knowable microscopic reality, suggesting instead a participatory reality that includes macroscopic observers.

13.2 Quantization

Certain experimental observations made in the late nineteenth century resisted traditional Newtonian explanations. We have already discussed Michelson and Morley's experiment (Section 10.5), which was explained only by Einstein's non-Newtonian proposals concerning light beams, time, and space.

Other phenomena indicated that new ideas were needed to understand the microscopic world. The photoelectric effect was one such phenomenon.

HOW DO WE KNOW? THE PHOTOELECTRIC EFFECT

Around 1890, scientists began noticing that when light and other electromagnetic radiation shine on a metal surface, they can eject electrons from the surface (Figure 13.1). This **photoelectric effect** is the basis of many practical devices, most notably the **photovoltaic cells** that can help resolve society's energy problems (Chapter 17). Here we are interested in the scientific, rather than the technological, impact of this effect.

It isn't difficult to explain the photoelectric effect in general qualitative terms, on the basis of the standard electromagnetic wave theory of radiation (Figure 13.2). According to this theory, light arrives at the metal surface in the form of a broad, continuous wave of electromagnetic field vibrations. The vibrating field exerts a vibrational force on all the charged particles (protons and electrons) in the metal surface. The field shakes these particles harder and harder until the most loosely bound electrons shake loose.

This explanation seems credible until one looks at the observed details. Here are two that are troubling:

1. There is no experimentally observable delay between the arrival of radiation at the surface and the emission of the first electrons. Even if the light is very dim, electrons are still ejected the instant that light strikes the surface.
2. Emission occurs only if the light's frequency is above a certain "threshold frequency."* Light whose frequency is below the threshold ejects no electrons, regardless of the light's intensity (brightness) and regardless of how long one waits. But light whose frequency is above the threshold ejects electrons immediately, even if the light is very dim.

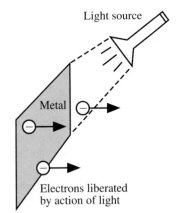

FIGURE 13.1
The photoelectric effect.

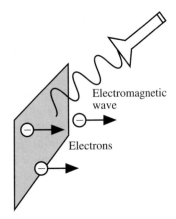

FIGURE 13.2
The photoelectric effect as described by the electromagnetic wave theory of radiation.

---

* This threshold frequency is different for different metals; that is, copper's threshhold differs from silver's, and so forth.

Both features contradict the electromagnetic wave theory of radiation. An electromagnetic wave arriving at the metal surface would be spread out over the portion of the surface illuminated by the light. Only a small portion of the wave's energy could be absorbed by any single atom in the surface. For sufficiently dim light, then, the energy absorbed by any single atom should be so small that a considerable time must elapse before an electron could acquire enough energy to shake loose. Calculations based on the electromagnetic wave theory show that in dim light, the delay should be as long as several seconds. Yet no delay has been observed, even though a delay as short as a few billionths of a second should be detectable.

There is nothing in the electromagnetic wave theory to suggest the second troubling feature. Since higher-frequency light has higher energy, we might expect that it would eject electrons sooner. But we would expect—and wave-theory calculations predict—that even low-frequency light would eventually transmit enough energy to shake an electron loose. Why the sudden cutoff at a specific frequency?

As an analogy to the first feature, imagine a string of ripples on a pond rolling in to a beach filled with small pebbles. We might expect the ripples to disturb the pebbles a little. But imagine our surprise if as soon as the ripples arrived at the beach, a few pebbles were violently knocked off the beach, landing hundreds of yards away! It is as though all the energy of the ripples were concentrated on just a few pebbles.

As an analogy to the second feature, suppose that the "ripple–pebble" effect occurred only if the ripple frequency were above some threshold frequency. At lower ripple frequencies, no pebbles would be ejected, even though the ripples might be huge (large amplitude), whereas at ripple frequencies above this threshold, pebbles would be violently ejected even by tiny ripples.

This was the paradox that faced anyone wanting to explain the photoelectric effect using the wave theory of radiation.

In 1900, German physicist Max Planck (Figure 13.3) had a new idea concerning vibrating objects. He had been thinking about the emission of radiation by vibrating charged particles. According to the traditional theory of radiation (Section 9.2), radiation was a wave emitted by charged objects when they vibrate. For example, the light from a flashlight was supposed to be an electromagnetic wave emitted by vibrating electrons and protons within the atoms of the flashlight bulb's filament.

Planck's study of radiation led him to the hypothesis that every vibrating object's energy was restricted in a way that had no explanation in Newtonian physics. He suggested that a vibrating object's energy must for some reason be chosen from a specific collection of allowed, or possible, values. To describe this restriction, we say that a vibrating object's energy is **quantized**, meaning "restricted to specific allowed quantities."

It is an odd hypothesis. Why should a vibrating object be restricted to just a particular set of allowed energies? It is as though you were in a child's swing and found that you could swing only at certain amplitudes (widths of vibration), such as 1 m, 2 m, or 3 m, but that for some reason the intermediate amplitudes such as 1.3 m or 2.7 m were not allowed. Your amplitude and your energy would have to increase or decrease in sudden "jerks," from

1 m to 2 m to 3 m, without going through any intermediate amplitude. Newtonian physics contains nothing like this.

Once you inject a really new idea into a preexisting scientific theory, there is no telling what the ramifications might be. Energy quantization was not much noticed for a few years, but it eventually swept across all of physics.

Einstein had a busy year in 1905. In that year, he published his special theory of relativity and his paper on Brownian motion that established the reality of atoms (Section 2.1), and he gave the now accepted explanation of the photoelectric effect by applying Planck's quantization idea.

Einstein reasoned that if a vibrating charge's energy is quantized, it can change its energy only by instantaneously jumping from one allowed energy to another, because it is never allowed to have any intermediate amount of energy. The conservation of energy then implies that the emitted radiation itself must emerge from a vibrating charge as an instantaneous burst of radiation rather than as the long continuous wave that the electromagnetic wave theory predicts.

Einstein concluded that radiation always appears as tiny bundles, tiny particles, but not particles of matter like protons, neutrons, and electrons. These new particles are made of radiation; they are light particles, infrared particles, X-ray particles, and the like. These particles of radiation are called **photons** (photo particles). Photons are different from material particles: They always move at lightspeed; they have a rest mass of zero; and vibrating charged particles create photons.

Einstein pointed out that if a light beam's energy is concentrated in tiny photons rather than spread out across an entire light wave, there is a simple explanation of the photoelectric effect (Figure 13.4): Photons simply rain down on the metal surface and knock out electrons the way machine-gun bullets can hit a concrete wall and knock out pieces of concrete. According to this view, we would expect electrons to come flying out of the metal as soon as the first photons hit.

But how can we explain the second observed feature of the photoelectric effect, the dependence of the effect on the light's frequency? Here Einstein turned to a formula that appeared in Planck's quantization hypothesis. Planck had devised a formula specifying the energy spacings, the gaps between the allowed energy values of a vibrating particle. Einstein reasoned that a photon must carry an amount of energy equal to one of these energy spacings, because a vibrating charge would have to suddenly jump downward in energy by at least one energy spacing whenever the charge emitted radiation. Planck's formula, as interpreted by Einstein, is

$$\text{energy of photon} = (6.6 \times 10^{-34}) \times (\text{frequency of radiation})$$
$$E = hf$$

where h equals 6.6×10^{-34} joule-seconds* and where the frequency f is measured in vibrations/second, or hertz. The very small number, 6.6×10^{-34} joule-seconds, obviously plays an important role here. Planck chose

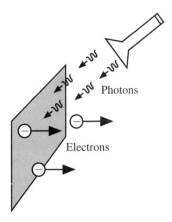

FIGURE 13.4
The photoelectric effect as described by the particle theory of radiation.

Photons

Electrons

---

* The units "joule-seconds" need to be attached so that when multiplied by a frequency, the result will be an energy measured in joules.

this particular number in order to get his theory to agree with the experimental data. This number came to be called **Planck's constant** and to be symbolized as h. Because h is so small, the energy of a photon is small too. The allowed quantity of energy carried by a photon is called a **quantum** of energy.

Since Einstein's idea is that radiation is made of photons rather than waves, you might wonder what f, the frequency of the radiation, can mean in the preceding formula. How can individual particles, or photons, have a frequency? It's an excellent question. We will discuss it in the next section.

Planck's formula tells us that a photon's energy is smaller or larger depending on whether the radiation's frequency is smaller or larger. So lower-frequency radiation should be accompanied by lower-energy photons. Suppose that in the photoelectric effect, the light's frequency is so low that individual photon energies are too small to jar loose even the most loosely bound electrons from the metal. Then no electrons at all will be knocked loose, no matter how long one waits and no matter how intense the low-frequency light is. So Einstein's photon idea explains the threshold frequency.

Once Einstein hit on using photons to explain the photoelectric effect, the explanation became surprisingly simple and obvious. Today, science students work through the details as an introductory exercise. But in 1905, scientists were uncomfortable with quantization. Just as in his invention of the theory of relativity, it was Einstein's originality, intellectual courage, physical insight, and search for clean and simple explanations that won the day. His quantitative predictions were beautifully confirmed by experiments more than a decade later, in 1916.

We summarize the ideas of Planck and Einstein as follows:

> THE PARTICLE THEORY OF RADIATION
> Electromagnetic radiation is created by vibrating charged particles. Because the energy of any vibrating particle is quantized, radiation appears as particles, called *photons*, each having energy hf, where h is Planck's constant and f is the radiation's frequency.

This theory of radiation is analogous to the atomic theory of matter. It says that radiation is grainy in the way that matter is grainy. The grains of matter are called *atoms* (or, more fundamentally, electrons and protons and neutrons), and the grains of radiation are called *photons*.

Neither Planck nor Einstein had any idea why a vibrating particle's energy must be quantized. Today we still do not know. For unfathomable reasons, nature requires a vibrating particle to choose its energy from among only particular allowed energies. So radiation must be emitted in lumps.

And we still do not know today why Planck's constant has the particular value it has. The small number h plays a role in quantum theory that is analogous to the role played by the large number c in relativity theory. Planck's constant shows up throughout quantum theory, just as c shows up throughout relativity theory, and affects all of the theory's quantitative predictions.

Physics is finished, young man. It's a dead-end street.
Advice to Max Planck from his teacher

And these fifty years of conscious brooding have brought me no nearer to the question of "What are light quanta [photons]?" Nowadays every clod thinks he knows it, but he is mistaken.
Einstein, near the end of his life

DIALOGUE 1   Light of two colors (two frequencies) shines on a metal surface whose photoelectric threshold frequency is 6.2×10^{14} Hz. The two frequencies are 5×10^{14} Hz (orange) and 7×10^{14} Hz (violet). Will this light eject electrons from this surface? If so, is it the orange light or the violet light or both that does the job?

DIALOGUE 2   Find the energies of the two types of photons in Dialogue 1. How much energy does a photon need to knock electrons out of this surface?

13.3 Radiation: waves or particles? *the duality of nature*

Section 13.2 presented evidence supporting a new particle theory of radiation and contradicting the older wave theory. You might suppose, then, that scientists would reject the wave theory and replace it with the particle theory.

But nature turns out to be more subtle than this. She won't let us reject the wave theory. Recall the evidence for the electromagnetic wave theory (Section 8.3). A wave-interference pattern like that in Figure 8.17 is direct evidence for light waves. Because nobody has found a way to explain such patterns except by assuming that radiation is made of waves, we cannot reject the electromagnetic wave theory. Yet the photoelectric effect requires a particle theory.

So we have two quite different theories of electromagnetic radiation: a wave theory and a particle theory. The wave theory is required by wave-interference experiments but is contradicted by the photoelectric effect, and the particle theory is required by the photoelectric effect but is contradicted by interference experiments. Although both theories are required, they seem mutually inconsistent. It is hard to imagine how radiation could be both waves and particles at the same time.

This paradoxical, wave-particle nature of radiation is evident in Einstein's proposal that light is made of photons whose energy is Planck's constant multiplied by their frequency. How can a photon have a frequency? A frequency is something that a wave has. A photon is not a wave. It simply moves in a straight line at lightspeed, without vibrating or waving in any way. What Einstein meant in his formula $E = hf$ is that f is the frequency that the radiation would have if we were thinking in terms of the old wave theory of radiation. But this still doesn't answer the question of how a particle, a photon, can have a frequency.

What is radiation? Is it waves or particles? Continuous or discrete? Spread out or concentrated in lumps? If it is waves, then how can it instantaneously knock electrons out of metals? If it is particles, then how can it interfere with itself?

To study this **wave–particle duality of radiation** in detail, we shall look closely at the evidence for waves: the **double-slit interference experiment** (Chapter 8). Double-slit interference is a fundamental experiment for investigating the concepts of quantum theory.

---

1. The violet light is above the threshhold frequency, so its photons have sufficient energy to eject electrons.
2. $E = hf = (6.6 \times 10^{-34}) \times (5 \times 10^{14}) = 33 \times 10^{-20}$ J (orange), $E = 46 \times 10^{-20}$ J (violet). For the photoelectric effect, a photon must have energy $E = hf = (6.6 \times 10^{-34}) \times (6.2 \times 10^{14}) = 41 \times 10^{-20}$ J.

Figure 13.5 reproduces the experimental arrangement. Light from a single source goes through two thin slits and then forms a wave-interference pattern of bright lines and dark spaces on a screen. The bright lines are regions of constructive interference, where wave crests coming from one slit meet crests from the other slit and valleys meet valleys. The dark spaces are regions of destructive interference, where crests from one slit meet valleys from the other slit and cancel each other out.

There seems to be no trace of photons here. But if we perform the experiment using very dim light and record the results on photographic film placed on the receiving screen and exposed very briefly to the light from the slits, a surprising result emerges. Instead of the wave-interference pattern, tiny individual flashes appear rather randomly all over the screen. Figure 13.6(a) is a drawing that simulates the results of an experiment of this sort. The flashes cannot be explained by the wave theory of light; they are in fact evidence that light is made of photons that create each flash as each photon hits the screen. If we expose the film for a longer time, an interference pattern begins to emerge, formed by the impacts of a large number of individual photons (Figure 13.6(b)). The pattern becomes clearer the longer we wait (Figure 13.6(c)).

HOW DO WE KNOW?

Ordinary photos are also made by photon impacts. Usually we think of photography in terms of light waves that come from the photographed object and that are focused onto the film by the camera's lens system. But dim light and a short exposure time allow us to see individual photon impacts. Figure 13.7 shows the photo emerging from individual impacts.

So the patterns made by light waves are due to large numbers of photons. In the double-slit experiment, each photon strikes the screen fairly randomly, the first hitting in one place, the second in quite another place, and so forth.

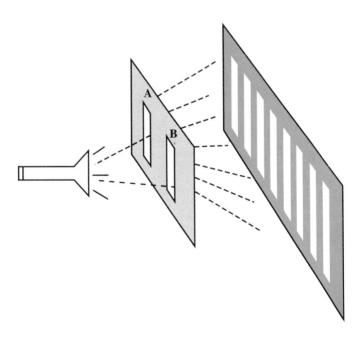

FIGURE 13.5
The double-slit experiment with light: the experimental setup and the results.

FIGURE 13.6

Close inspection shows that an interference pattern is formed by individual photons hitting all over a screen or film. In (a) a film briefly exposed shows 14 photon impacts. In (b) a longer exposure time allows about 150 photons to hit the film. In (c) a few hundred photons have hit.

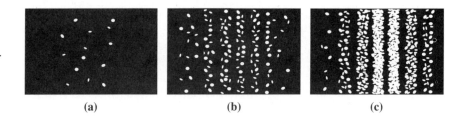

(a)   **(b)**   **(c)**

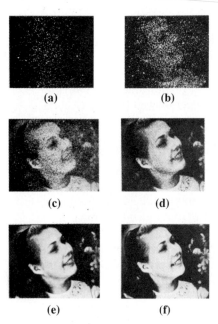

(a)   **(b)**

(c)   **(d)**

(e)   **(f)**

FIGURE 13.7

Light is made of photons striking randomly all over the lit surface. Here a photo emerges from individual photon impacts. Each photo in the sequence has a longer exposure time. The approximate number of photons in each photo is (a) 3×10^3, (b) 10^4, (c) 10^5, (d) 8×10^5, (e) 4×10^6, and (f) 3×10^7.

But there is a pattern in this randomness. A photon never hits in the regions that will emerge as dark regions in the interference pattern but instead strikes preferentially in the regions that will emerge as bright lines. The interference pattern is best described as a statistical pattern formed by large numbers of individual impacts. We will discuss the meaning of this in greater detail later.

We do not normally notice that light is made of photons, because each photon carries such a small energy. A light beam striking a surface seems to arrive continuously and seems spread out evenly over the surface, because it contains so many photons that individual ones cannot be detected without special apparatus. Because the energy of one photon is hf, the smallness of each photon's energy is a consequence of the smallness of Planck's constant. If Planck's constant were larger, light would be obviously lumpy.

These experimental results highlight, without resolving, the wave–particle duality of light. What is it that directs an individual photon to strike the screen at a point that will contribute to the appropriate interference pattern while still allowing each photon to strike rather randomly all over the screen? Although we will say more about this question later, we will find that ultimately science does not know the answer.

DIALOGUE 3   Name at least one common technological device whose operation is based on the wave aspect of light and at least one whose operation is based on the particle aspect of light.

DIALOGUE 4 MAKING ESTIMATES   On a clear day, the sunlight striking each square meter of Earth's surface during each second carries 1000 joules of energy. Estimate the number of photons striking 1 square meter during 1 second. Estimate the number of photons striking your hand during 1 second when you hold your hand open to bright sunlight.

---

3. Any device that uses a lens to focus light is based on the wave aspect of light: cameras, eyeglasses, microscopes, telescopes, and the like. Photovoltaic cells are based on the particle aspect of radiation. More generally, devices that convert light to an electric current are based on the photoelectric effect and so on photons: light meters and electric eyes, for example.

4. According to Chapter 9, sunlight is made mainly of radiation in or near the visible region, having a frequency around 10^{15} Hz (see Figure 9.7). The energy of one photon of this radiation is $(6.6 \times 10^{-34}) \times 10^{15}$, or 6.6×10^{-19} joules, or about 10^{-18} joules. To get 1000 joules of energy, we would then need $1000/10^{-18}$ or 10^{21} photons. This is a billion trillion photons in just 1 second. About how large is your hand? I measure my hand to be roughly 9 cm $\times$ 18 cm, or about 200 cm^2. One square meter is 100×100 cm, or 10,000 cm^2, in area. So a hand is about 200/10,000 m^2, or 0.02 m^2, and the number of photons falling on a hand in 1 second is about 0.02×10^{21}, or 2×10^{19}. This is 20,000,000,000,000,000,000 (20 billion billion) photons every second.

13.4 The electron wave-interference effect ___

FIGURE 13.8
Louis de Broglie.

As far as we know, everything is made of matter and radiation. The conventional view was that radiation was made of waves, but more recently we have found that radiation is in some sense also made of particles.

What about matter? Since the early Greeks, scientists have believed that matter is made of particles. Today we recognize many varieties of material particles: molecules, atoms, protons, neutrons, and electrons, for example.

Louis de Broglie (pronounced "de Broy"; Figure 13.8), a Ph.D. student at the University of Paris in 1923, felt that there should be a kind of symmetry, or balance, between matter and radiation. He thought it unsymmetric, unbalanced, that radiation should exhibit wave–particle duality, behaving sometimes like particles and sometimes like waves, and that matter exhibited no such duality, behaving always as particles. Believing that wave–particle duality was a universal feature of nature, he suggested that if radiation had a dual wave–particle nature, then so should matter. Matter, which had always been assumed to be made of particles, should also in some sense be made of waves. Despite the lack of experimental evidence at that time to support it, de Broglie found this bold idea so beautiful that he included it in his Ph.D. dissertation, submitted in 1924. De Broglie's Ph.D. committee didn't know what to make of it and sent the dissertation to Einstein for his opinion. Einstein was impressed and commented later that "it is a first feeble ray of light on this worst of our physics enigmas." The committee approved de Broglie's dissertation.

Waves of matter? De Broglie's idea was even odder than Planck's and Einstein's particles of radiation. How could individual particles of matter—particles such as electrons and atoms—also be waves? Nevertheless, de Broglie pursued his notion that the wave–particle duality of radiation implied a similar duality of matter. Based on the symmetry that he envisioned between radiation and matter and working from the formula $E = hf$ that connects the wave and particle aspects of radiation, he deduced a formula that predicted the wavelength of the wave that he believed was associated with any material particle:

$$\text{wavelength of material particle} = \frac{\text{Planck's constant}}{(\text{particle's mass}) \times (\text{particle's speed})}$$

$$\lambda = \frac{h}{ms}$$

This formula for waves of matter is analogous to the formula for particles of radiation. Both connect a particle property to a wave property. Planck's constant plays an important role in both formulas. The smallness of h implies that the wavelength λ of a material particle is very small, just as it implies that the energy E of a photon is very small. The smallness of λ means that the wave aspects of matter are difficult to detect, just as the smallness of E means that the particle aspects of radiation are difficult to detect. That is why we normally think of matter as made of particles but of radiation as made of waves.

If we apply de Broglie's formula to a typical macroscopic material object like a 1 kg baseball rolling across the floor at 1 m/s, we get a wavelength of

$$\lambda = \frac{6.6 \times 10^{-34} \text{ J} \cdot \text{s}}{(1 \text{ kg}) \times (1 \text{ m/s})} = 6.6 \times 10^{-34} \text{ m}$$

The baseball's wavelength is about a billionth of a trillionth of a trillionth of a meter. This is far smaller than an atom and far too small to be detected. It is no wonder that we have never noticed the wave aspects of ordinary objects.

The wavelengths of microscopic material particles are much longer. Since mass shows up in the denominator of de Broglie's formula, the least massive material particles generally have the largest wavelengths. The least massive material particle known today is the electron.* In experiments involving streams of electrons, such as the experiment described later, electrons typically move at speeds of 10^7 or 10^8 m/s. At these speeds, de Broglie's formula predicts an electron's wavelength to be about 10^{-11} m. Although this is very small, being about one-tenth the size of a typical atom, it is large enough to be detected in careful experiments.

HOW DO WE KNOW?

De Broglie's hypothesis was confirmed in two independent experiments in 1927. In one of these, British physicist G. P. Thomson passed a beam of electrons through a thin metal foil. The pattern formed by the impacts of large numbers of electrons on a screen behind the foil was a wave-interference pattern that could have been caused only by some unknown kind of wave accompanying the electrons as they passed through the foil. Apparently the gaps between the atoms in the foil were acting like the slits in a wave-interference experiment, causing the waves that passed through different gaps to interfere on the far side of the foil. We call this effect **electron wave interference**.

Although the physical nature of these mysterious waves accompanying the electrons was not known, it was possible to determine their wavelength by measuring the interference pattern on the screen. The experimentally measured wavelength agreed with de Broglie's formula.

It is interesting that G. P. Thomson, who showed that electrons are waves, was the son of J. J. Thomson (Section 8.6), who showed that they are particles.

We summarize de Broglie's idea as:

THE WAVE THEORY OF MATTER
Every material particle is accompanied by a wave. The wavelength of this wave is h/ms, where m is the particle's mass and s is its speed.

As if the wave-versus-particle confusion concerning radiation were not bad enough, de Broglie's idea now presents us with a paradoxical **wave–particle**

---

* It is possible that particles known as *neutrinos* are material particles with a rest mass far smaller than the electron's rest mass. Currently, however, neutrinos are believed to be nonmaterial, that is, to have no rest mass at all (Chapter 18).

duality for matter. How can a material particle such as an electron have a wavelength? An electron is not spread out in space, and it doesn't normally vibrate or wave as it moves. Rather, an electron is a single tiny object that moves in a straight line at a constant speed unless acted on by external forces. How can an electron have a wavelength? Let's turn to experiment, to see what nature says about this.

Figure 13.9 shows the experimental arrangement for a two-slit experiment, but using matter instead of light. The arrangement is identical to the two-slit experiment using light, Figure 8.16, except that instead of a light source this experiment uses a source of material particles. We will assume that the experiment uses electrons, although any other material particles such as neutrons, protons, atoms, or molecules could be used and the results would, as far as we know, be the same (although the wavelength would be different).

Figure 13.9 pictures electrons emerging from the source as tiny particles, since that is what we ordinarily suppose electrons to be. Electrons are sprayed all over a partition containing two thin parallel slits. We observe the pattern formed by the electrons as they strike a screen placed behind the partition. The experiment is similar to shooting machine-gun bullets at a pair of slits in a partition and observing the pattern made by the bullets as they hit a screen behind the partition (Figure 13.10). We will find this example, with bullets, helpful in understanding the experiment with electrons. The difference between the two experiments is that electrons are much smaller than bullets.

Suppose we begin by performing both experiments, with bullets and with electrons, with only slit A open. Figure 13.11(a) shows the experiment with bullets, viewed from above, looking at the thin dimension of the slits (compare Figure 13.10). Figure 13.11(b) shows the experiment with electrons, also viewed from above. We get just the result we expect from the assumption that

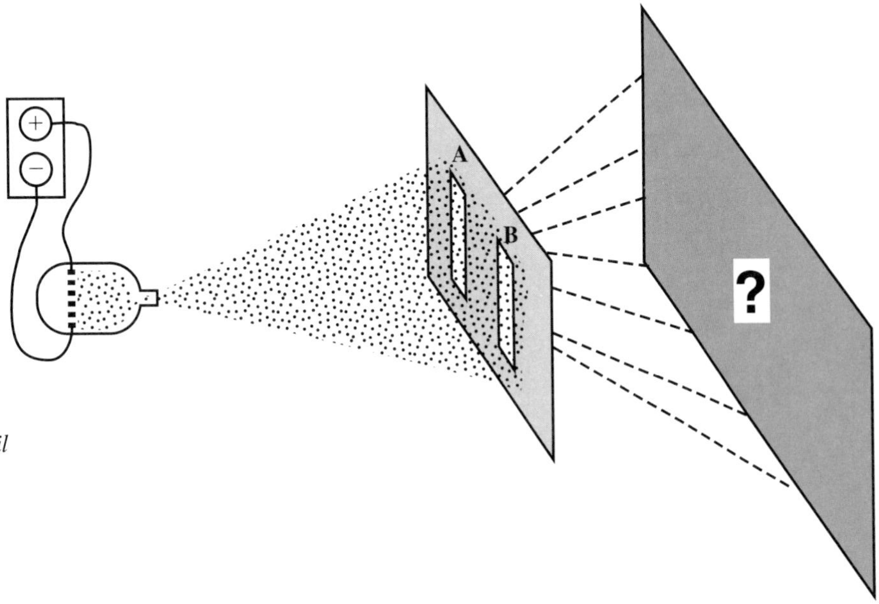

[The double-slit experiment is] a phenomenon which is impossible, absolutely impossible, to explain in any classical [Newtonian] way, and which has in it the heart of quantum mechanics. In reality, it contains the only mystery. We cannot make the mystery go away by explaining how it works.

Richard Feynman

FIGURE 13.9
The double-slit experiment with particles of matter such as electrons. The source of the electrons is a thin tungsten metal wire that is heated electrically until electrons "boil" off it. A similar electron beam is central to TV picture tubes. In the experiment shown, what will we see on the screen?

FIGURE **13.10**
*The double-slit experiment with
machine-gun bullets.*

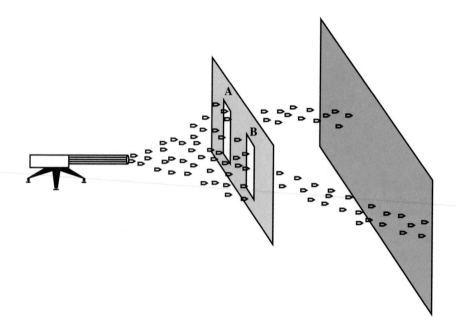

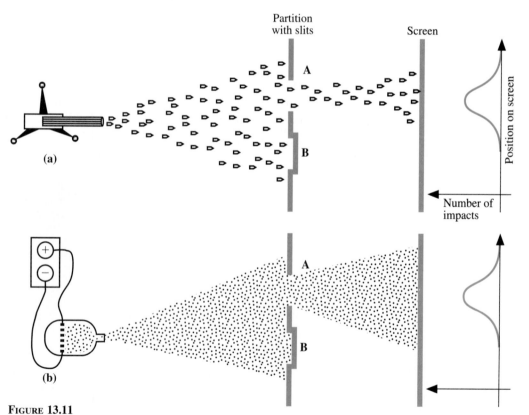

FIGURE **13.11**
*The double-slit experiment but with only slit A open, using (a) machine-gun bullets
and (b) electrons. The average number of impacts at each point on the screen is
graphed, versus the position on the screen.*

electrons are particles. Just like bullets, the electrons strike the screen in a somewhat spread-out pattern behind the slit. The results of firing a large number of bullets and a large number of electrons are shown in the graphs drawn behind the screen in the two experiments. The graphs show the average number of impacts (of bullets and electrons), plotted to the left on the page, versus the position on the screen, plotted upward on the page.

Each bullet hits somewhere within the pattern of Figure 13.11(a), and the overall pattern emerges only after many bullets have hit. In the same way, each electron hits somewhere within the pattern shown in Figure 13.11(b). It is possible to observe the individual flashes that each electron makes on the screen, showing directly that electrons are tiny particles. And like the bullets, if we allow many electrons to hit, we will find the pattern of hits shown in the graph. Both patterns are somewhat spread out behind the slit because of interactions with the sides of the slits or other disturbances.

Naturally, if we perform the two experiments with only the other slit, slit B, open, we will get the same result, only now the patterns will appear behind B instead of A (Figure 13.12).

What happens with both slits open? It is easy to predict the result of the experiment with bullets if the machine-gun sprays both slits. Any bullet that

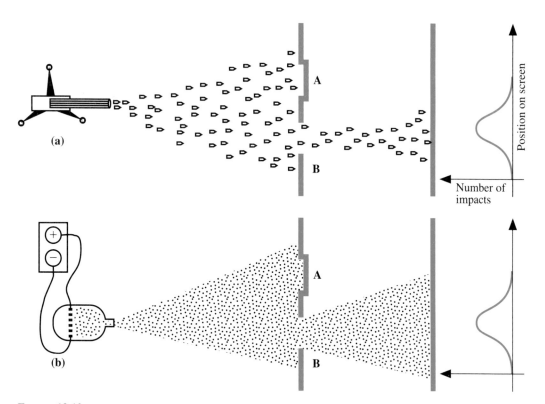

FIGURE 13.12
The double-slit experiment but with only slit B open, using (a) machine-gun bullets and (b) electrons.

passes through A will do precisely what it would do if B were closed, and any bullet that passes through B will do precisely what it would do if A were closed. Bullets passing through A form the impact pattern of Figure 13.11(a); bullets passing through B form the pattern of Figure 13.12(a); and the overall pattern of impacts is just the two single-slit patterns added together, as shown in Figure 13.13(a). This predicted result is exactly what we get when we actually do this experiment with bullets.

A complication would arise in the two-slit experiment with bullets if we fired them so frequently that the bullets passing through the two slits could occasionally hit one another. We are not interested here in how two bullets might interact with each other; we want to learn only about the behavior of individual bullets passing one at a time through the slits. So we suppose that bullets are fired so infrequently that only one bullet is in transit at a time. If we then record the positions of many impacts over a period of time, the overall pattern will be as shown in Figure 13.13(a).

Now we try the two-slit experiment with electrons. As with the bullets, we fire individual electrons at the slits one at a time so that the electrons can-

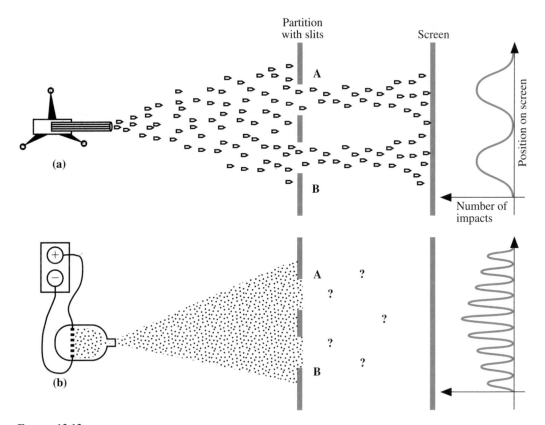

FIGURE 13.13
The double-slit experiment with both slits open, using (a) machine-gun bullets and (b) electrons. In (b), how can electrons create the wave-interference pattern shown in the graph? What is happening in the region between the slits and the screen?

not interact with one another. As in the single-slit experiment with electrons, we will observe individual electron impacts on the screen, showing again that electrons are tiny particles. Surely, then, each electron passes through only one slit, because an electron is far smaller than one slit. If it passes through slit A, it will contribute to the pattern of Figure 13.11(b), and if it passes through slit B, it will contribute to the pattern of Figure 13.12(b). The result is surely the two single-slit patterns added together, just as it was for bullets.

But as you may have suspected from the buildup we have given to this experiment, a most curious result emerges: The overall pattern made by many impacts is a wave-interference pattern (Figure 13.13(b)). The figure does not picture the electrons as they move through the space between the partition and the screen because, as we will see, there is a big question about how we should visualize the electrons in this region and even about whether we can visualize them.

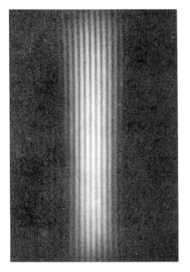

FIGURE 13.14
A wave-interference pattern made by electrons.

HOW DO WE KNOW?

The double-slit experiment using electrons was performed by Claus Jonsson in 1961, although electron wave interference was demonstrated much earlier by other experiments, such as G. P. Thomson's 1927 experiment. Figure 13.14 is a photograph of the pattern formed on the screen behind the slits. Comparison with a typical interference pattern for light, Figure 8.17, indicates that we are indeed dealing here with wave interference. But Figure 8.17 is made by light, a form of radiation, and Figure 13.14 is made by something quite different: electrons, a form of matter.

An experiment by A. Tonomura and other Japanese physicists, reported in 1989, shows in detail the manner in which the alternating bright and dark lines are formed in a pattern such as that in Figure 13.14. Figure 13.15 shows the gradual buildup of a wave-interference pattern from the impacts of individual electrons in a double-slit experiment. In the experiment, about 1000 electrons per second passed through the two slits and then hit a fluorescent surface similar to the inside of a TV tube. A time-exposure photograph was made of the results. The top photograph shows this time exposure after about 0.01 seconds when only 10 electrons had hit. The other photos show the time exposure after longer times. The electrons make individual impacts on the screen, confirming their particle nature. The first few electrons appear to arrive randomly. But it becomes apparent by the third photo that electrons are arriving preferentially in regions that turn out to be regions of constructive interference and are avoiding regions of destructive interference. This result is precisely like the buildup, from individual photon impacts, of an interference pattern for radiation, described in the preceding section; compare Figure 13.15 with Figure 13.6.

Other experiments indicate that a similar interference pattern would result if this experiment were performed using any type of material particle. Interference patterns have actually been observed using neutrons and entire hydrogen atoms.

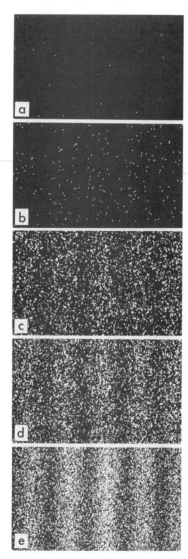

FIGURE 13.15

The buildup of an interference pattern in the electron wave-interference experiment, by individual impacts of electrons. The five photos use exposures of 0.01 s (when only 10 electrons have hit), 0.1 s (100 electrons), 3 s (3000 electrons), 20 s (20,000 electrons), and 70 s (70,000 electrons).

DIALOGUE 5   Arrange these from shortest to longest wavelength, assuming that all six have the same speed: helium atom, automobile, DNA molecule, electron, neutron, baseball.

13.5 Matter: particles or waves? *the quantum mystery*

Electron wave interference is strong evidence that matter is made of waves. These waves are exploited every day in such devices as the **electron microscope**. Using electromagnetic fields instead of the glass lenses used by visible-light microscopes, electron microscopes bend and focus the waves associated with electrons to form electron images of microscopic phenomena. Since electron wavelengths can be smaller than an individual atom, electron microscopes can form images of atoms (Figure 2.15), something that visible-light microscopes cannot do because visible wavelengths are thousands of times larger than an atom.

Even though matter acts as if it were made of waves, we know from the tiny flashes that material particles make on screens, as well as from such phenomena as Brownian motion (Chapter 2), that matter also acts as if it were made of particles. Like radiation, matter has a paradoxical wave–particle nature.

The similarity between wave–particle duality for radiation and wave–particle duality for matter goes fairly deep. Radiation is quantized into discrete (separated) photons; a beam of radiation makes tiny flashes on a screen; individual photons strike a screen in a random pattern that is somehow directed by an underlying wave; and the wave pattern emerges as a consequence of many individual impacts. All of this is true for matter, too, except that matter is quantized into electrons, protons, and other material particles, instead of into photons. Although the remainder of this chapter concentrates mostly on the wave and particle aspects of matter, using electrons as a primary example, nearly everything that is said here is equally true for radiation.

Let's look closely at the most paradoxical feature of electron wave interference. Figure 13.16 reproduces the graphs of the patterns formed in the two-slit experiment. The figure shows four graphs: (a) is the pattern of impacts formed with only slit A open; (b) is the pattern formed with only slit B open; (c) is the pattern we would expect to get with both slits open; and (d) is the wave-interference pattern that we actually get. In graph (d), look at the points of destructive interference, the points marked o in the figure. These are points where no electrons hit the screen when both slits are open. Electrons instead hit at points between the o's! It might be helpful to compare these o's with the capital O's of Figure 8.14. Remember that the pattern is made by electrons coming one at a time through the slits. Somehow, each electron "knows" that both slits are open and that it is supposed to avoid the places marked o. How could an individual electron "know" this? An electron is far smaller than either slit. It is hard to imagine how it could come through both slits or "sense" that the other slit is open.

---

5. Automobile, baseball, DNA molecule, helium atom, neutron, electron.

FIGURE 13.16

Four patterns of impacts in the double-slit experiment. In (a) only slit A is open, and in (b) only slit B is open. We expect (c) with both slits open, but we actually get (d) when using electrons or other microscopic particles.

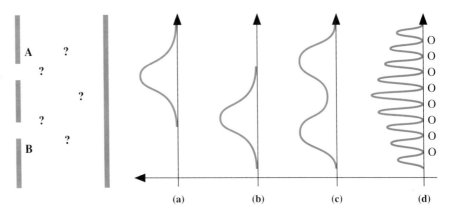

When we close one slit, however, we destroy the interference pattern. Electrons simply pass through the open slit and hit all over the screen, forming patterns (a) or (b), depending on which slit is open. With only one slit open, electrons strike at places like the o's as frequently as at any other points. Opening the other slit should *increase* the number of electron impacts at every point on the screen, since more electrons pass through two slits than pass through one. Yet the mere act of opening slit B somehow informs every electron—no matter which slit it goes through—that it is not to strike at the o's. And this happens even when the electrons go through the experiment one at a time—even if only one electron goes through every hour. Every electron avoids the points o, and after a large number of impacts, the overall pattern is the pattern graphed in (d).

We can conclude that when both slits are open and a single electron goes through the apparatus, something goes through both slits simultaneously. This seems to contradict the fact that every time a single electron is observed (for instance, by observing its flash on the screen), it is seen as a tiny particle, far too small to go through both slits. Something associated with each electron is spread out over both slits when both slits are open. Whatever this something is, it interferes in the way that a wave interferes after passing through the two slits. It is as though the electron were interfering with another electron—although there is no other electron. We describe this by saying that an individual electron interferes with *itself* when both slits are open.

This paradoxical phenomenon was predicted by de Broglie in 1923 and was first observed in 1927, yet physicists today are as perplexed by it as ever. If an electron (or a proton, atom, or whatever) is a particle, it must go through just one, and not both, of the two slits in the double-slit experiment. And if it goes through just one slit, it cannot form an interference pattern. But experiment shows that it does form an interference pattern. So it is not possible even to imagine that an electron passes through only a single slit.

Somehow, despite all the evidence showing that electrons are tiny particles, an electron is more than just a tiny particle during this experiment. As long as both slits are open, an electron acts like a spread-out wave right up until it actually hits the screen. Then when the flash appears on the screen, the electron "collapses" back into a tiny particle. An electron (or any other microscopic particle) seems to be a spread-out, wavy entity when it is not

All of modern physics is governed by that magnificent and thoroughly confusing discipline called quantum mechanics. . . .It has survived all tests and there is no reason to believe that there is any flaw in it. . . .We all know how to use it and how to apply it to problems; and so we have learned to live with the fact that nobody can understand it.
Murray Gell-Mann, physicist, in *The Search for Unity in Particle Physics*

being observed and seems to collapse into a tiny particle every time it is observed.

To try to understand this confused situation, we turn again to experiment. We would like to look more closely at what is actually coming through each of the two slits. So we set up a detector behind one or both slits, to see whether something comes through only one slit, or both. Experiments of this sort have been done. In every instance, the electron was observed to come through only one slit, not both!

At first, we might conclude that in the double-slit experiment, the electron really does come through only one slit. But this conclusion would be unwarranted, because in every instance in which we set up a detector of this sort, *the interference pattern is destroyed*, and we get just the simple pattern (c) of Figure 13.16. The mere act of observing the slits destroys the interference pattern! As soon as we activate a detector that can tell us whether the electron goes though one or both slits in the interference experiment, we change the experiment itself. The electron seems to "know" that a detector is present and that it (the electron) should go through only one slit instead of both. Putting the detector behind the slits does not give us the information we seek concerning what comes through the slits in the interference experiment, because the mere presence of the detector destroys the interference pattern.

In summary, an electron behaves like a tiny particle whenever a detector is present that can determine through which slit the electron came, but it behaves like a spread-out wave that goes through both slits whenever there is no detector.

This throws into question a basic premise of science: Science has always assumed that the real world is essentially the same when we are not observing it as it is when we are observing it. But wave–particle duality seems to imply that the real world is essentially different when detectors are absent than when they are present. The mere presence of detectors seems to change reality.

DIALOGUE 6   List several differences between particles and waves. Can you think of any ways in which particles and waves are similar?

13.6 The psi-field: *nature doesn't know what she will do next* _____

The formation of an electron wave interference *pattern* is significant. In order for a pattern like that graphed in Figure 13.13(b) to emerge from tiny particles striking a screen, different particles must strike at different places. You might suppose that electrons hit at different places because

_____

6. Some differences: A particle is limited in size, whereas waves are spread out and have no well-defined size; a particle has a well-defined mass, whereas we cannot define the mass of a wave; a particle is a thing or object, whereas a wave is a pattern; and a particle can exist by itself, whereas a wave can exist only in some medium. Some similarities: Both have a speed; both transport energy; and both can transport information.

they started out differently from the electron source. Could we then adjust the source so as to prepare every electron identically and make them all hit the same point on the screen? This would destroy the interference pattern, for we cannot have a pattern if all the electrons hit the same point. Experimentally, we find that no matter how finely we control the experiment, we get the same pattern. Even though electrons are prepared identically, they hit at different points; that is, identical preparations lead to different outcomes.

What causes different electrons to hit at different places? The answer must be that nothing causes it, because all the electrons are identical and all were prepared identically. This general rule is found, experimentally, to be true throughout the microscopic world. Given two particles of the same type (two photons, two helium atoms, two electrons, and so forth), no amount of careful identical preparation can make them behave identically. Despite our best efforts to control them, microscopic particles go their own way. "Randomness," or "uncertainty," is built into nature at the microscopic level. Nature herself doesn't know what she will do next. So wave–particle duality implies irreducible uncertainties, called **quantum uncertainties**, in nature.

This is quite a break with deterministic Newtonian physics. A few physicists disagree with the notion that the future is fundamentally uncertain, arguing instead that our current understanding (quantum theory) is simply not deep enough to penetrate the true principles governing the microscopic world and that these true principles would restore predictability to nature. Einstein argued forcefully during the 1930s that "God does not play dice," citing detailed examples to try to show that an irreducible element of uncertainty would be absurd. But despite such arguments, quantum theory as formulated during the 1920s continues to have a perfect record of success in predicting observed phenomena. Predictions that Einstein and others believed to be absurd have now been tested and found actually to occur.

Despite the randomness of individual impacts in the double-slit experiment, the overall pattern *is* predictable. That is, we get the same interference pattern every time we do the experiment. The pattern represents the overall statistics of many impacts. It tells us the approximate fraction of impacts in any portion of the screen. In Figure 13.17, we might predict, for example, that 15% of the impacts will appear between points a and b, that 10% will appear between b and c, and so forth. *The overall statistics are predictable, even though individual impacts are not.*

In 1926, German physicist Max Born (Figure 13.18) was the first to conclude that the interference pattern must be a *probability pattern* for each electron. That is, Born concluded that data such as the graph in Figure 13.17 show the probabilities for a single electron to strike at various points on the screen. Following up on this idea, Born concluded that de Broglie's wave that accompanies each particle is a **wave of probability** or "probability wave" for the presence of the electron. Similarly, the electromagnetic wave that accompanies each photon is a probability wave for the presence of the photon.

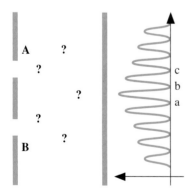

FIGURE 13.17
The pattern represents the statistics of the impacts—the number hitting between points a and b or b and c, for example.

FIGURE 13.18
Max Born.

The Born interpretation of the Schroedinger equation is the single most dramatic and major change in our world view since Newton.
Leon Lederman, physicist, in *The God Particle*

I think I can safely say that nobody understands quantum mechanics. ...I am going to tell you what nature behaves like. If you will simply admit that maybe she does behave like this, you will find her a delightful, entrancing thing. Do not keep saying to yourself, if you can possibly avoid it, "But how can it be like that?" because you will get down the drain, into a blind alley from which nobody has yet escaped. Nobody knows how it can be like that.
Richard Feynman, in *The Character of Physical Law*

Probabilities were invented long before quantum theory and usually have nothing to do with quantum theory. Probabilities are useful whenever the outcome of a particular experiment is uncertain, but the overall statistics of many repetitions are predictable. A simple example, having nothing to do with quantum theory, is the flip of a coin. It is common to speak of a 50% probability of heads and a 50% probability of tails for a fairly flipped coin. What this statement means is that in a long series of tosses, roughly 50% will be heads. This probability, 50% or 0.5, can be regarded as a statistic, a number representing the pattern that emerges in many repeated trials of the experiment.

But there is a difference between the probabilities observed in macroscopic experiments such as the flip of a coin and the probabilities referred to in quantum theory. A coin flip obeys Newtonian physics to an excellent approximation, and so the outcome is predictable in principle. That is, with enough information regarding the tension in the flipper's thumb, the height of the coin above the table, the elastic properties of table and coin, and so forth, it is possible to use Newtonian physics to predict the outcome. Our uncertainty about a coin flip arises only from our ignorance of the precise details. But quantum events are not predictable even in principle. Quantum uncertainties don't arise from our ignorance of the details. Rather, the microscopic world is inherently unpredictable: Nature herself doesn't know what she will do next.

The wave–particle duality of microscopic particles is just the kind of situation in which probabilities should be useful, because individual particles are not predictable but their statistics are predictable. So it is natural to interpret as probability waves the waves that accompany microscopic particles.

A wave of probability is a fairly abstract notion. Although each electron is described by a probability wave, that probability wave cannot be conclusively detected when just one electron goes through the double-slit apparatus, because the pattern emerges only after many electrons have gone through it. The probability wave for a single electron is a description of the possible places in which the single electron might be found if we decided to observe that electron's position. It is a wave of possible locations, a wave of possibilities.

Let's follow the probability wave for just one electron traveling through a double-slit apparatus. Figure 13.19 shows the probability wave at four different times, labeled (a), (b), (c), and (d). According to quantum theory, this single electron's probability wave undergoes the following development: (a) The wave leaves the source; (b) the wave spreads out so as to go through both slits; (c) the wave goes through both slits; and (d) the wave arrives at the screen in an interference pattern formed by the parts of the original wave going through each slit. The probability wave represents the likelihood of the electron's presence. For instance, at time (c), 50% of the probability wave is located just behind slit A, and 50% is located just behind B. The physical meaning of this is that if we observe the electron's position at time (c), we will have a 50% chance of finding the electron just behind A and a 50% chance of finding it just behind B.

The wave-interference pattern observed on the screen can be described by saying that the portions of the probability wave coming through each slit

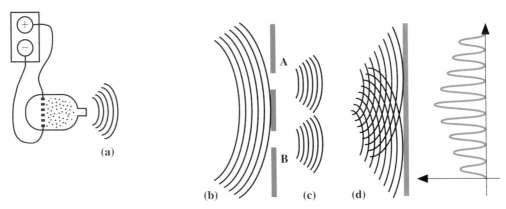

FIGURE 13.19

The probability wave for a single electron at four different instants: (a) as the wave (that is, the electron) leaves the source, (b) as it approaches the slits, (c) just after passing through the slits, and (d) just before arriving at the screen. The graph—of the probability wave at the position of the screen—shows the probabilities that the electron will hit at various positions.

FIGURE 13.20.
Erwin Schroedinger.

interfere with one another so as to reinforce one another at some places on the screen and cancel out one another at other places. Each individual electron must strike the screen at a point that agrees with this pattern.

The predictability of the overall statistical patterns shows that probability waves also are predictable. In 1926, Austrian physicist Erwin Schroedinger (Figure 13.20) invented a method of predicting probability waves. Schroedinger began with a well-known formula that had been used to describe waves in other situations not involving quantum theory. Into this wave formula, he inserted de Broglie's relation $\lambda = h/ms$, along with some judicious guesswork. The result was a formula, now called **Schroedinger's equation**,* that predicts the probability wave for electrons or any other material particles in a wide variety of situations. The quantity that does the waving in Schroedinger's equation is usually represented by the Greek letter Ψ (psi). We will call it the **psi-field** of the particle, to indicate that it is spread out over a region of space, like an electromagnetic field. A particle's psi-field is nearly the same thing as its probability wave: The probability of finding a particle at some point is simply the square of the particle's psi-field at that point. We often use "psi-field" and "probability wave" interchangeably.[†]

---

* Here it is: $-(h^2/8\pi^2 m)\,\nabla^2\Psi + V\Psi = E\Psi$, in which Ψ (psi) represents the wave and is a so-called probability amplitude rather than a probability. The meaning of this is that actual probabilities are found by squaring Ψ. Schroedinger's equation is a "differential equation" involving a "second derivative," $\nabla^2\Psi$. In addition, h is Planck's constant, m is the mass of the particle, and E is the numerical value of the particle's total energy. The symbol V represents the way that external forces act to alter the particle's energy. Because V is different in different physical situations, such as the double-slit experiment and within a hydrogen atom, Schroedinger's equation has a different form for each different situation.

† The psi-field goes by many other names in other books: psi-wave, matter-wave, and Dirac field. Einstein called it a ghost field, and physicist/author Nick Herbert calls it quantum-stuff.

Schroedinger's equation correctly predicts such phenomena as the electron wave interference pattern (Figure 13.16, pattern (d)), the pattern observed in a single-slit experiment (Figure 13.16, pattern (a)), and the interference patterns observed when electrons pass through thin sheets of solid material. Most important historically, Schroedinger showed that his equation could be applied to an electron within an atom and that the predicted results agree with experiment (Chapter 14).

Schroedinger's equation is the quantum equivalent of Newton's law of motion. Like Newton's law of motion, it predicts the behavior of a material particle subjected to external forces. But whereas Newton's law predicts a particle's precise motion, Schroedinger's equation predicts only a particle's psi-field and so the probabilities of various *possible* motions.

All of these ideas apply also to particles of radiation, photons. Like material particles, individual photons are unpredictable, whereas their statistical patterns are predictable. So an individual photon can be described by a psi-field that gives the probability of finding the photon at any particular place. But an Erwin Schroedinger was not needed to invent the wave equation that describes and predicts a photon's psi-field, because the waves themselves were discovered and described decades before quantum theory. The psi-field for photons is simply the electromagnetic field, and the probability waves for photons are simply electromagnetic waves. Because radiation's wave aspect is generally easier to observe than is its particle aspect, light waves were known long before photons were known and long before anybody realized that light waves are probability waves for photons. So the theory of radiation that parallels Schroedinger's theory of matter is simply the electromagnetic wave theory (Section 9.2).

13.7 The quantum theory of matter and radiation

Psi-fields, Schroedinger's equation, and electromagnetic theory help clarify and predict nature's microscopic behavior, but they do not resolve the quantum paradoxes. For instance, Schroedinger's equation predicts the probability pattern on the screen, but it doesn't tell us what happens to an electron as it goes through a double-slit apparatus. Which slit does it really go through? Which path does it really follow?

To highlight this question, Figure 13.21 shows the double-slit interference experiment once again, with the probability pattern drawn at the position of the screen. Suppose we perform this experiment with just one electron. The electron travels from the source, marked S, through the apparatus to the screen. It may hit anywhere in the probability pattern except at the places of destructive interference. Suppose that it actually hits at the point marked P. By what path did it get from S to P? Any possible path must go through one or the other slit. Figure 13.21 shows three such possible paths. Surely, it seems, the electron followed

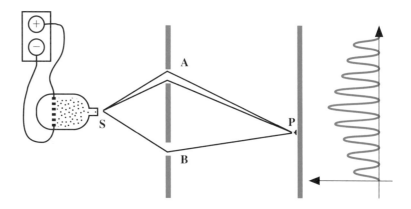

My own suspicion is that the universe is not only queerer than we suppose, but queerer than we can suppose....I suspect that there are more things in heaven and Earth than are dreamed of, or can be dreamed of, in any philosophy. That is...why I have no philosophy myself, and must be my excuse for dreaming.
  J. B. S. Haldane, philosopher and biologist

some such path from S to P. And yet we have already seen (Section 13.4) that the commonsense assumption that the electron came either through slit A or slit B implies that there is no interference pattern. The conclusion is that we are not allowed even to *imagine* that the electron moves from S to P along any single path. We are not allowed to imagine such things as "the electron really went through only one of the slits, but since we didn't observe it, we described it with probabilities for going through both slits." If we imagine the electron doing anything at all between S and P, we must imagine it moving along all the possible paths, such as the three shown.

What is really going on at the microscopic level? Because physics has always assumed that macroscopic phenomena such as human sense perceptions are determined by events at the microscopic level, the question of what the electron really is doing is fairly significant. When we ask what is going on, we really are asking: What is the nature of microscopic reality? We will probe this question more deeply in the next chapter.

We conclude this chapter by summarizing quantum theory in a way that emphasizes the dual wave–particle nature of both matter and radiation:

THE QUANTUM THEORY OF MATTER AND RADIATION
Both matter and radiation occur as particles such as electrons, protons, and photons. The behavior of a particle is unpredictable, but the statistical patterns formed by many particles are predictable. Such a pattern is called a psi-field. Since psi-fields are wavelike, both matter and radiation have both particle and wave characteristics.

A material particle's psi-field in any physical situation can be predicted from Schroedinger's equation for that particle in that situation. The psi-field's wavelength is related to the particle's mass and speed by means of de Broglie's relation, $\lambda = h/ms$, where h is Planck's constant.

The psi-field of a radiation particle (photon) can be predicted from electromagnetic theory. The frequency of a photon's psi-field (the frequency of its electromagnetic wave) is related to the photon's energy by Planck's relation, $E = hf$.

The discovery of quantum mechanics in the mid-1920s was the most profound revolution in physical theory since the birth of modern physics in the seventeenth century.
  Steven Weinberg, physicist, in *Dreams of a Final Theory*

For both matter and radiation, the underlying quantity that is waving is the psi-field. Underlying quantum theory is a field that is spread out in space. Yet the objects actually observed are particles. The connection between a particle and its psi-field is that the psi-field determines the probability of finding the particle at various places. This idea at least clarifies the paradox of wave–particle duality without exactly resolving it.

DIALOGUE 7   List at least two differences between Newtonian physics and quantum physics. Now list two similarities.

"ACTUALLY I STARTED OUT IN QUANTUM MECHANICS, BUT SOMEWHERE ALONG THE WAY I TOOK A WRONG TURN."

7. Differences: Newtonian physics is deterministic; it is objective; it assumes that matter is made of particles and that radiation is waves; and it is based on Newton's law of motion. Quantum physics is nondeterministic; it is dependent on the observer; it assumes that matter and radiation are made of particles that have a probabilistic wavelike behavior; and it is based on Schroedinger's equation. Similarities: conservation of energy, second law of thermodynamics, law of inertia, idea of universal natural laws.

Summary of Ideas and Terms

Quantum theory The currently accepted physical theory of the microscopic world.

Photoelectric effect When light and other radiation shine on a metal surface, they can eject electrons from the surface, thereby providing evidence that radiation is made of photons.

Quantized Limited to specific allowed quantities.

Photons Particles of radiation. A photon moves at speed c, has zero rest mass, and carries a **quantum** (a specific allowed amount) of energy.

Particle theory of radiation Radiation is created by vibrating charged particles and appears as photons carrying energy hf, where f is the radiation frequency and h is **Planck's constant**, 6.6×10^{-34} joule-seconds.

Wave theory of matter Every material particle is accompanied by a wave having wavelength h/ms, where m is the particle's mass and s is its speed.

Wave–particle duality of radiation and matter Both radiation and matter have wavelike and particlelike aspects.

Double-slit interference experiment When photons or material particles are sent through two narrow slits, the distribution of many impacts on a screen forms a wave-interference pattern.

Quantum uncertainties The irreducible uncertainties that arise because identical physical situations lead to different outcomes.

Probability An event's fractional number of occurrences in a long series of trials. Useful in situations involving uncertainties.

Probability wave A probability pattern that moves through space in a wavelike manner.

Electron microscope Uses electron probability waves to form images of microscopic objects.

Psi-field Roughly the same as "probability wave." The square of the psi-field at any point is the particle's probability of being found at that point. Quantum theory predicts a particle's psi-field.

Schroedinger's equation Predicts the psi-field for material particles.

The quantum theory of matter and radiation Both matter and radiation occur as particles whose behavior is unpredictable but whose probabilities form predictable wave patterns. For material particles, this pattern is predicted from Schroedinger's equation. For photons, it is predicted from electromagnetic theory.

Review Questions

QUANTIZATION

1. What is quantum theory?
2. Describe the photoelectric effect and its two troubling features. What does this effect tell us about radiation?
3. When a vibrating electron emits one photon, does the electron's energy increase, decrease, or remain unchanged? By how much?
4. Where do photons come from? What is their speed? Their rest mass?
5. Name a rather common device that utilizes the photoelectric effect.

RADIATION: WAVES OR PARTICLES?

6. What is a photon? What is the relation between its frequency and its energy? How can one photon have a frequency, and what does frequency mean here?
7. How do we know that radiation is made of waves? How do we know that it is made of particles?
8. If we perform the experiment shown in Figure 13.5 using a light source that emits only one photon per second, what will we observe on the screen? What if we perform it with a million photons per second?
9. Why don't we normally notice that light is made of photons?

MATTER: PARTICLES OR WAVES?

10. Can a single electron have a wavelength? How can you calculate its wavelength?
11. How do we know that material particles are accompanied by waves?
12. What will be observed on the screen in Figure 13.9? In Figure 13.10?
13. If one electron per second is sent through the experiment in Figure 13.13(b), what will we observe? What if we send through a million electrons per second?
14. Which detects the smallest objects: a visible light microscope or an electron microscope? Why? List some differences between the two devices.
15. What experimental arrangement would produce each of the four patterns graphed in Figure 13.16?
16. Of the four patterns shown in Figure 13.16, which one would be created by bullets with only slit A open?

By electrons with only slit A open? By bullets with both slits open? By electrons with both slits open but with a detector behind the slits capable of determining through which slit each electron will come? By electrons with both slits open and with no detector?

17. When a single electron travels through a double-slit apparatus, does anything go through both slits? How do we know?

18. What happens when we use a detector to tell us which slit each particle comes through in the double-slit experiment?

THE PSI-FIELD

19. What is Schroedinger's equation?

20. What are quantum uncertainties? How do they differ from the ordinary uncertainties in a coin flip?

21. The impact point of each electron is unpredictable in the double-slit experiment. What *is* predictable?

22. Explain in detail the meaning of each part of Figure 13.19 as it applies to a single electron.

23. If an electron traveling through a double-slit apparatus strikes directly behind slit A, is it correct to say that the electron came through slit A?

Home Projects

1. Observe the metal heating element on an electric stove as it changes color while heating up. As it heats up, what can you say about the number and kind of photons it is emitting? Before it begins to glow, what kind of photons is it emitting?

2. Find the probability pattern for tossing two dice. Record the number of twos, threes, and so on in 100 tosses, and graph the results. According to your experimental results, what is the percentage probability of a seven? A two? Compare this with the theoretical results obtained by assuming that each possible combination of two faces (such as a three on die 1 and a four on die 2) has a probability of one-thirty-sixth. Do these probabilities have anything to do with quantum uncertainties?

For Discussion

1. Discuss the similarities and the differences between matter and radiation.

2. Do electrons and other microscopic particles really exist? Are such particles more real or less real than, say, the chair you are sitting on? Less real? Is the "redness" of a red lightbulb more or less real than the atoms that physicists tell us are emitting the red light?

Exercises

QUANTIZATION

1. What energy transformation (from what to what) occurs in the photoelectric effect?

2. How can the photoelectric effect be used to open a door when someone approaches?

RADIATION: WAVES OR PARTICLES?

3. Which has greater energy: a radio photon or an infrared photon? Which has a longer wavelength? Larger frequency? Larger rest mass? Which moves faster?

4. What is meant by a red photon? A yellow photon? Which one has greater energy? Longer wavelength?

5. Explain why, in terms of photons, ultraviolet light can damage cells in your skin but visible light cannot.

6.*Which has greater energy, a microwave photon or a visible photon? About how many times greater (consult Figure 9.7)?

7.*You charge an object by rubbing it and then shake it at 1 Hz. Does it emit photons? How much energy does each photon carry?

8.*How much energy does one photon of 10^{24} Hz gamma radiation carry?

9.*Making estimates. About 10 visible photons are needed to cause a single photosynthesis reaction in living plants. About how much energy is carried by these 10 photons?

10.*Making estimates. The human eye can detect as few as 10,000 photons per second entering the pupil. About how much energy is this, per second?

MATTER: PARTICLES OR WAVES?

11. List the similarities and differences between a photon and an electron.

12. If electrons behaved only like particles and not like waves, what pattern would you observe in the double-slit experiment?

13. You don't notice the wave aspect of a pitched baseball. Is this because the wavelength is very long or because it is very short?

14. Which has the longer wavelength, an electron or a proton moving at the same speed?

15. Which has the longer wavelength, a slow proton or a fast proton?

16. If a "proton microscope" could be devised, how would you expect its wavelength to compare with the wavelength of an electron microscope?

17. Suppose that protons rather than electrons were used in the double-slit experiment. Would we still get an interference pattern on the screen? How would this pattern differ from the pattern for electrons?

18.*Suppose that we fire a high-velocity pellet gun that accelerates 1-gram (10^{-3} kg) pellets to speeds of 1000 m/s (three times the speed of sound). Find the wavelength of the pellet's probability wave.

19.*If you double a hydrogen atom's speed, how will this affect its wavelength? What if you double its mass (by attaching two hydrogen atoms) without changing its speed?

20. What is the percentage probability of getting two heads in a row in fair coin tosses? Of getting five heads in a row? How could you experimentally test your prediction about getting five heads in a row?

21. Do quantum uncertainties have anything to do with the probabilities involved in a coin toss? Explain.

22. In the double-slit experiment with electrons, is the impact point predictable? Are there any points where we can predict that an electron will certainly not hit?

23. What is predictable in the double-slit experiment with electrons?

24. Would the answers to the preceding two questions be different if we were talking about photons instead of electrons?

25. Suppose that two electrons travel separately through a double-slit apparatus such as shown in Figure 13.19. The first electron hits at the upper end of the interference pattern, and the other hits at the lower end. Is it correct to say that the first electron came through slit A and the second came through slit B? Explain.

14

THE UNIVERSE ACCORDING TO QUANTUM THEORY

Recall Democritus's words (Chapter 2): "By convention sweet is sweet, by convention bitter is bitter—the objects of sense are supposed to be real, and it is customary to regard them as such, but in truth they are not. Only the atoms and empty space are real."

The Newtonian worldview also argues that sense impressions are not real but are merely a reflection of the real atomic events. Since human experience consists entirely of sense impressions, it seems important to ask what is going on in the supposedly real world of the atoms. The previous chapter found that world to be paradoxical, unpredictable, and quite non-Newtonian, all of which raised some deep questions about the meaning of quantum theory and even the meaning of the scientific enterprise. This chapter looks at some specific quantum principles that can help us understand these questions.

But first we devote two sections to one of the theory's many successful practical applications: the atom. We have already studied two theories of the atom: the Greek atom and the planetary atom. We saw that the failure of the Greek atom to explain electrical phenomena led to the planetary model. Similarly, certain failures of the planetary atom (Section 14.1) lead to the quantum atom. The quantum atom (Section 14.2) accounts for every atomic observation to date and, as far as we know, needs no further improvements. That is saying a lot because the observations have been many and they have been extraordinarily precise. For example, a certain radiation frequency for the hydrogen atom, known as the *Lamb shift*, is predicted by quantum theory to be 1057.860 ± 0.009 megahertz and has been measured as 1057.845 ± 0.009 megahertz. It seems miraculous that an abstract theory about unseen atoms can predict a seven-digit number that is then verified in the natural world. Quantum theory might be odd, but there is something right about it.

We then return to the problem of explaining the meaning of quantum theory. Section 14.3 presents the uncertainty principle, one of the theory's most meaningful and useful ideas. Sections 14.4 and 14.5 present two topics that are closely related to the uncertainty principle and that could lay claim to being the oddest ideas that have cropped up yet in physics or in all of science: quantum jumps and the nonlocality principle. We bolster these ideas with several recent experiments that support them. Section 14.6 discusses the meaning to humans of all of this. Section 14.7 returns to one of this book's themes: comparisons between Newtonian and post-Newtonian physics. Quantum theory is the central item in any such comparison and is perhaps a central item guiding us toward finding a philosophical framework that can sustain humankind in the twenty-first century.

14.1 How do we know? *observing atomic spectra*

The most accurate scientific measurements known are made with **spectroscopes**, devices that measure the frequencies or wavelengths present in radiation. Figure 14.1 shows how a spectroscope studies the visible radiation emitted by a light source such as a heated, glowing gas. Radiation from the source passes through a single thin slit and emerges as a narrow beam. This beam passes through a glass prism or some other device that can separate the light beam's different frequencies (colors). Light beams bend when they pass from one medium into another, for instance, from air into glass. You might have noticed this effect in a pool of water, where partly submerged objects appear to bend at the water's surface. The reason that a prism separates a light beam's frequencies is that different frequencies bend by different amounts at each glass surface. This separation of frequencies is also seen in a rainbow, where each raindrop acts like a small prism for sunlight.

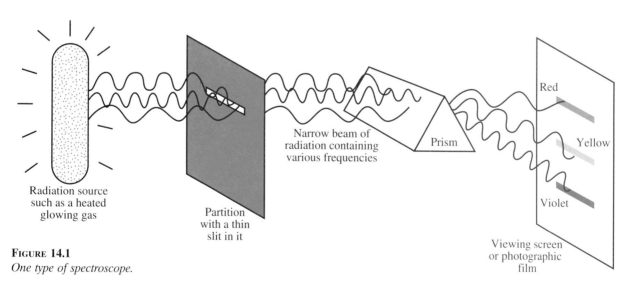

FIGURE 14.1
One type of spectroscope.

By the time the light exits the far side of the prism, it has separated into many beams, each with its own wavelength. In terms of photons instead of waves, photons with different energies are now moving in different directions. A screen or photographic film intercepts all the light beams and displays their various colors. Each beam's frequency can be determined by measuring the position at which it strikes the screen. The set of frequencies measured in such an experiment is called the **spectrum** of the source that emitted the radiation.

Different kinds of spectroscopes operate in every part of the electromagnetic spectrum. For example, a radio receiver is a spectroscope for detecting the frequencies of the radio radiation present in a room.

Spectral measurements yield an enormous quantity and variety of information. Most of our data about the microscopic world come from spectral measurements. For instance, telescopes gather data about the mass, temperature, motion, chemical composition, and other properties of stars and galaxies by means of a spectroscope at the viewing end.

A glowing solid or liquid, such as a tungsten lightbulb filament heated to around 3000°C, emits a **continuous spectrum**, one that contains an unbroken range of visible frequencies and is spread out in a continuous band of color across the spectroscope screen. But surprisingly, if a dilute gas is heated until it glows, it will emit a spectrum that is not continuous. Instead, it is "quantized," or restricted to a limited number of precise frequencies, each frequency appearing on the screen as a narrow slit-shaped line (Figure 14.1). This collection of frequencies is called a **line spectrum**. Figure 14.2 (see color insert) shows a continuous spectrum and line spectra from several gases. As you can see, the line spectra for different gases are different. Since each gas has its own spectrum, spectroscopy can identify different gases.

Heating is one way to **excite** a gas, in other words, to cause it to emit radiation. Most gases glow once they reach temperatures above about 2000°C. Flames are glowing gases of this sort, heated by combustion. The sun's light comes from hot gases on its visible surface, at a temperature of 5500°C.

A second way to excite a gas is to send an electric current through it. This process, called **electric discharge**, creates the light seen in neon tubes, fluorescent bulbs, mercury or sodium vapor bulbs, sparks, and lightning strokes. Electric discharge tubes containing a dilute gas can be used to study the gas's spectrum (Figure 14.3).

FIGURE 14.3
An electric discharge tube containing a dilute gas. With a large enough charge on the two "electrodes" at the ends of the tube, the electrodes "discharge" by forcing electrons off the negative electrode. These electrons then excite the gas by colliding with it as they move through the tube toward the positive electrode.

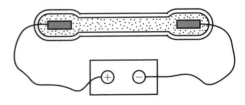

Atoms are completely impossible from the classical [Newtonian] point of view.

Richard Feynman

How can we explain the observed spectra? As we know, when any substance is heated, the random kinetic energy of its atoms increases. The Greek model of the atom offers no obvious reason that this should cause materials to glow, but the planetary model does: The subatomic parts of atoms are electrically charged; these charged particles move more energetically when the material is heated; and these vibrating and orbiting charged particles should send out electromagnetic radiation.

But why are only some wavelengths emitted, rather than all wavelengths? What determines which wavelengths are emitted? The planetary atom does not answer these questions.

There is an even more glaring problem with the planetary atom. As explained in Figure 14.4, an orbiting electron can be thought of as vibrating along two directions at once. So the electromagnetic wave theory predicts that an orbiting electron should radiate electromagnetic energy all the time! But observations show that atoms do not radiate all the time. Worse yet, if an electron did radiate all the time, it would have to continually lose energy, and as it lost energy, it should spiral into the nucleus and cease orbiting. So when coupled with the electromagnetic theory of radiation, the planetary model is not self-consistent, because it predicts that atoms should collapse.

It is possible to imagine a universe in which Newtonian physics would be correct, even down to the smallest sizes. But atoms could not exist in such a universe, and so there could be no chemistry, life would be impossible, and things would be pretty boring.

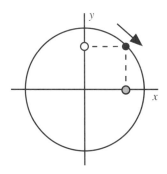

FIGURE 14.4
An orbiting electron (black circle) can be thought of as making two vibrational motions: When viewed from below, it appears to be vibrating along the x axis (green circle), and when viewed from the side, it appears to be vibrating along the y axis (white circle).

DIALOGUE 1   You might have noticed that as you heat a metal hot plate, it first glows dark red and then becomes brighter and whiter. If you observe this process through a spectroscope, what will you see?

DIALOGUE 2   Instead of the thin slit in a spectroscope, could a round hole be used? How would a "line" spectrum then appear? Why is a slit better?

14.2 The quantum atom

We will examine the simplest atom, hydrogen, made of one proton and one electron. Because the electron is some 2000 times less massive than the proton, it does nearly all the moving, orbiting in the electromagnetic field of a nearly stationary proton. To a good approximation, we can ignore the proton's motion, treating it as a tiny material particle at rest.

The electron's behavior is described by its psi-field, that is, by its probability wave. Imagine a hydrogen atom at rest and isolated. Since there is no reason for anything to be physically changing in such an atom, we would expect the electron's probability wave to be stationary, unchanging.

1. At first, you will see a dim spectrum extending from the longest visible wavelengths (red) through a portion of the red part of the spectrum. As the plate gets hotter, the spectrum will get more intense, and it will extend through a larger portion of the visible spectrum.
2. Yes. The spectrum would appear as a series of circles instead of a series of lines. Two neighboring lines on the screen are more distinctly separated than are two neighboring circles.

Stationary wave-patterns are observed in many phenomena in nature, phenomena usually having nothing to do with quantum theory. Such patterns are called **standing waves**. A vibrating violin string is an example. Figure 14.5 shows four different standing waves on a string. Each photo is a time exposure over the duration of a few vibrations.

As you can see, each standing wave is formed by vibrations of the entire string, but the resulting wave shape is not moving in either direction along the string. Nevertheless, the string has a wave shape at any particular instant. For example, Figure 14.6 shows what the third standing wave of Figure 14.5 would look like if it were photographed in a short-exposure snapshot. Each end of the string must be a point of nearly zero vibration. Within these fixed end points, the string is vibrating in one, two, three, and four loops in each of the successive photographs of Figure 14.5. Notice that only certain standing waves can fit onto the string, namely, those for which an integral number (1, 2, 3, 4, and so on) of loops fit onto the string's length.

Returning to the psi-field of an electron in a hydrogen atom, since the probability wave is expected to be stationary, we expect that the electron's psi-field must form a standing wave pattern. But this standing wave will not stretch along a straight line as in Figure 14.5; instead the hydrogen atom's spherical shape causes the psi-wave to form a circular standing wave, as shown in Figure 14.7(a). Just as the standing waves of Figure 14.5 must just fit onto the string, the standing psi-wave must just fit around the nucleus.

FIGURE 14.5

Four standing waves on a string. In the first state of vibration, the string vibrates in one segment; in the second state, it vibrates in two segments with one stationary point (which does not vibrate) in the middle; and so forth.

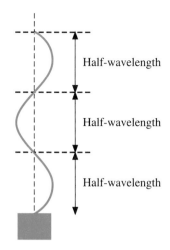

FIGURE 14.6
A "snapshot" of the third standing wave in Figure 14.5, at a single instant.

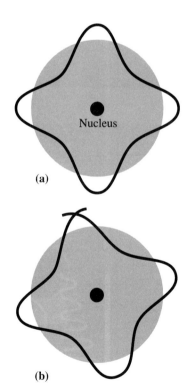

(a)

(b)

FIGURE 14.7
(a) A standing wave that just fits around the nucleus. (b) A standing wave that doesn't fit; its wavelength is a little longer than the wavelength in (a).

Figure 14.7(a) shows a psi-wave that fits properly, and Figure 14.7(b) shows one that does not and so is not allowed.

It isn't surprising that the Schroedinger equation for a hydrogen atom predicts that the electron's psi-field can have any one of several different allowed standing-wave shapes. These, and only these, wave patterns are allowed. At any one time, the electron is represented by one, and only one, of these **quantum states** of the hydrogen atom.

Figure 14.8 is one way of picturing a few of these quantum states. Each of the ten patterns shown is one of the allowed stationary probability patterns of the single electron in a hydrogen atom. To visualize the full three-dimensional probability patterns, imagine rotating the two-dimensional diagram around the vertical z axis shown in each diagram.

As we know, a single electron is always observed as a tiny particle, not a spread-out wave. The pattern shows only the probability that the electron will be observed to be at various places. The diagram represents this probability by darkening, and so the electron is more likely to be found at places where the diagram is darker. Physicists find it convenient to label these quantum states by means of three numbers called **quantum numbers**, indicated as N, L, and M in the figure.

For example, the state labeled $N = 1$, $L = 0$, $M = 0$ occupies a smaller volume than does any other state. In this state, the electron is highly likely to be found close to the nucleus and is equally likely to be found in any direction out from the nucleus (upward, downward, to the left, and so forth).

The state $N = 2$, $L = 0$, $M = 0$ is larger, so the electron is likely to be found farther from the nucleus than is an electron in the state $N = 1$, $L = 0$, $M = 0$. This state has an interesting gap partway out from the nucleus, representing a distance from the nucleus at which the electron will never be found. It is interesting that an electron in this state can be found inside or outside this distance but never at this distance. How can the electron get from the inside to the outside without sometimes being at this distance? Paradoxical questions like this are related to wave–particle duality: The electron is in some sense spread out on both sides of the gap, even though the electron is always observed at only one point.

The state $N = 3$, $L = 0$, $M = 0$ is larger still. The electron is likely to be found still farther from the nucleus. There are now two gaps where the electron will not be found.

Unlike the states mentioned so far, states having $L = 1$ or $L = 2$ are not the same in every direction. The state $N = 2$, $L = 1$, $M = 1$ is shaped like a fat doughnut circling the z axis and is reminiscent of the planetary model of the atom. The state $N = 2$, $L = 1$, $M = 0$ is shaped like a dumbbell (two spheres) along the z axis. It is separated into two parts, between which the electron is not found.

Figure 14.8 shows all the states having N values equal to 1, 2, or 3. There is another set of states having $N = 4$, another set having $N = 5$, and so forth. There is an infinite number of quantum states of hydrogen.

These diagrams represent the patterns of hydrogen, nature's simplest and most prevalent atom. Each pattern represents one way that a hydrogen atom can exist. Atoms with more than one electron have more complex quantum states, but they all are found by solving the Schroedinger equation in the

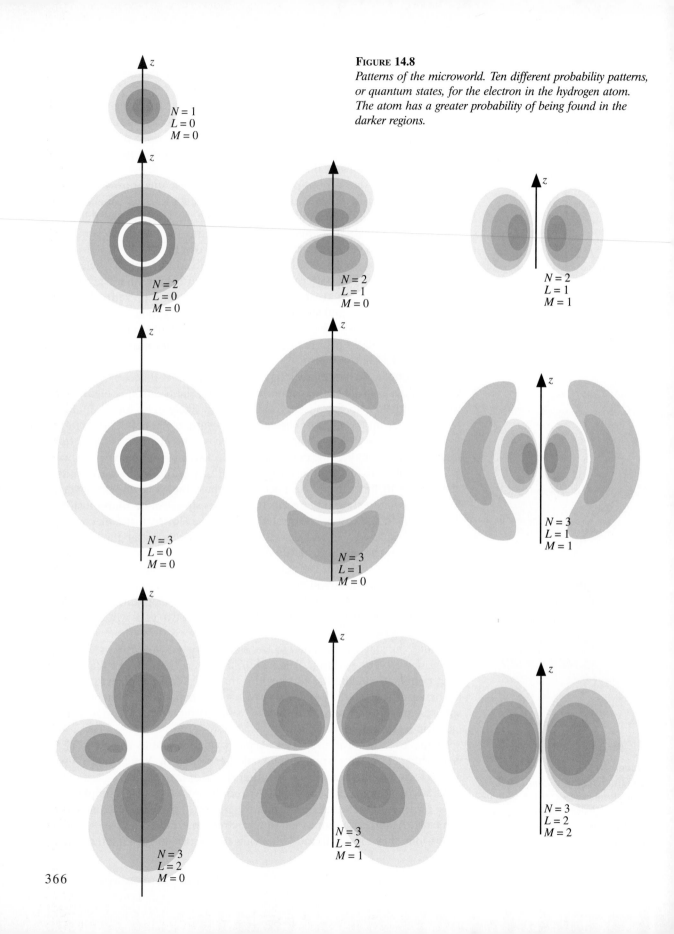

FIGURE 14.8
Patterns of the microworld. Ten different probability patterns, or quantum states, for the electron in the hydrogen atom. The atom has a greater probability of being found in the darker regions.

$N = 1$
$L = 0$
$M = 0$

$N = 2$
$L = 0$
$M = 0$

$N = 2$
$L = 1$
$M = 0$

$N = 2$
$L = 1$
$M = 1$

$N = 3$
$L = 0$
$M = 0$

$N = 3$
$L = 1$
$M = 0$

$N = 3$
$L = 1$
$M = 1$

$N = 3$
$L = 2$
$M = 0$

$N = 3$
$L = 2$
$M = 1$

$N = 3$
$L = 2$
$M = 2$

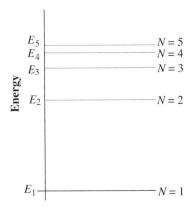

FIGURE 14.9

The lowest five energy levels for the electron in the hydrogen atom. When measured in joules, these atomic energy levels are quite small: The energy difference, $E_2 - E_1$, between the lowest two levels is only 1.6×10^{-18} joules.

You surely must understand, Bohr, that the whole idea of quantum jumps necessarily leads to nonsense. . . .If we are still going to have to put up with these damn quantum jumps, I am sorry that I ever had anything to do with quantum theory.

    Schroedinger, during a conversation
                    with Niels Bohr

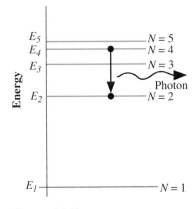

FIGURE 14.10

A symbolic representation of a quantum jump from an N = 4 quantum state to an N = 2 quantum state. A photon, carrying energy $E_4 - E_2$, is given off at the instant the quantum jump occurs.

form appropriate to that particular atom and requiring that the psi-field just fit around the nucleus.

Since each of these states is a standing wave with one particular frequency and since the frequency of a wave is related to its energy, we expect each state to have just one specific energy. That is, we don't expect the energy of any one of these states to have any quantum randomness. From Figure 14.8, we can make some educated guesses about the amount of energy present in each state. Because the force by the proton on the electron is attractive, one would have to do work to pull an electron outward, away from the nucleus. So the electromagnetic energy of the electron goes higher as the electron travels farther from the nucleus. This is just like gravitational energy: Because Earth exerts an attractive gravitational force on a rock, the rock's gravitational energy increases as the rock travels farther from Earth.

Judging from Figure 14.8, the N = 1 state has the lowest energy. Because of the gravitational analogy, this is called the **ground state**. It is the state in which the electron is as close to the nucleus as it can be. Other states are called **excited states** because they are more energetic. Because all the N = 2 states have about the same average distance from the nucleus, we expect them to have about the same energy. Similarly, it appears that all the N = 3 states have about the same energy, an energy that is higher than that of the N = 2 states.

The precise energy of each quantum state can be calculated using Schroedinger's equation, which shows that the energies increase with increasing N, as depicted in Figure 14.9. The energy values E_1, E_2, and so on are labeled by their N values 1, 2, and so forth. A diagram like this, showing the energies of various quantum states, is called an **energy-level diagram**.

Each of these standing-wave states represents an isolated hydrogen atom that isn't changing. What happens when something does change? What happens, for example, when a hydrogen atom emits radiation?

As we know, radiation comes in bundles of energy, photons. An atom must emit at least one photon whenever it radiates. In emitting a photon, an atom must lose energy, so it must be in an excited state to begin with, and it must change to a lower-energy state. This transition from one quantum state to another must be instantaneous, because the conservation of energy demands that as soon as the photon's energy is emitted, the atom's energy must drop by exactly this amount. Such an instantaneous transition from one quantum state to another is called a **quantum jump**.

Figure 14.10 shows a common way of representing quantum jumps. The transition is shown as an arrow on an energy-level diagram stretching from the initial to the final energy. The diagram represents an atom making a transition from the N = 4 to the N = 2 energy level and also indicates that a photon is emitted, carrying away the energy. In Figure 14.8, imagine that an N = 4 (not shown) pattern vanishes instantaneously, to be replaced by an N = 2 pattern. The hydrogen atom truly jumps from one pattern to another.

Now we can understand, in terms of quantum theory, the **emission of radiation** by atoms. Atoms emit radiation when they quantum-jump to a lower energy level, creating and emitting a photon in the process.

When a hydrogen atom quantum-jumps from some higher-energy state to a lower-energy state, it emits a photon of radiation having energy hf. Because of the conservation of energy, the emitted energy must equal the energy difference in the quantum jump; that is,

hf = (energy of high-energy state) − (energy of low-energy state)

So if you know the two energy levels, you can find the frequency of the photon emitted in a quantum jump between these two levels. Physicists can predict the frequencies that hydrogen atoms can emit, by finding the allowed energies of hydrogen, as predicted by the Schroedinger equation, and then finding the frequency of the photon emitted in each possible quantum jump between pairs of energy levels.

HOW DO WE KNOW? SCHROEDINGER'S EQUATION AND THE SPECTRUM OF HYDROGEN

Figure 14.11 shows the ten downward quantum jumps that are possible between the lowest five energy levels for hydrogen. Since the photon's energy is equal to the atom's energy change, the length of the arrow representing each quantum jump is proportional to the frequency of the radiation emitted in that quantum jump. So a hydrogen atom can emit ten different frequencies by quantum-jumping from the N = 2, 3, 4, or 5 levels downward into a lower level. Figure 14.12 shows these ten frequencies quantitatively. Since Schroedinger's equation can determine the energy levels (Figure 14.11), it can also determine these frequencies.

FIGURE 14.11

The ten possible downward quantum jumps between the lowest five energy levels of the hydrogen atom.

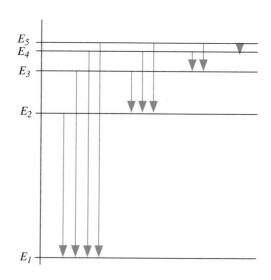

FIGURE 14.12

The frequencies of the photons that are given off during the ten quantum jumps shown in Figure 14.11.

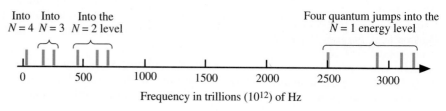

If one uses a spectroscope (Section 14.1) to measure the spectrum of atomic hydrogen gas, the frequencies of the photons will turn out to be precisely the frequencies indicated in Figure 14.12 and predicted by quantum theory. Schroedinger's equation first gained fame because Schroedinger was able to show that it correctly predicted these frequencies. Physicists knew that there must be something right about it if it could predict such precise numbers.

The absorption of radiation is similar to emission, only backward. A photon from the environment enters an atom. If the photon's energy just matches one of the atom's possible quantum jumps, the atom may quantum-jump upward to a higher energy level, absorbing the photon in the process.

The quantum theory of more complex atoms and of molecules is similar. The carbon atom, the hydrogen molecule, and indeed a DNA molecule all could be studied by starting from the Schroedinger equation for this atom or molecule and finding the allowed quantum states and their energies. This procedure gets difficult for more complex molecules, but the same quantum principles apply.

Quantum theory deals very simply with the collapsing-atom paradox posed by the planetary model. Once it enters its lowest-energy state, $N = 1$ in Figure 14.8, an atom cannot emit radiation because it has no lower-energy state to quantum-jump to. The reason is that no smaller standing waves will fit, so this is the smallest the electron's psi-field can become.

DIALOGUE 3   Describe the three-dimensional shapes of the three quantum states for hydrogen with $N = 3$ and $L = 2$.

DIALOGUE 4   Among the ten quantum jumps between the five energy levels of hydrogen shown in Figure 14.11, which one will create the photon with the highest frequency? Which will create the lowest frequency?

DIALOGUE 5   How many different frequencies can be created by quantum jumps among only the lowest six energy levels of hydrogen?

14.3 The uncertainty principle: *a particle's realm of possibilities*

Quantum theory's central idea is that everything is made of unpredictable particles whose statistical behavior follows a predictable wave pattern. In 1927, German physicist Werner Heisenberg (Figure 14.13) found that this wave–particle duality implies that the microworld has an inherent uncertainty that can be quantified.

Let's consider just one electron moving through empty space, along a direction that we will call the x axis. Since the electron's position x is unpredictable, its psi-field must be spread out somewhat along the x axis, as indicated in Figure 14.14. The spread in possible positions, called the **uncertainty**

There is no part of chemistry that does not depend, in its fundamental theory, upon quantum principles.
Linus Pauling, chemist

FIGURE 14.13
Werner Heisenberg.

3. A small doughnut circling the nucleus, plus a dumbbell along the z axis. Two doughnuts circling the z axis, above and below the nucleus. A fat doughnut circling the nucleus.
4. $N = 5$ to $N = 1$. $N = 5$ to $N = 4$.
5. 15.

in position, is indicated by the symbol Δx (delta x), the range of possible places where we might find the electron if we precisely measured its position. A traveling psi-field like Figure 14.14, that extends over only a limited distance, is called a **wave packet**. It is a standard way of representing a moving particle whose position is somewhat uncertain.

Uncertainty is a common notion, usually having nothing to do with quantum theory. For example, suppose you hike along a straight road for 10 kilometers, give or take 1 kilometer. Then your uncertainty in your position x (measured perhaps from your starting point) is $\Delta x = 2$ km.

What kinds of traveling waves does quantum theory allow? Since the electron is moving freely, it is not surprising to find that Schroedinger's equation allows psi-fields of the type that de Broglie had in mind when he first proposed that freely moving particles had waves associated with them. Several such allowed psi-fields are shown in Figure 14.15. Each has a different wavelength λ, and different wavelengths correspond to different electron speeds because of de Broglie's formula $\lambda = h/ms$, so each wave corresponds to a different speed for the electron moving along the x axis.

Each allowed wave in Figure 14.15 is infinitely spread out along the x axis, so each psi-field corresponds to an electron for which Δx is infinite! So none of these waves represents a wave packet. Fortunately, Schroedinger's equation also allows combinations of the solutions shown in the figure. By judiciously combining different single-wavelength solutions, it is possible to get the different waves to cancel out one another everywhere along the x axis, except in a limited region Δx, and so to obtain physically plausible wave packets like the one shown in Figure 14.14, representing an electron known to lie within a range Δx along the x axis.

But now look at what has happened: In order to get a wave packet representing an electron whose position is localized along a portion of the x axis, we had to combine waves having a range of different wavelengths, and different wavelengths correspond to different speeds for the electron.

FIGURE 14.14
The psi-field representing a single particle whose uncertainty in position is Δx, moving along the x axis. A psi-field like this, which is spread out over only a limited distance, is called a wave packet.

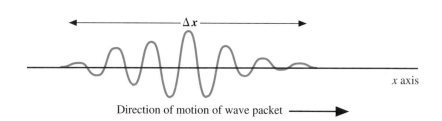

FIGURE 14.15
Four different psi-fields, each representing a single freely moving particle. The different wavelengths correspond to different speeds of the particle.

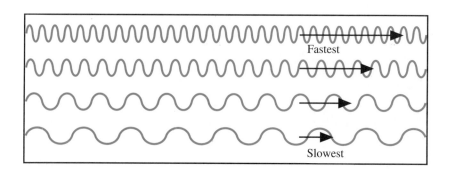

So a wave packet represents an electron whose speed is uncertain. A particle's range of possible speeds is called its **uncertainty in s**, abbreviated Δs.

Here is a simple example, again having nothing to do with quantum theory: If you glance at your car's speedometer and read it as "75 km/hr, give or take 2 km/hr," then your uncertainty in speed is $\Delta s = 4$ km/hr.

So every wave packet representing a moving electron has two kinds of uncertainties: Δx and Δs. Let's compare one wave packet A with another wave packet B that has been squeezed into half of A's length (Figure 14.16). As you can see, B's wavelengths are shorter, and we know that this means larger speeds (Figure 14.15). It seems plausible (and it can be shown*) that larger speeds mean a larger *uncertainty* in speed and that in fact the halving of Δx implies a doubling of Δs.

This illustrates a general feature of wave packets: Whenever Δx is squeezed by some amount, Δs expands by the same amount. And vice versa: squeezing Δs expands Δx. Quantitatively, this means that the *product* of the two uncertainties, $\Delta x \cdot \Delta s$, remains unchanged. We refer to a particle's Δx and Δs, considered together, as the particle's **realm of possibilities**.[†]

Working through these ideas in detail, Heisenberg found that this rule holds for every particle (not just electrons) in every physical situation (not just when moving freely). This idea is called:

HEISENBERG'S UNCERTAINTY PRINCIPLE

Every material particle has inherent (unavoidable) uncertainties in position and in speed. Although either one of these uncertainties can take on any value, the two uncertainties are related by the fact that their product must approximately equal h/m. In symbols,

$$(\Delta x) \cdot (\Delta s) \approx h/m\text{[§]}$$

where h is Planck's constant and m is the particle's rest mass.

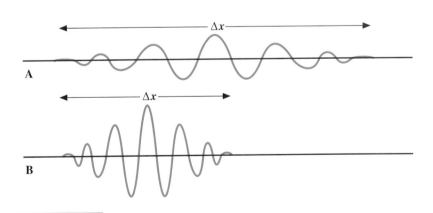

FIGURE 14.16
Two wave packets, having different Δx's. Packet B can be constructed by squeezing packet A to half its size. In this process, all of A's wavelengths get squeezed to half their original length, which means that the speeds and also the uncertainty in speed get doubled.

---

* Since B is squeezed to half of A's length, B's wavelengths are half as long as A's. So B's component speeds are twice as big as A's, because wavelength and speed are inversely proportional. So the range of speeds, Δs, is twice as big for B as for A.

† This nice phrase is from Nick Herbert's wonderful nontechnical account of the meaning of quantum theory, *Quantum Reality* (New York: Doubleday/Anchor, 1985).

§ More precisely, $(\Delta x) \cdot (\Delta s) \geq h/4\pi m$. The product can be greater than $h/2\pi m$ but not less.

The uncertainty principle tells us that any particle preserves its overall realm of possibilities in position and speed; any reduction in one uncertainty must expand the other by an equal amount.

We can visualize a particle's realm of possibilities in a diagram such as Figure 14.17(a), where a single point represents a particle's precise position x and speed s. Newtonian physics assumes that every object has a precise x and s. For example, the location and motion of the center of a baseball can be described, according to Newtonian physics, by a particular x representing the center's location and a particular s representing the center's speed. Newton's law of motion is basically a method for predicting an object's future x and s from its present x and s. For example, given the position and speed of the center of a falling baseball at one time, we can predict its position and speed at any later time during the fall.

But quantum theory does not allow such precision. Quantum theory demands that an object's x and s have uncertainties Δx and Δs whose product is roughly h/m. In an x-versus-s diagram, this product is the area formed by the rectangle whose sides are Δx and Δs, as shown in Figure 14.17(b). If for any reason Δx is reduced, then Δs must expand to fill up the same overall

FIGURE 14.17
Position and speed uncertainties.

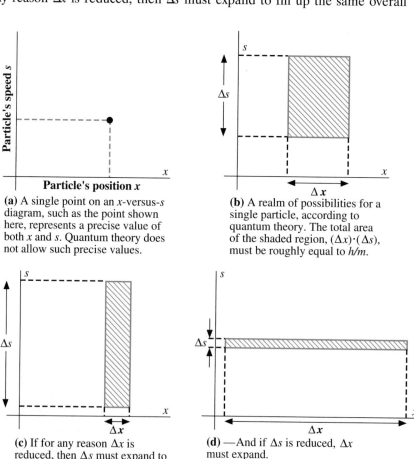

(a) A single point on an x-versus-s diagram, such as the point shown here, represents a precise value of both x and s. Quantum theory does not allow such precise values.

(b) A realm of possibilities for a single particle, according to quantum theory. The total area of the shaded region, $(\Delta x) \cdot (\Delta s)$, must be roughly equal to h/m.

(c) If for any reason Δx is reduced, then Δs must expand to fill up the same overall realm of possibilities.

(d) —And if Δs is reduced, Δx must expand.

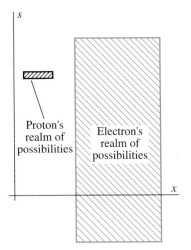

FIGURE 14.18
Because of its larger mass, a proton's uncertainties are much smaller than an electron's.

realm of possibilities, as shown in (c). And if Δs is reduced, Δx must expand as in (d). Either x or s can be as highly predictable as you like, but they both cannot be highly predictable. If one is highly predictable, the other must be highly uncertain.

Since the uncertainty principle says that $(\Delta x) \cdot (\Delta s)$ is inversely proportional to m, more massive particles have smaller realms of possibilities. A proton's realm, for example, is 2000 times smaller than an electron's realm (Figure 14.18). Because x and s are what we need to predict an object's future behavior, a proton is much more predictable than an electron is.

A 1-kilogram baseball, one million trillion trillion times more massive than an electron, is so predictable that quantum uncertainties can be neglected. That is why Newtonian physics works just fine for large objects like baseballs. Even a grain of sand is so massive (it contains some 10^{18} atoms) that quantum uncertainties are negligible.

Suppose a particle's Δx has been squeezed into a very small uncertainty. This particle must then have a large Δs. But you can't have a large Δs without at the same time having a large s; for instance, if Δs were 1000 km/s, the lowest (slowest) realm of possibilities for s would be 0 to 1000 km/s, so the average s must be at least 500 km/s. So when Δs is large, s must be large too. This means that a highly confined particle (Δx small) must move fast. The smaller the confinement, the faster the motion. The microscopic world cannot sit still!

For example, the protons and neutrons in a nucleus are confined to a region 10,000 times smaller than the atom itself, by the strong nuclear forces that hold the nucleus together. So these protons and neutrons must move rapidly, which is why nuclear energies must be large.

The uncertainty principle prevents atoms from collapsing. As we have seen, the planetary atom has the grave defect that it should collapse. The uncertainty principle prevents this: As an electron radiates energy and falls inward, it is confined to a smaller region. This process of losing energy cannot continue indefinitely because the particle would eventually not be able to fill out its realm of possibilities: The particle's uncertainty Δx would eventually become so small that the uncertainty principle could not be satisfied.

Quantum uncertainties lie at the heart of the nuclear phenomenon known as radioactive decay (Chapter 15) and cause this process to be fundamentally unpredictable.

When a child is conceived, the DNA molecules of each parent are randomly combined in a process in which the quantum features of the DNA's chemical bonds play a role. So quantum uncertainties operating at the microscopic level play a role in our genetic inheritance.

In these and other ways, we are in the hands of the god who plays dice.

DIALOGUE 6   Arrange these objects in order, beginning with the object having the largest realm of possibilities and ending with the one having the smallest: proton, glucose molecule $C_6H_{12}O_6$, helium atom, baseball, electron, grain of dust, water molecule, automobile.

---

6. Electron, proton, helium atom, water molecule, glucose molecule, grain of dust, baseball, automobile.

14.4 Quantum jumps

Quantum theory continues to surprise and perplex physicists. Einstein was among those who found the theory too counterintuitive to believe. He and two other physicists showed in 1935 that because of quantum uncertainties, quantum theory predicts some phenomena that are, as he put it, so "spooky" that "no reasonable definition of reality could be expected to permit this." Einstein and others took these predictions as evidence that a correct theory would not contain quantum uncertainties. However, Einstein did not suggest a way to put quantum theory's spooky predictions to an experimental test.

Because quantum theory proved so gloriously successful in practice, few physicists worried much about such untested objections. Among those who did worry were David Bohm and John Bell (Figure 14.19). Bohm began publishing his theoretical analysis of quantum theory during the 1950s. Working from these ideas, Bell showed in 1964 that some of quantum theory's spooky predictions are experimentally testable. John Clauser (Figure 14.19) and four collaborators* carried out the first such test in 1972 and found that contrary to the expectations of Einstein and others, the spooky phenomena actually occur! In 1982, Alain Aspect (Figure 14.19) refined Clauser's test so as to leave little doubt that the real world is stranger than Einstein and others had thought.

All the spooky predictions are related to the psi-field, especially its sudden changes. Suppose, for example, that a freely moving electron's psi-field is a wave packet (Figure 14.14), with a position uncertainty Δx. Suppose we then measure the electron's position to within an accuracy better than Δx, perhaps $\Delta x/2$. Then just after this measurement, the electron must be represented by a new wave packet having position uncertainty $\Delta x/2$. The measurement caused the electron's psi-field suddenly to change. Such sudden changes in a psi-field are called **quantum jumps.**

For another example, we return to the double-slit experiment using electrons.[†] In Chapter 13, Figure 13.19(d) shows an electron's psi-field just before the electron hits the screen, along with the possible impact points and their probabilities. But once the particle actually hits the screen, it is suddenly "localized" at the impact point. So its psi-field jumps instantaneously, from the spread-out pattern in Figure 13.19(d) (this is what is graphed on the screen in the figure) to a small region at the impact point.

Quantum jumps can affect quite large regions. For example, it happens that the psi-field for each photon from any very distant star is spread out over many kilometers by the time it reaches Earth. British physicist Robert Hanbury Brown confirmed this prediction in 1965 by measuring, for the light from an individual star, interference patterns that were over 100 meters in diameter. Despite the psi-field's large size, it instantaneously contracted to a point when the photon hit a detector.

For another example of quantum jumps, activation of a particle detector behind one slit in the double-slit experiment destroys the interference pattern

---

[*] Michael Horne, Abner Shimony, and Richard Holt collaborated on the design of the experiment, and Stuart Freedman collaborated on its execution.

[†] Most of the following experiments have been done using photons rather than material particles such as electrons, because the experiments are easier with photons. But the quantum predictions are the same for electrons or other material particles, and physicists have little doubt that the results would be the same for electrons.

FIGURE 14.19

Four explorers of quantum theory: Clockwise from upper left: David Bohm, John Bell, Alain Aspect talking with Bell (r.) and physicist Albert Messiah, John Clauser.

(Chapter 13). Merely switching on the detector causes the psi-field to jump from the interference pattern of Figure 14.20(a) to the noninterference pattern of Figure 14.20(b). As a normal or Newtonian explanation of this strange effect, perhaps the detector interacts with each particle as it comes through a slit, forcing the particle to follow pattern (b) rather than pattern (a). To check this, experimenters placed the detector far behind the slits, near the screen, as in Figure 14.21. With the detector far from the slits, it is implausible that the detector could exert forces on the particle as it comes through the slits. And yet when the detector is on, we again get pattern (b) instead of (a).

FIGURE 14.20

Merely switching on a particle detector at a point such as D causes the psi-field to jump from the interference pattern (a) to the noninterference pattern (b).

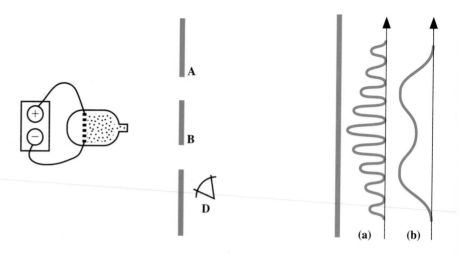

FIGURE 14.21

Even if the detector is placed far behind the slits, near the screen, the pattern still jumps from pattern (a) to (b) whenever the detector is activated.

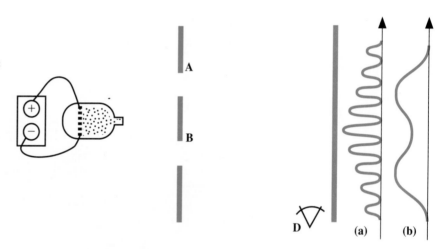

The detector can even be switched on after the particle is well past the slits, and the interference pattern still will jump to the noninterference pattern at the instant the switch is thrown! It is as though the particle comes through both slits (at least, its psi-field comes through both slits) right up until the experimenter decides to switch on the detector; then the particle "changes its mind" and decides that it actually came (in the past!) through only one slit.

We see here hints of an effect known as *nonlocality*. The psi-field changes instantaneously, all over the screen, the moment that the detector is switched on. The detector seems intimately connected with the particle's psi-field, instantaneously, everywhere on the screen. Such an instantaneous connection, across a distance, is called a **nonlocal effect**.

HOW DO WE KNOW? MANDEL'S EXPERIMENT ON THE NONLOCAL EFFECT OF DETECTORS

To what extent does the mere detection of an event cause changes in related events some distance away? In 1991 Leonard Mandel and coworkers conducted a striking test of quantum theory's predictions about the effect that detectors have on distant events.

Figure 14.22 shows Mandel's experimental arrangement. A source emitted a single particle whose psi-field moved along two paths labeled path A and path B. This single particle was then converted into *two* particles, 1 and 2, each moving along two possible paths, labeled 1A, 2A, 1B, 2B. Particle 1 moved along either path 1A or path 1B through a double-slit apparatus having slits A and B, and particle 2 moved along either path 2A or path 2B toward a detector. As long as the experimenters had no way of knowing which slit (A or B) particle 1 came through, the probability-wave for particle 1 came through both slits and so the impact points on the screen formed the interference pattern (a), just as in the double-slit experiment with electrons (Figure 13.13(b)).

The unique feature of this arrangement is that the two particles must be *either* on paths 1A and 2A, *or* on paths 1B and 2B. That is, the two particles' psi-fields were connected with each other or "entangled" (see the following section) in such a manner that, if particle 1 was on path 1A, then particle 2 was necessarily on path 2A rather than 2B; and if particle 1 was on path 1B, then particle 2 was necessarily on path 2B.

Now suppose that a barrier is placed in the way of path 2A, as shown, and that a single particle is emitted from the source. Then if we see a flash on the screen and also detect a particle hitting the detector, we can conclude that particle 2 came along path 2B, so the flash on the screen was caused by a particle moving along path 1B. And if we see a flash on the screen but we do *not* detect a particle hitting the detector, we can conclude that particle 2 came along the blocked path 2A, so particle 1 must have moved along path 1A.

In other words, with the barrier in place, the detector placed along one of the paths of particle 2 can determine which slit particle 1 came through. On the other hand, if the barrier is removed, then the detector in the path of particle 2 no longer gives any information about particle 1. Note that the barrier and detector interact only with particle 2, and could be quite

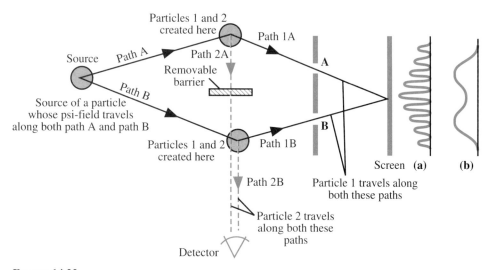

FIGURE 14.22

Mandel's experiment. A barrier placed in path 2A causes the pattern on the screen to quantum-jump from pattern (a) to pattern (b), even though the barrier does not directly disturb the particle that impacts the screen, and even though the barrier and detector can be quite distant from the screen.

distant from the path of particle 1—they could be many kilometers away, or even on the moon or in a distant galaxy. Thus, the detector and barrier constitute an extremely remote and noninterfering detection scheme for particle 1. It is hard to believe that the placement or removal of the barrier along path 2A could affect what happens to particle 1 at the screen.

Yet quantum theory predicts, and the experiment confirms, that when the barrier is placed in path 2A, the interference pattern (a) quantum-jumps to the noninterference pattern (b). Furthermore, the effect persists even when the detector is removed—the mere blocking of path 2A destroys the interference between paths 1A and 1B! Apparently the mere possibility that an observer *could* insert a detector and thus determine whether path 1A or 1B was taken causes the interference pattern to switch to noninterference. We conclude that detectors that can provide information about distant events can change those events even without directly physically interfering with those events.

14.5 The nonlocality principle: *spooky action at a distance*

Except for Mandel's experiment, the quantum uncertainties discussed so far have involved only one particle at a time. For example, in the double-slit experiment with electrons, the essential uncertainty concerned whether a single electron goes through slit A or slit B.

We now consider the possibility of quantum uncertainties involving two or more particles at a time. If two particles are involved in the same microscopic event, for example, if they are created together in a single event or if they physically interact with each other, their psi-fields can become intimately connected to each other. The two particles can become a single quantum system so that there is no psi-field for either particle separately but only a single inseparable psi-field that the two particles share. In a sense, the "two" particles are not really two particles at all but are one object, even though they are in two different places. The particles are then said to be **entangled.** Figure 14.23 shows the entanglement of two particles that start out separately and then interact to become entangled.

HOW DO WE KNOW? THE POSITION-ENTANGLEMENT EXPERIMENT*

This experiment begins with the creation of two entangled particles that move directly away from each other (Figure 14.24). Each particle then passes through a double-slit apparatus. Because of their opposite directions, if particle 1 goes through slit A, particle 2 must go through slit B (the dashed double arrow in the figure), and vice versa (solid arrow). The experiment could be described as a "duplicate double-slit experiment" with two entangled particles. If the two particles were not entangled, each half of the experiment (the left side and the right side of the figure) would yield the usual double-slit interference results discussed previously.

*The experiment was suggested in 1986 by Michael Horne and Anton Zeilinger and performed in 1990 by John Rarity and Paul Tapster; a similar experiment was performed by Z. Y. Ou and Leonard Mandel in 1989. The description presented here simplifies some of the experimental details without changing the physical essentials.

No theory of physics that deals only with physics will ever explain physics. I believe that as we go on trying to understand the universe, we are at the same time trying to understand man. The physical world is in some deep sense tied to the human being.

John Wheeler, physicist

In some strange way, the universe is a participatory universe.

John Wheeler

One is led to a new notion of unbroken wholeness which denies the classical idea of analyzability of the world into separately and independently existing parts. We have reversed the usual notion that the independent "elementary parts" of the world are the fundamental reality. Rather, we say that the interconnectedness of the whole universe is the fundamental reality, and that the "parts" are merely particular and contingent forms within this whole.

David Bohm

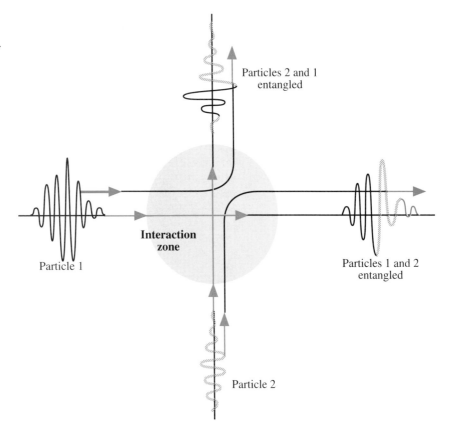

Figure 14.23
When two particles interact and then separate, it is possible for their psi-fields to become entangled.

Figure 14.23
When two particles interact and then separate, it is possible for their psi-fields to become entangled.

Particles 2 and 1 entangled

Particle 1

Interaction zone

Particles 1 and 2 entangled

Particle 2

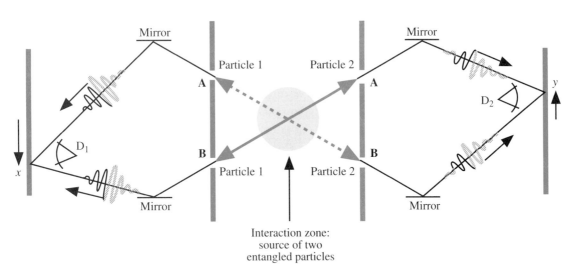

Mirror

Particle 1

Particle 2

Mirror

A

A

D_2

y

D_1

x

B

B

Particle 1

Particle 2

Mirror

Mirror

Interaction zone: source of two entangled particles

Figure 14.24
The position-entanglement experiment. Because of their entanglement, particles 1 and 2 coordinate their impact points x and y instantaneously, regardless of the distance between them.

In this experiment, the entangled particles must separate in opposite directions. Because of this, there are so-called *statistical correlations* between the impact point of particle 1 on its screen and the impact point of the particle 2 on its screen.

Statistical correlations are common in situations involving uncertainties. For a typical example—having nothing to do with quantum theory—suppose your friend tells you that he has sealed a gold coin and a silver coin in separate envelopes and mailed one to you in Tokyo and one to Betty in Paris. Even though you have not yet received your envelope in Tokyo, you do know of a statistical connection between your envelope and Betty's envelope: If your envelope contains gold, Betty's will contain silver, and if your envelope contains silver, Betty's will contain gold. Such a connection between the probabilities of different events is called a **statistical correlation**. Two different events are correlated if the outcome of one affects the probabilities associated with the other. Note that in this example, the correlation is not the result of any actual physical interaction between the two coins in the two cities; that is, neither coin causes the other actually to change from gold to silver. The correlation is due only to the fact that your friend put a gold coin in one envelope and a silver coin in the other.

In the position-entanglement experiment, detectors (D_1 and D_2) were placed at each screen. Each detector monitored one point on the screen and registered a hit whenever a particle struck at (or very near) that point. The experimenters measured the degree of correlation between simultaneous hits and misses on the two detectors. Given a hit on, say, D_1, how likely is a simultaneous hit on D_2?

Because the two particles separate in opposite directions, we expect some correlation. For instance, if the distances x (of D_1 below the midpoint of its screen) and y (of D_2 above the midpoint of its screen) are equal to each other, then hits on D_1 should usually (but not always, because of quantum uncertainties) occur simultaneously with hits on D_2, because the two particles have opposite directions. Such a correlation is similar to the gold and silver coin correlations and is due simply to the previously opposite directions of the two particles.

The interesting correlations occur when x and y differ from each other. Quantum theory predicts an interference pattern for these correlations: If D_1 is held fixed at any point x while D_2 is moved from one position y to another, quantum theory predicts positions where hits on D_2 are particularly likely to occur simultaneously with hits on D_1, and other positions where misses on D_2 always occur simultaneously with hits on D_1. The second case is the most interesting: Misses on D_2 always occur when x and y differ by certain fixed amounts, such as 0.5, 1.5, or 2.5 mm. That is, the two particles somehow "know" that they are not supposed to impact at points x and y that differ by these particular amounts. *How can they "know" that?* Each particle's interference pattern, for a fixed impact position of the other particle, is dependent on where the other particle impacts its screen. As another way of putting this: Each particle's psi-field quantum-jumps into a new pattern when the other particle impacts at a particular point—the two particles instantaneously adjust their psi-fields to each other's impact point! The two particles are truly entangled with each other—anything that happens to one of them affects the two-particle psi-field and hence simultaneously affects the other particle.

The two screens can be as widely separated as you like; they could be in different galaxies, yet quantum theory predicts the same results. Each entangled particle "knows" instantaneously what the other is doing. They coordinate their impact points so as not to hit when the difference between x and y is, say, 0.5 mm. If particle 1 happens to strike at $x = 0.3$ mm, particle 2 will instantaneously "know" that it must not hit at $y = 0.8$ mm, 1.8 mm, 2.8 mm, and so on, and if particle 1 happens to hit at $x = 0.4$ mm, particle 2 will not hit at $y = 0.9$ mm, 1.9 mm, 2.9 mm, and so forth. How can they cooperate this way when they are far apart?

Maybe this cooperation is not spooky. Maybe it is merely of the "gold-and-silver-coin" variety, due entirely to the prior connection. John Bell analyzed this question in 1964 and proved that the correlations are not of this variety. In other words, Bell proved that this cooperation is due to a real, instantaneous, physical contact between the particles—each particle really does "know" what the other particle is doing from one instant to the next. The cooperation occurs because the two particles behave as a single entity; they are in a single quantum state so that they both must quantum-jump at the same time. We summarize Bell's idea as follows:

> **BELL'S NONLOCALITY PRINCIPLE**
> Quantum theory predicts that entangled particles exhibit correlations that can be explained only by the existence of real nonlocal (that is, instantaneous and distant) connections between the particles.*

Bell was able to discover quantum predictions involving correlations of entangled particles that required a nonlocal explanation and that could be experimentally tested. As just mentioned, Clauser and then Aspect were the first to carry out such tests. Aspect's experiment included a refinement that showed that the connections really were instantaneous, or at least faster than light. The position-entanglement experiment is an example of such nonlocal connections between entangled particles, whose results fully confirm the "spooky" quantum predictions.

Such a conclusion might seem to contradict relativity theory's prohibition on faster-than-light motion. But relativity says only that energy (in the form of material objects or of radiation) cannot travel faster than light. The statistical correlations referred to in Bell's principle cannot transfer energy, so Bell's principle does not contradict relativity.

DIALOGUE 7   Can two particles be entangled even when they are not exerting forces on each other?

14.6 What does it mean? *quantum theory and reality*

It isn't easy to say what all of this means. Danish physicist Niels Bohr (Figure 14.25) was the leading contributor to our understanding of quantum theory. In

---

* The proof of this is surprisingly nontechnical. See, for instance Nick Herbert, *Quantum Reality* (New York: Doubleday/Anchor, 1985), pp. 215–223.

7. Yes, once entangled particles have separated sufficiently widely that they exert no forces on each other.

FIGURE 14.25
Niels Bohr's engagement photograph, taken in 1911. Two years later he developed the first accurate theory of atomic structure.

There is no quantum world. There is only an abstract physical description. It is wrong to think that the task of physics is to find out how nature is. Physics concerns what we can say about nature.

Niels Bohr

There are two sorts of truth: trivialities, where opposites are obviously absurd, and profound truths, recognized by the fact that the opposite is also a profound truth.

Niels Bohr

When it comes to atoms, language can be used only as in poetry. The poet, too, is not nearly so concerned with describing facts as with creating images.

Niels Bohr

1913, he invented a partially quantized version of the planetary atom, a useful stepping-stone to the later full quantum theory of the atom (Sections 14.1 and 14.2). During the 1920s and 1930s, Bohr, along with Heisenberg, Born, and others, developed a philosophy of quantum theory, a view of what the theory means, that has withstood the test of time and new experiments such as those showing nonlocal connections. This view is generally, but not universally, accepted among physicists today. The remainder of this chapter presents an updated version of this now standard **Copenhagen interpretation**.

Perhaps the most characteristic feature of quantum theory is its uncertainties. The essence of quantum uncertainty is that identical physical situations lead to different outcomes. For example, if we send 10 electrons, one at a time, through a double-slit experiment, they will hit at 10 different places, even though the 10 trials are identical. The Copenhagen interpretation concludes that microscopic events really are unpredictable, that God really does play dice.

When we say that the position of a microscopic particle is uncertain, we do not mean simply that we lack knowledge of its position. Rather, we mean that the particle actually *has* no definite position. A particle described by a wave packet (Figure 14.14), for example, has no definite position and no definite speed. The particle is in some sense all over its entire psi-field; it fills up its entire realm of possibilities $\Delta x \cdot \Delta s$. As Heisenberg put it, a particle's psi-field "introduces something standing in the middle between the idea of an event and the actual event, a strange kind of physical reality just in the middle between possibility and reality."

Now consider the effect of a precise position measurement on a particle described by a wave packet. Such a measurement finds the particle to be at some specific position x. It would be a mistake, however, to conclude that the particle was at or even near x just before the measurement. Instead, we must visualize the particle as residing all over the range Δx before the measurement and that the measurement actually *creates* the particle's position rather than simply discovering it. Measurements to some extent create the properties they detect. A position measurement creates a position, and a speed measurement creates a speed (Figure 14.26).

The uncertainty principle prevents the simultaneous existence of both a precise x and a precise s. It is as though a baseball could be either white or spherical but not both at once. Because the existence of either property precludes the existence of the other, these two properties are said to be **complementary** to each other. When Bohr was knighted as an acknowledgment of his achievements in science and his contributions to Danish culture, he chose as a suitable motif for his coat of arms the Chinese symbol representing a similar complementary relationship of the archetypal opposites yin and yang (Figure 14.27).

The meaning of quantum theory is bound up with observation. Observations can instantaneously and radically change the observed system. When we see a flash on an observing screen, we should not imagine that just before the flash a particle was nearing that point. Before the impact, all the particle's possibilities are live possibilities. If we must talk about it at all, just before its impact the particle is everywhere at once. The impact causes the particle to quantum-jump to a new state and gives the particle a location. We could not predict the particle's position before impact because there was nothing there to predict. The particle *had* no position.

FIGURE 14.26

The effect of a position measurement or of a speed measurement is to create a position and speed for the measured particle.

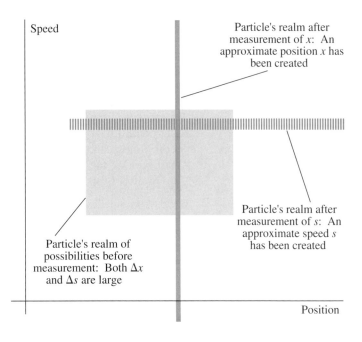

Speed

Particle's realm after measurement of *x*: An approximate position *x* has been created

Particle's realm after measurement of *s*: An approximate speed *s* has been created

Particle's realm of possibilities before measurement: Both Δ*x* and Δ*s* are large

Position

FIGURE 14.27

Niels Bohr's coat of arms. The inscription above the yin–yang symbol reads "opposites are complementary." In this way quantum theory's most profound philosopher acknowledged the harmony between seeming opposites, a harmony contemplated in the ancient Eastern teaching that the essence of all natural and human phenomena lies in the dynamic interplay of complementary opposites.

But "observation" must be understood broadly. Observations are possible without physical disturbances and without human observers. Indeed, we have seen that even the threat of possible detection can cause a quantum jump.

The properties of particles have meaning only in the context of the entire experimental environment, because that environment helps determine a particle's psi-field. An electron's position, for example, refers to the position as it would be defined by some particular position-measuring device. It is improper to think of the electron as having a position in the absence of such a measurement. As Bohr often said, attributes of microscopic particles do not belong to the particle itself but reside in "the entire measurement situation." This view that a particle's properties are created by the experimental context is called **contextual reality**, or **observation-created reality**.

Microscopic particles do not have the same reality status as, say, a penny does. A macroscopic object like a penny is well described by Newtonian physics, so it makes sense to speak of its position and speed in the ordinary sense. But to the extent that an object's behavior must be described by quantum theory, the object has, as Heisenberg put it, "a strange kind of physical reality just in the middle between possibility and reality." But microscopic particles are by no means subjective or only in the mind. For example, the flash of an electron on a screen is real, and it occurs even when no observer is looking (it could, for example, be recorded by an automatic camera). The microscopic world is real, but its reality status is not what we are used to.

Entanglement represents an extreme form of contextual reality. When two particles are entangled, each particle's nature is bound up with the other particle. Each particle becomes the context for the other particle. The two particles form a single experimental situation, even though they may be in different galaxies. They are a single object, only in two different places. Just as we must not think of an electron in a wave packet as really being at one point *x*

To what appear to be the simplest questions, we will tend to give either no answer or an answer which will at first sight be reminiscent more of a strange catechism than of the straightforward affirmatives of physical science. If we ask, for instance, whether the position of the electron remains the same, we must say "no"; if we ask whether the electron's position changes with time, we must say "no"; if we ask whether it is in motion, we must say "no." The Buddha has given such answers when interrogated as to the conditions of a man's self after his death; but they are not the familiar answers for the tradition of seventeenth and eighteenth century science.

J. Robert Oppenheimer

It moves.
It moves not.
It is far,
and it is near.
It is within all this,
And it is outside of all this.
The Upanishads, Hindu sacred script,
ca. 1000 B.C.

[W]e shall always be able to imagine other [false] theories—like the boring world of particles governed by Newtonian mechanics.
Steven Weinberg, physicist, in *Dreams of a Final Theory*

and we must not think of an electron in the double-slit experiment as really coming through one slit, we must not think of entangled particles as separate. Any attempt to do so will run into contradictions with experiment.

This seems to invert the conventional relationship between the microscopic and macroscopic worlds. In the usual view, macroscopic objects such as tables are made of atoms. But quantum theory provides a sense in which tables are more fundamental than atoms, because macroscopic detectors define the conditions of existence for atoms. There is a sense in which atoms depend on macroscopic objects such as screens or tables.

Again quoting Heisenberg:

Some physicists would prefer to come back to the idea of an objective real world whose smallest parts exist objectively in the same sense as stones or trees exist independently of whether we observe them. This however is impossible. . . .Materialism rested upon the illusion that the direct "actuality" of the world around us can be extrapolated into the atomic range. This extrapolation, however, is impossible—atoms are not things.

14.7 Toward a post-Newtonian worldview

Scientifically, humankind is now nearly one century into the post-Newtonian age, but it is unclear where we are philosophically. What is our metaphysical direction today? Let us recall four key ideas of the Newtonian worldview (Section 5.5):

Atomism. Atoms, rather than macroscopic phenomena or human perceptions, form the fundamental reality. Newton called them "solid, massy, hard, impenetrable particles" that "never wear or break in pieces." Tastes, colors, and the like are mere names. For example, a napkin's redness is due to motions of atoms in the napkin and in the observer. Things are not really red in themselves.

Objectivity. Galileo wanted to remove all human influence from his experiments, in order to study nature itself without human influences. It was assumed that objectivity actually was possible, at least in principle. If everything is caused by atoms and if atomic behavior is determined by physical laws, then what is real does not depend on humans. So it should be possible to observe the universe, uninfluenced by humans.

Predictability. Every physical system is entirely predictable. Once started, the clockwork universe was required to do precisely what it has done and what it will do.

Analysis. Ever since Galileo, science has progressed by separating phenomena into their simplest components and studying those components. This led to a focus on the simplest and smallest components: atoms.

Today's physics denies all four of these ideas:

Atomism. Atomism was first contradicted by the electromagnetic field, which is physically real but not made of atoms. The microscopic world, including atoms, appears to be made of psi-fields, but a psi-field is certainly not made of atoms. Atoms themselves are made of energy. Far from being solid, hard, and impenetrable, they are empty. They are made of fields, and their rest mass, their matter, is a consequence of the energy of these fields.

Far from never wearing or breaking, atoms can be annihilated. Although energy is indestructible, matter can be destroyed and created.

Atoms were once viewed as tiny things in the traditional sense. For example, when Ernest Rutherford (Chapter 8) was asked whether atomic nuclei really existed, he replied, "Not exist—not exist! Why I can see the little beggars there in front of me as plainly as I can see that spoon!" But atoms are not things in the same way that a spoon is a thing. An atom's nature is dependent on its surroundings. Descartes argued that secondary qualities such as a napkin's redness are dependent on primary realities such as atoms. Turning this order around, quantum theory views macroscopic objects as the experimental context for the atoms.

Objectivity. The idea of objectivity was crucial to scientific progress in Galileo's and Newton's time. But over time, objectivity took on a mythical status as scientists came to believe that natural phenomena could always be divorced from their surroundings.

Beginning at least with Einstein's relativity in 1905, the observation process intruded on this belief in the possibility of complete objectivity. It became essential to specify exactly how quantities like time and space can be measured. In quantum theory the entire experimental context becomes essential to the defining properties of the objects under study. Does an electron have a position? The answer depends on the experimental context. This does not make physics subjective, because every observer using the same observational equipment still sees the same result. Reality is not quite dependent on the observer, but it is dependent on observation, which is different for every different type of observation.

Predictability. Identical causes no longer lead to identical effects. Individual events are not predictable, even in perfectly controlled experiments. A single radioactive decay, the flash of a photon, and individual chemical reactions such as those that determine a person's genetic inheritance are unpredictable quantum events. The universe is not like a clock. But statistical patterns are predictable, even though single events are not. Chance rules individual events, but nature determines the odds.

Analysis. The analytic process assumes that it is possible to divide a phenomenon into parts without changing it. Quantum theory contradicts this notion. The analytic process works quite well for macroscopic systems. For instance, it is useful to separate the solar system into the sun, planets, and so forth and to consider the ways that each part interacts with each other part. But quantum theory says that we cannot necessarily consider a microscopic system as made of separable parts. The electron double-slit experiment cannot be broken down into an electron, on the one hand, plus the double-slit apparatus, on the other, because the electron is in part defined by the apparatus. Whereas Mars would still be Mars even if Venus did not exist, an electron changes its nature when its environment changes. Quantum entanglement is the most striking example. Two entangled particles are so closely connected that it is not possible even to think of them as independent particles, even though they are on separate planets. There is a microscopic wholeness that is not obvious to our macroscopic eyes.

Despite nearly a century of post-Newtonian physics, a post-Newtonian worldview is still not in sight. What grand metaphors will we eventually adopt

I don't think there's one unique real universe. . . .Even the laws of physics themselves may be somewhat observer dependent.

Stephen Hawking

It would be a poor thing to be an atom in a universe without physicists. . . .A physicist is an atom's way of knowing about atoms.

George Wald, biologist

to replace the Newtonian metaphors? Or do we even need grand metaphors? The Newtonian age certainly had its grand metaphor, one to which nearly all physical scientists subscribed, from Kepler through Maxwell, namely, the clockwork or mechanical universe. Today this metaphor continues to deeply influence our culture's view of physical reality.

Perhaps a biological rather than mechanical analogy would be more fitting. It appears from quantum physics that natural systems are closely tied to their surroundings, that they really possess the macroscopic contextual properties such as redness or hotness that they appear to possess, that they are partly undetermined, and that they cannot be separated without essentially changing their nature. All of this seems similar to our perceptions of biological organisms in interaction with their environments. Perhaps we live in an "organic" universe rather than a "mechanical" universe.

Will we construct a scientifically accurate and humane worldview that can sustain us in the post-Newtonian age, and regain our philosophical roots? Humankind has barely scratched the surface of this task. It might be the critical issue for the modern world.

"I THINK YOU SHOULD BE MORE EXPLICIT HERE IN STEP TWO."

Summary of Ideas and Terms

Spectroscope A device that measures the **spectrum,** or set of frequencies, emitted by a radiation source.

A **continuous spectrum** covers a range of frequencies; a **line spectrum** contains only precise frequencies.

Standing wave A wave in which the medium vibrates in a wave pattern but the pattern does not move.

The **planetary model of the atom,** in which electrons move in planetlike orbits around the nucleus, cannot explain line spectra and predicts that atoms will lose energy until they collapse.

The **quantum model of the atom,** in which electrons are described by standing psi-waves surrounding the nucleus, agrees with all experiments so far.

The **quantum states** of the hydrogen atom are the various possible psi-fields for its electron. Each quantum state is a standing-wave pattern that obeys Schroedinger's equation and represents a possible probability pattern for the atom's electron.

Quantum numbers The numbering system used to label an atom's quantum states. The states of hydrogen are labeled N, L, and M.

The **energy level** of a quantum state is the precise, predictable energy the atom has when it is in that state. An **energy-level diagram** shows the collection of energy levels for an atom.

The **ground state** of an atom is its quantum state of lowest possible energy. Higher-energy states are **excited states.** A gas can be excited by heating and by an **electric discharge** or electric current flowing through it.

Quantum jump An instantaneous change of an entire psi-field. When an atom quantum-jumps, the psi-field for its electron changes to a new quantum state.

Emission of radiation occurs when an excited atom quantum-jumps into a lower energy level, emitting a photon.

Since the photon's frequency is determined by the atom's energy levels, an atom's spectrum can be predicted from Schroedinger's equation.

Heisenberg's uncertainty principle Every material particle has an inherent **uncertainty in position (Δx)** and **uncertainty in speed (Δs).** Although either one of these uncertainties can take on any value, the two uncertainties are related through the fact that their product must approximately equal h/m.

Wave packet A psi-field moving through space and spread out over only a limited distance Δx.

Nonlocal effect An instantaneous connection between two separated points in space.

Mandel's experiment Demonstrates that a detector can have nonlocal effects even if the detector is remote from the particle being detected.

Two particles are **entangled** when they interact and then separate in such a way that they form a single inseparable psi-field.

Two events are **statistically correlated** if the outcome of one affects the probabilities of the other.

The position-entanglement experiment Demonstrates nonlocal connections between two entangled particles, each of which goes through a double-slit apparatus. The statistics of the impact points of the two particles are instantaneously correlated across a distance.

Bell's nonlocality principle Entangled particles cooperate in a way that can be explained only by the existence of real nonlocal connections. Entanglement of this sort is predicted by quantum theory and has been confirmed by experiments.

Copenhagen interpretation The consensus interpretation among physicists. Uncertainties, complementarity, and nonlocality are inherent in nature; observations help create the observed properties; microscopic reality is **contextual.**

Review Questions

OBSERVING ATOMIC SPECTRA

1. The theoretical value of the Lamb shift (in the chapter's introduction) is 1057.860 ± 0.009, and the measured value is 1057.845 ± 0.009 megahertz. Do these two results agree?

2. What is the purpose of the prism in a spectroscope? What is the purpose of the thin slit?

3. Exactly what is measured by a spectroscope?

4. Describe two ways to excite a gas. What does "excited" mean, in microscopic terms?

5. Describe one way in which the planetary model disagrees with observations of atomic spectra. Describe one problem with the concept of the planetary atom.

THE QUANTUM ATOM

6. What are standing waves? How are they related to the quantum theory of the atom?

7. Describe the three-dimensional shapes of some of the states in Figure 14.8. Exactly what does one of these states represent?

8. Which state(s) in Figure 14.8 has the lowest energy? The highest? Which is a ground state? An excited state?

9. Consider any one of the states in Figure 14.8. In this state, does the electron have a predictable energy? A predictable position? A predictable speed?

10. Describe the process by which atoms create radiation.

11. What is meant by a quantum jump in an atom?

12. How many different radiation frequencies are emitted in the quantum jumps shown in Figure 14.11?

THE UNCERTAINTY PRINCIPLE

13. Does the particle represented by the psi-field of Figure 14.14 have a precise position? A precise speed?
14. Do the four particles represented by the psi-fields of Figure 14.15 have precise positions? Precise speeds?
15. Which wave packet of Figure 14.16 has the more precise position? The more precise speed?
16. Does the uncertainty principle say that a particle must have a Δx that is larger than some prescribed value? What does it say?
17. Does a baseball have large quantum uncertainties or small ones? Why?

QUANTUM JUMPS

18. What happens to a particle's wave packet when a new and more accurate position measurement is performed?
19. Is it possible for a single microscopic particle's psi-field to be spread out over macroscopic dimensions, such as several meters or larger? Give an example.
20. Give an example in which the switching on of a detector causes a particle's psi-field to quantum-jump.
21. What is meant by a nonlocal effect?
22. What does Mandel's experiment demonstrate?

THE NONLOCALITY PRINCIPLE

23. What is a statistical correlation? Give a common example of statistically correlated events.
24. What is entanglement, and what does Bell's principle tell us about entangled particles?
25. What does the position-entanglement experiment demonstrate?
26. Can the nonlocal connections described by Bell's principle transfer energy instantaneously?

QUANTUM THEORY AND REALITY

27. According to the standard interpretation of quantum theory, which of the following are actually inherent in nature: uncertainties, wave–particle complementarity, nonlocal connections, contextual reality?
28. Describe at least two key ideas of the Newtonian worldview that are contradicted by quantum theory. Are some aspects of the Newtonian worldview contradicted by Einstein's relativity theory?

Home Projects

1. Burn different substances and observe the colors of their flames. Try scattering a little salt into a candle flame. Give a microscopic explanation, based on the quantum theory of the atom, of what you are seeing.
2. Use your radio as a spectroscope to determine the frequencies of the radio radiation present in your room. Record the frequencies and each frequency's relative intensity (don't touch the volume knob) on a scale of 1 to 10 as your ear perceives it. How long are the wavelengths your radio is detecting (consult Figure 9.7)?
3. Observe a mercury-vapor (gas made of mercury) and a sodium-vapor streetlight. What color dominates mercury's spectrum? The sodium atom's spectrum? What are you seeing, according to the quantum theory of the atom?

For Discussion

1. Is the moon still there when you are not looking at it? How do you know? How could you experimentally determine whether or not it is still there when nobody is looking at it?
2. Does social materialism (the desire to possess TV sets, cars, and so forth) have anything to do with philosophical materialism (the belief that reality consists only of matter in motion)?
3. Are there senses in which contemporary U.S. culture might be said to be Newtonian? Pre-Newtonian? Post-Newtonian?

Exercises

OBSERVING ATOMIC SPECTRA

1. In what ways is your radio a type of spectroscope? In what ways does it differ from the spectroscope described in the text?
2. Why, when different materials burn, do they often create flames of different colors? How might the chemical composition of a burning substance be determined?
3. If you compared the spectra from two sodium vapor bulbs, would they be the same? What if you compared a sodium vapor bulb with a mercury vapor bulb?

THE QUANTUM ATOM

4. Explain, in terms of inertia, why the electron does nearly all the moving in a hydrogen atom.

5. Assuming that the string's length in Figure 14.5 is 2 m, could the string vibrate in a standing wave whose wavelength is 2.1 m? 1.9 m? 0.5 m? Explain.

6. Describe the three-dimensional shapes of the two quantum states shown in Figure 14.8 having $N = 3$ and $L = 1$.

7. If a very accurate measurement of an atom's mass could be made in an excited state and in its ground state, would any difference be found? (*Hint*: Remember $E = mc^2$). What happens to an atom's mass when it emits a photon?

8. In Figure 14.11, which quantum jump creates the higher frequency photon, $N = 4$ to $N = 3$ or $N = 4$ to $N = 2$? Which of the two photons has the longer wavelength?

9. In Figure 14.11, which quantum jump creates the highest frequency, $N = 5$ to 4, 4 to 3, 3 to 2, or 2 to 1? Which creates the longest wavelength?

10. The four spectral lines of hydrogen photographed in Figure 14.2 have wavelengths and frequencies that agree precisely with the four lowest-energy transitions into hydrogen's $N = 2$ energy level. Which three of these four lines are graphed in Figure 14.12? Give the initial and final N values for the red line and for each of the other three lines in the photo.

THE UNCERTAINTY PRINCIPLE

11. If Planck's constant were smaller than it is, how would the uncertainty principle be affected? What if Planck's constant were zero?

12. How would it affect you if Planck's constant were 1 J-s instead of 6.6×10^{-34} J-s?

13. If Planck's constant were smaller than it is, would this affect the sizes of atoms? How? What would become of atoms if Planck's constant were zero?

14. Think of a few common situations, unrelated to quantum theory, in which observation changes reality. Would public opinion polls be an example? Would looking at the moon be an example?

15. One everyday example in which a measurement disturbs the measured object is the measurement of the temperature of a pan of water using a thermometer. How does this disturb the temperature? Is this a quantum effect?

16.*Making estimates. The electron in a ground-state hydrogen atom is confined by electric forces to remain in a sphere measuring roughly 10^{-10} meters in diameter. An electron's mass is about 10^{-30} kilograms. Use these data along with the uncertainty to estimate the speed of this electron. (*Hint*: In the ground state, the electron's speed should be roughly equal to the uncertainty in its speed, in other words, $\Delta s = s$. See the discussion at the end of Section 14.3.) What fraction of lightspeed is this?

QUANTUM JUMPS

17. What would happen to the wave packet of Figure 14.14 if an accurate speed measurement were performed? How would the measurement affect Δx and Δs?

18. Your friend flips a coin but covers it up so that neither of you can tell whether it is heads or tails. What odds (probability) would be fair to put on heads? Suppose he uncovers it and you see that it is tails? What odds should you now assign to heads? Does this sudden shift in the probabilities have anything to do with quantum theory?

19. The graph behind the screen in Figure 13.12(b) shows the pattern formed by the psi-wave on the screen in a single-slit experiment. Is this a graph of a single electron's psi-wave just before or just after the electron hits the screen? What happens to the psi-wave when the electron hits the screen?

THE NONLOCALITY PRINCIPLE

20. Suppose two dice are physically connected in such a way that an "even" outcome (2, 4, or 6) on either die always occurs with an even outcome on the other, and odd outcomes also always occur together. Could we then say that the outcome on one die entirely determines the outcome on the other die? Could we say that the two dice are statistically correlated? If the first die comes up 2, what are the probabilities for each of the six sides of the other die?

21. Is the statistical correlation described in the preceding exercise a quantum effect?

22. What experiment demonstrates statistical correlations that are not explainable in terms of Newtonian concepts but that can be explained in terms of quantum theory?

QUANTUM THEORY AND REALITY

23. List several general ways in which nature is non-Newtonian and several specific phenomena (such as radioactive decay) that are non-Newtonian. In what ways is nature Newtonian? List several specific phenomena (such as the fall of a rock) that are, to a very good approximation, Newtonian.

PART **V**

WITHIN THE ATOM
*fire of the nucleus, fire
of the sun*

15

THE NUCLEUS AND RADIOACTIVITY:
a new force—

One way that science has expanded human awareness is by expanding the "distance scales" that humans comprehend. Prescientific cultures were aware of what they could see, down to the smallest dust particles (10^{-5} m) and up to the distance a person might see on Earth (100 km, or 10^5 m). Today, telescopes have extended our awareness to other galaxies at over 10^{21} m and to the edge of the observed universe at some 10^{26} m. At the other end of the scale, microscopes of various sorts, including the giant accelerators that investigate subatomic objects, have extended our awareness down to atoms (10^{-10} m in size), the nucleus (10^{-14} m), and subnuclear particles (10^{-19} m and still counting).

The nucleus is the star of this and the next chapter. Although the nucleus might seem remote from human concerns, technology has exploited it as a source of great power. Power always has both dangers and opportunities. It is up to all of us to see that when used, it is used wisely. So we study nuclear physics not only for its intellectual significance but also because you and I must figure out how to use this powerful knowledge beneficially and nondestructively.

Quantum principles are important to any object as small as the nucleus. And we have seen that processes within such a small region of space must occur at high energies, so high that relativistic principles are important. Nuclear physics is an offspring of both of the post-Newtonian physical theories.

This chapter explores the structure of the nucleus and the nuclear reaction known as radioactive decay. Chapter 16 then studies the other two major types of nuclear reactions, fusion and fission. Section 15.1 of this chapter discusses the forces acting within the nucleus, and Section 15.2 explores the energetics of the nucleus, a topic crucial to our further discussions of nuclear physics. Then we present the physics of radioactive decay (Section 15.3), including the concept of half-life (Section 15.4). The next two sections explore two radioactivity-related cultural and societal topics: radioactive dating (Section 15.5) and human exposure to radioactivity (Section 15.6). Section 15.7 expands on Section 15.6 in regard to the question of technological risk.

DIALOGUE 1   Where should nuclear physics be placed in Figure 5.17?

15.1 Nuclear forces: *the third glue*

Scientists are like detectives, forming hypotheses and deductions from observed clues. Let's do some nuclear detective work.

The main clue is Rutherford's evidence that the atom is mostly empty space. An atom's protons and neutrons are tightly bunched in a tiny region 10^{-14} m across (Figure 15.1).* Since protons are positively charged and neutrons are not charged, there must be an attractive force between these nuclear particles to bind them to each other, because otherwise the repulsion between the protons would blow the nucleus apart.

This attractive force must be strong enough to overcome the great repulsion that exists between two protons as close as 10^{-15} m. Gravity is always attractive. Could it hold the nucleus together? As you might guess, gravitational forces between nuclear particles are insignificant compared with electric and other forces, so gravity can't do the job. Now recall that the forces experienced in your daily life can be explained in terms of only two fundamental forces: gravity and electricity. Since the force that holds the nucleus together cannot be either of these, we deduce that there must be a third fundamental force that holds the nucleus together.

This third force must strongly attract protons to one another when they are separated by about 10^{-15} m. It must also attract neutrons to one another and to protons, since otherwise neutrons would fall out of the nucleus. But despite its great strength at short separations, this force cannot extend very far. It certainly cannot extend from one nucleus to the next in solid matter (about 10^{-10} m), because if it did, all the nuclei would clump together. In fact, we could hypothesize (remember, this is a detective story) that this force extended only far enough to hold the largest nuclei together, because if it extended much farther, still larger nuclei would be possible and would be observed in nature.

This force, which is strongly attractive between protons and neutrons at around 10^{-15} m and negligible at much larger distances, is called the **strong nuclear force**, or simply the strong force. It holds the nucleus together.

We can distinguish among three kinds of "glue" that bind things together. At the subatomic level, the strong force holds the nucleus together. At the atomic level, the electric force binds orbital electrons to their nuclei, binds atoms into molecules, and holds solids and liquids together. And at the as-

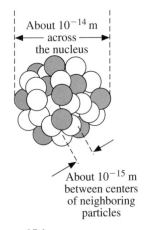

About 10^{-14} m
across
the nucleus

About 10^{-15} m
between centers
of neighboring
particles

FIGURE 15.1

A schematic representation of a nucleus. The green spheres represent protons, and the white spheres represent neutrons. Diagrams like this, showing subatomic particles as if they were ordinary Newtonian objects, are useful aids to thought, even though they are highly simplified. Nature at this level is quantum mechanical, having both wave and particle properties, and cannot be properly represented with any picture.

---

1. In "quantum + special relativity."
* Nuclear diameters range from about 3×10^{-15} (hydrogen) to 20×10^{-15} m (uranium).

tronomical level, the gravitational force holds planets, stars, solar systems, and galaxies together and reaches across clustered galaxies to determine the shape of the universe.

Every force ever observed in nature can be reduced to the action of these three forces plus one other: the **weak nuclear force** (Section 15.3). Together, these **four fundamental forces** determine the structure of our universe. We don't know why they have the properties that they do, properties that make the universe such an interesting and varied place. What if the properties of any one of them were different? For example, what if the strong force had a longer range, or what if it were weaker? What if the electric or gravitational force were stronger or weaker than it actually is? What if there were three kinds of electric charge or only one? What if gravity were repulsive instead of attractive? Unless the changes were only slight, the universe as we know it could not then exist.

15.2 Nuclear energy and nuclear structure ____

The strong force is by far the strongest of the four fundamental forces. Since this force is so strong, the energy changes accompanying individual nuclear processes are large, compared, for example, with chemical processes (dominated by the electric force). Nuclear weapons explosions and nuclear reactors yield huge energies per kilogram of material, and gamma radiation from the nucleus is the most powerful form of electromagnetic radiation.

There is a fundamental quantum reason that nuclear energies must be large. Because the strong force has a short range, any nuclear particle is trapped in a small region of space. The uncertainty principle tells us that any such well-localized particle must have a highly uncertain speed and that the speed itself must be large (Section 14.3). Since nuclear particles must move quickly, they must have high energies. When this argument is worked out quantitatively, the conclusion is that any nuclear particle trapped in a region as small as 10^{-14} m must have an average speed of at least 3×10^7 m/s, or about 10% of lightspeed. Relativistic phenomena are important at such high speeds, and so nuclear processes necessarily involve both quantum and relativistic physics. This is indicated in Figure 15.2, which is a more detailed version of Figure 5.17. The x at the edge of the region that is forbidden by the quantum uncertainty principle indicates a typical nuclear size and speed.

Since nuclei are made of parts, we suspect that it is possible to alter their structure. Any process that does so is called a **nuclear reaction**. Most nuclear reactions fall into one of three types: radioactive decay, fusion, and fission.

A favorite enterprise of the medieval scientist-magicians known as *alchemists* was to try to transform one chemical element into another, especially lead into gold. Neither their chemistry nor their magic produced any gold, but the dream has turned out to be achievable. Although magic doesn't help, chemical reactions are on the right track. The problem is that chemical energies are far too low to do the job. Today, nuclear chemists routinely turn one element into another. It is even possible to convert lead to gold, but you can't turn a profit this way: The needed energy costs much more than the gold is worth.

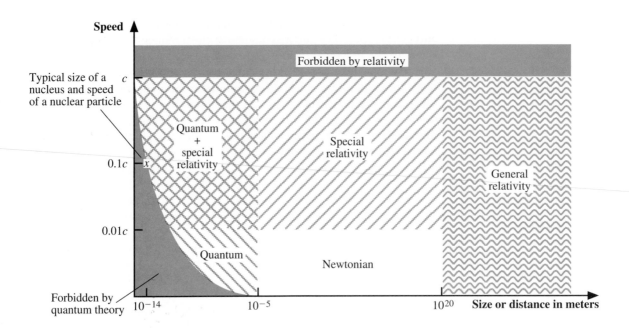

FIGURE 15.2
*The domains of Newtonian, relativistic, and quantum physics. The numbers and
boundaries are only representative; there are no definite borders between the domains
of the theories.*

Nuclear reactions are a little like chemical reactions. The chemical properties of an atom are determined by its number of orbital electrons, that is, by the **element** to which the atom belongs, specified by its atomic number. The elements and their atomic numbers are listed in the periodic table (inside back cover). The atomic number is important to nuclear reactions, too, because it is the number of protons in the nucleus. But neutrons are also significant in determining nuclear behavior. Nuclei with identical numbers of both protons and neutrons are said to belong to the same **isotope**. Just as atoms of a particular element have identical properties in chemical reactions, nuclei of a particular isotope have identical properties in nuclear reactions.

The numbering system for isotopes is only slightly more complicated than that for elements. An isotope is numbered by its **atomic number**, the number of protons it contains, and also by its **mass number**, the total number of protons and neutrons it contains. The mass number is simply the total number of particles in the nucleus. Since the proton's mass is nearly identical to the neutron's mass (they differ by less than 1%), the mass number is also nearly proportional to the mass of the nucleus, which is why it is called the mass number.

For example, the element carbon has atomic number 6, so every carbon nucleus has 6 protons. Because some carbon nuclei contain 6 neutrons, others contain 7, and still others contain 8, there are three different isotopes of carbon. These three isotopes have the same chemical properties but different nuclear properties. Their mass numbers are 12, 13, and 14.

We indicate specific isotopes by their chemical symbol preceded by their atomic number as a subscript and their mass number as a superscript. For example, the three isotopes of carbon are written $^{12}_{6}C$, $^{13}_{6}C$, $^{14}_{6}C$. But we often drop the atomic number because the chemical symbol actually specifies it and simply write ^{12}C, ^{13}C, ^{14}C, pronounced "carbon-12, carbon-13, carbon-14."

DIALOGUE 2   Compare and contrast nuclear reactions with chemical reactions: (a) Which of the four fundamental forces is involved in each? (b) What is changed, or rearranged, by each?

DIALOGUE 3   How does the mass of a ^{14}C nucleus compare with the mass of ^{12}C nucleus? How do their electric charges compare?

DIALOGUE 4   How many protons and how many neutrons are in a nucleus of each of the following isotopes: $^{1}_{1}H$, $^{2}_{1}H$, $^{3}_{1}H$, $^{235}_{92}U$, $^{238}_{92}U$?

DIALOGUE 5   In what way are ^{3}H and ^{3}He similar? In what way are they different?

15.3 Radioactive decay: *spontaneous nuclear disintegration*

In 1896, French physicist Henri Becquerel finished his research for the week and stored a certain uranium compound away in a drawer for the weekend. By chance, an unexposed photograph plate was stored in the same drawer. When he returned the following week, Becquerel found to his surprise that the film had been exposed, despite having been kept in a dark drawer. A lesser scientist might have shrugged his shoulders and tossed out the ruined film, but Becquerel suspected a connection between the uranium and the exposure. He discovered that he could reproduce the effect whenever he placed the uranium near photographic film. Apparently, the uranium radiated something that could expose a photographic plate. The process was called **radioactivity**. When Becquerel subjected the uranium to various chemical treatments, they produced no change in the effect. So radioactivity had little to do with chemistry. Science had had its first brush with the nucleus, even though the nucleus had not yet been discovered.

Two years later, French physicist Marie Curie (Figure 15.3) and her husband Pierre detected radioactivity in pitchblende, a tarry black substance.

FIGURE 15.3
Marie Curie shared, with Pierre Curie and Henri Becquerel, the 1903 Nobel Prize for physics. Pierre died in a traffic accident in 1906. Marie Curie continued her research, confident that she could succeed despite the widespread prejudice against women in physical science. She did succeed, becoming the first woman to teach at the Sorbonne. In 1911, she became the first person to receive a second Nobel Prize. Her daughter Irène became a nuclear scientist, winning the 1935 Nobel Prize in chemistry. In 1934 Marie Curie developed cataracts and lesions on her fingers and died of radiation-induced leukemia. Irène Curie also died of leukemia.

2. (a) Chemical reactions involve electric forces, whereas nuclear reactions involve nuclear forces (and electric forces also, because protons are charged). (b) Chemical reactions alter the structure of molecules (by altering the electron orbits), whereas nuclear reactions alter the structure of nuclei.
3. One-sixth larger mass (the mass ratio is $14/12 = 7/6$). Same charge.
4. 1 and 0, 1 and 1, 1 and 2, 92 and 143, 92 and 146.
5. They both have 3 nuclear particles. ^{3}H has 1 proton and 2 neutrons, and ^{3}He has 2 protons and 1 neutron.

They were not surprised at the radioactivity because pitchblende is a known ore of uranium, but they were surprised to find that the radiation was more intense than the radiation from pure uranium, even though pitchblende contains only low concentrations of uranium. Apparently some other substance, much more radioactive than uranium, was present in pitchblende. The Curies then performed the monumental task of chemically separating this substance from eight tonnes of pitchblende. This involved some real detective work, because the new substance and its chemical properties were unknown. They managed to get only a bare powdery pinch, 0.01 grams, of the stuff. Like uranium, it radiated spontaneously, but it gave off rays at a much higher rate than did an equal mass of uranium. Because of its powerful radiation, they named the new element *radium*.

Marie Curie is, of all celebrated beings, the only one whom fame has not corrupted.
Albert Einstein's eulogy upon the death of Marie Curie in 1934

Since the discovery of uranium and radium, scientists have found that there are very many radioactive substances. Every isotope heavier than bismuth (atomic number 83, just above lead in the periodic table) is radioactive, and there are many **radioactive isotopes** of the other elements. For example, among the three isotopes of carbon, ^{14}C is radioactive, but ^{12}C and ^{13}C are not.

Experiments (Figure 15.4) show that radioactive materials emit three distinct types of rays, known as **alpha**, **beta**, and **gamma** rays. The way that these rays respond to electric or magnetic fields shows that alpha rays are positively charged, beta rays are negatively charged, and gamma rays are not charged. These rays come from the nuclei of various isotopes.

Closer examination reveals these rays are created in one of two types* of spontaneous nuclear processes, known as **radioactive decay** processes. In **al-**

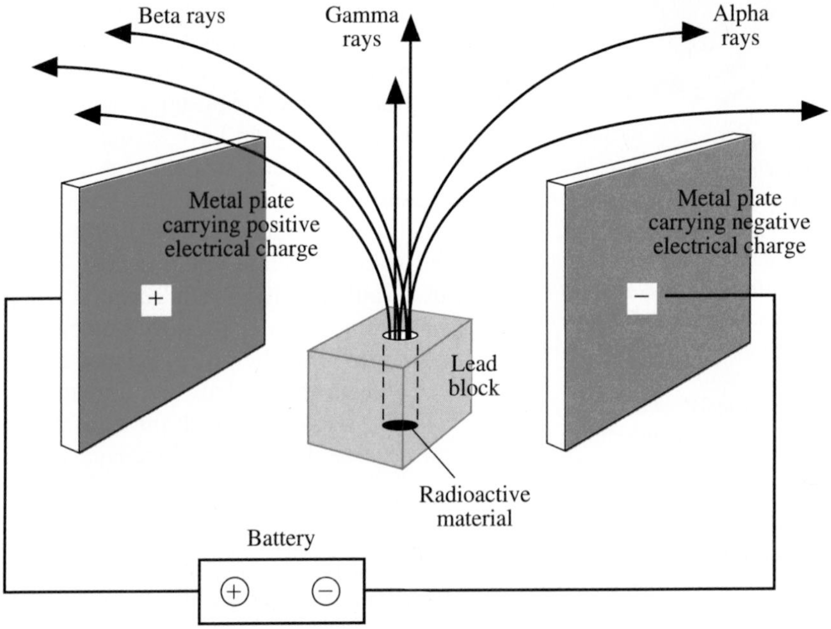

Figure 15.4
An experiment showing that radioactive materials can emit three different types of rays, alpha, beta, and gamma. The two metal plates create an electric field in the space between the plates. The experiment shows that because they are attracted to the negative plate and are repelled by the positive plate, alpha rays carry a positive electric charge. Similarly, beta rays are negatively charged, and gamma rays are not charged.

* In addition to the two types discussed here, there are other less common types of radioactive decay processes, such as positron decay.

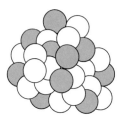

Daughter nucleus after
radioactive decay

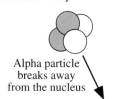

Alpha particle
breaks away
from the nucleus

FIGURE 15.5
Alpha decay.

Beta particle is
created in the
nucleus and
immediately
ejected

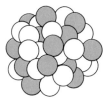

Daughter nucleus after
radioactive decay

FIGURE 15.6
Beta decay.

Alpha particle about
to break away

FIGURE 15.7
*During alpha decay, an alpha
particle becomes separated from
the rest of the nucleus and is then
pushed rapidly away by the
repulsion from the protons in the
daughter nucleus.*

pha decay, a radioactive nucleus spits out a particle called an **alpha particle** that is identical with the nucleus of helium: $_2^4\text{He}$, two protons and two neutrons bonded together by the strong nuclear force (Figure 15.5). Once an alpha particle escapes into its surroundings, it soon slows down, owing to collisions with surrounding molecules, and picks up two electrons from nearby atoms to become a normal atom of helium.

In **beta decay**, a radioactive nucleus spits out an electron (Figure 15.6). This is surprising because there are no electrons in the nucleus! We'll straighten out this mystery later. Once out into the open, this electron soon slows down, owing to collisions, and is captured by some nearby atom to become an ordinary orbital electron. But when it is emitted by a radioactive nucleus, we call it a **beta particle** to indicate the process that created it.

Most radioactive isotopes decay by only one of these two processes. For example, uranium and radium are alpha emitters, and ^{14}C is a beta emitter. Because alpha and beta decay violently disturb the charged particles of the nucleus, both cause the nucleus to emit high-energy electromagnetic radiation. So each alpha or beta decay is normally accompanied by a high-energy photon from the nucleus: a gamma-ray photon. Although alpha rays and beta rays are often called radiation because they radiate outward from the nucleus, they are not electromagnetic radiations. They are, instead, streams of material particles. Gamma rays, however, are a form of electromagnetic radiation.

Radioactivity occurs because some nuclei do not hold together very well and eventually fall apart spontaneously. Such nuclei are said to be **unstable**. In a **stable nucleus**, on the other hand, the balance of nuclear forces holds the nucleus together forever, or at least until some outside influence breaks it apart. One source of instability is simply large size: A very large nucleus has a hard time sticking together because each proton tends to be pushed out of the nucleus by the repulsive force of all the many other protons. This is why all isotopes heavier than bismuth are radioactive.

In alpha decay, a small part of the nucleus is pushed out (Figure 15.7). Because the alpha particle, $_2^4\text{He}$, is one of nature's most stable structures, it is this combination of protons and neutrons that breaks away. Once separated from the nucleus, an alpha particle is pushed strongly away by the electric force from the protons in the remaining nucleus, called the **daughter nucleus**, which causes the nucleus to lose two protons and two neutrons.

Whereas alpha decay is the sort of falling apart that one might expect in an unstable nucleus, beta decay is more surprising. It is caused by the weak nuclear force, which actually causes a neutron to transform spontaneously into a proton while simultaneously creating a high-energy electron (Figure 15.8). When this process occurs within a nucleus, the nucleus loses a neutron and gains a proton. At the same time, the electron created in the process is spewed out of the nucleus because it is created at high energy. This is the source of beta particles.

The nucleus itself is transformed into a different isotope during radioactive decay. It is the alchemist's dream, and it has been occurring spontaneously all the time! For example, when $_6^{14}\text{C}$ beta decays, it loses a neutron

Before

After

FIGURE 15.8
The details of beta decay: A neutron is transformed into a proton and an electron, and the electron moves away with high energy. Compare this with Figure 15.6.

Now, the special interest of radium is in the intensity of its rays, which is several million times greater than the uranium rays. And the effects of the rays make the radium so important. . . .These effects may be used for the cure of several diseases. . . .What is considered particularly important is the treatment of cancer. . . .Radium is more than a hundred thousand times dearer than gold.
Marie Curie, in a talk at Vassar College, 1922

and gains a proton, so its atomic number increases by 1 and its mass number remains unchanged. The daughter nucleus has atomic number 7 and mass number 14; in other words, it is $^{14}_{7}$N.

We represent nuclear reactions the way we represent other processes: with an arrow from the initial to the final situation. For example, the beta decay of ^{14}C is represented by

$$^{14}_{6}\text{C} \rightarrow {}^{14}_{7}\text{N} + \text{beta}$$

As another example, $^{238}_{92}$U is an alpha emitter. Alpha decay reduces the atomic number by 2, from 92 to 90 (thorium), and reduces the mass number by 4. This nuclear reaction is represented by

$$^{238}_{92}\text{U} \rightarrow {}^{234}_{90}\text{Th} + \text{alpha}$$

It always is useful to view processes in energy terms. What energy transformation occurs in radioactive decay? Alpha and beta particles carry microscopic kinetic energy (thermal energy) into the environment around the nucleus, and gamma photons carry radiant energy. These energies came from the nuclear structure of the radioactive material. So the universe loses nuclear energy and gains thermal and radiant energies, and the energy transformation is

$$\text{nuclear energy} \rightarrow \text{thermal energy} + \text{radiant energy}$$

Radioactive decay is like a landslide, but caused by nuclear forces instead of gravitational forces. In a landslide, gravity pulls part of a hill downward into a more stable, compact configuration. The slide transforms the gravitational energy of the elevated land into kinetic energy during the slide and finally into thermal energy. In radioactive decay, the forces in the nucleus cause a similar spontaneous "sliding" of an unstable nucleus into a more stable configuration, causing the nucleus to fall apart instead of downhill. In the process, nuclear energy (instead of gravitational energy) is converted to other forms.

Radioactive isotopes exist—or can be manufactured artificially in nuclear reactors—for nearly every chemical element. Because they have chemical properties identical to the chemical properties of the stable isotopes of the same element, radioactive isotopes have many uses. For example, thyroid cancer may be treated by injecting a small amount of a radioactive isotope of iodine into the patient. Iodine's chemical properties cause it to migrate to the thyroid where the isotope's radioactivity destroys cancer cells.

Radioactive isotopes make excellent tracers because their presence can be detected by detecting their radiations. For example, the progress of a lubricant moving through an industrial process, of a chemical moving through the body, or of fertilizer moving through a growing plant may be traced by "doping" that chemical with a radioactive form of the same chemical. The radioactive form behaves chemically just like the stable form, but it can be monitored by detecting its radioactivity.

Radioactive dating is another significant application of radioactive isotopes, but for this we need to discuss an idea called half-life.

15.4 Half-life: *when does a nucleus decay?*

The central feature of quantum theory is that individual microscopic events are unpredictable, but the statistics of these events are predictable. The situation is like flipping coins, only for a different reason. The outcome of one coin flip is unpredictable, but the statistics of many flips (50% heads) is approximately predictable. It is the same with microscopic processes. The difference between a macroscopic process like a coin flip and an unpredictable microscopic process is that "nature knows" how the coin will fall. You could use Newtonian physics to predict the outcome of a coin flip, although the measurements and calculations would be very complex. But microscopic processes are non-Newtonian and inherently unpredictable.

I hope that no one discovers how to release the intrinsic energy of radium until man has learned to live at peace with his neighbour.

Ernest Rutherford, in 1915

Given the submicroscopic nature of the nucleus, it is not surprising to learn that radioactive decay is an inherently unpredictable event. It is one of nature's clearest examples of quantum uncertainties.

To be specific, consider the alpha decay of one uranium nucleus. Because of nature's preference for alpha particles, you can think of the surface of a big nucleus like uranium as a collection of alpha particles that drift near the surface, attracted to the nucleus by the strong nuclear force but also repelled by the nucleus's protons (Figure 15.9). The motion of each alpha particle is determined by a quantum probability wave of the same sort that is responsible for the wave properties of electrons. So there is a certain chance that any particular alpha particle will find itself outside the nucleus, far enough outside that the strong nuclear force (which, remember, has a short range) can no longer be felt. Such an alpha particle will be immediately pushed out of the nucleus by electric forces.

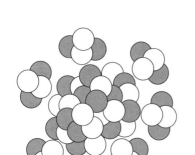

FIGURE 15.9
Think of the surface of the nucleus as a collection of alpha particles that drift near the surface. When, by chance, one of them finds itself too far from the rest of the nucleus, it is immediately pushed away by electric forces.

Because of quantum uncertainties, nature knows only the chances that an alpha particle will be ejected, not when it will be ejected. A particular uranium nucleus might decay during the next second, in 5 minutes, or 20 billion years. A nucleus gives no warning of when it will decay.

But like all quantum events, we can predict the statistics of radioactive decay. For example, given a large number of uranium nuclei, it is possible to predict roughly what fraction of them will decay in any particular period of time, even though it is impossible to predict exactly which nuclei will decay and which ones won't. Coin tosses are statistically predictable in this way, too: If we toss a large number of coins a single time, we can predict that roughly 50% will come up heads, but we can't predict which particular coins will be heads and which will be tails.

---

6. $^{131}_{54}$Xe (xenon). $^{222}_{86}$Rn (radon).

TABLE 15.1

Half-life and Decay Process of Several Radioactive Isotopes

Isotope	Name of element	Decay process	Half-life (approx.)
$^{14}_{6}\text{C}$	carbon	beta	6000 yr
$^{90}_{38}\text{Sr}$	strontium	beta	30 yr
$^{131}_{53}\text{I}$	iodine	beta	8 days
$^{137}_{55}\text{Cs}$	cesium	beta	30 yr
$^{214}_{84}\text{Po}$	polonium	alpha	0.00016 s
$^{222}_{86}\text{Rn}$	radon	alpha	4 days
$^{226}_{88}\text{Ra}$	radium	alpha	1600 yr
$^{234}_{90}\text{Th}$	thorium	beta	24 days
$^{235}_{92}\text{U}$	uranium	alpha	0.7×10^9 yr
$^{238}_{92}\text{U}$	uranium	alpha	4.5×10^9 yr
$^{239}_{94}\text{Pu}$	plutonium	alpha	24,000 yr

For any particular radioactive isotope, the most important statistic of this sort is the **half-life**, the time during which 50% of a large collection of those nuclei will decay. Although any particular isotope has a definite half-life, different isotopes have widely different half-lives. The reason is that some radioactive nuclei are highly unstable and others are much more stable. A highly unstable nucleus is like a landslide waiting to happen: It will probably decay soon, just as the waiting landslide will probably slide soon. A more stable nucleus will probably take longer to decay. Table 15.1 gives the half-lives of several radioactive isotopes.

As an example, start with 1 gram of pure ^{14}C. This radioactive isotope has a half-life of about 6000 years and decays by beta decay to the common stable form of nitrogen, ^{14}N. After 6000 years, 50% of the ^{14}C nuclei will have decayed, and you (or rather, your descendants) will have just half a gram of ^{14}C. The other half-gram will have been transformed into nitrogen.

How much ^{14}C is left after another 6000 years? The answer is half of the half-gram, or 0.25 gram, because the half-life statistic still applies. The 6000-year half-life is the time during which 50% of *any* macroscopic amount of ^{14}C decays. So there will be 0.50 grams of ^{14}C after 6000 years, 0.25 grams after 12,000 years, 0.125 grams after 18,000 years, and so forth. These values are graphed in Figure 15.10 and connected with a smooth curve showing the amount remaining at any time. This is the **decay curve** for ^{14}C. The decay curve for any radioactive isotope is identical to that for ^{14}C, only with a different half-life (Figure 15.11).

Radioactive decay is often called **exponential decay** because it is similar to exponential growth (Section 7.7). Exponential decay has a fixed halving time (half-life), just as exponential growth has a fixed doubling time.

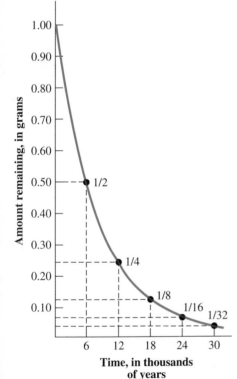

FIGURE 15.10

Radioactive decay curve for 1 gram of ^{14}C.

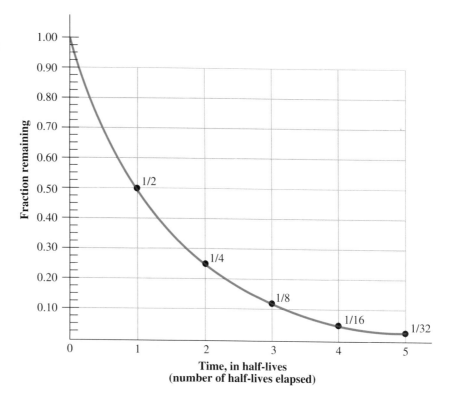

FIGURE 15.11
The exponential decay curve of any radioactive isotope.

DIALOGUE 7 MAKING ESTIMATES   Suppose you have 16 pennies and you toss all of them and then remove the ones that come up tails. You toss the remaining coins and again remove the tails. If you continue the process, about how many tosses must you make before you get down to a single remaining penny? (Try it!) What if you started with 100 pennies? What is the half-life of pennies in this game?

DIALOGUE 8   If you start with a gram of pure ^{14}C, how much will remain after 36,000 years? Approximately how much will remain after 10,000 years?

15.5 Radioactive dating: *when did we come from?*

Radioactive decay is a kind of clock. If you know how much of a substance has decayed, you can read the elapsed time from the decay curve (Figure 15.11). For instance, if a radioactive substance has decayed away to 25% of its original amount, you will know that it has been two half-lives since it started decaying.

---

7. After 4 tosses, you will (on the average) be down to 1 penny. If you start with 100 pennies, you will be down to about 1 after 6 to 7 tosses. The half-life of pennies is 1 toss.
8. 1/64 of a gram, or 0.016 gram. Either Figure 15.10 or 15.11 shows that after 10,000 years, which is 1.67 half-lives, about 30%, or 0.30 gram, will remain.

Carbon dating is one example. Essentially all of Earth's carbon is one of the two stable carbon isotopes, ^{12}C and ^{13}C. But in Earth's atmosphere, about 1 carbon atom in a trillion is the radioactive form, ^{14}C. Since the half-life of ^{14}C is only 6000 years and Earth is far older than 6000 years, you might wonder how any ^{14}C could still be in the atmosphere. The answer is that *cosmic rays*, high-energy particles that travel through outer space, continually replenish it. They occasionally enter the atmosphere, collide with atmospheric nitrogen, and transform the nitrogen nucleus into ^{14}C.

Because the carbon in all biological organisms comes ultimately from the atmosphere, ^{14}C is distributed throughout the biological world at 1 ^{14}C atom per trillion carbon atoms, about the same as the atmospheric ratio. This ratio of radioactive to stable carbon is maintained until a living organism dies. Then the ^{14}C gradually decays. The time elapsed since death can be determined by measuring the amount of ^{14}C remaining compared with the amount of stable carbon. For instance, if an old ax handle has only a quarter of its normal amount of ^{14}C, the tree from which the ax handle was made must have died two half-lives ago, or 12,000 years.

Such a measurement can determine the time elapsed since death if one knows how much ^{14}C was present when the organism died. The usual assumption is that the fraction of ^{14}C in the atmosphere in the past was nearly the same as it is today, so the fraction in a long-dead tree when it died would have been 1 ^{14}C per trillion carbon atoms, the same as today. But is this assumption correct? We will discuss this in a moment.

To measure the ^{14}C content in, for example, an old ax handle, a small amount of the carbon in the ax handle is separated chemically, and the ratio of ^{14}C to total carbon is measured by measuring the carbon's radioactivity and weighing the carbon sample. One problem is that the ^{14}C content in fresh wood is low to begin with and it decreases with time. Consequently, its radioactivity level is so low that it is hard to measure. For example, every gram of carbon in your body contains 5×10^{22} carbon atoms, of which only 50 billion are ^{14}C, of which only an average of 10 decay every minute.

But ^{14}C isn't the only radioactive isotope in town. Uranium, for example, is the basis for several dating methods. The uranium isotope ^{238}U has a half-life of 4.5 billion years and decays to thorium. But the daughter nucleus, thorium, also is radioactive, so it decays too. And its daughter nucleus is radioactive, so it decays. And so forth. An entire **radioactive decay sequence** of lighter and lighter daughter nuclei results from the decay of each ^{238}U nucleus until finally the sequence reaches a stable isotope. In the case of ^{238}U, this final stable isotope is the lead isotope ^{206}Pb. Since the common natural form of lead is ^{208}Pb, any ^{206}Pb can be identified as originally coming from uranium. Uranium is surprisingly common throughout Earth's crust and can be found in most rocks. The date at which a rock's uranium was first locked into position—the date at which the rock crystallized—can be determined by comparing the amounts of ^{238}U and ^{206}Pb at the same location in the rock.

Many different radioactive decay processes are used for a variety of dating methods (Table 15.2). Each method compares a radioactive isotope with

TABLE 15.2
Isotopes Used for Radioactive
Dating

Isotope	Comparison nucleus
^{238}U	^{206}Pb
^{238}U	^{234}U
^{235}U	^{207}Pb
^{234}U	^{230}Th
^{187}Re	^{187}Os
^{147}Sm	^{143}Nd
^{87}Rb	^{87}Sr
^{40}K	^{40}Ar
^{14}C	total C
^{10}Be	total Be

In all of us there is a hunger, marrow deep, to know our heritage, to know who we are, where we have come from. Without this enriching knowledge, there is a hollow yearning. No matter what our attainments in life, there is a vacuum, an emptiness, and a most disquieting loneliness.

Alex Haley, in *Roots*

a second "comparison nucleus" to determine what fraction of the radioactive isotope has decayed and how long it has been decaying.

Radioactive and also nonradioactive methods have taught us "when we came from" (Table 15.3). Some spans of years are so large that it is difficult to relate them to anything meaningful. To put them in perspective, the table's third column compresses Earth's history into 12 hours, beginning with Earth's formation. This perspective can be an eye-opener (Figure 15.12). Throughout most of Earth's history, the dominant life-forms have been simple organisms such as algae. On the 12-hour clock, complex animals appear only late in the evening; the earliest humans evolve at less than a minute before midnight; and our species (*Homo sapiens*) appears at 1 second before midnight. All of human culture spans far less than 1 second. Great human movements that changed the face of our planet, events such as the spread of agriculture, the human population explosion, the Industrial Revolution, and the information revolution, have occupied only a moment on the world stage. To paraphrase Norman Mailer (Section 1.1), the itch has been to accelerate.

HOW DO WE KNOW? CROSS-CHECKING RADIOACTIVE DATING

How far can we trust the time spans given in Table 15.3? Science is a "web of consistency." Scientists are always checking different results against one another. For instance, if several independent methods of dating agree, our confidence in all of them is increased. It would be quite a coincidence if they were all wrong in exactly the same way.

TABLE **15.3**

When We Came From: Some Approximate Dates Relevant to Our Species

Event	Years before present	On a 12-hour clock
Creation of Earth	4.6 billion	noon
Life		
Earliest known fossils (single-celled)	3.3 billion	3:00 P.M.
First vertebrates (backboned animals)	500 million	10:45 P.M.
First reptiles	300 million	11:15 P.M.
First mammals	200 million	11:30 P.M.
First primates (monkeylike animals)	70 million	11:50 P.M.
Humans		Seconds before midnight
Earliest hominids[a]	4 million	40 s
Use of stone tools	2 million	20 s
Genus *Homo*	1 million	10 s
Homo sapiens (modern humans)	100,000	1 s
Culture		
Invention of agriculture	10,000	0.1 s
First cities, earliest writing	5,000	0.05 s
The scientific age (Copernicus)	500	0.005 s
The industrial age	250	0.002 s
Twentieth century	100	0.001 s

[a] Humanlike creatures that were direct human forebears.

FIGURE 15.12

The cosmic clock. During the final second of this 12-hour "day," agriculture began to develop at 0.1 s before midnight, and the industrial age began at 0.002 s before midnight.

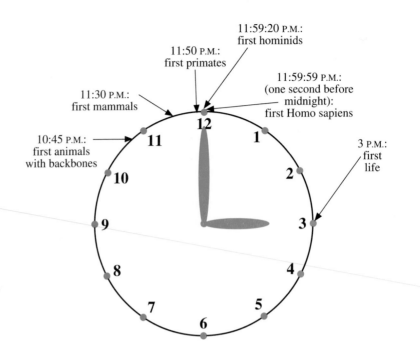

There is a reason to expect errors in carbon-inferred dates: If the amount of cosmic rays hitting the atmosphere were different in the past than it is today, the ^{14}C content in the biological world would have been different in the past than it is today. For this reason, scientists have cross-checked the carbon-based dates and have found that carbon dating actually is in error, giving dates that are 10 to 20% too young.

One method of cross-checking is by studying ^{14}C in the annual growth rings of trees. We can date living trees (up to 5000 years old) by simply counting these growth rings, and we can date dead wood of unknown age by matching the rings' growth patterns (variations in ring thicknesses, caused by varying climate) with the growth patterns in dated specimens. Tree specimens up to 10,000 years old can be dated in this way. There is little reason to doubt these dates, for they are obtained by simple counting. The results show that the older rings are as much as 10% older than is indicated by their carbon dates.

Another method of cross-checking studies the ^{14}C content in coral reefs. These reefs can be dated by the carbon method and also by a uranium method that is thought to be reliable to within 1%. Figure 15.13 shows the results of this cross-check (dots on the graph) and also the tree-ring cross-check (crosses). The dots show the uranium age (along the vertical axis) at which a given carbon-inferred age (horizontal axis) occurs. If both dating methods were accurate, these dots would lie on the straight line shown, along which the uranium age equals the carbon age. For example, the data point circled shows that a carbon age of 16,000 years corresponds to a uranium age of 19,000 years, a difference of about 15%.

All three methods agree to within 15%. Tree-ring dating is entirely independent of radioactive methods, and uranium dating is unrelated to cosmic rays striking the atmosphere. It is hard to believe that all three

FIGURE 15.13

Checking carbon dating by using the tree-ring method (black crosses) and a uranium method. Apparently carbon dating gives ages that are a few thousand years too young for objects that are around 10,000 to 20,000 years old. The greatest discrepancy between carbon dating and the other two methods is 15%.

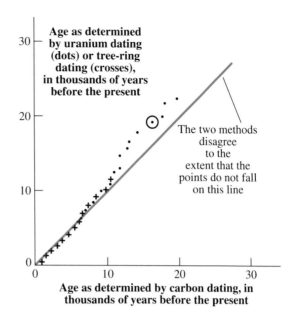

Age as determined by uranium dating (dots) or tree-ring dating (crosses), in thousands of years before the present

The two methods disagree to the extent that the points do not fall on this line

Age as determined by carbon dating, in thousands of years before the present

About 43% of Americans, for example, doubted that humans are descended from other animal species and 45% said that early humans lived at the same time as dinosaurs.
National Science Board, *Science and Engineering Indicators*, 1989

This is of course a drastic suggestion—orthodox geologists indeed reject it out of hand. However, there is no other alternative. If the Bible is the Word of God—and it is—and if Jesus Christ is the infallible an omniscient Creator—and He is—then it must be firmly believed that the world and all things in it were created in six natural days and that the long geological ages of evolutionary history never really took place at all
Henry M. Morris, Director of the Institute for Creation Research, in *Scientific Creationism*, Creation-Life Publishers, San Diego, 1974

We do not look upon the Bible as an authority for science or history. We see truth in the Bible as not to be reduced solely to literal truth, but also to include salvation truths expressed in varied literary forms.
Pastoral Statement for Catholics on Biblical Fundamentalism, National Conference of Catholic Bishops, 1987

could be far wrong—a different something wrong in each case—and yet that they all would arrive at about the same answer. Nature would really have to be conspiring against us for that to happen.

Many cross-checks are possible using radioactive methods (Table 15.2) and many other methods (Table 15.4). Each of the radioactive and nonradioactive methods is a kind of natural "clock," all of which give about the same dates, producing a strong scientific consensus in support of such results as Table 15.3.

In the past, many scientists and nonscientists believed that Earth was only a few thousand years old. For example, Kepler suggested (Chapter 1) that "God waited six thousand years" for an observer such as Tycho Brahe. Such estimates were based on counting backward through the generations of the Old Testament and so calculating a date of about 6000 years ago for Adam and Eve.

Creationism (see also Sections 7.8 and 12.7) is the belief that the Old Testament, including Genesis, can be read as science and that Earth and the major types of biological organisms, including humans, all were created separately (that is, they were not connected by evolution) and at roughly the

TABLE 15.4
Nonradioactive Dating Methods

Method	Property that is measured
Tree rings	precise calendar year of ring growth
Relative dating by rock strata	location of sedimentary rock layers
Astronomy	age of universe, galaxy, solar system
Electron spin resonance	radiation-induced changes in rock crystals
Thermoluminescence	radiation-induced changes in rock crystals
Mitochondrial DNA	mutation rate of DNA molecules
Amino acid analysis	gradual change in amino acids over time

same time, just a few thousand years ago.* Many Americans continue to hold creationist beliefs. Questions about human origins remain controversial because of their religious significance, tending to divide Christians into those who interpret the Bible "liberally," as a spiritual and moral guide but not as an authority for science or history, and those who interpret it "literally," as an authority for science and history.

There is a debate in the United States today about the teaching in public schools of topics such as Earth's age and the age and origin of different species, including humans. This question is perhaps the most important of the "pseudoscience" issues (astrology, extrasensory perception, unwarranted faith in extraterrestrial visitations, and so forth). These debates are really about scientific methodology, that is, about the validity of science itself.

Today, the hypothesis that Earth is only a few thousand years old conflicts with data and principles from astronomy, physics, chemistry, geology, biology, paleontology, archaeology, and history. There is a broad scientific consensus in favor of an Earth that is far older than a few thousand years. Scientific results are always open to question, and the estimates in Table 15.3 will doubtless be revised as new results come in, but the general picture of an Earth measured in billions of years, a long-term evolution toward more complex organisms, and human forebears measured in millions of years appears to be both durable and fruitful.

DIALOGUE 9   A future archaeologist digs up a body and measures an average of only about 1 ^{14}C decay per minute in 1 gram of the body's carbon. This is about 10% of the decay rate when the person was alive. About how long has this person been dead (when she digs up the body)?

DIALOGUE 10 MAKING ESTIMATES   How radioactive are you? The human body is 18% carbon by weight. The biological world contains some 50 billion ^{14}C atoms per gram of carbon, of which some 10 atoms decay every minute. Estimate the number of radioactive carbon atoms in your body and the number inside you that decay in 1 minute.

15.6 Human exposure to ionizing radiation ____

The walls of your room, the air you breathe, and even your body are radioactive (Dialogue 10). What are the effects? Are the effects different today than they were before the nuclear age? What are the risks? Can or should anything be done?

---

* This is essentially the definition of creationism given by one of its major proponents, Henry M. Morris, in his book *Scientific Creationism* (San Diego: Creation-Life Publishers, 1974).
9. The decay curve, in Figure 15.10 or 15.11, is down to 10% after about 19,000 years.
10. If your body's mass is 60 kg (130 pounds), you have about 12 kg, or 12,000 gm, of C. The number of ^{14}C atoms in your body is about $(50 \times 10^9) \times (12,000) = 600 \times 10^{12}$, or 600 trillion. The number of atoms that decay in 1 minute is $10 \times 12,000 = 120,000$. Your body's total radioactivity is actually twice this large, owing to other radioactive isotopes in your body.

The alpha, beta, and gamma particles emitted by radioactive materials damage biological cells. The damage is done when these high-energy particles pass, like little bullets, through a cell, ionizing (knocking electrons out of) some of the cell's molecules. Because they can do this, alpha, beta, and gamma rays are called **ionizing radiations.** Since X rays also have enough energy to ionize biological material, they are classified as ionizing radiation, although they do not come from the nucleus.

Ionization changes a molecule's chemistry. Because nature has spent several billion years perfecting the biological cell, any change is likely to be for the worse. Any of the cell's functions can be altered, depending on which molecules are ionized. The most important effects result from damage to a cell's DNA molecules, because DNA carries the biological information inherited by other cells.

The biological damage done in humans by ionizing radiation is measured in a unit called the **rem** (for Roentgen Equivalent in Man). If a person receives ionizing radiation, the number of rems received is a direct measure of the number of damaged cells. A quantitative feel for the rem is best obtained by looking at examples.* For instance, the amount of radiation that an average person in the United States receives every year from all sources is about 0.3 rems, or 300 **millirems**. At the other extreme, a sudden dose of 1000 rems causes death within 30 days. So 0.3 rems during a year is normal, and 1000 rems in a short time is catastrophic.

There are three main types of biological damage to humans. The most immediately obvious is **acute short-term effects** to the red blood–forming cells of the bone marrow and to the cells that line the intestinal wall. A sudden dose of 25 to 100 rems to the whole body causes short-term changes in the blood that the person might not be aware of. At 100 to 300 rems, the effects on the blood and the intestines produce typical symptoms of radiation sickness: fever, vomiting, damaged red blood cells, reduced white blood cells and platelets, loss of hair, spontaneous internal and external bleeding from weakened blood vessels, and small hemorrhages beneath the skin. A dose of 500 rems produces 50% fatalities; 1000 rems causes death within 30 days; and 10,000 rems causes death within hours.

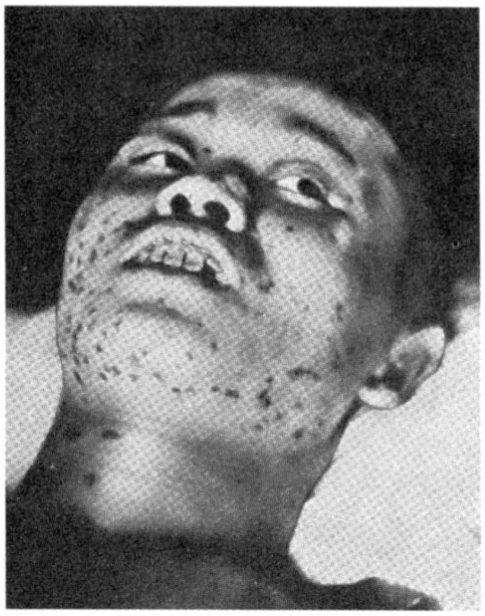

Fortunately, humans have seldom experienced doses large enough to produce acute effects. The most severe examples have been the nuclear bombs dropped by the United States in 1945 on the Japanese cities of Hiroshima and Nagasaki (Figure 15.14) and the 1986 Chernobyl (in Ukraine) nuclear reactor accident. Most of our knowledge of radiation damage to humans has come from these events, and from the medical uses of radiation.

The second form of radiation damage is **mutation**, an inheritable alteration of the genetic material in a sperm or egg cell. Radiation-caused mutations have been observed in experiments with fruit flies and other species. Mutations can produce successive generations of altered offspring. They are almost always harmful, but occasionally a mutation is advantageous, and this effect is in fact essential to the process of biological evolution. To date,

FIGURE 15.14

A victim of radiation sickness, 23 days after the Hiroshima bombing. The spots on his face are hemorrhages beneath the skin, caused by a weakness of the blood vessels and blood-clotting defects. He died a few days later. These acute, short-term effects are quite different from the long-term effects, primarily cancer, caused by low levels of ionizing radiation.

---

* Here is the precise definition of the rem: For gamma rays and X rays, the rem is the amount of radiation that would produce 0.01 joule of absorbed ionizing energy in 1 kilogram of biological material. The definition is a little more complicated for alpha and beta rays.

studies of radiation survivors and their descendants at Hiroshima and Nagasaki have found no observable inherited mutations, but this does not guarantee that no mutations have occurred. Since recessive mutations can lie dormant for several generations before becoming apparent, it will be some time before mutations can be completely ruled out among descendants of nuclear bomb survivors.

The third form of damage is **cancer** in ordinary body cells. The rate of certain cancers observed in nuclear bomb survivors is far above normal. Among the Hiroshima survivors, the leukemia rate between 1950 and 1985 was four times the normal rate for an unexposed population. Many other forms of cancer were double the normal rate. Based on the available statistics, the National Academy of Sciences estimates that a sudden radiation dose of 50 rems produces about a 4% probability of eventual death by radiation-caused cancer. This means that out of every 100 people exposed to 50 rem, an average of 4 will ultimately die of radiation-caused cancer.

The principal open question about radiation and cancer concerns the effects from lower exposures, below 50 rem. Here, statistical evidence gives very uncertain conclusions. The working hypothesis is that smaller exposures do have effects and that these effects are proportional to the exposure level. In other words, since 50 rems causes an estimated 4% probability of cancer, 25 rems is assumed to cause a 2% probability, 12.5 rems is assumed to cause a 1% probability, and so forth.

As an example, one cancer-causing isotope produced by nuclear weapons explosions and nuclear reactors is radioactive strontium, $^{90}_{38}$Sr. After a nuclear explosion, ^{90}Sr and other radioactive isotopes attach themselves to atmospheric dust particles that eventually fall to Earth as **radioactive fallout**. As you can see from the periodic table, strontium is chemically similar to calcium. If you breathe or eat ^{90}Sr, it will do what calcium does, namely, migrate to your bone marrow. Since the bone marrow is where red blood cells are created, the connection between radioactivity and leukemia (a blood condition) is not surprising. Isotopes such as ^{90}Sr that are both radioactive and chemically active in the human body are especially dangerous, because once inside the body they combine with other chemicals and stay there, rather than being quickly released. Because of isotopes like ^{90}Sr, radioactivity is much more dangerous inside the body than outside it.

There is a lot of disagreement about the long-term effects of low radiation doses, comparable to the amounts people receive every day. Because the effects are difficult to observe among human populations, predictions also are difficult to check. A prudent working hypothesis is that even a single microscopic decay process has a small probability of causing a cancer or mutation and that at these low levels the probability of harm is proportional to the amount of radiation received. This hypothesis is considered prudent or conservative in the sense that the facts might be less severe than this hypothesis predicts.

Table 15.5 shows what kinds of radiation most Americans receive. As you can see, people generally receive about one-third of a rem during each year of life. In addition to the scientific uncertainty about the damage caused by such low doses, the doses in Table 15.5 are themselves uncertain estimates.

Figure 15.15 shows the relative significance of each source of radiation. Most of it is natural, from sources that have been present in nature for thou-

TABLE **15.5**

Ionizing Radiation Received by the U.S. Population. Estimated annual effective dose received in one year from various sources, averaged (per person) across the U.S. population.

Source	Annual dose per person (millirem)
Natural	
Radon from the ground	200
Cosmic rays	27
Rocks and soil	28
Internal consumption	40
Subtotal	295
Artificial	
Medical and dental X rays	39
Nuclear medicine	14
Consumer products	10
All other[a]	1
Subtotal	64
Grand total	359

[a] Occupational, nuclear weapons, nuclear power and its fuel cycle, and miscellaneous.

SOURCE: National Research Council's fifth committee on the biological effects of ionizing radiation (BEIR V), 1990

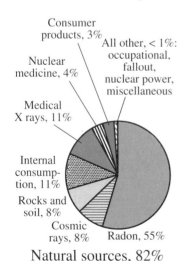

Nonnatural sources, 18%

Consumer products, 3%

All other, < 1%: occupational, fallout, nuclear power, miscellaneous

Nuclear medicine, 4%

Medical X rays, 11%

Internal consumption, 11%

Rocks and soil, 8%

Cosmic rays, 8%

Radon, 55%

Natural sources, 82%

FIGURE 15.15

Sources of radioactive risk: the relative contributions from various radiation sources that the average U.S. citizen commonly receives.

sands of years. Two widely discussed artificial sources, nuclear power and nuclear weapons tests, contribute only a tiny fraction. By far the largest artificial source is medical.

Most of the average dose is thought to come from a single source, **radon**, which comes from the alpha decay of radium in the ground. Because radon is a gas, it escapes into the atmosphere shortly after it is created underground. It is not really radon itself that is dangerous, because it is chemically inert and is breathed in and right back out before it can decay (its half-life is 4 days). But radon's decay products are both radioactive and chemically active, and they attach themselves to microscopic airborne particles, are breathed in, become lodged in the lungs, and can lead to lung cancer. Because radon collects inside closed houses, it is five times more concentrated in the average U.S. home than in outdoor air. Most of the radon in homes enters through the substructure. Radon levels vary widely among different houses, with some not much above outdoor levels and others a hundred or a thousand times greater.

You can check the radon level in your home by buying at a hardware store a small device called a track detector. Alpha particles leave a track as they pass through the detector. After the detector has been in your house for a few months, you can mail it to a laboratory where the tracks are counted and used to calculate your home's radon level. If your home has a high level, you might want to seek professional assistance in reducing it. Sealing cracks doesn't help much, because air pressure differences suck air from the ground through the substructure and up into the house, even in well-sealed houses. Simple ventilation can help, though. The most effective solution is a blower inserted between the house's concrete base and the gravel on which the concrete is laid, to change the pressure system beneath the house so that the house will no longer suck air from the ground.

15.7 Risk assessment: *dealing with risk in a technological society*

To live is to risk: There is risk in drinking a cup of coffee, there is risk in going out of the house, and there is risk in not going out of the house (such as breathing in radon).

The nuclear accident at **Chernobyl** provides a good starting point for thinking about risk. In 1986, an explosion and fire at the former Soviet Union's Chernobyl nuclear power plant threw massive amounts of radioactive isotopes into the air. It was history's worst nuclear power accident. The immediate toll was 237 cases of acute radiation sickness, with 31 fatalities within a month, all among the reactor's personnel or emergency response personnel such as fire fighters.

It is possible to estimate the number of long-term cancer deaths that may be caused by Chernobyl during the succeeding several decades. Radioactive fallout from the accident spread over much of Europe, producing an irregular fallout pattern (Figure 15.16). Many Europeans received total individual doses ranging from 10 millirem (equivalent to one diagnostic X ray) to over 1 rem. The fallout doses were much larger closer to Chernobyl. Over 100,000 people were evacuated from a 30-kilometer zone around the accident, after receiving some 20 rems per person within a few days. Significant effects persist today in Ukraine and Belarus in the former Soviet Union.

The two most harmful isotopes were radioactive iodine and cesium, ^{131}I and ^{137}Cs. Both are biologically active. Although ^{131}I has a half-life of only 8 days and so soon decayed to harmless levels, it was inhaled during the first 8 days after the accident, and it also fell onto grass that cows ate, so it entered milk supplies. In addition, ^{137}Cs has a half-life of 30 years and so will remain dangerous in the soil for decades, although it will sink into the ground and after a few years offer less surface radiation. Both isotopes are expected to cause future cancer deaths.

Some 18,000 cancer deaths are expected from the accident (Table 15.6). This is certainly a large number of deaths. Nevertheless, it is not large enough to be measurable by any feasible method.* The problem is that there is no feasible way to tell whether a particular cancer was caused by Chernobyl or by something else. And the Chernobyl numbers are so small, relative to the total cancer rate, that the excess cancers outside the 30-kilometer zone around Chernobyl will not be observable in any future statistics. For example, the 7000 Soviet deaths will add only 0.02% to the normal 35 million can-

---

* This has even caused some people to argue that the excess cancers don't exist because they can't be measured, since science deals with only observable effects. But this point about the scientific method applies only to effects that cannot conceivably be observed because of the principles of physics. There are, however, conceivable (but impractical) ways to determine how many cancers are caused by Chernobyl, for instance, by directly monitoring every cell in every person's body.

FIGURE 15.16

The spread of radiation following the Chernobyl nuclear power plant accident. The three shaded regions represent three different levels of exposure to ^{131}I. During the first four days following the accident, persons in the three gray areas received the iodine radiation exposures indicated.

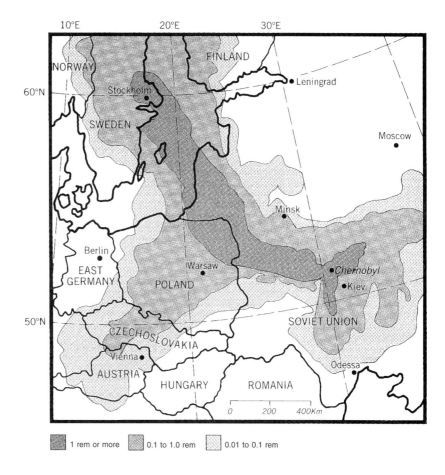

cer deaths (Table 15.6). This increase is far smaller than the unpredictable natural variations (the "noise") in the measured cancer rate.

What are we to make of an estimated 18,000 deaths that cannot be observed? One way to make sense of such numbers is by making numerical comparisons with other risks. For example, the 18,000 long-term worldwide deaths from Chernobyl are fewer than half the number of people killed by automobiles in one year alone in the United States alone (nearly 50,000—a number that I like to remember for comparison with other statistics that I

TABLE 15.6

Projected Cancer Deaths Caused by Chernobyl, in Different Parts of the World, Compared with Total Populations and with Expected Total Cancer Deaths

Region	Population (M = million)	Cancer deaths (lifetime) Natural	Cancer deaths (lifetime) Caused by Chernobyl
Former Soviet Union (FSU)	280 M	35 M	7,000
Europe (outside FSU)	490 M	88 M	10,000
Asia (outside FSU)	1900 M	342 M	500
U.S.A. and Canada	250 M	48 M	20

SOURCE: Lynn R. Anspaugh et al., *Science*, December 16, 1988, pp. 1513–1519.

come across in the news) and about the same as the number of murders in the United States in one year (20,000). Comparisons such as this can help put abstract numbers into perspective.

For another example, radon exposure causes between 5000 and 20,000 (the number is highly uncertain) U.S. lung cancer deaths per year. Since this is comparable to (although somewhat less than) the annual numbers of U.S. car deaths and murders—and these are generally considered serious problems—radon should be considered serious, too.

One general way to comprehend such risks is to look at probabilities of death for one person. For example, the fallout from Chernobyl might kill 7000 citizens of the former Soviet Union, out of a total population of 280×10^6. So the fraction of the total population that will die from this cause is $7000/280 \times 10^6$, or about 25/1,000,000. Another way of looking at this is that each of these people has, on the average, a lifetime risk of 25 in 1 million of dying from Chernobyl-caused cancer. That is, in a randomly chosen group of 1 million people in the former Soviet Union, an average of 25 will die from this cause.

Now compare this with the yearly risk of a U.S. citizen dying from radon-caused cancer. The fraction dying each year from radon is 12,000 (the expected yearly deaths) divided by 250 million (the U.S. population). The answer is 50/1,000,000, twice the lifetime Chernobyl risk to former Soviet citizens. Again, radon is significant.

Quantitative risk estimates such as these give us a rational way to compare risks. Table 15.7 lists several activities, all of which carry the same quantitative risk of death, 1 in 1 million. All these risks are equivalent in the sense that if everybody in a population of 1 million were to perform any one of the indicated activities, one of them (on the average) would eventually die from it. From an individual person's point of view, every time he or she does one of these activities, there is 1 chance in 1 million (still on the average) that it will eventually kill that person. It's a little unnerving.

To say that everything has its risks is really another way of saying that we all must die. In the long run, the total risk of death is 100%. It is sad but true, and one might as well try to deal with it sensibly. Many risks, such as radon exposure, are impossible to reduce to negligible proportions, and it is usually impossible to reduce any particular risk to absolutely zero. Furthermore, attempts to reduce risk can sometimes be counterproductive. For example, if you don't get that chest X ray, you might die from tuberculosis. You could significantly reduce your radon risk by living outdoors, but that would expose you to other risks. If you really wanted to reduce your radon risk, you could live outdoors and breathe only once every hour. But then you would suffocate. You could decide to do nothing that is nonessential if it carried any risk. But then you would probably die of boredom.

Radiation safety specialists have thought a lot about radiation risk, and they offer two general pieces of wisdom: First, every radiation dose may carry some risk, so no exposure is permissible unless it carries a compensating benefit. Second, the dose to any person should be kept as low as is reasonably possible, taking into consideration all other factors (social, economic, and so forth). In other words, be aware, and balance the risks against the benefits. Similar principles apply to all the risks of living.

TABLE **15.7**

Activities Carrying an Average Risk of Death of One Part per Million

Activity	Cause(s) of death
Ionizing radiation	
One chest X ray at a good hospital	Cancer from ionizing radiation
Traveling cross-country once by jet	Cancer from cosmic ionizing radiation
Living 1 week in a building	Cancer from indoor radon
Living 5 weeks outdoors	Cancer from outdoor radon
Living 2 months in Denver	Cancer from cosmic ionizing radiation
Living 5 years next to nuclear power plant	Cancer from ionizing radiation
Living 50 yrs within 5 mi of nuclear power plant	Accident
Internal consumption	
Smoking 1.4 cigarettes	Cancer and heart disease
Living 2 months with a cigarette smoker	Cancer and heart disease
Drinking 0.5 liters of wine	Cirrhosis of the liver
Normal consumption of tap water for 1 year	Cancer from chloroform
Drinking 30 12-ounce cans of diet soda	Cancer from saccharin
Eating 40 tablespoons of peanut butter	Liver cancer from aflatoxin B
Eating 100 charcoal broiled steaks	Cancer from benzopyrene
Travel	
3 miles by motorcycle	Accident
10 miles by bicycle	Accident
30 miles by car	Accident
800 miles by train	Accident
1000 miles by commercial airplane	Accident
6000 miles (cross-country) by jet	Cancer from cosmic ionizing radiation
Work	
Spending 1 hour in a coal mine	Black lung disease
Spending 3 hours in a coal mine	Accident
Other	
Living 2 days in New York or Boston	Air pollution

SOURCE: Richard Wilson, "Comparing Risks," *Physics and Society*, October 1990, pp. 3–5.

DIALOGUE 11   Given that ^{131}I and ^{137}Cs have half-lives of 8 days and 30 years, respectively, how much time must pass after the Chernobyl accident before the radiation from each isotope will have decreased to 1% of its original level?

DIALOGUE 12 MAKING ESTIMATES   Your doctor recommends a diagnostic X-ray exposure of 1000 millirem (1 rem). The lifetime risk of death from a single 1 rem dose is 0.1%. In a population of 1 million, how many would be killed by this exposure? Suppose the disease is detected, thanks to the X rays, and then treated. If the disease had a 10% chance of killing you, did you reduce your risk by undergoing the X-ray diagnosis?

11. According to Figure 15.11, after 6 to 7 half-lives, the decay curve is down to about 1%. For ^{131}I, 6 half-lives is 48 days, and for ^{137}Cs, 6 half-lives is 180 years.

12. Because 0.1% is 1 part per 1000, or 1000 parts in 1,000,000, about 1000 persons would be killed. You reduced your risk from 10% to 0.1% (a risk-reduction factor of 100).

Summary of Ideas and Terms

Strong nuclear force Holds the nucleus together, acts between nuclear particles (protons, neutrons), is strongly attractive at separations of around 10^{-15} m, and is negligible at larger distances.

The four fundamental forces Gravitational, electromagnetic, strong nuclear, and weak nuclear.

Nuclear reaction Any process that alters the structure of a nucleus.

Isotope A particular type of nucleus. Specified by its **atomic number** (number of protons) and **mass number** (total protons and neutrons). A symbol like $^{14}_{6}C$ represents an isotope with the atomic number 6 and the mass number 14.

Radioactivity or **radioactive decay** The spontaneous emission, by a nucleus, of an **alpha particle** (a helium nucleus) or a **beta particle** (an electron created in the nucleus). Any nucleus that can do this is **radioactive**, or **unstable**. At the time of decay, a **gamma ray** (a high-energy photon) is also emitted.

Radioactive isotope An isotope that is radioactive. An isotope that is not radioactive is **stable**.

Half-life of a radioactive isotope. The time during which half of a macroscopic amount of the isotope will decay.

Decay curve A graph of the amount of a radioactive material remaining, versus time. Decay curves follow **exponential decay**, meaning that they have a fixed halving time (the half-life).

Radioactive dating Determining the ages of old objects by means of radioactive methods. In **carbon dating**, radioactive ^{14}C in a dead organism is measured as a fraction of the total carbon in order to determine how long the ^{14}C has been decaying and when the organism died.

Radioactive decay sequence A sequence of radioactive decay processes that occur when a radioactive isotope decays to become another isotope that is also radioactive.

Ages Some approximate ages determined by several radioactive and other methods are Earth, 5 billion years; life, 3 billion years; humans (hominids), 4 million years.

Creationism The belief that the biblical book of Genesis can be read as scientific truth and that Earth and the major biological organisms all were created only a few thousand years ago.

Ionizing radiation Any atomic or subatomic emissions that are capable of ionizing biological molecules. Alpha particles, beta particles, gamma rays, and X rays are the main types. The biological damage is measured in a unit called a **rem**. The main types of damage are radiation sickness, mutations, and long-term cancers.

Radioactive fallout Dust that falls to the ground carrying radioactive isotopes from a nuclear explosion or nuclear accident.

Natural radiation is ionizing radiation from natural sources, and **artificial radiation** is from human-made sources. Natural sources include **radon gas**, cosmic rays, the ground, and internal consumption. The main artificial source is medicine.

Chernobyl Site of history's worst nuclear power accident, in 1986. A few tens were killed from short-term effects, and thousands more are expected to die of long-term cancers.

Quantitative risk estimates The quantitative evaluation of human-made and natural risks, especially for the purpose of comparing different risks.

Review Questions

NUCLEAR FORCES

1. In what ways is the nucleus non-Newtonian?
2. Name the four fundamental forces, and describe the main function of each of the three that act as "glues" to hold things together.
3. Which of the four forces holds the nucleus together, and which tends to push it apart?
4. What property of the strong force causes nuclei to be so small? Why can't much larger nuclei exist?

NUCLEAR ENERGY AND NUCLEAR STRUCTURE

5. Explain why nuclear particles must move rapidly and so must have high energy.
6. How are nuclear reactions similar to alchemy?
7. What is an isotope, and what quantities need to be known in order to specify a particular isotope?
8. What are the differences among $^{12}_{6}C$, $^{14}_{6}C$, and $^{14}_{7}N$? Compare their numbers of protons, numbers of neutrons, chemistry, nuclear chemistry (behavior in nuclear reactions), numbers of orbital electrons in the neutral atom, and mass.

RADIOACTIVE DECAY

9. What kinds of rays, or particles, are emitted by radioactive nuclei? Describe each kind.
10. What happens to the atomic number and mass number during alpha decay? During beta decay?
11. What energy transformation occurs during radioactive decay?
12. List some useful applications of radioactivity.

HALF-LIFE

13. In radioactive decay, what quantity cannot be predicted? Of what basic principle is this an example?
14. We know that ^{131}I has an 8-day half-life. If you start with 100 grams, how much will remain after 24 days?

15. If you started with 100 grams of ^{14}C (6000-year half-life) and only 3 grams remains, about how time has elapsed?

RADIOACTIVE DATING

16. Explain how carbon dating works. What kinds of objects does it work on? Where does radioactive carbon come from?
17. We know that $^{238}_{92}U$ is an alpha emitter and that its daughter nucleus is a beta emitter. What are the atomic numbers and mass numbers of the next two isotopes in the decay sequence that begins with ^{238}U?
18. According to science, about how old is Earth: a few thousand years, a few million years, a few hundred million years, a few billion years, or a few trillion years? How about the human race (hominids)?
19. Describe at least one way to cross-check a particular dating method such as carbon dating.

HUMAN EXPOSURE TO IONIZING RADIATION

20. Which of the following are ionizing radiations: radio, ultraviolet, gamma, alpha rays, X rays?
21. What is a rem?
22. Name and describe the three main types of biological damage caused by ionizing radiation.
23. List two natural sources of ionizing radiation and one artificial source.

RISK ASSESSMENT

24. Will Chernobyl cause long-term cancers? Do scientists expect to observe these effects? Why?
25. What is fallout, and where can it come from?
26. What is the meaning of a risk (of cancer, for example) of 3/1,000,000 per person?
27. Where do we find radon in our normal living environment? How does it get there?

Home Projects

1. The decay game. Get 100 pennies, perhaps from a bank. Think of them as a new radioactive isotope, "coinium." One coin "decays" when it comes up tails. Start with your sample of 100 coins, and toss them all on the floor. Remove the ones that decayed. Toss the remainder, and remove the ones that decayed. And so forth. Graph the number remaining versus the number of "years" (throws). What is the half-life? How many years do you predict it will take until your coinium sample has decayed to just one coin? Predict the amount of coinium left after 3 years (it's 12.5); after 5 years. Let your coinium decay all the way to zero, several times; find the experimental average amount remaining after 3 and 5 years; and compare this with your predictions. Note the in-creased importance of random errors as the number of coins decreases. (Compare this with "the growth game," Home Project 5 in Chapter 7.)

2. You can see individual radioactive decays in watch dials. Luminous dials are coated with either a radium and zinc sulfide mixture or a nonradioactive substance. You can tell the difference because nonradioactive dials get progressively dimmer in the dark whereas radium dials stay bright. Look at a radium dial, in the dark, with a magnifying glass. With a strong magnifying glass, you can see individual tiny flashes. Each flash occurs when an alpha particle from a single radium nucleus strikes a zinc sulfide molecule. These flashes are unpredictable: You are seeing quantum uncertainty!

For Discussion

1. Discuss the issue of evolution versus creationism. Do you think that science's conclusions about Earth's age and other ages are correct? If not, why not? If so, do these conclusions contradict the 5000-year age of Earth that is sometimes deduced from the Bible? If there is a contradiction or disagreement between science and the Bible, how do you propose to resolve it? Is science correct, is the Bible correct, is neither correct, or is there some way in which they both might be correct?

2. Discuss the ethical implications of Tables 15.5, 15.6, and 15.7 by comparing some of the risks. Compare and contrast the number of deaths at Chernobyl with the number of U.S. auto deaths, in view of the fact that the number of long-term cancer deaths from Chernobyl is about the same as the number of deaths on U.S. highways in 4 months. Is there any sense in which the radioactive risks in Table 15.7 are more, or less, significant than the other risks? Compare and contrast some of the quantitatively equivalent risks in Table 15.7: Are they really equivalent, or are there significant differences? Which are avoidable? Of the avoidable risks, which should you try to avoid or reduce? Can, or should, governments help reduce any of these risks?

Exercises

NUCLEAR FORCES AND STRUCTURE

1. Which force is stronger between two protons separated by 10^{-15} m (the size of a small nucleus), the electric or the strong force? What evidence do you have for your answer?

2. Which force is stronger between two protons separated by 10^{-10} m (the size of an atom), the electric or the strong force? What evidence do you have for your answer?

3. How many protons and neutrons are there in these nuclei: $^{13}_{6}C$, $^{56}_{26}Fe$, $^{90}_{38}Sr$, $^{3}_{1}H$?

4. How do the masses of ^{1}H, ^{2}H, and ^{3}H compare? How do their charges compare?

5. How do the masses of ^{3}H and ^{3}He compare? How do their charges compare?

RADIOACTIVE DECAY

6. Why are radioactive materials often warm?

7. What do you suppose heated the water in a naturally heated hot spring?

8. Radioactive materials often glow in the dark. But the gamma radiation emitted by a nucleus during decay is not visually detectable, so where does the light come from?

9. Which one is most similar to X rays: alpha, beta, or gamma rays?

10. Can a hydrogen nucleus emit an alpha particle?

11. Can an element decay "forward" in the periodic table to a higher atomic number?

12. Use the periodic table to find the residual nucleus in each of the following: beta decay of ^{3}H, beta decay of ^{90}Sr, alpha decay of ^{222}Rn.

HALF-LIFE

13. If a radioactive isotope has a 1-year half-life, what fraction will remain after 5 years?

14. Elements above uranium (larger atomic numbers) don't exist in nature because they have short half-lives and have decayed away since Earth was created. But there are several elements below uranium with short half-lives that do exist in nature. How can this be?

15. Radon has a 4-day half-life. Starting with 1 gram of radon, how much will remain after 4 days? Starting with 10 radon atoms, how many will remain after 4 days? Is this a precise prediction? Why? If you have just 1 radon atom, will you still have it after 4 days?

16.*You have a gram of ^{131}I, a gram of ^{222}Rn, and a gram of ^{234}Th. Use Table 15.1 to predict how much of each you will have after 24 days.

17.*You start with 5 grams of ^{131}I. How much will remain after 20 days? How long will it take the radiation to decline to 5% of its original value?

RADIOACTIVE DATING

18. Can we carbon-date ordinary rocks? Why? Can we carbon-date objects that are a million years old? Why?

19. How does the impact of cosmic rays on the atmosphere affect carbon dating?

20. Carbon dating was used to find the age of the Dead Sea Scrolls. Would this method have worked if the scrolls had been carved in stone?

21.*The carbon radioactivity of an old wooden ax handle has declined to 20% of its original value. Estimate the age of the ax handle.

22.*The ^{238}U-to-lead ratio in a rock is found to be 70% of the ratio in similar rock that recently crystallized. How long ago did this particular rock crystallize?

RISK ASSESSMENT

23. At a party, Edgar drinks half a liter of wine and smokes four cigarettes. According to Table 15.7, which was more risky, the drinking or the smoking? How much more dangerous? (Assume that Edgar doesn't drink and drive!)

24.*You travel 1500 miles by car. Use Table 15.7 to find your risk of death by accident, assuming that you are an average driver. If you had flown by jet, what would have been your risk of death by accident or by cancer from the cosmic radiation at high altitudes? What if you had traveled by train? Motorcycle?

25.*The number of people killed worldwide by volcanoes has increased from 315 per year in 1800 to 845 per year in the mid-twentieth century. But the population has also increased during that time, from 750 million to 3 billion. Has the yearly risk per person from volcanoes increased or decreased?

26.*Making estimates: The average risk of daily life. Everybody dies precisely once. Average this one death over a typical life of 70 years, and show that the average risk of death per day from all causes is 1 chance in 25,000, or 0.00004, or 40 chances in a million.

27.*Making estimates: the meaning of one in a million. Table 15.7 says that there is a death risk of 10^{-6} in smoking 1.4 cigarettes. Suppose you smoke an average of 1.4 cigarettes daily for 40 years. How large will your overall total risk be from smoking? What if it's a pack (20 cigarettes) per day instead?

16

FUSION AND FISSION
—and a new energy

We get useful energy from each of the three fundamental "glues": gravity, electromagnetism, and the strong nuclear force. For example, we get electric power from the gravitational forces acting on a lake held up behind a hydroelectric dam and from microscopic electric forces that cause molecules to combust chemically in a coal-burning plant. As we will see, we can also get electric power from the strong nuclear force.

In our familiar macroscopic world, the gravitational force is the most obvious, and the strong nuclear force the least obvious of the three. In the microscopic world, this order is reversed. Acting between such subatomic particles as neighboring protons within a nucleus, the strong nuclear force is by far the strongest of the three; the electric force is next in strength; and gravity is by far the weakest.

Because of the different strengths of the three forces, equal amounts of energy output from each of the three require quite different amounts of fuel. For example, a 1000 MW hydroelectric power plant uses the gravitational energy of some 60,000 tonnes of water every second; a 1000 MW coal-burning power plant requires about 10,000 tonnes of coal (150 truckloads or 1 trainload) every day; and a 1000 MW nuclear power plant uses only some 100 tonnes of uranium (a few truckloads) every year. These three forces also differ in their destructive power. Towns of a few thousand people have been leveled by the gravitational energy in millions of tonnes of Earth or water in a landslide or flood. A town of this size can also be leveled by perhaps 1000 tonnes of a high explosive in several hundred chemical bombs. But Hiroshima, a city of a quarter of a million people, was leveled by one nuclear bomb carrying 40 kilograms of uranium.

The strongest of the three forces is also the least understood. Humans have probably had some intuitive understanding of gravity for millions of years. Electromagnetism began to be understood in the eighteenth century, although magnetic effects in "lodestones" (the naturally magnetic mineral magnetite) were observed many centuries ago. We became aware of the strong nuclear force only during the twentieth century and are still far from a fundamental scientific understanding of it.

Fusion and fission, the topic of this chapter, are nuclear reactions in which the strong nuclear force comes fully into play. The most useful way for us to think about the physics of these reactions is in terms of energy, a recurring theme of this book. Section 16.1 continues the discussion of nuclear energy that was begun in Section 15.2. Section 16.2 looks at the physics of fusion and why the stars shine. Section 16.3 applies these ideas to tell one of nature's most fascinating tales: how the atoms in your body and elsewhere were forged in stars. Section 16.4 presents the physics of fission, and Section 16.5 discusses the method, called a chain reaction, by which fission is used to get useful energy from the nucleus. Unfortunately, history's first large-scale application of nuclear physics was to building history's most terrible weapon (Section 16.6). Because we all must learn the lessons of this tale, Sections 16.4, 16.5, and 16.6 are presented in their historical context. Section 16.7 presents another grim nuclear technology, the fusion bomb.

16.1 The energy of the nucleus

Radioactive decay is one way to get energy from a nucleus. Radioactivity converts nuclear energy into other forms:

$$\text{nuclear energy} \rightarrow \text{thermal energy} + \text{radiant energy}$$

There are two other ways to get nuclear energy, fusion and fission. Briefly, in **nuclear fusion**, two nuclei combine (fuse) to form a single larger nucleus, and in **nuclear fission**, a single nucleus splits (fissions) roughly in half to form two smaller nuclei.

Continuing the nuclear detective story begun in Section 15.1, let's see what we can deduce about the energetics of fusion and fission. As detectives, we must look closely at certain interesting details. We need two such details: (1) The strong nuclear force that holds the nucleus together is strongly attractive but short ranged, and (2) protons exert a repulsive electric force on one another that is weaker but much longer ranged than the strong force.

We begin with the simplest multiparticle nucleus, the hydrogen isotope ^2_1H, made of one proton and one neutron. Suppose you tried to separate this nucleus into an isolated proton and neutron. Because the proton and neutron are held together by the strong nuclear force, this would not be easy; you would have to do a lot of work to pull them apart (Figure 16.1).

FIGURE 16.1

You would have to do work to pull apart the two particles in the nucleus of ^2H.

(a) In which case does the system have greater gravitational energy: when the rock is at a lower or a higher altitude?

(b) Which has greater nuclear energy: a proton and neutron that are closer together or farther apart?

FIGURE 16.2

(a) If you begin with a carbon nucleus —

(b) and put energy into the system by pulling it apart into two fragments—

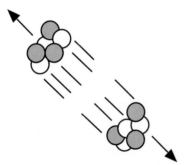

(c) you will get some of the energy back when the fragments are pushed apart by the electric forces between them.

FIGURE 16.3

DIALOGUE 1   Which situation has more energy, the $^2_1$H nucleus before separation or the separated proton and neutron?

The separated proton and neutron have more energy than does the $^2_1$H nucleus. This excess energy is a form of **nuclear energy**, since it is due to nuclear forces. The energy relationship here is like that of two objects, such as Earth and a rock, that are attracted by gravity: The gravitational energy is larger when the rock is held higher above Earth's surface, that is, when the two objects have a larger separation (Figure 16.2).

A similar argument applies to every nucleus. Pulling apart any nucleus into individual protons and neutrons requires work. So the energy of any nucleus is always less than the total energy that its protons and neutrons would have if they all were separate from one another. This is a basic feature of nuclear energy.

Now let's consider a larger nucleus, say $^{12}_6$C, which contains six protons and six neutrons. Imagine pulling it apart into two equal fragments:

$$^{12}_6C \rightarrow {}^6_3Li + {}^6_3Li$$

To do this, you must do work to overcome the strong nuclear forces in the carbon nucleus. But as soon as you get the two fragments far enough apart to be out of range of the strong nuclear attraction between the fragments, something new happens: The two fragments now repel each other because they continue to feel an electric force from the protons in the other fragment. So the fragments are now pushed away from each other (Figure 16.3). That is, some of the energy that you put into the $^{12}_6$C nucleus to separate it is retrieved in the form of an electric push that sends the fragments flying apart.

If you start instead with a larger nucleus, say $^{24}_{12}$Mg, and pull it apart into two fragments, you must again do work to pull the nucleus a short distance apart, and you retrieve some of this energy from the electric push that then sends the two fragments flying apart. Look at this more closely: To pull the $^{24}_{12}$Mg nucleus apart, you must do only a little more work than you did to pull the $^{12}_6$C nucleus apart. The reason is that the fragments feel only the nuclear forces of their nearest protons and neutrons, because the nuclear force has a short range. But once the two fragments are separated, the electric push between them is much stronger than it was when you started with $^{12}_6$C. The reason is that all the protons in each fragment feel the electric force from all the protons in the other fragment, because the electric force has a long range.

To summarize this important point: Whereas it takes only a little more work to separate a larger nucleus into two fragments, we get much more energy back from the electric push between its fragments than we get back when we separate a smaller nucleus into fragments.

This leads to an interesting deduction. As we consider separating larger and larger nuclei into two fragments, we should reach a point at which the

---

1. Because you must do work on the nucleus to separate it, the work–energy principle says that the system's energy increases, so the separate proton and neutron have more energy than does the $^2_1$H nucleus. But the nuclear *force* is larger when the proton and neutron are closer together.

energy we get back from the electric push between the fragments is actually larger than the energy we had to put in to separate the nucleus.

This turnaround point in the energy balance occurs when the atomic number reaches 26 (iron). The energetics of separating a nucleus that is lighter than iron is quite different from the energetics of separating a nucleus into two fragments that are themselves heavier than iron.* In separating the lighter nucleus, we must put net energy into the nucleus because we get back less energy than we put in. This energy shows up as a larger total nuclear energy of the two separated nuclei. But in separating the heavier nucleus, we get net energy from the nucleus because we get back more energy than we put in. This energy shows up as a smaller nuclear energy of the two separated nuclei.

Figure 16.4 is a useful way to represent these deductions. The nuclear energy per nuclear particle is graphed versus the mass number (the number of nuclear particles). This **nuclear energy curve** shows the two features that we have deduced: First, the entire curve lies below the straight line representing the energy of the separated nuclear particles. This must be so because it takes work to separate any nucleus into its component protons and neutrons. Second, the curve decreases until we reach about mass number 55, corresponding to iron (atomic number 26), after which it increases. So nuclear energy is gained if a nucleus lighter than iron is separated into two smaller parts (we move "uphill" on the left side of the nuclear energy curve), and nuclear energy is lost if a nucleus much heavier than iron is separated into two parts (we move "downhill" on the right side of the nuclear energy curve).

The shape of the nuclear energy curve reveals two important facts about possible ways of obtaining nuclear energy: First, if we could find a practical way to combine (fuse) nuclei, we could transform nuclear energy into other forms by fusing elements much lighter than iron to make heavier elements,

FIGURE 16.4

The nuclear energy curve. The solid curve shows the energy per nuclear particle for nuclei of various mass numbers. Iron, with a mass number of about 55 and an atomic number of 26, has the least nuclear energy per particle, so it is the most stable nucleus.

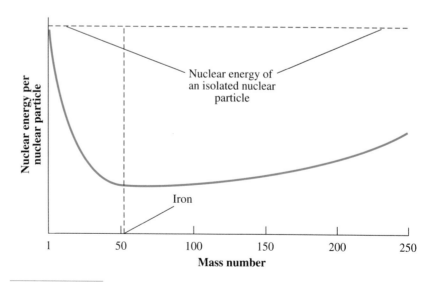

_____

* If the original nucleus is heavier than iron and either fragment is lighter than iron, the situation is more complicated.

because we would then be working "downward" on the nuclear energy curve. We can "release" nuclear energy (transform it into other forms) by fusing light nuclei.

Second, if we could find a practical way to fission nuclei, we could transform nuclear energy into other forms by fissioning elements much heavier than iron into lighter elements, because we would again be working downward on the nuclear energy curve. We can release nuclear energy by fissioning heavy nuclei.

This concludes our detective work. It appears that energy can be obtained from light nuclei by fusion and from heavy nuclei by fission if there is a feasible method of actually causing either reaction.

DIALOGUE 2   Suppose you fused two $_2^4$He nuclei. What nucleus would this create? Would this process transform nuclear energy into other energy forms, or would it transform other energy forms into nuclear energy?

DIALOGUE 3   Suppose you fused three $_2^4$He nuclei into a single nucleus. What nucleus does this create?

DIALOGUE 4   Suppose that you fissioned a sulfur nucleus into two fragments. Does this process convert nuclear energy to other forms, or is it vice versa? Suppose that instead you fissioned a lead nucleus. Now which direction does the energy transformation go?

16.2 Fusion: *the fire in the sun*

We saw in the preceding section that the fusion of two light nuclei, such as two hydrogen nuclei, transforms nuclear energy into other, possibly useful, forms. But is there a practical way to do this?

The way to do this would be to bring the two hydrogen nuclei so close together that they are within each other's strong nuclear force range (10^{-15} meters). Then they would suddenly snap tightly together because they would be attracted strongly by each other's strong nuclear force. But it would be hard to get them this close because while still far outside each other's strong nuclear force range, they feel an electric repulsion owing to their positive charges. In order to get them within each other's strong nuclear force range, this repulsion must be overcome. Nature manages to carry out this feat at the centers of stars, where stars get their energy.

Stars are made mostly of hydrogen, and they are millions of degrees hot at the center. Atoms at the center are moving so rapidly that they strip off

---

2. $_4^8$Be. Nuclear energy is transformed into other energy forms.
3. $_6^{12}$C.
4. Since sulfur is lighter than iron, other energy forms are transformed into nuclear—you would have to put net energy into the system to cause the reaction. But if you fissioned lead, which is much heavier than iron, nuclear energy would be transformed into other forms.

one another's orbital electrons when they collide, so that a star's center is an ionized gas of hydrogen nuclei plus unattached electrons. The nuclei are moving so rapidly that they are able to move through one another's region of repulsive electric force and get close enough together to feel one another's strong nuclear force. Then they fuse.

Our sun is a typical example (Section 5.3). It was born about 5 billion years ago from a large gas cloud. During a period of only 100 million years, this cloud collapsed down to nearly its present size, drawn together by its own gravitational attraction. As it fell inward, it warmed because the inward-falling gas and dust particles sped up and then collided with one another. It continued warming until the cloud's center reached millions of degrees, a temperature high enough to initiate the fusion process just described. Then the sun turned on, and it has kept itself turned on ever since.

The sun gets its energy by the fusion of hydrogen into helium, mainly via the reaction*

$$\mathrm{^1_1H + {}^2_1H \rightarrow {}^3_2He}$$

This reaction works downward on the nuclear energy curve, converting nuclear energy to other forms. Since the sun is hot and bright, it is not hard to guess that two of the major other forms are thermal and radiant energy. So the dominant energy transformation is

nuclear energy → thermal energy + radiant energy

Ever since fusion began, the sun has created thermal energy from the fusion reaction itself. This keeps the sun hot enough to maintain fusion at its center, which is why it and other stars shine.

Fusion in stars is similar to chemical combustion. Once initiated by some external process, both processes sustain themselves by creating the thermal energy they need. Because of the central role of thermal energy, a self-sustaining fusion reaction is called a **thermonuclear reaction**.

Nuclear reactions are good examples of mass–energy equivalence (Section 11.4), because they usually produce measurable mass changes.[†] The reason is that the nucleus's small size makes it inherently relativistic—its parts must move fast in order to satisfy the uncertainty principle.

HOW DO WE KNOW? CHECKING $E = mc^2$

Nuclear reactions are one type of process in which Einstein's famous mass–energy equivalence formula is readily validated. As an example, let's consider the hydrogen-to-helium fusion reaction. The nuclear energy curve tells us that a helium nucleus has less energy than do two hydrogen nuclei. So the principle of mass–energy equivalence predicts that the helium nucleus should also have less mass (Figure 16.5).

---

[*] More precisely, this nuclear reaction is one step of a three-step process that begins with single protons ^1_1H and creates the common isotope of helium ^4_2He as the output of the process. The reaction given above is the helium-creating step of this process.

[†] There is a popular belief that Einstein's ideas apply only to nuclear energies and that Einstein's relativity played a crucial role in the development of the theory of fission and of nuclear weapons. Actually, $E = mc^2$ applies to every energy transformation, and Einstein's ideas played only a small role in the development of fission and nuclear weapons.

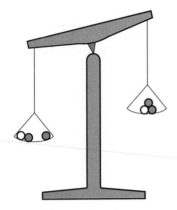

FIGURE 16.5
The mass of the whole does not equal the sum of the masses of its parts: When ^1H and ^2H fuse to form ^3He, the ^3He has less total energy when it is at rest, so it must have less mass.

The masses of all three nuclei are known:

$$\text{mass of } {}_1^1H = 1.6727 \times 10^{-27} \text{ kg}$$

$$\text{mass of } {}_1^2H = 3.3437 \times 10^{-27} \text{ kg}$$

$$\text{mass of } {}_2^3He = 5.0066 \times 10^{-27} \text{ kg}$$

The first two masses add up to 5.0164×10^{-27} kg, which is 0.0098×10^{-27} kg more than the mass of the helium nucleus. Just as Einstein predicted, the mass becomes smaller when the system loses energy.

To validate mass–energy equivalence quantitatively, the energy loss during the reaction (the energy that is transformed into radiant and thermal energy) must be directly measured. This can be done by using the transformed energy to warm up water and measuring the resulting temperature change of the water. The measured energy "released" (in other words, transformed) per individual fusion reaction turns out to be 8.815×10^{-13} joules. Let's see whether this does equal the known mass difference times the square of the speed of light:

$$
\begin{aligned}
0.0098 \times 10^{-27} \times c^2 &= 0.0098 \times 10^{-27} \times 9 \times 10^{16} \\
&= (0.0098 \times 9) \times (10^{-27+16}) \\
&= 0.08815 \times 10^{-11} \\
&= 8.815 \times 10^{-13} \text{ J}
\end{aligned}
$$

It checks.

DIALOGUE 5   The sun's total power output of 400 trillion trillion watts (400×10^{24} joules per second) comes from three types of fusion reactions whose net effect is to convert the sun's hydrogen to ${}_2^4He$. How much mass does the sun lose every second? Despite this enormous mass loss, the sun will end its "life" 5 billion years from now with only 1% less mass than it had when it turned on, 5 billion years ago!

16.3 Creation of the universe and the atoms: *we are star dust*

Where did I come from? is a question we all ask. One answer is that we came from our parents. But there is a shorter-term perspective: We create ourselves at the molecular level day by day, from the food we eat and the air we breathe. There is an evolutionary perspective: Our genetic inheritance evolved from hominid species millions of years ago and from the earliest life billions of years ago. And there is a cosmological perspective. As we will see, the atoms of our bodies came from the stars.

The scientific consensus, since at least 1965, has been that our universe began in a single event some 15 billion years ago, a violent explosion that created space and time themselves, that created the different forms of energy and matter, and that hurled matter and energy outward. The creation event is called the **big bang** (Section 11.7).

---

5. $E/c^2 = (400 \times 10^{24})/(9 \times 10^{16}) = (400/9) \times 10^{24-16} = 44 \times 10^8 \text{ kg} = 4.4 \times 10^9$ kg, or 4.4 million tonnes every second.

Four independent lines of evidence support the big-bang theory:

1. Astronomers proposed the theory in 1929 because they discovered evidence that all the galaxies throughout the universe are receding from one another just as if they had been driven apart by an explosion. There is now considerable evidence, based on observation of the radiation from distant galaxies, confirming this expansion of the universe. Judging from the separation speeds and distances that we see today, the galaxies should have been together at the same place some 15 billion years ago.

2. In 1964, radio astronomers detected the "cosmic background radiation," the faint remnant of the hot initial explosion, that still fills the universe. The radiation has now cooled all the way down to $-270°C$.* This cooled radiation has too little energy to be visible and is now in the radio region of the spectrum. Its observed characteristics, such as its temperature, agree with the big-bang theory's predictions.

3. In 1992 an observing satellite mapped the high-frequency radio radiation from all directions in space. The results showed that the background radiation contains large but very subtle "ripples" of just the sort expected if the initial big bang did indeed develop into the structured universe that we see today (Figure 16.6, see color insert).

4. The fourth line of evidence concerns nuclear fusion and creation of the elements. During the first moments following the big bang, the only kinds of ordinary matter that could exist were unattached protons, neutrons, and electrons. The universe was too hot for protons and neutrons to stick together and form nuclei or for electrons to go into orbit around the protons. It was not until 3 minutes after the beginning that the universe had cooled enough for protons and neutrons to remain attached to one another. According to well-developed theories of nuclear physics, conditions during about the next 1 minute were right for the nuclei to fuse. After this single minute, the universe was too cool and too dilute for fusion. Nuclear physics predicts that during this fourth minute of the universe, about 25% of the protons ($_1^1H$) and neutrons fused into just four different isotopes: $_1^2H$, $_2^3He$, $_2^4He$, and $_3^7Li$. Nuclear physics predicts that by the end of the fourth minute, the universe was made of just five isotopes, in the proportions stated in Table 16.1.

Astronomers have made spectroscopic measurements of the light from the oldest stars, stars that presumably formed from the original material created in the big bang and that have changed little since that time. These measurements show relative amounts of $_1^1H$, $_1^2H$, $_2^3He$, $_2^4He$, and $_3^7Li$ that are in excellent agreement with the theoretically predicted amounts (Table 16.1). This quantitative agreement with the big-bang theory's predictions for so many different isotopes is the most important evidence in support of the theory.

If you're religious, it's like looking at God.

George Smoot, leader of the team that announced in 1992 the discovery of the structure of the cosmic background radiation that remains from the creation of the universe

TABLE 16.1
Predicted Nuclear Composition of Universe at 4 Minutes After Start of Big Bang. Current spectroscopic observations of the oldest material in the universe agree well with these predictions.

Isotope	Relative concentration by mass
$_1^1H$	75%
$_1^2H$	5–10 parts in 100,000
$_2^3He$	2–5 parts in 100,000
$_2^4He$	25%
$_3^7Li$	2–5 parts in 10 billion

* This is just 3 degrees above absolute zero, the lowest possible temperature, the temperature at which all microscopic motion is the least it can be without violating the uncertainty principle.

The universe is still made mostly of hydrogen and helium, although heavier elements created since the big bang now contribute a small percentage. Nearly all the hydrogen and helium can be traced back to the big bang. Although our bodies contain no helium, the hydrogen made in the big bang is one of the four common elements (C, O, H, N) present in living organisms.

Once matter began gravitating together, nuclear fusion turned on the stars and began transforming lighter nuclei into heavier ones. Stars such as the sun convert hydrogen to helium during most of their history. But a star cannot burn (fuse, actually) forever. After a few billion years, a star such as the sun uses up the available hydrogen at its center. Then the fusion ceases; the star collapses further; this collapse causes an increase in temperature; and the higher temperature ignites new fusion processes such as

$$_2^4He + {}_2^4He \rightarrow {}_4^8Be$$

$$_2^4He + {}_4^8Be \rightarrow {}_6^{12}C$$

These hotter thermonuclear reactions now cause the star to expand. It becomes a "red giant," far larger than the original star, destroying any life that might have existed on its planets. The star converts its helium to heavier elements such as carbon. Eventually, all possible fusion material is used up; fusion ceases; and gravity reasserts itself. The star collapses once again, but there is now nothing to stop it until it has shrunk all the way down to become a **white dwarf** (Section 5.3). The elements, such as carbon and oxygen, created by such a star remain bound up in the white dwarf.

Where, then, did all the various elements that we see on Earth come from? The answer is that they came from stars that exploded. Most stars settle down without too much fuss to become white dwarfs. But a few, namely, those that are much more massive than the sun, go through violent death throes. High-mass stars are pulled strongly inward by their own gravity, and so they become extremely hot during the collapse that follows the consumption of the star's usable hydrogen. The high temperatures ignite thermonuclear processes that create heavier and heavier elements, all the way up through iron, with atomic number 26. But the nuclear energy curve (Figure 16.4) tells us that thermonuclear reactions cannot create elements heavier than iron, because such reactions would consume thermal energy, thereby cooling the star and ending the reaction. Iron continues forming at the star's core until the core becomes so massive that it cannot hold itself up against its own gravity. The solid iron core then suddenly collapses, reducing the core's radius from 1000 km to 20 km in just 1 second, an unimaginably violent process. The energy sent outward by this collapse heats and lights the star dramatically. All of this literally rips apart the star in a **supernova explosion** and throws much of its material out into space. The remaining core material collapses further, becoming either a spinning, superdense **neutron star** or a **black hole** (Section 5.4).

What about the elements heavier than iron? No self-sustaining process can create these elements because they require a net increase of nuclear energy and this energy must come from some outside source. This outside source is, again, supernova explosions. Some of the energy of the explosion itself is transformed into the nuclear energy needed to fuse together the elements heavier than iron, from lighter ones.

In our galaxy, only five supernovae have been visible to the naked eye during the past 1000 years. In addition, one was visible in a neighboring galaxy only a few years ago, in 1987 (Figure 16.7). From Earth, a supernova explosion looks like a new (*nova*) star because of the star's sudden brightening; hence its name. The 1987 supernova occurred in a neighboring galaxy, the Large Magellanic Cloud, 160,000 light-years away from Earth.

Without supernova explosions, only the material listed in Table 16.1 could be spread throughout the universe; all stars and planets would be formed from that material; and nobody would be here to tell this story. Our sun is a "second-generation" star that formed several billion years after our galaxy first formed, and so it incorporated heavier elements from supernovae that exploded before the sun was born. The inner planets—Mercury, Venus, Earth, and Mars—were close enough to the sun that their higher temperatures caused the lighter elements, hydrogen and helium, to remain mostly in a gaseous form that then escaped from these planets, leaving only the heavier elements behind along with some hydrogen that was bound chemically in water (H_2O) and other compounds.

We can literally thank our stars for the rare, ancient, and distant supernova explosions that sent forth the elements heavier than helium to places such as this one, where a new star and a new planetary system eventually formed and where a tiny fraction of the fused star stuff has been borrowed, for a few years, by our bodies.

DIALOGUE 6   Because the supernova shown in Figure 16.7 occurred in a galaxy 160,000 light-years away, when did it actually occur?

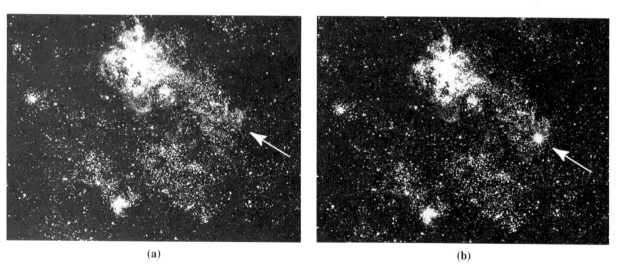

(a)   (b)

FIGURE 16.7
The supernova of 1987, the brightest supernova in 400 years. Its light reached Earth on February 23, 1987. "Before" (a) and "after" (b) photos show the star as it looked before and shortly after the explosion.

DIALOGUE 7   During a supernova explosion, if an iron nucleus captured a neutron and then the nucleus beta-decayed, into what element would it be transformed? What if the iron nucleus captured three neutrons and then beta-decayed twice? This is the type of process that created the elements heavier than iron.

The unleashed power of the atom has changed everything save our modes of thinking, and we thus drift toward unparalleled catastrophes.
Einstein

16.4 The discovery of fission: *passage to a new age**

It has been said that those who cannot learn from history are condemned to repeat it. The saga of the discovery and first use of nuclear fission is a powerful case study in science and society, a story that we must assimilate if we want to use science and technology without, in one way or another, destroying ourselves. It is a heroic story of a job that needed to be done in the face of a world-threatening enemy. Once these events were set in motion in the early days of World War II, the story's conclusion at Hiroshima and Nagasaki may have become, like all high tragedy, inevitable.

Scientists discovered radioactive decay, the first known nuclear reaction, in 1896. During the next several decades, they studied the new phenomenon intensely. Using a simple technique now common in high-energy physics, scientists bombarded tiny things by throwing other tiny things at them, and they observed what happened.

In 1933, Irène Joliot-Curie (daughter of Marie and Pierre) and her husband, Frédéric Joliot (Figure 16.8), bombarded a thin aluminum foil with alpha particles. This created a previously unknown isotope of phosphorus. Although natural phosphorus is stable, the new isotope was radioactive. It was the first creation of a radioactive isotope and the first artificial release of nuclear energy in radioactive decay. Joliot foresaw the potential consequences: "We are entitled to think that scientists, building up or shattering elements at will, will be able to bring about transmutations of an explosive type. . . .If such transmutations do succeed in spreading in matter, the enormous liberation of useful energy can be imagined."[†]

At about the same time, scientists bombarded beryllium nuclei with alpha particles and detected a previously unknown particle that was ejected from the beryllium during the collision. The new particle was electrically neutral. Science had discovered the third fundamental constituent of matter, the neutron.

The subtle new particle played a key role in the application of nuclear energy. Being uncharged, neutrons can sneak into a nucleus without having to overcome the electrical repulsion that protons feel when they approach the

FIGURE 16.8
Irène and Frédéric Joliot-Curie at the Radium Institute in Paris, in about 1935. The institute was founded by Irène's mother, Marie Curie.

6. 160,000 years ago.
7. Atomic number 27: cobalt. Atomic number 28: nickel.
* I am indebted to Richard Rhodes's *The Making of the Atomic Bomb* (New York: Simon & Schuster, 1986) for most of the historical details and quotations in Sections 16.4, 16.5, and 16.6. Rhodes's book is the definitive work on the topic, a beautiful and intelligent account.
† Rutherford and others had pursued the release of such energies since the beginning of the century.

positively charged nucleus. And unlike the electron, which does not feel the strong force, the neutron interacts strongly once it sneaks inside.

Hungarian physicist Leo Szilard was a lifelong admirer of the visionary science fiction of H. G. Wells, whose 1914 novel *The World Set Free* predicted nuclear energy, nuclear bombs, nuclear war, and world government. Szilard saw the possibility that neutrons emitted in a nuclear reaction would then bombard other nuclei in the same material and so create a series, a "chain," of nuclear reactions that could release nuclear energy in a large mass of material. Perhaps neutrons offered a way to extract useful amounts of nuclear energy. It was a double-edged vision, hopeful and fearful.

In 1934, Enrico Fermi (Figure 16.9) began bombarding various nuclei with neutrons to induce radioactivity, just as the Joliot-Curies had done. He worked his way through the elements to uranium, creating forty new radioactive isotopes in the process. One element that Fermi bombarded was uranium, element number 92 and the heaviest natural element. The results were ambiguous.

German chemist Ida Noddack, codiscoverer of the element rhenium, published an interpretation: "One could assume. . .that when neutrons are used to produce nuclear disintegrations, some distinctly new nuclear reactions take place which have not been observed previously. . . .[Perhaps] when heavy nuclei are bombarded by neutrons. . .the nucleus breaks up into several large fragments." In 1934, her prophetic words were universally ignored.

In 1938, Irène Curie reported that the slow neutron bombardment of uranium had produced a mysterious element that she could chemically separate from the uranium target by using lanthanum (atomic number 57) as a "carrier" to pick up the new element and carry it away. But the mysterious element was chemically inseparable from lanthanum. Surely it could not actually *be* lanthanum. How could the bombardment of uranium, element 92, create an element that was thirty-five places away from uranium in the periodic table? Otto Hahn, who led a uranium research group in Berlin, was skeptical of such unusual results and accordingly directed his research toward what he believed were needed corrections in Curie's work.

1938 was a fateful year. Austria was annexed by Hitler's Germany; the Munich Agreement allowed Hitler to occupy part of Czechoslovakia; and anti-Jewish mobs in Germany torched synagogues and beat Jewish families in the streets. Lise Meitner (Figure 16.10), an Austrian Jew, had worked with Hahn since 1907. She had been protected from German anti-Semitism by her Austrian passport, but the annexation of Austria made her a German citizen. In July, with Hahn's help, daringly traveling on her now invalid Austrian passport, she fled from Germany to Stockholm, Sweden.

As the clouds of war gathered across Europe, Hahn and his coworkers in a laboratory in a peaceful suburb of Berlin bombarded uranium with neutrons, trying to find the error in Irène Curie's work. The mystery only increased. Not only lanthanum but also a second element, barium (element 56), made a good carrier for some of the radioactive isotopes created by the bombardment. Once again, it proved impossible to separate chemically the mysterious isotope from its barium carrier. Hahn communicated the results to Meitner in Stockholm. Far from exposing any error in Curie's earlier work, Hahn's results were similar to Curie's.

FIGURE 16.9
Enrico Fermi in 1930.

FIGURE 16.10
Lise Meitner, around 1930.

FIGURE 16.11

The fission of a U nucleus after being struck by a neutron.

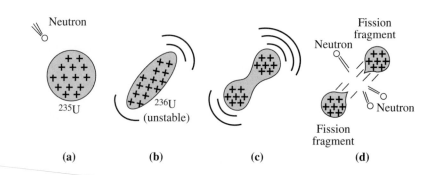

(a)   (b)   (c)   (d)

Meitner enjoyed frequent 10-mile hikes "to keep me young and alert." On Christmas Eve 1938, she and her nephew Otto Frisch took a long walk on cross-country skis through the Swedish countryside. Frisch was a colleague, a physicist, and together they pondered Hahn's data. Niels Bohr had suggested that a nucleus could be viewed as a liquid drop. With this picture in their minds, they debated whether a neutron added to a uranium nucleus might cause the nucleus to oscillate and elongate. Electric forces would then push the two ends away from each other, and two smaller nuclei would appear where the one had been before (Figure 16.11). It was like a drop of water that elongates and splits in two. One of the smaller nuclei might be barium or lanthanum. Taking a cue from microbiology, they named the new process **nuclear fission**.

Using the liquid drop model, Meitner calculated that the energy with which the two **fission fragments** should be pushed apart by electric forces should be about 3×10^{-11} joules. She then recalculated the energy in an entirely different way, by applying Einstein's $E = mc^2$ to the experimentally known mass difference between uranium and the fragments (Figure 16.12). That mass difference was known to be about one-fifth of a proton's mass. Multiplying this mass by c^2, she got about 3×10^{-11} joules. It checked.

Meitner's calculations were solid evidence of fission. The nuclear age had begun.

DIALOGUE 8   When the Joliot-Curies bombarded $^{27}_{13}$Al with alpha particles, the aluminum nuclei absorbed the alpha particles and emitted one neutron. Use the periodic table to determine the isotope created by this process.

FIGURE 16.12

The uranium nucleus is more massive than the pieces into which it splits (see the nuclear energy curve). According to Einstein, the mass difference multiplied by the square of light speed should equal the energy released during fission.

235U

Fission fragments, including the neutrons created during fission

16.5 The chain reaction: *unlocking the strong force*

By 1939, leading scientists such as Einstein, Fermi, and Szilard had fled to the United States from Hitler's Europe. When Szilard, in New York, heard that uranium could absorb a neutron and then break into two parts, he foresaw a feasible way to realize H. G. Wells's dream of nuclear energy. Because

8. $^{30}_{15}$P.

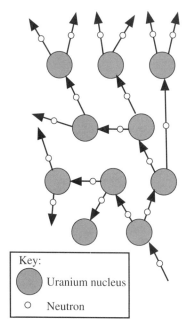

Key:
⬤ Uranium nucleus
○ Neutron

FIGURE 16.13
A chain reaction. At the lower right, a single neutron strikes a uranium nucleus. When the uranium nucleus fissions, it emits two neutrons that then fission two other uranium nuclei, and so forth. In this way, a large mass of uranium can be quickly fissioned.

I was shocked and depressed beyond measure. The thought of the unspeakable misery of countless innocent women and children was something that I could scarcely bear. . . .By the end of a long evening of discussion, attempts at explanation, and self-reproaches, I was so agitated that others became seriously concerned on my behalf.

Otto Hahn, codiscoverer of fission, while interned with other German nuclear scientists in England after the war in 1945

neutrons, being uncharged, are an especially good nuclear "glue," heavier elements contain many more neutrons than protons. The lighter fragments formed when uranium splits should have many more neutrons than is normal for their atomic number. Consequently, individual neutrons should split off during the reaction, which could then fission other uranium nuclei. As this **chain reaction** proceeded from one uranium nucleus to the next, a large mass of uranium might be fissioned (Figure 16.13). The number of neutrons would multiply quickly as the chain reaction spread, causing a large mass of uranium (several kilograms, for example—enough to destroy the center of a city) to fission in a few millionths of a second.

Szilard thought that if neutrons were in fact emitted during fission, this fact should be kept secret from the Germans.

Within a week of the announcement of fission, Fermi and others had independently hit on the idea of a chain reaction using neutrons and were making estimates of the energy that might be released. Once, standing at his Columbia University office window overlooking the bustling streets of New York City, Fermi cupped his hands as if he were holding an orange. "A little bomb like that," he said, "and it would all disappear."

Szilard devised a simple experiment to detect directly the neutrons that he suspected were released when one neutron fissioned a uranium nucleus. From the data, he estimated the average number of neutrons released per fission. A number larger than one could be enough to allow a chain reaction to build up quickly and fission a large mass of uranium. The number turned out to be about two. Szilard immediately telephoned a fellow Hungarian physicist now living in the United States, Edward Teller. Szilard said only one thing: "I have found the neutrons." That night there was little doubt in Szilard's mind that the world was headed for grief.

Hitler ordered the invasion of Poland on September 1, 1939, starting World War II. German scientists were aware of the weapons potential in fission. Accordingly, the German government banned the sale of uranium and in 1939 started a secret nuclear weapons program. It was the beginning of the international nuclear arms race, which climaxed with about 50,000 nuclear weapons by the end of the cold war in around 1990 and which continues today among several rival nations in the world.

Physicists such as Szilard and Teller understood the possibilities of fission. Fission bombs could be winning weapons for Hitler if he were allowed to build them sooner than the United States could. They discussed their fears with Einstein, who agreed to lend his prestige to an effort by scientists to alert the U.S. government to the problem. Together they drafted a letter from Einstein to President Franklin D. Roosevelt that they delivered in October 1939. Einstein suggested that the U.S. government stay informed of further developments and financially support fission research. The final paragraph noted, "Germany has actually stopped the sale of uranium from the Czechoslovakian mines which she has taken over," and "in Berlin. . .some of the American work on uranium is now being repeated."

As it turned out, Germany did try to develop nuclear weapons throughout the war but never got close to its goal. A U.S. project, on the other hand, was successful, although not in time for use against the enemy whom Einstein had feared.

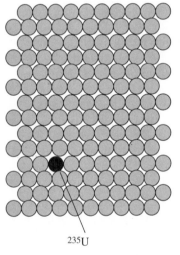

^{235}U

FIGURE 16.14
*In natural uranium, only 1 atom
in 140 is* ^{235}*U. The others are* ^{238}*U.*

Einstein's letter had little effect. There was a meeting and a committee report. Nothing more. America did not enter the war until December 1941, and in 1939 only Szilard, Einstein, and other knowledgeable physicists took nuclear dangers seriously.

In 1940, Germany invaded and occupied most of Europe and bombed Britain in preparation for an invasion. Both sides began bombing cities, and massive civilian casualties became a reality of modern warfare.

Although the U.S. government declined an active role, U.S. uranium research proceeded between 1939 and 1941. It gradually became clear that there was an important difference between the two uranium isotopes, $^{235}_{92}$U and $^{238}_{92}$U. When a neutron strikes uranium, only ^{235}U has much chance of fissioning; ^{238}U just absorbs the neutron to become a new radioactive isotope, ^{239}U. This means that a nuclear bomb requires nearly pure ^{235}U to sustain the rapid chain reaction needed to fission a large mass of uranium. If much ^{238}U is present, it will absorb most of the neutrons, and the bomb will fizzle.

Natural uranium has very little ^{235}U, less than 1%. To make a bomb, this 1% must be separated from the 99% that is ^{238}U (Figure 16.14). To many scientists, this appeared essentially impossible. The problem was that two isotopes of the same element must behave identically in every chemical reaction, so chemistry cannot separate them. The difficulties of extracting enough ^{235}U to build a bomb seemed so great that Niels Bohr insisted that "it can never be done unless you turn the United States into one huge factory." Bohr believed that it would therefore not be done. His words proved prophetic, although not in the way he imagined.

In 1940, scientists created the first nonnatural chemical element, that is, one not found naturally on Earth. They found evidence that when $^{238}_{92}$U is bombarded with neutrons, it absorbs a neutron to become $^{239}_{92}$U, which then quickly emits a beta particle to become element number 93. Its discoverers named the first element beyond uranium *neptunium*, for Neptune, the planet lying just beyond Uranus, the planet for which uranium was named.

Like all elements heavier than lead, neptunium is radioactive. Because it is a beta emitter, it decays to a higher atomic number, creating yet another nonnatural element, number 94 (Figure 16.15). Early in 1941, scientists detected the new element, which turned out to have an important property: Like ^{235}U, element 94 fissions readily when struck by a neutron. Furthermore, the new element can be chemically separated from the uranium in which it was created, thereby avoiding the difficulties of separating two isotopes of the same element.

It was not until 1942 that its discoverers proposed a name for the new element that fissions like ^{235}U but that can be chemically separated from uranium. They called it **plutonium**, for the outermost planet, which had in turn been named for Pluto, the Greek god of the underworld.

In October 1941, scientists convinced President Roosevelt that fission weapons could work.

The Japanese attacked the United States at Pearl Harbor on December 7, 1941. In 1942 a U.S. nuclear weapons program, code-named Manhattan Engineer District and known as the **Manhattan Project**, began in earnest.

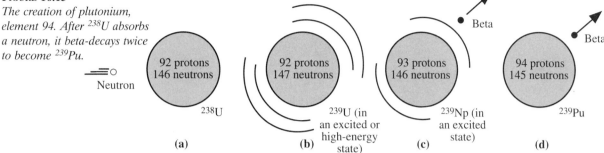

FIGURE 16.15

The creation of plutonium, element 94. After ^{238}U absorbs a neutron, it beta-decays twice to become ^{239}Pu.

Neutron

92 protons
146 neutrons

^{238}U

(a)

92 protons
147 neutrons

^{239}U (in an excited or high-energy state)

(b)

93 protons
146 neutrons

Beta

^{239}Np (in an excited state)

(c)

94 protons
145 neutrons

Beta

^{239}Pu

(d)

The fateful question of the human species seems to me to be whether and to what extent the cultural processes developed in it will succeed in mastering the derangements of communal life caused by. . .aggression and self-destruction. In this connection, perhaps, the phase through which we are at this moment passing deserves special interest. Men have brought their powers of subduing the forces of nature to such a pitch that by using them they could now very easily exterminate one another to the last man. They know this; hence arises a great part of their current unrest, their dejection, their mood of apprehension.

Sigmund Freud

DIALOGUE 9   When a ^{235}U nucleus is struck by a neutron, it splits into two large fragments and emits two to four neutrons. Although the fragments are different in different instances, one typical fragment is $^{142}_{56}Ba$. Use the periodic table to deduce the chemical name of the other fragment in this particular fission process.

DIALOGUE 10   In the preceding dialogue, suppose the second fragment is $^{91}_{36}Kr$. Write a "reaction equation" showing the isotopes and other particles that go into and come out of this reaction.

DIALOGUE 11   By roughly what percentage do the masses of one atom of each of the two uranium isotopes differ?

16.6 The Manhattan Project and fission weapons

Constructed on a monumental scale, the U.S. project to build a **fission bomb** or "A-bomb"* was a measure of U.S. fears about a German fission bomb. In December 1942, a research group under Fermi at the University of Chicago created the first self-sustaining chain reaction (Figures 16.16 and 16.17). To achieve this, Fermi's group constructed the world's first **nuclear reactor**, a device to transform, in a controlled way, nuclear energy into other energy forms. We discuss nuclear reactors in Chapter 17.

The world's first reactor illustrates several facts that are relevant today to a possible chain reaction of a different kind: **nuclear weapons proliferation** to countries around the world. Just as Fermi's reactor was a crucial stepping-stone to a U.S. fission bomb, countries seeking to build their own

9. The other fragment must have atomic number $92 - 56 = 36$. This is krypton.

10. $142 + 91 = 233$. Two additional neutrons $(235 - 233)$ must have been released. The overall reaction equation, including the neutron that struck the ^{235}U nucleus, is

$$^{235}_{92}U + \text{neutron} \rightarrow \ ^{142}_{56}Ba + \ ^{91}_{36}Kr + 3 \text{ neutrons}$$

11. They differ by 3 parts $(238 - 235)$ in 238, or a little more than 1%.

* *Atomic bomb* or *A-bomb* is an inappropriate name, because only the nucleus, rather than the entire atom, is the energy source. The term *nuclear weapon* is appropriate, but it applies equally to fission and fusion weapons. We use *fission bomb* and *fusion bomb*.

FIGURE 16.16

A painting of the opening ceremony of the world's first nuclear reactor beneath the football field (now removed) at the University of Chicago. Photographs were not allowed because of wartime security. Note the "suicide squad" of three young physicists in the back; they were holding jugs of neutron-absorbing liquid to pour into the reactor in case something went wrong.

FIGURE 16.17

The bronze plaque at the University of Chicago commemorating Fermi's achievement.

nuclear weapons are likely to begin by building a reactor. Although the primary purpose of nuclear reactors today is to provide peaceful electric power, reactors can also provide research relevant to nuclear weapons. An apparently peaceful nuclear power program can be used as a cover for a secret weapons program, as was demonstrated by South Africa in the clandestine development of its nuclear weapons arsenal between 1974 and 1990, as well as by other nations such as Iraq and North Korea.

Another problem is that reactors can produce militarily useful plutonium by converting ^{238}U to ^{239}Pu, providing one possible path to a fission bomb. Since uranium is common in Earth's crust, the fuel is easy to obtain. During a few weeks of operation, a reactor can create enough plutonium for a fission bomb.

Near Hanford, Washington, the Manhattan Project engineers built three large natural-uranium reactors for the purpose of making plutonium. These reactors created large quantities of Earth's new element ^{239}Pu. The engineers also built a plant to carry out the difficult operation of chemically extracting the plutonium from the highly radioactive used fuel.

Among the vast assortment of natural and artificial isotopes, ^{235}U and ^{239}Pu are essentially the only ones that will produce a chain reaction. There is no obvious reason that there should be two chain-reacting nuclei rather than some other number such as one or zero. It seems ironic, like a kind of test for the human race, that nature gave us two combinations of protons and neutrons that could be used for bombs and energy.

The two isotopes offer two paths to fission bombs: uranium and plutonium. The Manhattan Project pursued both paths. Along the uranium path, the key problem was the **isotope separation** of the 1% of natural uranium that is ^{235}U from the 99% that is ^{238}U. Any separation technique must be based entirely on the tiny mass difference between the two isotopes because they are identical chemically.

It was not very clear that the job of separating large amounts of uranium 235 was one that could be taken seriously.

Enrico Fermi, 1939

A technique used during the Manhattan Project, and still widely used, is **gaseous diffusion**.* Recall (Chapter 2) that gas molecules that are initially concentrated in one region will spread out as a result of their thermal motion, a phenomenon known as **diffusion.** Lighter molecules, which have less inertia, have more thermal motion than heavier molecules, so lighter molecules diffuse faster than heavier ones do. If uranium is put into a gaseous form, molecules containing the lighter isotope ^{235}U will diffuse faster than molecules containing ^{238}U, and this difference can be exploited to separate gaseous uranium into a part that is slightly **enriched** in ^{235}U and a part that is less enriched. After many such enrichment steps, **highly enriched uranium** is obtained. If the enrichment exceeds a 90% ^{235}U content, the material is considered to be weapons grade.

Physicists at Columbia University studied gaseous diffusion between 1941 and 1943. In 1943, in Oak Ridge, Tennessee, army engineers began constructing a gaseous diffusion isotope separation plant. It was of monumental size (Figure 16.18), to house the thousands of diffusion containers needed.

---

* There are several other isotope separation methods: electromagnetic, liquid thermal diffusion, centrifuge, nozzle, and laser separation.

This war, in contradistinction to all previous wars, is a war in which pure and applied science plays a conspicuous part.
Sir William Ramsay in 1915, commenting on World War I

The nuclear proliferation lesson is that it is not easy to enrich uranium. Acquiring the tens of kilograms of highly enriched uranium needed for a bomb is the key obstacle to building a uranium bomb.

In October 1942, the U.S. Army selected physicist J. Robert Oppenheimer (Figure 16.19) to direct the laboratory that would design and build the world's

(b)

FIGURE 16.19
J. Robert Oppenheimer was one of the United States' most brilliant physicists, politically left-wing and controversial, a wide-ranging intellectual and sensitive man given to literary pursuits. (a) In the 1930s. (b) During the Manhattan Project.

438

(a)

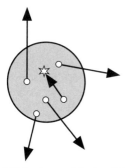

(a) If the lump of fissionable material is too small, most neutrons emitted during fission will escape through the surface without striking any nuclei.

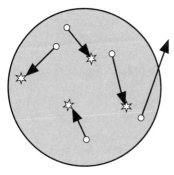

(b) But if the lump of material is large enough, most neutrons will strike other nuclei before they escape through the surface.

FIGURE 16.20
The concept of critical mass.

FIGURE 16.21
Little Boy: the gun design of the Hiroshima bomb.

first nuclear weapons. It proved to be an inspired choice. Oppenheimer chose a high, isolated desert mesa in New Mexico as the site of the new laboratory. They named the place after the boys' school that had been on the mesa: Los Alamos.

One of the first things that the Los Alamos scientists did was calculate the minimum amount of fissionable material needed to sustain a chain reaction. If the amount is too small, most of the nuclei will be close to the material's surface, and so most of the neutrons created during fission will simply pass right out through the surface without hitting anything. But if the amount of fissionable material is large enough, most of the neutrons created during fission will fission another nucleus (Figure 16.20). This minimum amount is called the **critical mass**. For ^{235}U, the critical mass is a grapefruit-sized 15 kilograms. For ^{239}Pu, it is 5 kilograms—about the size of an orange, as Fermi had guessed. For a bomb, facilities such as Hanford and Oak Ridge would need to produce this much material.

The basic principle of the fission bomb is simple. Start with a subcritical (smaller than critical) mass of fissionable material; quickly add enough material to it to make it critical; and start the chain reaction by showering the critical mass with neutrons. The additional mass must be added quickly and the neutrons injected immediately, because otherwise stray neutrons from the environment could pass through the material and start the chain reaction prematurely. A critical mass of fissionable material is not something you can safely keep around. The design chosen for the uranium bomb was so straightforward that there was little need to test it before use. But simplicity came at the price of inefficiency: The design needed 42 kilograms of precious ^{235}U, nearly three times the minimal critical mass. There is a proliferation lesson here: A country need not test a fission bomb in order to have one. If enough highly enriched uranium can be made, begged, or stolen, building a crude bomb is fairly simple.

Figure 16.21 shows the **gun design** of the uranium bomb. An explosive charge slams a subcritical ^{235}U "bullet" into a subcritical target to quickly form a critical mass. At this instant, a neutron source showers the critical mass with neutrons, thereby initiating a chain reaction.* A massive steel

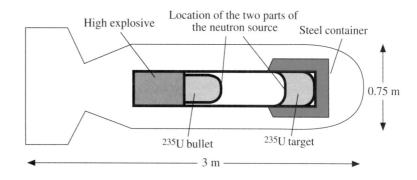

High explosive   Location of the two parts of the neutron source   Steel container

0.75 m

^{235}U bullet ^{235}U target

3 m

---

* The neutron source was a radioactive isotope that emits alpha particles, coupled with the element beryllium. When an alpha particle strikes a beryllium nucleus, it knocks a neutron out of that nucleus. In the bomb, the alpha emitter was in one of the two subcritical pieces, and the beryllium was in the other piece, so that the two pieces created neutrons when they came together.

container confines the reaction until a large fraction of the ^{235}U nuclei have fissioned. They called this first bomb Little Boy.

The plutonium bomb was much more complex. The gun method will not work with plutonium because early ignition of the chain reaction by stray neutrons occurs much more quickly, making it impossible to slam the two pieces together fast enough. Instead, researchers designed a faster way of putting together a critical mass. It is based on the fact that a subcritical mass can be made critical by simply squeezing it into a small volume, without adding any additional mass. When the material is more compact, the fissionable nuclei are closer together, and so it is more likely that a neutron released during fission will hit another nucleus and continue the chain reaction.

Figure 16.22 shows the **implosion design** of the plutonium bomb. The high explosive is arranged around a 5-kilogram sphere of plutonium, and the sphere is surrounded by metal to confine the reaction. The high explosive is detonated from the outside, and a wave of exploding material moves inward. That is, the explosive "implodes," inward, squeezing the plutonium sphere into a small enough volume to become critical. At this instant, a neutron source releases neutrons at the center. The scientists named this bomb Fat Man. Although this design is technologically demanding, plutonium is not as hard to obtain as is the highly enriched uranium needed to construct a uranium bomb.

Highly enriched uranium for Little Boy began arriving at Los Alamos in late 1944, and by the following summer there was enough for a bomb. By May 1945, enough plutonium had arrived to allow final critical-mass experiments for Fat Man.

In May 1945, Germany surrendered; the German bomb had never approached completion. Nonetheless, the United States did not halt work on the Manhattan Project, even though the feared German bomb had been its initial driving force. There are lessons here. Once we find that something can be done, there is often a drive, a **technological imperative**, to do it. And once a project is started, it often develops a self-justifying **technological momentum**. It seems that what can be done must be done and what is being done should be done. On the other hand, there were good military arguments for continuing the project. Nuclear weapons might shorten the war that still continued against Japan.

FIGURE 16.22

Fat Man: the implosion design of the Nagasaki bomb.

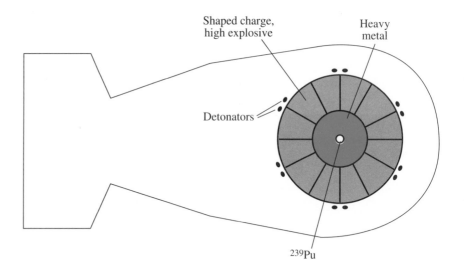

There floated through my mind a line from the Bhagavad-Gita in which the Krishna is trying to persuade the Prince that he should do his duty: "I am become death, the shatterer of worlds." I think we all had that feeling more or less.
J. Robert Oppenheimer, at the test of the first atomic bomb

Oppenheimer helped choose the test site for the plutonium bomb, in a barren landscape south of Los Alamos. He code-named the site Trinity, referring to a line by poet John Donne: "Batter my heart, three person'd God." Just before dawn on July 16, 1945, a burst of brilliant purple never seen before lit up the desert, and a small part of Earth was brought to temperatures that were unprecedented within the solar system, save at the center of the sun.

The energy released was the same as would be released by 18,000 tons, or 18 "kilotons," of a chemical explosive such as TNT. A **kiloton** is a unit of energy,* the energy that would be released by 1000 tons of exploding TNT. For comparison, typical chemical bombs carry perhaps one-quarter ton of high explosive.

Little Boy, the untested uranium bomb, was loaded onto a B-29 bomber on August 5. Most Japanese cities had been firebombed to ashes by this time. But Hiroshima, an industrial city with an army depot, an ocean port, and 400,000 people, was still untouched.

The B-29 arrived over Hiroshima by 9 A.M. on August 6. As the world's first combat nuclear weapon fell toward its target, the airplane quickly turned and dove away to escape the blast. By the time its crew looked back, the city was hidden by an awful cloud (Figure 16.23). Little Boy released 12 kilotons of nuclear energy; 140,000 lay dead with an equal number wounded; and the city, 5 kilometers (3 miles) across, lay in ruins (Figure 16.24). The dying continued: By 1950, the total had reached 200,000, 50% of the city's population.

The war continued. A debate raged among Japanese leaders about whether it was time to surrender. The Soviet Union, no longer fighting against Germany, was poised to go to war against Japan.

On August 9, the United States dropped Fat Man. The target was Nagasaki, a city somewhat smaller than Hiroshima. The bomb released

FIGURE 16.23
By the time the B-29's crew looked back, Hiroshima was hidden by an awful cloud.

FIGURE 16.24
Hiroshima, August 1945. Little Boy released 12 kilotons of nuclear energy; 140,000 people lay dead, with an equal number wounded; and the entire city, 5 kilometers (3 miles) across, lay in ruins. The dying continued: By 1950, the total had reached 200,000, 50% of the population.

* One kiloton = 4.2×10^{12} joules.

22 kilotons of energy and killed 70,000 outright and 140,000 all together by 1950, again a death rate of 50%.

On August 14, 1945, Japan surrendered.

DIALOGUE 12   About how many 500-pound chemical bombs would need to be dropped to release the energy of Fat Man? How many large highway trucks (30-ton freight capacity) would be needed to carry this much high explosive? How many World War II heavy bombers (5-ton bomb capacity) would be needed to carry this much high explosive?

DIALOGUE 13   It was discovered during the 1980s that North Korea had built two large nuclear reactors and a chemical processing plant for nuclear fuel. Which, if either, of the two kinds of fission bombs might North Korea have been developing?

16.7 Fusion weapons: *star fire on Earth*

The sequel to the Manhattan Project also has instructive lessons for us today.

Following World War II, the U.S.–Soviet cold war rivalry commenced. Like the rest of the scientific world, Soviet physicists learned of fission in early 1939, and by 1940 they had concluded that a chain reaction could be established in uranium. Fifty to 100 Soviet scientists had worked on isotope separation and uranium and plutonium bomb designs during the war, and the bombings of Hiroshima and Nagasaki only caused them to redouble their efforts after the war. The Soviet Union achieved a chain reaction in a reactor in 1946 and tested a fission bomb in 1949.*

In 1941, Edward Teller began calculations about a bomb based on the energy released when hydrogen fuses to form helium. Whereas the research during World War II focused on a fission bomb, Teller concentrated his research on fusion. The efforts of his fusion group expanded after the war, and by 1949 it appeared that they might eventually succeed. But the United States had not yet decided to build a fusion weapon.

The Soviet fission bomb test, coming as cold war tensions were building, surprised and alarmed the United States. After the Soviet test, President Harry Truman and a small circle of officials and scientists debated whether to proceed with an all-out U.S. effort to build a hydrogen bomb, or **fusion bomb**. Their fear was similar to earlier fears about Germany: The Soviets might build the winning weapon first. Teller and others felt that this would be intolerable, and in 1950 Truman decided to develop a fusion bomb.

---

12. 500 pounds = 0.25 tons. 22,000/0.25 = 88,000 bombs. 22,000/30 = 730 trucks. 22,000/5 = 4,400 bombers.
13. Plutonium, since an enrichment plant is not mentioned.
 * Physicist Klaus Fuchs, a researcher at Los Alamos between 1942 and 1949, later confessed to being a Soviet spy. Some observers believe that his reports on the United States' fission bomb gave little real assistance to the Soviets, but others estimate that he might have shortened the Soviet project by one or two years. Fuchs also passed on information about U.S. fusion bomb research, but this information probably actually hindered the Soviet fusion project because U.S. fusion research was headed along a wrong path up until 1951.

In 1950, the U.S. fusion bomb project was, according to Oppenheimer, "a tortured thing that you could well argue did not make a great deal of technical sense." But during 1951, Teller and mathematician Stanislaw Ulam devised a clever design that solved many of the project's problems. In a revealing comment on the allure of high technology, Oppenheimer stated, "By 1951 the program was technically so sweet that you could not argue about that."

In October 1952, the United States exploded the world's first thermonuclear fusion device. Figure 16.25 shows a typical fusion bomb design. To bring the hydrogen fuel to the multimillion-degree temperatures needed for fusion, an implosion-type fission device is used as a "trigger." Inexpensive natural uranium is used at several places in the design. Under the conditions that exist while the hydrogen is fusing, even natural uranium can be made to fission, yielding a larger blast that is also higher in radioactive fallout.

Although the critical mass limits the size of fission weapons, there is no natural upper limit to a fusion weapon. The yield of the world's first fusion explosion was 10,000 kilotons, or 10 **megatons**, equivalent to 10 million tons of TNT. A single fusion weapon such as this releases a thousand times more energy than did the Hiroshima bomb, or roughly twice the total explosive energy released by all combatants during all of World War II.

The success of the Soviet fission bomb test and reports that the U.S. government was considering a fusion program stimulated the Soviets to move ahead rapidly with their own fusion program. In November 1955, the Soviet Union also tested a thermonuclear fusion weapon.

We see here, once more, the action–reaction cycle. This dangerous and expensive process is hard to control as long as nations feel threatened, and so similar arms race cycles, involving nuclear and other sorts of weapons, are likely to continue among many nations.

The U.S.–Soviet nuclear arms race was, by any measure, extreme. By the mid-1980s each side possessed about 25,000 nuclear weapons, enough to destroy the other side as a functioning society many times over. Each side held its weapons out of fear of what the other side might do if it had a significant

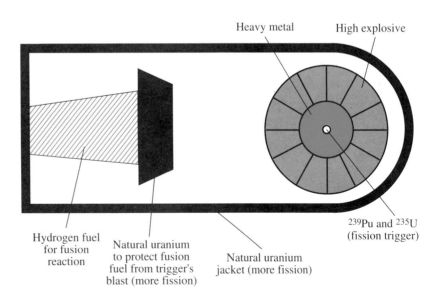

FIGURE 16.25
The design of a fusion bomb.

nuclear advantage. Each side's arsenal was designed to "deter" the other side from initiating an attack. Any significant increase in either stockpile always produced a compensating, or overcompensating, increase in the other. Each side's nuclear arsenal contained about 10,000 strategic fusion weapons, each one of them capable of destroying the center of a large city such as New York City or Chicago. Weapons like this still exist.

DIALOGUE 14 MAKING ESTIMATES: HOW BIG IS 1 MEGATON?
Estimate the number of large highway trucks (capacity 30 tons) needed to haul 1 megaton of TNT and the length of a single line of trucks carrying such a load on the highway. What about 10 megatons?

TRYING TO DESCRIBE THE SIZE OF THE BIG BANG

---

14. 1,000,000/30 = 33,000 trucks. A large highway truck is about 15 m long, so the length of 33,000 trucks is 500,000 m, or 500 km, with no spacing. If trucks are separated by one truck length, the line is 1000 km long. For 10 megatons, it is 10,000 km, or twice across the United States.

Summary of Ideas and Terms

Nuclear fusion A nuclear reaction in which two nuclei combine to form a single larger nucleus.

Nuclear fission A nuclear reaction in which a large single nucleus splits into two roughly equal smaller nuclei.

Nuclear energy Energy resulting from a nuclear structure.

Thermonuclear reaction A fusion reaction that creates the thermal energy needed to sustain itself.

Big bang The event about 15 billion years ago that created our universe.

White dwarf A small, burned-out star; the end state of most stars, such as the sun.

Supernova explosion The explosion of a giant star. Supernova explosions spread the chemical elements into space and so are the source of the elements heavier than helium in our solar system.

Mass–energy equivalence, and the nucleus When a nucleus releases nuclear energy, the total rest mass of the participants must decrease. The energy release is related to the mass decrease by $E = mc^2$.

Chain reaction A series of neutron-induced fission reactions that proceed from one nucleus to the next by means of the neutrons released during each fission reaction.

Action–reaction cycle in arms races A mutually reinforcing cycle of increased armaments by two or more hostile nations.

Manhattan Project The U.S. project during World War II to build fission bombs.

Hiroshima and **Nagasaki** The Japanese cities that were fission-bombed near the end of World War II.

Isotope separation, or **enrichment**, of uranium Any process that increases the percentage of ^{235}U relative to ^{238}U. One method is based on **gaseous diffusion**. Weapons-grade uranium is **highly enriched** to about 90% ^{235}U.

Critical mass The minimum amount of fissionable material that will sustain a chain reaction.

Nuclear reactor A device that controllably transforms nuclear energy into other energy forms.

Fission bomb, or atomic bomb (A-bomb) A bomb based on a fission chain reaction in ^{235}U or in Pu. In the **gun design**, two subcritical masses are brought together. In the **implosion design**, a subcritical mass is made critical by squeezing it to high density.

Fusion bomb, or hydrogen bomb (H-bomb) A bomb based on the fusion of hydrogen, triggered by a fission bomb.

Nuclear weapons proliferation The spread of nuclear weapons to more and more nations.

Technological imperative, technological momentum The tendency to build whatever technology is possible and to continue a technological project once it is started.

Kiloton, megaton The amount of energy that would be released in the explosion of 10^3 tons and 10^6 tons, respectively, of TNT.

Review Questions

THE ENERGY OF THE NUCLEUS

1. What energy transformation occurs in radioactive decay? In nuclear fusion? Nuclear fission?
2. Of the four fundamental forces, which ones are important inside the nucleus?
3. Which has more nuclear energy, a separated proton and neutron or a ^2H nucleus? How do you know? Which has greater mass? How do you know?
4. If you separate a nucleus into its individual protons and neutrons, does its nuclear energy increase or decrease, or does the answer depend on which nucleus you started with?
5. Suppose you pull a nucleus in half. Does the system's nuclear energy increase or decrease, or does the answer depend on which nucleus you started with?
6. Sketch the nuclear energy curve (per nuclear particle). What does this curve tell us about the possibilities of getting useful energy from fusion or fission?

FUSION AND CREATION OF THE ELEMENTS

7. Give the reaction formula for one fusion reaction.

8. What isotope is created when ^{12}C fuses with ^4He? When ^{14}N fuses with ^2H? (Use the periodic table.)
9. List two pieces of evidence for the big bang.
10. List two isotopes created in the big bang.
11. Where did your body's hydrogen atoms originate? The oxygen atoms?
12. Where do stars get their energy?

FISSION AND THE CHAIN REACTION

13. Why do neutrons work better than protons do in causing fission? Why do neutrons work better than alpha particles do? Why do neutrons work better than electrons do?
14. Describe what happens during a chain reaction.
15. When $^{238}_{92}U$ absorbs a neutron, what isotope does it become? What new element is then created by a single beta decay of this isotope?
16. Where does plutonium come from?

NUCLEAR WEAPONS

17. Can natural uranium fuel a fission weapon? Why?

18. Why must uranium be enriched before it can be used as a weapon material?
19. Name the two chain-reacting materials that can be used in fission weapons.
20. Explain how a gun-type fission bomb works.
21. Explain how an implosion fission bomb works.
22. How is a fusion bomb heated to fusion temperatures?
23. What is a kiloton, in reference to a weapon? A megaton?

Home Project

Ask each of your classmates to get three sheets of paper and bunch up each sheet tightly. Gather the class closely together in the center of the room. Ask them to toss their three bunched sheets in the air when they are struck with one sheet. Toss one bunched sheet into the crowd, and watch the chain reaction. Add a few "control rods"—people who throw no bunched sheets into the air. Can you see the difference? How many control rods are needed to stop the reaction? How many control rods are needed to make the reaction "barely critical," neither expanding nor dying out?

For Discussion

1. Science and the search for truth. Is it always better to know than not to know? And is it always better to know sooner rather than later? The contrasting opinions of scientists Pierre Curie and Ernest Rutherford are of interest here: "Humanity will obtain more good than evil from future discoveries" (Curie, in 1903); "I hope that no one...release(s) the intrinsic energy of radium until man has learned to live at peace" (Rutherford, in 1915). Consider three examples: the 1930 discovery of chlorofluorocarbons (Chapter 9), the 1938 discovery of fission, and the 1951 discovery of a way to initiate a fusion bomb explosion. Would we be better off if any of these had not been discovered or had been discovered later? If scientists had discovered fission not in 1938 but in 1934, as they could have done by following Ida Noddack's suggestion, would the world have been better off?

2. Science and democracy. Science-related decisions that affect citizens can be difficult in a democracy, because the questions are sometimes highly technical and understood only by specialists, yet they affect all of us. Examples include radioactive waste, nuclear weapons, and global warming. How should democracies deal with such questions? If all such questions were left to specialists such as scientific experts, would it still be a "democracy"? What role should be played by specialists? By political leaders? By citizens?

Science-related decisions are especially difficult when they also involve secrecy. For example, the general public was excluded from the decision in 1941 to build a fission bomb and the decision in 1950 to build a fusion bomb. Does the public's right to know extend to such cases? If a few officials and experts make these decisions, is this significantly nondemocratic? Who should make such decisions?

An extreme example occurred in 1942. Oppenheimer gathered a few scientists to discuss nuclear weapons. One speaker suggested that a fusion bomb might inadvertently trigger fusion reactions in the atmosphere and destroy Earth. Other scientists made calculations to see whether this was possible. Their calculations showed that even under worst-case assumptions, the temperature needed to ignite the atmosphere exceeded by a factor of 100 the calculated temperature of the fusion bomb. So the scientists felt certain that fusion bombs would not ignite the atmosphere. Should the public have been told about this possibility and been allowed to come to its own conclusions, before the first fusion test in 1952?

What about decisions by one nation that affect the entire planet? Should all people have a say about the United States' decisions concerning chlorofluorocarbons or carbon-dioxide emissions? About Chinese decisions concerning birth control? About Russian decisions concerning nuclear power safety?

Exercises

THE ENERGY OF THE NUCLEUS

1. An alpha particle is removed from $^{16}_{8}O$. Show this process as a nuclear reaction, with the proper symbol for the remaining nucleus. Which side of this reaction, the left or the right, represents more nuclear energy?

2. How does a nucleus's mass compare with the sum of the masses of its protons and neutrons? Does the answer depend on which nucleus you are considering? If so, explain.

3. Would fusion or fission or neither release nuclear energy from carbon? From gold? From iron?

4. Would you expect radioactive materials to be warmer than most objects? Explain.

5. In which of the following processes is there a change in rest mass due to the mass–energy equivalence principle? In which ones is the change in rest mass actually measurable? Warming a cup of coffee, operation of a nuclear reactor, explosion of TNT, explosion of a fission bomb, explosion of a fusion bomb, lifting a book.

FUSION AND CREATION OF THE ELEMENTS

6. The fusion of helium nuclei into heavier nuclei occurs only during the later stages of a star's history, when the star has reached a higher temperature, owing to its collapse. Why would you expect helium fusion to occur only at higher temperatures than those needed for hydrogen fusion?

7. In what sense have we always been sustained by the energy from nuclear fusion?

8. Before it exploded, the 1987 supernova fused many elements. In one reaction, ^{12}C fused with ^{4}He. What nucleus did this create?

9. Another reaction occurring in the 1987 supernova, before the explosion (see the preceding exercise), was the fusion of two ^{12}C nuclei. What nucleus did this create?

10. Can matter be destroyed? If so, give an example. Can energy be destroyed? If so, give an example.

FISSION AND THE CHAIN REACTION

11. A neutron strikes a ^{235}U nucleus and creates lanthanum. What other element is created?

12. List one or more similarities between combustion and fission. List one or more differences.

13.*When ^{235}U fissions, it loses about 1% of its rest mass. Suppose that 10 kg of ^{235}U fissions. How much rest mass is lost? How much nuclear energy is released?

14.*Making estimates. A large nuclear power plant supplies energy at a rate of about 1000 megawatts, or 10^9 joules/second. The energy comes from ^{235}U. About how much rest mass vanishes in one day? If the energy came instead from coal, still at 1000 megawatts, would any rest mass vanish? If so, how much in one day?

NUCLEAR WEAPONS

15. Is plutonium radioactive? Defend your answer.

16. Can natural uranium metal explode spontaneously? Why?

17. Is a chain reaction possible in a substance that emits no neutrons when it fissions?

17

THE ENERGY FUTURE

The way that a society organizes energy to do work is a key determinant of its character. Most of the differences among primitive, agricultural, and industrial societies stem, directly or indirectly, from their different ways of using energy. Energy is even more significant in today's technology-based society than it was in the past. Its many societal connections include, for example, global warming, nuclear power, the quality of life in our cities, and the structure of our economy.

Although we will be discussing the future, we will not try to make predictions. In most respects, the energy future is not predictable. It is, rather, something that we will determine by our own actions. It is precisely because we have the power to choose that we have an obligation to study that future, and so we study possible energy futures, "scenarios," rather than predictions.

Section 17.1 takes both a long-term and a short-term view of the history of energy use. Section 17.2 looks at where we are now and outlines our future options. Because some of our most important traditional energy resources are either running out or getting us into trouble with the environment, the present era is a watershed in our use of energy. Perhaps the most important issue is the balances and trade-offs between traditional nonrenewable sources—natural gas, oil, coal, and nuclear—and the alternative renewable options, especially solar energy and energy efficiency. Sections 17.3, 17.4, and 17.5 look at the traditional energy sources, especially nuclear and coal. Section 17.3 studies how nuclear power works, and Section 17.4 makes a comparative assessment of nuclear and coal power; such technology assessments are important tools for making technology-related decisions. Section 17.5 considers some of the most important issues for nuclear power.

Section 17.6 explores the physics and a little of the economics of renewables, especially the solar-related energy sources. Finally, Section 17.7 looks at the fascinating economics and physics of energy conservation.

A personal note: Most people who have spent much time studying society's energy problems have formed opinions about the topic. I am no exception to this. However, I have made every effort to keep my own opinions out of this chapter and to present fairly all of the reasonable options together with some of their pros and cons. Do not, for example, assume that the emphasis given to nuclear power in this chapter means that I am either especially friendly, or especially hostile, to nuclear power. Rather, I have tried to emphasize those topics that are important for all of us to think about and that have a strong physics component.

17.1 A brief history of energy

Before deciding where you want to go, it's a good idea to know where you've been. Figure 17.1 presents an approximate long-term view of where our energy use has been. The earliest humans used only their own bodily energy, obtained by respiration of food and oxygen. A typical person puts some 2000 Calories, or 8 million joules (8 MJ), of energy to work every day, roughly equivalent to a single 100-watt bulb burning continuously.

During the agricultural age, beginning some 10,000 years ago, humans used the chemical energy of farm animals to do work and the chemical energy released in fires for cooking and warmth. This might have increased the

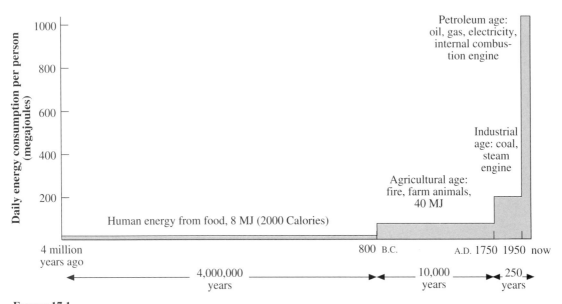

FIGURE 17.1

A brief and approximate history of the use of energy resources: individual daily consumption versus the approximate date. Note that the time axis is not drawn to scale: The preagricultural period is really four hundred times as long as the agricultural period, which is in turn forty times as long as the industrial period.

energy put to work by one person about fivefold, to 40 MJ per day, equivalent to five 100-watt bulbs burning continuously.

Heat engines fueled by coal, the first fossil fuel, ushered in the industrial age. Fossil fuels store the chemical energy created by millions of years of accumulating layers of energy-rich plant and animal remains. Time and pressure transformed these remains into great beds of coal, pools of oil, and pockets of gas. In only 250 years of intense carbon burning, we have sent much of this carbon—removed from the atmosphere over millions of years—back into the atmosphere. This enabled people in industrialized countries to increase their personal energy use by another factor of 5. From about 1750 to 1950, each person in the industrialized world put about 200 MJ of energy to work every day, equivalent to twenty-five 100-watt bulbs.

During the past fifty years, oil and natural gas have dominated in the industrialized world. The intense use of fossil-fueled heat engines (mostly for transportation and electricity) has pushed up industrialized nations' energy use by yet another factor of 5, to 1000 MJ per person per day, equivalent to 125 100-watt bulbs.

Figure 17.2 takes a closer look at the recent past in one industrialized country: the United States. The numbers plotted along the vertical axis are in units of 10^{18} joules, or exajoules ("exa" means 10^{18}). Each of the six portions marked Wood, Hydro, Coal, Oil, Gas, and Nuclear represents the energy provided by one particular resource, and the upper boundary represents the total energy provided by all six resources. Note the dominance and growth of fossil fuels since 1900, the growth of oil and gas since World War II (about 1940), and the large overall rise in energy consumption. The nation's total energy consumption rose by much more than a factor of 5 during these 150 years because although per capita energy consumption was rising about fivefold

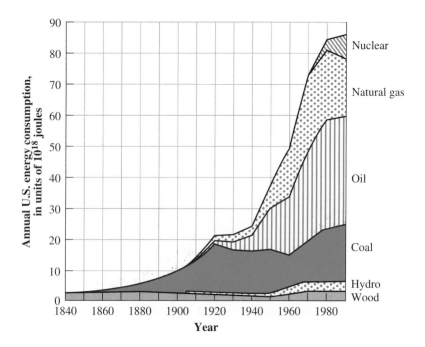

FIGURE 17.2
History of U.S. energy use since 1840: total annual U.S. energy resource consumption from various sources.

(Figure 17.1), the nation's population was increasing about tenfold, for a nearly fiftyfold overall increase in energy use. Resource use results from two distinct driving factors: individual (per capita) use and population size.

DIALOGUE 1   According to Figure 17.2, in approximately what year did the United States begin getting as much, or more, energy from fossil fuels as from wood? In approximately what year did the combined energy from oil and natural gas begin to exceed that from coal?

DIALOGUE 2   How much total energy did the United States use in each of the years 1880, 1900, 1920, 1940, 1960, and 1980? By about what factor did the energy consumed in 1980 exceed that consumed 100 years earlier?

DIALOGUE 3 MAKING ESTIMATES   Use Figure 17.2 to document roughly the estimate shown in Figure 17.1 for the period from 1950 to the present. The U.S. population is about 250 million.

17.2 Energy use today and future options: *an overview*

Fossil fuels dominate the industrialized world's energy today, with smaller contributions from nuclear power, hydroelectric, and wood. Figure 17.3 shows the current U.S. energy mix.

Figure 17.4 shows annual energy flows through the U.S. economy, the amounts supplied from each of today's six primary (or natural) energy resources, the amounts of each resource going to electric and nonelectric uses, the amounts from electric and nonelectric to each of the three main economic sectors (industry, residential–commercial, transportation), and the amounts used and wasted by each economic sector.

About one-third of the nation's energy goes into electricity generation. Hydroelectric and nuclear resources go entirely to electricity, and coal resources go mainly to electricity, whereas oil, natural gas, and wood go nearly entirely to nonelectric uses.

Figure 17.4 reflects the two great principles of energy. We see the law of conservation of energy in the fact that the energy flows (the pipe widths and the numbers) match: Energy in always equals energy out. We see the second law's effect in the transformation of energy from 85 exajoules of high-quality chemical, nuclear, and gravitational energy into 56 exajoules of waste thermal energy and 29 exajoules of useful energy, some of which again produces thermal energy for heating. The second law's rigorous restrictions on heat engine efficiencies shows up in the lower efficiency of electric power generation and transportation, both powered mainly by heat engines.

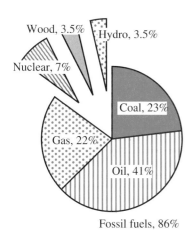

FIGURE 17.3
The U.S. energy mix today.

Wood, 3.5%   Hydro, 3.5%
Nuclear, 7%
Coal, 23%
Gas, 22%
Oil, 41%
Fossil fuels, 86%

---

1. About 1880. About 1940 or 1950.
2. 5, 10, 21, 25, 50, and 85 exajoules. 85/5 = 17.
3. In 1980 the total annual energy consumption was 85×10^{18} J, so the annual consumption per person was $85 \times 10^{18}/250 \times 10^6 = 340 \times 10^9$ J, and the daily consumption per person was about 340×10^9 J/365 $\approx 10^9$ J = 1000 MJ.

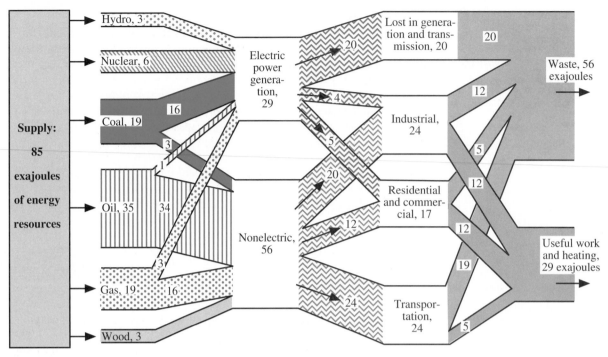

FIGURE 17.4
Approximate energy flows in the U.S. economy in 1992, in exajoules (10^{18} J).

As we will see, resource depletion and environmental problems will probably drive the world away from fossil fuels during the next few decades. This raises a serious question: What will be our energy future?

One option is an expansion of nuclear power, which could take several forms. The most immediate possibility is a new generation of light-water reactors (Section 17.3), improved versions of today's uranium-fueled reactors. Another possibility is new and safer types of uranium-fueled reactors. A third option is a new kind of reactor called a *breeder reactor* (Section 17.5), designed to produce plutonium for use in plutonium-fueled fission reactors. Breeder reactors were once regarded as the main future source of energy, but many factors have changed that outlook dramatically during the past two decades. Breeder reactors have not been a major factor in U.S. energy planning since the 1970s, although a few other countries such as Japan have retained a serious interest in them. We will discuss nuclear fission power further in Sections 17.3 and 17.5.

A final nuclear option is the development of fusion reactors. The energy-releasing fusion reaction that would be used is the same reaction that fuels a fusion bomb:

$$^2H + {}^3H \rightarrow {}^4He + \text{neutron}$$

In a **fusion reactor** this reaction would take place nonexplosively, yielding a continuous source of thermal energy that could boil water to make power-plant steam. The ^2H would be extracted by isotope separation from ordinary water, and the ^3H, a radioactive isotope of hydrogen that is not found naturally, would be made in nuclear reactors from a naturally occurring isotope of lithium.

It isn't easy to make a fusion reactor. One must heat the fuel to millions of degrees to start a self-sustaining thermonuclear reaction (Chapter 16). In one of the most-studied fusion reactor designs, a current is sent through the gaseous fuel to heat it to fusion temperatures. The problem is that in order for there to be a net energy gain, three conditions must be satisfied: The fuel must be compressed to a sufficiently large number of atoms per cubic centimeter; the temperature must be sufficiently high; and these two conditions must be sustained for a sufficiently long time (typically a tenth of a second). It is very difficult to meet all three conditions simultaneously because the gaseous fuel is highly ionized (charged) at these temperatures, which makes it difficult to confine it at the required high compressions. For instance, if the fuel touches the walls of its container, the walls will immediately cool the fuel far below fusion temperatures.

Despite steady progress during the decades of intense research that have gone into the fusion option, the conditions for creating usable energy have not yet been met, and a commercial fusion reactor could not be ready until at least the middle of the twenty-first century. But if fusion reactors do become operational, they might solve many energy problems and completely change the energy picture. Fusion-generated electricity might, for example, turn out to be inexpensive or to produce very little radioactive waste. Despite its possible future importance, we will not discuss the fusion option further here, simply because there are so many nearer-term decisions that need to be made.

Renewable energy resources are widely discussed future options. An energy resource is renewable if it is continuously available, like sunlight, or if it can be replaced within a human lifetime, like wood. Other resources, such as fossil fuels (which are really forms of solar energy stored over long periods) and uranium, are **nonrenewable**.

Sunlight energized the agricultural revolution thousands of year ago, and renewable wood and nonrenewable coal energized the beginning of the Industrial Revolution a few centuries ago. Industrial societies are now energized mainly by nonrenewable fossil fuels and uranium, but renewables still do play a significant role in the form of wood (3.5% of today's total) and hydroelectric (3.5%).

Future renewable energy options take many forms. Hydroelectric could expand a little. Biomass—biological products including wood, waste paper, garbage, agricultural wastes, sugar crops, and grains—can be burned for thermal energy or transformed into liquid or gaseous fuels such as alcohol and methane. Wind can generate electricity. The sun can heat water or some other fluid that can then be used to generate electric power, to create electricity directly in photovoltaic cells, and to provide direct heating for warmth and hot water. All of these renewable options are related, directly or indirectly, to the sun. A related nonsolar option is the geothermal energy of Earth's hot steam, hot water, and hot rock. Because there is so much geothermal energy, we classify it as renewable, even though strictly speaking, it is replaced over time periods much longer than a human life. We will discuss renewable energy further in Section 17.6.

Another widely discussed future option is greater conservation. Although not really an energy resource, conservation acts in many ways like an energy resource. **Conservation** refers to all those measures that reduce our energy consumption so that fewer resources are needed. Conservation includes **energy efficiency** measures, such as home insulation, energy-efficient lighting, and energy-efficient automobiles, that reduce energy consumption without altering the services provided by that energy. Conservation also includes saving energy by changing or reducing energy services, for example, switching from automobiles to mass transit, living in smaller homes, and building more compact communities. This distinction between efficiency and changed services is important, because many studies show that energy consumption can be reduced dramatically by efficiency alone, with no need for controversial changes in the amount or quality of the services we get from energy. We will discuss conservation further in Section 17.7.

Table 17.1 summarizes the main current and future options for resolving the energy problem. It is convenient to group them into four categories: fossil, nuclear, renewables, and conservation.

Figure 17.5 documents an important point concerning the history of energy use since 1950. This graph is similar to that of Figure 17.2, but with more detail. The curve marked "Total annual energy" traces the total consumption of natural energy resources. Fossil fuels dominate, with a small long-term contribution from renewable wood and hydro and an expanding contribution from nuclear beginning in the early 1970s.

The line marked "U.S. annual inflation-adjusted GNP" traces the nation's annual gross national product (GNP), its annual output of goods and services. The dollar GNP amounts are not shown. Instead, the scale of the GNP graph is chosen to make the GNP and energy graphs coincide in 1950. Since the GNP graph is a good measure of the services that the nation actually receives from the energy it consumes, it is not surprising that the United States' energy consumption rose almost exactly in step with its GNP between 1950 and 1973: As energy services increased, so did energy consumption.

The first international oil crisis, in 1973, changed this. Energy prices increased rapidly in response to decreased foreign oil supplies, and the nation looked for ways to save money by conserving energy. The effect of this crisis, and of the second crisis in 1979, is evident in the graph. Although the GNP fell briefly following each crisis, it soon resumed its upward trend. But the total energy use declined further than the GNP did after each crisis and did not resume its upward trend. Apparently the conservation measures, promoted by increased prices, became permanent. By using energy more efficiently, the nation obtained a steadily increasing level of goods and services out of a roughly constant supply of energy between 1974 and 1990.

Without conservation, it is likely that the United States' total energy consumption would have stayed in step with its GNP. So the difference between the two graphs is appropriately labeled "Conservation." By 1990, conservation contributed 40 exajoules to the nation's energy mix, more than did any other energy resource.

DIALOGUE 4 From Figure 17.4, find the overall energy efficiency of each of the following: electric power generation, the industrial sector, the residential–commercial sector, the transportation sector.

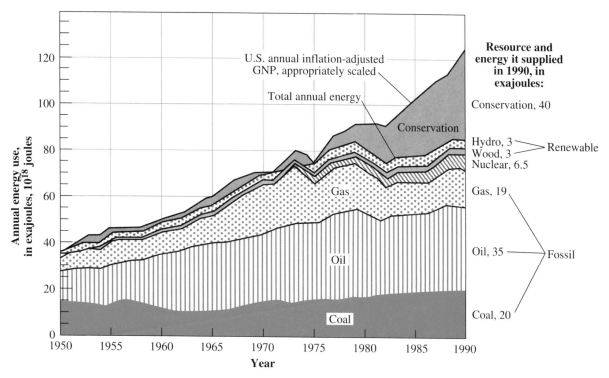

FIGURE 17.5

A more detailed view of the history of U.S. energy consumption since 1950. The upper boundary traces the GNP, one measure of the goods and services provided by energy. The gap that opens between the total energy consumed and the GNP, beginning at the time of the 1973 energy crisis, is a measure of the effect of conservation since 1973.

DIALOGUE 5 MAKING ESTIMATES   In Dialogue 3, we used the 85 exajoules of annual U.S. energy consumption to verify that the daily energy consumption per person is 1000 MJ (Figure 17.1). Now use this figure, 1000 MJ, to estimate the average person's power consumption, in watts. Express this in kilowatts. If all this energy were in the form of electricity, how many 100-watt bulbs could it light up?

17.3 Nuclear power: *how it works* _____

Most nuclear power comes from uranium chain reactions that provide energy for steam-generated electricity. Nearly every U.S. power reactor is a so-called light-water reactor, most of them of the type shown in Figure 17.6. Except for the reactor itself, nuclear power plants operate the

---

4. Electric power 9/29, or 31%; industrial 50%; residential–commercial 71%; and transportation 21%.

5. Power consumed is energy consumed per second (Chapter 6). Accordingly, divide the 1,000 MJ of daily energy consumption by the number of seconds in a day: $1{,}000 \times 10^6$ J/$(60 \times 60 \times 24$ s$) \approx 12{,}000$ watts, or 12 kilowatts. This is the equivalent of 120 100-watt bulbs.

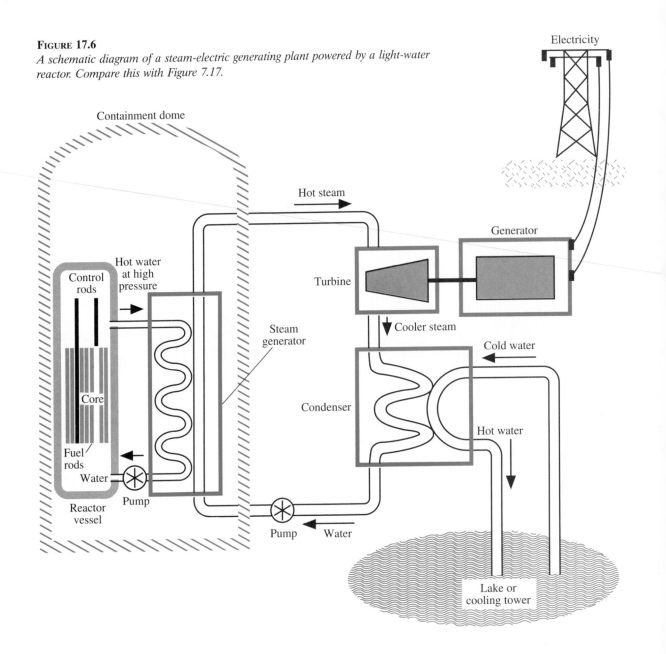

FIGURE 17.6

A schematic diagram of a steam-electric generating plant powered by a light-water reactor. Compare this with Figure 7.17.

We cannot control atomic energy to an extent which would ever be of any value commercially, and I believe we are not likely ever to be able to do so.

Prediction made by physicist Ernest Rutherford in 1933

same way as does every other steam-electric generating plant, whether powered by coal, oil, natural gas, wood, or the sun: Thermal energy from the primary energy source (nuclear fission, in this case) heats water to make high-temperature steam that turns a turbine that turns a generator that makes electricity. A **light-water reactor** uses natural water, called *light water* to distinguish it from the nonnatural *heavy water* that is used in power reactors in a few countries. **Heavy water** is water in which the two hydrogen atoms in each H_2O molecule are the rare "heavy" isotope $_1^2H$.

All power reactors have four essential ingredients: **fissionable fuel** to provide energy; **coolant** to transfer thermal energy away from the fuel; neutron-absorbing **control rods** that can be moved in or out to slow down, stop, or speed up the chain reaction; and a **moderator** to slow down the neutrons to make it more likely that they will cause fission.

The coolant—natural water in light-water reactors—circulates through the center of the reactor and removes the thermal energy created by the chain reaction. The same water acts as the reactor's moderator, by slowing down the neutrons to their most efficient speeds when these neutrons collide with the hydrogen nuclei in the water.

The fuel in light-water reactors is uranium, slightly enriched to 3%. The energy comes only from the ^{235}U nuclei (3%), and not from the nonfissionable ^{238}U nuclei (97%). Natural uranium, made of roughly 99% ^{238}U, won't work in light-water reactors, because both ^{238}U and $^1_1$H can absorb neutrons and spoil the reaction. Natural uranium can, however, be used as fuel if heavy hydrogen, $^2_1$H, replaces light hydrogen in the moderator, because $^2_1$H does not absorb neutrons. Consequently, countries with heavy-water reactors can use natural uranium as fuel, making it possible to obtain plutonium for nuclear weapons without having to enrich uranium.

Since power reactors use slightly enriched or natural uranium rather than highly enriched (90%) bomb-grade uranium, it is impossible for a power reactor to explode the same way that a full-fledged nuclear weapon would. And it is impossible to use the uranium directly in a bomb.

The uranium fuel is shaped into small cylindrical pellets and stacked in long thin metal tubes. Some 40,000 of these fuel rods, 100 tonnes of fuel, form the power-producing core of the reactor. The core is enclosed in a heavy steel "reactor vessel" (Figure 17.7) designed to withstand high pressures, to shield against radiation, and to absorb the many neutrons that escape from the core.

Water circulates through two physically separated loops (Figure 17.6). The first loop circulates through the core, removing heat from the fuel rods. Although this water is heated to above 300°C, the high pressure maintained inside the reactor vessel keeps it from boiling. This water circulates around to a "steam generator" where it heats the water in the second loop. This thermal energy is exchanged by placing the first loop's pipes in contact with the second loop's pipes, so that thermal energy flows from the hot loop to the cooler one without allowing their water to mix.

Because the reactor's operation causes the core, the reactor vessel, and the first water loop to become highly radioactive, these elements are encased in thick concrete. For further shielding and protection against accidents, a large airtight **containment dome**—made of steel-reinforced concrete about a meter thick and built to absorb an impact as great as that of a jetliner crash—surrounds the radioactive elements (Figure 17.8).

Since the secondary loop is not kept at high pressure, its water immediately turns to steam in the steam generator. The rest of the operation is identical with every other steam-electric power plant (Figure 17.6, also see Section 7.6).

FIGURE 17.7
Reactor vessel.

FIGURE 17.8
Nuclear power plant under construction. This view shows a cooling tower and a reactor containment dome under construction.

DIALOGUE 6   Uranium is one possible fission-reactor fuel. Name another. How could this second fission reactor fuel be obtained?

17.4 Technology assessment: *assessing the nonrenewables*

We nuclear people have made a Faustian bargain with society. On the one hand, we offer an inexhaustible source of energy. But the price that we demand of society for this magical energy source is both a vigilance and a longevity of our social institutions that we are quite unaccustomed to.
   Alvin Weinberg, former director of Oak Ridge National Laboratory, 1972

How should we evaluate our energy options (Table 17.1)? From energy to ozone, science-related social issues such as this are part of today's landscape. Should we turn over these issues to the experts? To the government? Beneath the technical details, these issues concern human values, and neither experts nor governments are necessarily wise about values. In democratic nations, citizens must decide these issues because they are primary determinants of the kind of society we live in. In fact, citizens do decide these issues—every time they use technology, purchase consumer items, or vote. Should everybody become experts, then, on all these questions? Obviously that is impossible. But at the other extreme, thoughtlessly following our beliefs and intuitions is likely to be a prescription for disaster.

During the past few decades, citizens groups, governments, businesses, and individual persons have clarified these issues through **technology assessment**, which encourages open-minded, rational decisions about technology. It is not, however, a complete prescription for finding answers, because the questions concerning values generally go beyond technical assessments.

To assess a technological issue fairly, one should begin with a clear and unbiased statement of the question at hand. Second, one should list all of the

---

6. Plutonium, obtainable from nuclear reactors themselves. Plutonium is produced from the nonfissionable isotope ^{238}U that is present in the reactor fuel.

plausible solutions, without bias. Third, one should make a fair and balanced consideration of the costs and benefits of each option, realizing that all options have costs, even the option of doing nothing. Finally, one should fairly weigh both costs and benefits of the options and be willing to accept trade-offs between them. Rather than looking for zero-cost solutions, it is more constructive to look for the costs of each solution and then compare them with the benefits. Although few persons or groups are able to carry out a complete assessment of all options, these guidelines form a useful framework to help all of us organize our thinking about technology-related issues.

In assessing our energy options, we begin by looking at the options already in wide use. These are primarily the nonrenewables: the three fossil fuels and nuclear energy (Figure 17.3). The industrialized nations have already used up much of the world's oil and natural gas, and so these cannot sustain us in the future in the way that they have in the recent past. But there is a lot of coal remaining in the world, and so we will compare coal with nuclear power. These two resources produce 76% of the United States' electricity, with coal supplying 56% and nuclear power 20%.

DIALOGUE 7   Draw a line down the center of a single sheet of paper. Write "coal power" at the top of one side and "nuclear power" on the other side. List as many possible problems (costs) and strengths (benefits) as you can think of under each heading. If you intuitively favor one of the options, compensate for this possible bias by looking especially hard for the costs of that option and the benefits of the other option. After each entry on either side, consider whether there should be a comparable entry on the other side. For example, nuclear power plant accidents might be listed opposite coal mine accidents.

Table 17.2 is one possible list of problems for coal and nuclear power.

Because both resources are used to generate electricity, the two lists are similar. In fact, there are parallel entries for all but two problems: global warming caused by coal, and nuclear weapons proliferation caused by nuclear power. But this similarity of problems does not necessarily mean that the evaluations will be similar. For example, the amount of land degraded per unit of energy produced is far greater for coal than for nuclear power, because the volume of fuel is so much greater. For another example, terrorists are more of a threat to nuclear power than to coal power, because they could release radiation or steal materials to help make nuclear weapons.

One of the more fundamental issues listed in the table is resource availability. Table 17.3 shows the resource picture for coal and uranium and also for oil and natural gas. The situation is quite different for the four resources. Only 2% of U.S. coal has been consumed, whereas oil and

7. There is no single answer to this question, although several items should appear on any reasonably complete list: greenhouse effect (coal only), accidents (both), waste disposal (both), nuclear weapons proliferation (nuclear only), chemical pollution such as acid rain (coal only), radioactivity in the environment (nuclear only, although coal plants release some radioactivity). See Table 17.2 for suggestions.

TABLE **17.2**
Problems for Coal Power and Nuclear Power

	Examples or comment	
Problem	Coal power	Nuclear power
Land degraded	Mining, especially strip mining	Mining
Pollution	Acid rain, SO_2, NO_2, ash	Radioactive isotopes
Worker's health	Miners' black lung disease	Radiation in mines and power plants
Accidents	Mine explosions and collapses	Power plant accidents
Thermal pollution	Heating of lakes and rivers	Heating of lakes and rivers
Solid-waste disposal	Ash, sludge	Used fuel rods, low-level radioactive waste
Available resources	Coal resources	Uranium resources
Economics	Cost of plant, fuel, operations	Cost of plant, fuel, decommissioning
Global warming	Due to CO_2	—
Nuclear weapons proliferation	—	Due to Pu, enrichment facilities, knowledge
Public perceptions	Acid rain, strip mining	Radiation risks, waste disposal, accidents
Sabotage, terrorists	Shut off electricity	Shut off electricity, release radiation, build bomb

natural gas are about half gone, and uranium stands between these extremes. A similar picture emerges when one looks at the number of years before each resource would be used up if the current annual rate of U.S. production of that resource remained unchanged (second column).* The last column of the table shows how many years each resource would last if it supplied all of the United States' energy. Apparently the United States has a lot of coal and not so much oil or gas, with uranium lying somewhere in the middle.

TABLE **17.3**
Percentage Consumed and Years Remaining at Present Rates of Use for Four Nonrenewable Energy Resources in the United States

	Percentage of the total U.S. resource consumed by 1992[a]	Years remaining until total resource is gone at present annual rate	Years remaining if used to provide total U.S. annual energy consumption
Coal	2	3600	805
Oil	70	25	10
Natural gas	40	69	15
Uranium (without breeder reactors)	8	200	14

[a] The "total resource" means the amount estimated to be recoverable by known technical means, perhaps at increased but not unreasonable prices.

---

* The rate will not remain unchanged, however. If a resource's use increases, its remaining years must decrease. As oil and natural gas approach depletion, they will become more difficult and thus more expensive to produce, and so production will decline. Oil is already past the high point of the bell-shaped curve (Figure 7.24) that is typical of nonrenewable resource consumption. Despite these qualifications, the years remaining at current rates of use (Table 17.3) is a useful measure of a resource's remaining "lifetime."

Although acid rain and the disposal of large volumes of ash and sludge are serious problems for coal, global warming is coal's biggest drawback (Chapter 9). Carbon-dioxide emissions cause 55% of the global warming problem, and essentially all of this comes from the three fossil fuels. The consensus among energy analysts is that coal burning must decline over the coming decades because of global warming. Switching from coal to natural gas can help, because for equal amounts of energy, natural gas produces only about half as much carbon dioxide as coal does. But because of global warming and also limited gas supplies, this can be at most only a temporary solution. Because global warming is pushing the world toward a nonfossil future, energy analysts have been looking more seriously at nuclear power, renewable resources, and conservation.

DIALOGUE 8 MAKING ESTIMATES   U.S. coal plants collect 80 million tons of unburnable ash every year, most of which is disposed of in ponds, where it settles to the bottom. How much is 80 million tons? For instance, how many large highway trucks would it fill, at a 30-ton load per truck? If these trucks formed a bumper-to-bumper caravan across the United States, how long would the line be? (Highway trucks are some 15 m long.)

17.5 Issues for nuclear power

There are many lessons, for all nations and for all technologies, in the history and current problems of nuclear power in the United States.

The first commercial nuclear plant opened in 1957. By 1975, with fifty-four plants providing 10% of the nation's electricity, the nuclear age seemed firmly established. But then the nation hesitated. The last firm order for a new plant was placed in 1974. Despite the growing number of operating plants, it is now clear that the nuclear power era ended, for now at least, during the 1970s (Figure 17.9).

Nuclear power has always attracted strong proponents and opponents. Today, proponents argue that global warming, fossil-fuel shortages, fossil-fuel pollution, and the prospect of a safer generation of nuclear reactors call for a revival of nuclear power, the only nonfossil resource that has been proved capable of producing large amounts of energy. Opponents contend that nuclear power is unacceptable because of safety concerns, waste disposal problems, economics, and nuclear proliferation, that conservation plus natural gas can gain time for a few more decades, and that renewable resources and conservation will then be able to meet our needs. There is, however, agreement by energy analysts on both sides that it is time to start disengaging from fossil fuels.

---

8. $80 \times 10^6/30 = 2.7 \times 10^6$: nearly 3 million trucks. $(15 \text{ m}) \times (2.7 \times 10^6) = 40 \times 10^6$ m, or 40,000 km, or about eight times across the United States.

FIGURE 17.9
The United States' commitment to nuclear power, 1960–1990: total megawatts installed and under construction.

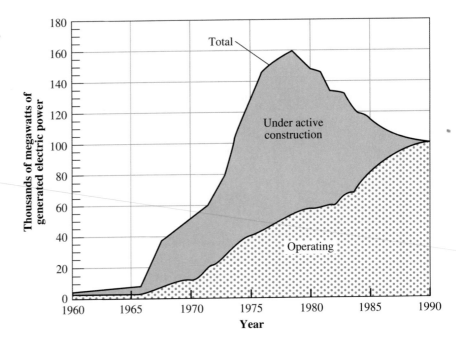

Nuclear power must play a significant role in a sustainable future, but it must be carefully administered. Safety standards, design features, waste storage and the entire fuel cycle must be under the jurisdiction of an international regulatory agency. Only then can the full potential of nuclear power be realized. In view of the worldwide population growth and its accompanying problems, I sincerely doubt that turning backward will allow us to master the future. Instead we have to go forward.
Wolf Hafele, physicist, energy expert, and former scientific adviser to the West German government

The "electricity-specific" uses that need, and economically justify, electricity's high quality are only 8 percent of U.S. end-use needs. Positing huge new markets for coal or uranium is thus like pushing on a string—imaginative but ultimately unsatisfying.
Amory Lovins, physicist, energy consultant, and director of research at the Rocky Mountain Institute at Snowmass, Colorado

Radioactive waste. . .is a rather trivial technological problem.
Bernard L. Cohen

The decline of nuclear power was caused by several factors, factors that are still important:

· Public support for nuclear power has always been weak because of fears of radiation and nuclear weapons, and doubts about the honesty and propriety of the nuclear enterprise.*
· Business and public confidence were shaken by the accidents at Three Mile Island in 1979 and at Chernobyl in the Soviet Union in 1986.
· The demand for electricity has not grown as much as expected.
· The costs of building nuclear power plants have risen dramatically.
· No solution to the problem of disposing of nuclear waste has been found that is acceptable to the public.

Next we will discuss four interrelated questions: nuclear waste, accidents, uranium resources, and weapons proliferation.

Although lower-level forms of nuclear waste pose some problems, the intensely radioactive **high-level nuclear waste** that comes from used fuel rods is the most serious problem. Used fuel rods are currently stored on site in large pools of water inside the containment domes. Because they must be stored safely for hundreds or perhaps thousands of years, a long-term solution is needed.

Used fuel rods contain the fission fragments and other radioactive isotopes created during the chain reaction (Table 17.4). Shorter-lived isotopes dominate the radioactivity for the first few centuries, so that the waste's radiation level decreases by a factor of 300 during the first 10 years, by a fac-

---

* Physicist and historian Spencer Weart gives an excellent historical analysis of these issues, in *Nuclear Fear: A History of Images* (Cambridge, MA: Harvard University Press, 1988).

TABLE **17.4**

Some Important Isotopes in
High-Level Nuclear Waste and
Their Half-lives

Fission products	Half-life (yr)
^{90}Sr (strontium)	29
^{93}Zr (zirconium)	950,000
^{99}Tc (technetium)	210,000
^{135}Cs (cesium)	2,300,000
^{137}Cs	30

Other isotopes created during the chain reaction	
^{238}Pu (plutonium)	88
^{239}Pu	24,000
^{241}Am (americium)	430
^{244}Cm (curium)	18
^{245}Cm	8700

No scientist or engineer can give an absolute guarantee that radioactive waste will not someday leak in dangerous quantities from even the best of repositories. . . .Should repository construction be authorized? Or should it be delayed. . .? These are not technical questions, but questions for the man in the street and his elected representatives.
Konrad Krauskopf, geologist, Stanford University

The proposed storage of high-level radioactive wastes in the Nevada desert poses no apparent risk of groundwater contamination.
From a newspaper report on the findings of a National Research Council study on the suitability of Yucca Mountain, in Nevada, as a high-level nuclear-waste disposal depository

People living near the Three-Mile Island reactor get more radiation exposure from radon in their homes every day than they got from the 1979 accident.
Bernard L. Cohen

tor of 100,000 during the first 1000 years, and by a factor of 1,000,000 during the first 10,000 years. Many experts believe that safe burial for 1000 years is sufficient, although Congress requires that any burial site be designed to be secure for 10,000 years. Experts believe that such long-term burial can be achieved by incorporating the wastes into glassy materials, putting these materials into steel canisters with a lifetime of 10,000 years, and storing the canisters in human-made caverns in geologically stable sites several hundred meters below ground. Various sites have been proposed, most recently at Yucca Mountain in Nevada, but there is opposition to each one.

The accident in 1979 at **Three Mile Island** illustrates the kinds of things that can go wrong in complex technologies. It began with a fairly routine event: A pump failed in the water loop that carries steam to the turbine (Figure 17.6). Control rods immediately and automatically dropped into the reactor, and fission ceased. But stopping fission in a reactor doesn't stop all the heating, because some 5% of a reactor's power comes not from fission but from radioactivity, and this 5% dies away very slowly. So water must be kept moving through a "shut-down" reactor to prevent overheating.

A backup pump took over to prevent overheating. But a valve in this backup pump had been left closed by mistake, and a warning light that should have alerted operators to the closed valve was obscured by a tag, so this water did not get into the reactor for another 8 minutes. During this time, the temperature and pressure rose in the primary water loop, forcing a pressure-relief valve to open. Unfortunately, this valve stuck open and, for 2 hours, was not discovered to be open. The open valve allowed water to escape from the reactor, flooding the containment building floor with radioactive water. Some of this water was automatically pumped to an adjoining building, releasing radioactivity.

Water in the core evaporated and escaped. Because of the threat of overheating due to the loss of water, every reactor contains a tank full of emergency cooling water for use in an accident. The reactor's declining water level automatically triggered the release of this water. But unfortunately again, the operators interpreted their control room dials to mean that there was too much water in the core, rather than too little. So they shut off the emergency cooling water. The water level dropped below the top of the fuel rods, and the fuel heated until much of the core had melted, an event known as a **melt-down** because it causes the fuel to slump downward. This permanently destroyed the reactor. Only the dwindling water still in the bottom of the core prevented the entire reactor vessel from melting and spilling molten fuel into the containment building.

Unusual chemical reactions in the hot reactor created hydrogen gas that remained in the reactor vessel for several days, causing concern that a hydrogen explosion might rip open the reactor vessel. The explosion did not happen because of a lack of oxygen in the reactor vessel.

There were no immediate deaths, although the excess radiation might cause one extra long-term cancer death among the surrounding population. The total radioactivity released was small. But the accident cost the utility company $1 billion to clean up, plus the loss of the reactor and the containment dome. Most of the cleanup was finished by 1990, eleven years after the accident.

History's worst nuclear power accident happened at **Chernobyl**. Unlike the current U.S. power reactors, the Chernobyl reactor was moderated by graphite rather than water, and it lacked a containment dome.* Although low-enriched reactor fuel cannot explode in the same way that a full-fledged nuclear weapon can, graphite-moderated reactors can suffer a runaway chain reaction that releases thermal energy rapidly enough to create a low-grade nuclear explosion, similar in its energy release to that of a conventional chemical bomb. Such a runaway chain reaction occurred at Chernobyl, causing a melt-down, contaminating a large area, and causing many short-term and long-term casualties (Section 15.6). The fuel melted through the containment vessel and partly penetrated the plant floor. If it had melted entirely through the floor, it could have contacted ground water and set off a massive radioactive steam explosion.

Although these two accidents led to many new safety procedures, experts estimate that with some 400 power reactors operating worldwide today, the chances of another core melt in one of them during the next 10 years is between 4 and 40%. If there is to be a revival of nuclear power, it will probably need to be based on new safer reactors. For example, one reactor being developed uses fuel formed into tiny grains of uranium individually encased in ceramic shells. The grains are so small and so widely dispersed that they cannot become hot enough to melt even if all the coolant is lost from the core.

Some energy scenarios envision a fivefold to tenfold increase in nuclear power during the next few decades. Any such scenario must take into account the uranium resources (Table 17.3). Under a fivefold expansion, uranium resources would run out in 40 years.

There is a way to solve the uranium resource problem. Recall (Section 16.5) that uranium's nonfissionable ^{238}U is transformed into fissionable ^{239}Pu during a chain reaction. This plutonium can then be used as fuel for another reactor. With proper design, it is possible to create more than one ^{239}Pu nucleus for every ^{235}U nucleus fissioned. Such **breeder reactors**, which create more ^{239}Pu fuel than they consume in ^{235}U fuel, could eventually convert much of the ^{238}U in any country's uranium resources to fissionable plutonium, greatly extending those resources.

Although breeder reactors can solve the resource problem, they make the **nuclear weapons proliferation** problem worse. As we saw in Chapter 16, there are several connections between nuclear power and nuclear weapons. For example, South Africa's nuclear weapons program demonstrates that an apparently peaceful nuclear power program can provide cover for a secret weapons program. And Pakistan's illegal purchase of nuclear materials from several European nations demonstrates that one country's peaceful nuclear power program can aid another country's nuclear weapons program.

Furthermore, uranium-fueled power reactors can produce militarily useful plutonium, although the plutonium must first be processed to separate it from the remaining uranium and from the highly radioactive isotopes in the used fuel rods. India followed this route when it utilized the used fuel rods from a nuclear power reactor, along with a supposedly peaceful processing plant,

---

* However, U.S. military reactors of this type operated until 1987.

to obtain plutonium for nuclear weapons. The ongoing nuclear arms race between Pakistan and India is intertwined with nuclear power issues.

Breeder reactors only exacerbate all of these proliferation problems. Reactor-grade plutonium can fuel a nuclear weapon, whereas reactor-grade uranium cannot. Widespread breeder reactors and reprocessing will create large amounts of already separated plutonium to be stored, shipped to other reactors, and used. An expanded nuclear power industry could put more than a million kilograms of plutonium into global commerce per year. Because fewer than 10 kilograms are needed for a fission bomb, it might be difficult to keep significant amounts of this plutonium from going into weapons.

There are lessons in all this. Many observers believe that the nuclear power industry should have planned more thoroughly and more realistically decades ago. Carroll Wilson, the first general manager (from 1947 to 1951) of the U.S. Atomic Energy Commission, writing in 1979 shortly after the Three Mile Island accident, recalled that during the early development of nuclear power, "No one spent much time. . .looking at the total system of the nuclear fuel cycle from the fuel enrichment through radioactive wastes. . . .There was no awareness that the whole system must function or none of it might be acceptable." He stated that the decision to adopt light-water reactors did not follow from a careful technology assessment but instead evolved historically out of the U.S. Navy's submarine reactors. The nuclear industry rapidly scaled up these small navy reactors to much higher powers, partly because the people who ran the industry had been trained on submarine reactors. And according to Wilson, "one of the grievous errors. . .was the failure to carry out the repeatedly recommended experiment of running a reactor to destruction [by purposely allowing it to lose all of its coolant] to find out what would really happen instead of depending on studies based on computer models for such vital information."

One might say that in regard to the questions of both scaling up the small submarine reactors and finding out what happens when a reactor loses all of its coolant, the nuclear industry ignored the basic question of the scientific method: How do we know—what is the evidence? Whether our energy future is to be nuclear, renewable, or something else, it would be wise to heed such lessons.

DIALOGUE 9   Does a coal-burning generating plant have combustion products that are analogous to a reactor's fission products? What is done with them? Is this harmful?

17.6 Future energy options: *renewables*

Among the renewable resource options (Table 17.1), only hydroelectric and biomass contribute significantly to today's total U.S. energy budget. However, other renewables are expanding, and their future possibilities are significant.

---

9. The main product of combustion is carbon dioxide, which goes into the atmosphere where it contributes to global warming.

Hydroelectric and biomass each contribute 3.5% of the U.S. energy budget (Figure 17.5). **Hydroelectric energy,** the gravitational energy of water that the sun's warmth has raised by evaporation, goes entirely to centralized electricity. Although more energy could be squeezed out, it is close to its practical limit.

Biomass energy, the chemical energy of organic substances such as wood, sugar crops, grains, and trash is available from many sources and can be transformed into many forms. There are three general ways of using biomass for energy. First, it can be burned for heating or to make power-plant steam for electricity; these are the main uses today. Second, bacteria can decompose biomass and convert it to alcohol liquid fuels for transportation and to methane gas for heating and other purposes. Third, biomass can be chemically processed to create gaseous or liquid fuels to replace natural gas and oil. All together, an estimated 20% of current annual energy consumption could be supplied in these ways. Biomass makes no long-term net contribution to global warming, provided that the trees or crops grown for energy are replanted.

The remaining renewables in Table 17.1 contribute very small amounts today, but each has substantial future potential. Of these renewables, geothermal, wind, solar–thermal, and photovoltaic cells are used to make electricity. The remaining two, active and passive solar, are used for direct heating.

Geothermal energy is the thermal energy that radioactivity and pressure create in Earth's molten core. The easiest way to tap it is by drilling underground and directly removing naturally hot water or steam (Figure 17.10). The country's largest installation, at The Geysers in northern California, provides the equivalent of two large coal or nuclear generating plants. But geothermal hot water and steam are very limited resources. The large, essentially unlimited geothermal resource is **hot dry rock** lying several kilometers underground. The thermal energy that might be recovered from the hot dry rock within a few kilometers of the surface amounts to many times more energy than is contained in coal deposits and could provide a large proportion of the nation's electricity. The drilling and underground fracturing technology to recover this energy is highly demanding and is as yet unproved (Figure 17.11).

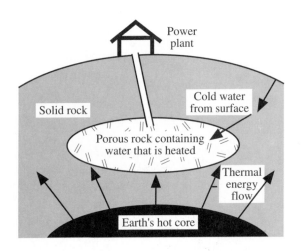

FIGURE 17.10
A model of a hot-water geothermal system. The water in the porous rock is heated by Earth's hot core. The hot water escapes through a well, boiling near the top.

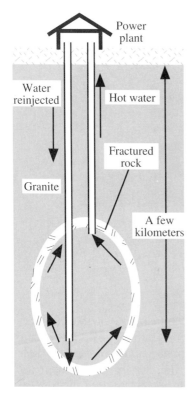

FIGURE 17.11
Thermal energy might someday be extracted from dry hot rocks by circulating water through large cracks created by hydraulic fracturing.

Wind energy is the kinetic energy of air set into motion when the sun warms the daylight side of Earth. Since antiquity, it has driven sailing ships and turned windmills for grinding and pumping. Today, it also generates electricity. The world's largest installation is at Altamont Pass in California, where small electric generators attached to 7500 **wind turbines** have a capacity equivalent to two large power plants (Figure 17.12). This is a proven technology: The Altamont installation is 50 years old. In regions where wind turbines are feasible, wind power is already financially competitive with coal for large-scale electric power generation. New designs should reduce costs further, potentially making twelve U.S. Great Plains states major wind-power producers. At the present time, 300-kilowatt wind turbines, each one powerful enough for 300 homes, are under development. As a simple but unrealistic illustration of wind energy's potential, 2 million of these turbines, sited on perhaps 5% of the Great Plains land area, could provide all of the nation's electricity.

Solar–thermal electricity is generated from thermal energy created by the sun. This resource generates the equivalent of one large power plant today, using two types of technologies. In the first, reflective solar collectors track the sun and focus it on a liquid such as oil that is then piped to a central location where it produces steam to drive an electric generator (Figure 17.13). The second uses sun-tracking mirrors to reflect solar energy to a central boiler that produces steam (Figure 17.14). A more speculative scheme uses the warm water near the ocean's surface and the colder water deeper down as the hot and cold ends of a heat engine that turns an electric generator (Figure 17.15).

Photovoltaic cells use the photoelectric effect (Section 13.2) to transform solar radiation directly into electric current. The cells are made of **semiconducting materials**, such as silicon, that have electrical properties lying

FIGURE 17.12
Wind generators, each one producing 100 kilowatts of electricity (enough for about 100 households), in northern California.

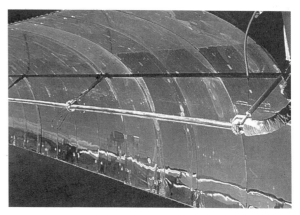

FIGURE 17.13
Solar collectors for a solar–thermal power plant. The mirrored faces focus sunlight onto the pipe running the length of the collectors. The pipes transfer thermal energy to a steam turbine coupled with an electric generator.

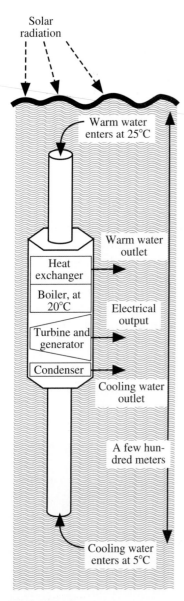

Solar
radiation

Warm water
enters at 25°C

Warm water
outlet

Heat
exchanger

Boiler, at
20°C

Electrical
output

Turbine and
generator

Condenser

Cooling water
outlet

A few hun-
dred meters

Cooling water
enters at 5°C

FIGURE 17.15
*Solar–sea power plant. The heat
engine's "working fluid" must boil
at around 20°C in order to change
from liquid to gas within the
narrow temperature range available
in the ocean's top few hundred
meters. Ammonia is one possible
working fluid.*

FIGURE 17.14
*A 10-megawatt solar–thermal test facility in Barstow,
California. The mirrors reflect sunlight onto a boiler on
top of the tower, and the boiler makes steam for electricity
generation.*

midway between conductors and insulators. Normally, semiconductors behave like insulators: It is difficult to make their electrons flow. But if some
of a semiconductor's electrons are given a small amount of energy, these
electrons can flow easily, as in a conductor. Because of these properties,
semiconductors are the basis of the solid-state revolution that is the basis
for modern electronic technology. In addition to being a promising energy
technology, photovoltaic cells are an instructive example of this electronic
revolution.

In photovoltaic cells, light provides the energy that puts electrons into the
conducting state. A typical cell is made of two thin layers of silicon placed
on top of each other (Figure 17.16). These two layers are constructed differently so as to have different electrical properties, called *n-type* (negative)
and *p-type* (positive). The difference is that nonsilicon "impurity" atoms of
different types are introduced into the two different layers. N-type semiconductors are made of silicon containing impurity atoms that have more
semiconducting electrons (per atom) than does silicon. P-type semiconductors contain impurities that have fewer semiconducting electrons than does
silicon.

Suppose the two layers are simply placed together, without any external
electrical contacts. Because of the impurities, microscopic forces acting
within each layer cause electrons to quickly flow from the n-type side (where
there were more electrons to begin with) to the p-type side (where there
were fewer). Within a small fraction of a second, this process builds up a
sufficiently strong electric field at the "p–n boundary" (the junction between
the two layers) to stop the flow of electrons.

If light now shines on the two layers—still with no external electrical
contacts—the photoelectric effect will energize some of the electrons, causing those to cross back across the p–n boundary from the p side to the n
side. At this point, the n side is like the negative terminal of a battery that
is not connected to anything, and the p side is like the positive terminal.

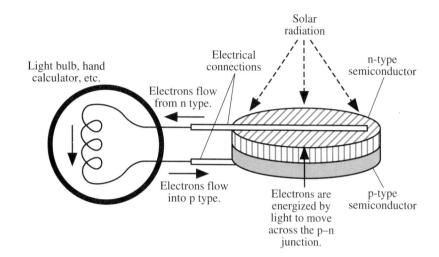

FIGURE 17.16

How a photovoltaic cell works. The cell is almost as thick as a dime (1 mm). Because of the differing microscopic properties of the two semiconductors, electrons that are energized by light cross the p–n junction from the p to the n side. These electrons can then flow through an external circuit. The junction acts like a battery.

Now suppose we connect this tiny "battery" to an external device, as shown in Figure 17.16, where a metal strip is attached to each side and a wire attaches each strip to an electrical device such as a light bulb. Then the energized electrons that were pushed to the n side will flow through the wires back around to the p side. As electrons arrive at the p side, light continues to energize electrons to move across the junction from the p side to the n side, creating a complete loop of flowing electrons. The p–n junction acts like a battery energized by light.

Arrays of such cells (Figure 17.17) can provide electricity for centralized electric power (Figure 17.18), or they can power individual buildings or appliances, especially in remote locations that are difficult to reach with centralized power. This resource is under intense development and should become increasingly competitive for large-scale electric power.

FIGURE 17.17

A series of photovoltaic cells, attached together, can provide a large electric current.

FIGURE 17.18

A large photovoltaic array near Davis, California, part of a public–private partnership to demonstrate the viability of utility-scale photovoltaic systems.

Finally, solar energy can be used directly for **solar heating**. Figure 17.19 shows a rooftop "flat-plate" collector where a pumped liquid is heated and then is circulated back indoors to be used for space or water heating. Solar collectors can also use forced air, instead of a liquid, to transfer thermal energy. Collectors using pumps or blowers are **active** forms of solar heating. Solar energy use can also be **passive**, based on natural flow. The methods are older than humankind (most animals and plants seek the sun at times), as simple as the backyard clothesline (Figure 17.20) and as new as the latest triple-pane argon-filled multilayer-coated windows. Figure 17.21 illustrates several passive concepts. Even in cold climates, these methods can reduce heating needs by 60 to 80%, and so it is surprising that during the 1980s, fewer than 10% of new houses incorporated passive features.

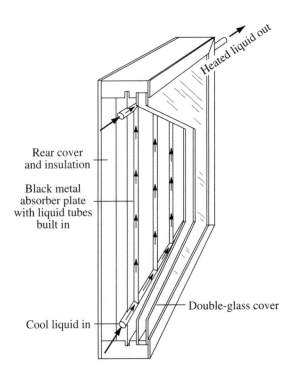

FIGURE 17.19

A flat-plate solar collector that uses forced circulation of a liquid to collect and transfer solar thermal energy.

FIGURE 17.20
A solar-fueled clothes dryer.

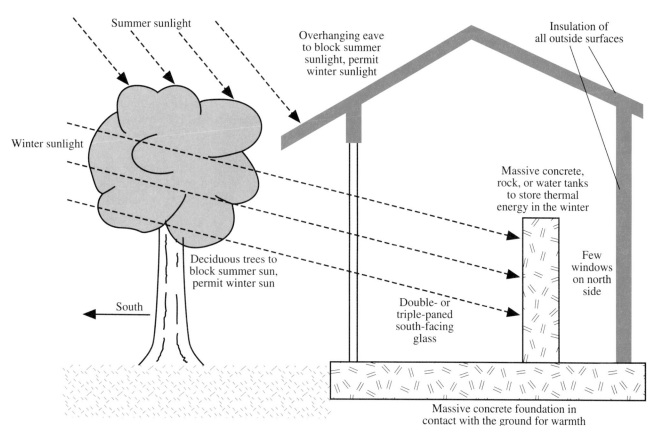

Summer sunlight

Overhanging eave
to block summer
sunlight, permit
winter sunlight

Insulation of
all outside surfaces

Winter sunlight

Massive concrete,
rock, or water tanks
to store thermal
energy in the winter

Few
windows
on north
side

Deciduous trees to
block summer sun,
permit winter sun

South

Double- or
triple-paned
south-facing
glass

Massive concrete foundation in
contact with the ground for warmth
in winter, coolness in summer

FIGURE 17.21
*Several passive solar energy concepts for keeping houses
warm in cold weather and cool in hot weather.*

The economics of these renewable resources will be critical in determining which, if any, of them will be widely used. Figure 17.22 graphs the past and expected future costs of electricity from three renewable sources and compares them with electricity from new coal-fired "base-load" plants that are operated continuously to provide most of the electricity consumed, and with natural gas-fired "peaking-load" plants that are operated only intermittently to provide electricity at times of high demand. Peak-load electricity costs more than twice that of base-load electricity. This cost means that efficiency or alternative energy measures that reduce the need for power at times of high demand are highly competitive economically. Wind and solar thermal electricity are competitive for peaking power today, and in some locales wind is competitive for base-load power. Early in the twenty-first century, all three might compete with coal for base-load electricity.

Many energy analysts have suggested that if a free-market nation is to preserve its environment and resources, it must incorporate "externalities" such as pollution and resource depletion into the price of goods. According to this view, coal, for example, is being subsidized by the environment because the price of coal does not reflect its full pollution costs, and this hidden subsidy distorts the market, leading to the overuse of coal because it is underpriced. Many analysts have suggested that such externalities be incorporated into the economics of fossil fuels. Several methods have been suggested, the simplest being a tax on the carbon content of all fuels, with the proceeds used to clean up the environment or to support development of less harmful energy resources. If this were done, the cost of fossil fuel–generated electricity would rise, and renewables would be more competitive. On the other hand, more traditional economists regard such measures as unhealthy manipulations of the free market.

DIALOGUE 10   Which of the following are heat engines; that is, which ones convert thermal energy into other forms: hydroelectric power plant, geothermal power plant, wind turbines, solar–thermal electricity generation, photovoltaic cells?

Substantial increases in importation of oil. . .could restore control of the market to OPEC by the year 2000 or before. Unless there are economic incentives for improving energy efficiency, conservation, and substitution, and for increasing domestic supplies, the United States will become increasingly vulnerable to a great, multicomponent energy crisis.

Philip Abelson, physicist, past editor of *Science*

FIGURE 17.22
Solar electricity costs compared with fossil-fueled electricity. The electricity costs of renewables (vertical axis) are based on actual (solid lines) or projected (broken lines) costs for centralized power generation. (Source: Carl Weinberg and Robert Williams, Scientific American, September 1990)

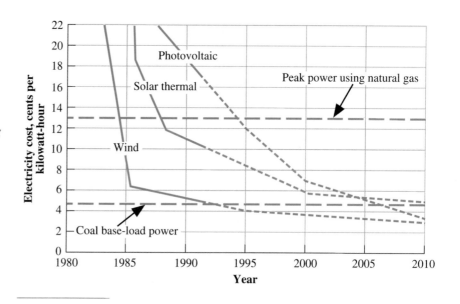

10. Geothermal and solar–thermal.

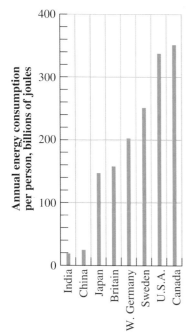

FIGURE 17.23
Per capita energy resource consumption in several nations in 1988.

By 1990, conservation contributed the equivalent of 40 exajoules annually to the nation's energy services (Figure 17.5), but with the bonus of not causing the pollution and the resource depletion that accompanies actual energy use. Since the 40 exajoules were saved with no change in the upward trend of the GNP, they involved no reduction in actual energy services. That is, these energy services were "bought" not with energy but with energy efficiency. This 40-exajoule contribution from efficiency exceeded even the 35-exajoule contribution of the nation's most-used energy resource, oil. Without the efficiency measures that have been taken since the 1973 oil embargo, the nation would have used more than 120 exajoules instead of 85 exajoules in 1990. And this saved a lot of money: 40 exajoules of energy was worth well over $200 billion in 1990, although the net savings were less than this because of the cost of the efficiency measures.

What further efficiency gains might be possible while maintaining the upward trend in goods and services? One indication comes from comparisons among different nations with similar economies. Figure 17.23 compares the energy consumed per person in the United States with that in five other industrial democracies whose industrial outputs (GNP per person) are similar to the United States'. For comparison, two less-developed countries, India and China, are also shown. Although Canada consumes a little more energy per person than does the United States, the other four industrialized countries listed consume far less. Germany, for example, consumes about 60% as much energy per person as does the United States, suggesting that the United States might be able to reduce its energy use to 60% of present consumption without reducing services.

Perhaps the most direct measure of a nation's energy efficiency is the amount of energy it must expend to produce a unit (a dollar's worth, say) of goods and services. This is known as the nation's **energy intensity.** It can be measured in joules per dollar of GNP. More efficient nations have lower energy intensities. Figure 17.24 compares the energy intensities between 1965 and 1990 of three national groupings: the United States, the western European countries plus Japan (that is, most of the other industrialized countries), and the less-developed countries. Because their GNP is so low, the less-developed countries have high energy intensities despite their low energy consumption (see India and China in Figure 17.23). Although all industrialized nations have reduced their energy intensity greatly during the past twenty years, the United States has consistently used about twice as much energy as the others have to produce a unit of goods and services as measured by GNP. This could be caused in part by inherent differences such as the United States' large area, but it might also indicate a potential 50% gain in the services that the United States obtains per joule of energy.

Chapter 9 (Figure 9.26) listed several specific energy-efficiency measures, along with fossil-fuel reductions totaling about 50% that might be obtained from these measures, indicating again that the United States might be able to halve its energy use without reducing energy services.

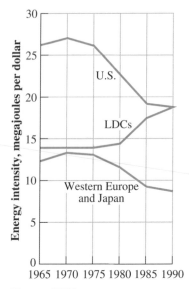

FIGURE 17.24
Energy intensity: energy use divided by gross national product for the United States, other industrialized countries (western Europe and Japan), and the less-developed countries (LDCs), in megajoules per 1985 dollar, between 1965 and 1990. An economy's "energy intensity" is the amount of energy it must consume to produce one dollar's worth of goods and services.

FIGURE 17.25
Compact fluorescent bulbs.

We will look at only one specific example of the surprising savings that are possible when energy efficiency is taken seriously: recent advances in efficient lighting.

Traditional incandescent bulbs create light by simply heating a thin wire until it glows. This produces a lot more heat than light. **Fluorescent bulbs** operate on a more energy-efficient principle, one that does not involve heating. The bulb's glass tube is filled with dilute mercury vapor or some other gas. An electric current (moving electrons) passes through the tube, ionizing some of the mercury atoms by collision and putting many atoms into excited quantum states (Sections 14.1 and 14.2). The excited mercury atoms radiate invisible ultraviolet photons which are absorbed by a powdery material called the *phosphor* which coats the inside of the glass tube which in turn causes the phosphor to radiate visible light. This process is five times more efficient than incandescent bulbs, so the same amount of lighting can be obtained for one-fifth as much energy. Even the older fluorescent bulbs provided enormous energy savings.

Fluorescent bulb efficiency and performance nonetheless have improved recently because of the development of high-frequency *ballasts*. A fluorescent bulb's ballast is the device that controls the current flowing through the tube. In standard electrical circuits in homes, electrons don't keep flowing in one direction but instead vibrate back and forth 60 times per second. Older fluorescent bulbs operate at this 60 Hz frequency. But both bulb quality and efficiencies would be higher if the frequency within the bulb were much higher than 60 Hz. The reason is that at 60 Hz, electrons and ions have a full one-sixtieth of a second for each back-and-forth motion—a long time in the world of electrons and ions. This is time enough for large numbers of electrons and ions to run into the ends of the tube, creating heat instead of light. This inefficiency is also unsightly because it causes the light to dim near the end of the tubes.

New electronically controlled ballasts cause the current to oscillate at up to 50,000 Hz. This reduces energy losses at the ends of the tube by restricting the electrons and ions to much shorter distances of vibration. Along with improving the efficiency and solving the end-dimming problem, high-frequency ballasts make possible shorter "compact fluorescent" bulbs. Fluorescent bulbs can now be screwed into ordinary incandescent light sockets (Figure 17.25).

The consequences of this single efficiency improvement are surprising for both consumers and the nation's energy budget. For one person, one compact fluorescent bulb, purchased at $18 and used over its normal 10,000-hour lifetime, saves about $40 (see Dialogue 11). For the nation, it is estimated that in a few decades when the market is saturated, the electricity savings from compact fluorescents will equal the output of 55 large power plants, cutting home electricity consumption by 8%, producing net savings (after paying for the bulbs) of $5.5 billion annually, and reducing carbon-dioxide emissions by 55 million tons per year.

This is just one example of the potential of energy efficiency. Economics is closely entwined with energy efficiency. Whenever energy efficiency saves money, it is possible for governments or companies to stimulate large energy-saving programs that reward consumers financially. For example, some electric companies and some governments are making large-volume low-cost purchases of compact fluorescent bulbs and financing them to consumers below cost to the company, with the difference recovered in electricity charges to those consumers. The consumers save money because of reduced electricity consumption, which more than compensates for the extra charges paid for the bulbs. Everybody wins: the company, the consumer, and the environment.

Because energy efficiency has been overlooked for so long, small investments can result in large savings. For example, a $7.5 million compact-fluorescent lamp factory saves as much electricity as a $1 billion power plant makes while also avoiding the power plant's fuel cost and pollution. A $10 million "superglass" factory making windows that block heat but permit light can produce the comfort that would be provided by the air conditioners run by $2 billion worth of generating stations.

The price of energy strongly affects the amount and types of energy consumed. It is instructive to note that the lock-step link between growth in GNP and in energy was broken only when the 1973 oil embargo raised energy prices. Beginning in 1973, as energy efficiency began to take hold, GNP grew but energy use did not (Figure 17.5). As a result of the higher energy prices resulting from the oil embargo, an enormous amount of energy has been saved.

Many studies suggest energy taxes as an incentive for conservation. Energy taxes could be made "revenue neutral" by lowering other taxes in compensation, to discourage energy use while not increasing overall taxes. Or the tax income could be used to develop further efficiency measures or for other purposes. Incentives could encourage energy-conscious shopping. For example, the U.S. Office of Technology Assessment suggests imposing gas-guzzler taxes on inefficient automobiles and gas-sipper rebates on efficient ones. An underlying view is that global warming, resource depletion, and other environmental concerns must be incorporated into the costs of doing business and consuming goods if a market economy is to address such concerns.

DIALOGUE 11   A typical 18-watt compact fluorescent bulb costs $18, lasts 10,000 hours, and provides as much light as does a 75-watt incandescent bulb that costs $0.75 and lasts 750 hours. How much does it cost for each type of bulb to provide 10,000 hours of light? Do not forget to include energy costs, at about 9¢ per kilowatt-hour.

DIALOGUE 12   In Dialogue 11, how many joules of electrical energy are saved by using the compact fluorescent rather than the incandescent bulb for 10,000 hours? (Recall that 1 watt is 1 joule per second.)

11. Compact fluorescent: One bulb is needed, at $18. The cost of the energy is $0.09 × 0.018 × 10,000 = $16.20. Total cost = $18 + $16.20 = $34.20. Incandescent bulb: The number of bulbs needed is 10,000/750 = 13.3 (14 will be needed). Cost of bulbs = $0.75 × 14 = $10.50. The cost of the energy is $0.09 × 0.075 × 10,000 = $67.50. Total cost = $10.50 + $67.50 = $78.

12. Each type of bulb burns for 3,600 × 10,000 = 36 × 10^6 seconds. The 18-watt bulb consumes 18 × 36 × 10^6 = 650 million joules. The 75-watt bulb consumes 75 × 36 × 10^6 = 2,700 million joules. Energy savings = about 2,000 million joules.

"THE ONLY OTHER SOLUTION IS THAT WE MAY EVOLVE
INTO A SPECIES IMMUNE TO ALL THIS JUNK."

Summary of Ideas and Terms

Energy resources Natural resources containing useful energy. The major U.S. resources today are fossil (coal, oil, natural gas), nuclear (uranium), hydroelectric, and wood. A resource is **renewable** if it can be replaced within a human lifetime; otherwise it is **nonrenewable**.

Energy conservation Measures to reduce energy consumption, including **efficiency** measures to save energy while providing the same services, and changes in the quality of services.

Nuclear power reactor A device in which chain-reacting nuclei transform nuclear energy into thermal energy for electric power. Its main components are **fuel** to provide energy, a **moderator** to slow the neutrons so that they will split the nuclei more efficiently, neutron-absorbing **control rods** to control the reaction, and a **coolant** to transfer thermal energy from the fuel.

Light-water reactor A uranium-fueled reactor that is cooled and moderated by natural water. Nearly all U.S.

power reactors are of this type. **Heavy water** is H_2O in which the hydrogen atoms are the rare heavy isotope ^2_1H.

Breeder reactor A reactor that creates more than one ^{239}Pu nucleus (from ^{238}U) for each ^{235}U nucleus it fissions and so creates more plutonium fuel than the uranium fuel it uses.

Fusion reactor A nuclear reactor that obtains its thermal energy from fusion rather than fission. Now under development, it could be commercially viable by the middle of the twenty-first century.

Global warming Greenhouse-effect warming of Earth's atmosphere, due mainly to fossil fuels. Many experts believe that because of this problem, coal-fired electricity generation must decline over the coming decades.

High-level nuclear waste Used reactor fuel rods containing highly radioactive fission products.

Three Mile Island Site of the most significant U.S. nuclear power plant accident. Although the fuel melted down, little radioactivity escaped.

Chernobyl Site of world's worst nuclear power accident. The fuel melted down; the reactor suffered a "slow nuclear explosion," comparable to a chemical bomb; and a large amount of radioactivity escaped.

Nuclear weapons proliferation The spread of nuclear weapons. Many experts believe that the use of plutonium as reactor fuel, and breeder reactors, will worsen this problem.

Renewable energy forms:

· **Hydroelectric**, the gravitational energy of raised water.
· **Biomass**, the chemical energy of organic substances.
· **Geothermal**, the thermal energy of hot underground steam, water, or rock.
· **Wind**, the kinetic energy of moving air.
· **Solar–thermal electricity**, generated from thermal energy created by the sun.
· **Photovoltaic electricity**, generated from solar radiation using the photoelectric effect. Two thin layers of silicon, constructed somewhat differently, cause electrons to move from one layer to the other when illuminated with light. The device acts like a low-power battery.
· **Active solar**, in which a solar-heated liquid or gas is pumped elsewhere for space or water heating.
· **Passive solar**, in which solar radiation, natural air flows, energy storage, and insulation provide direct heating and cooling.

Energy intensity The amount of energy a nation consumes to produce a unit of goods and services. It can be measured in joules per dollar of gross national product.

Fluorescent bulb An electric current flows through a gas that fills the bulb, exciting the gas, which emits ultraviolet radiation, which is absorbed by the phosphor coating, which emits light. New **high-frequency ballasts** have recently been developed for increased efficiency.

Review Questions

ENERGY HISTORY AND FUTURE

1. How does a "scenario" differ from a "prediction"?
2. List the six main U.S. energy resources today. Which three are the most used?
3. Which of the six main U.S. energy resources are renewable?
4. List the four broad categories of options for resolving the energy problem. Which one is dominant today?
5. Explain why, in reference to Figure 17.5, it is appropriate to label the gap between the top two graphs "Conservation." What caused these two graphs to separate?

NUCLEAR POWER

6. The United States uses mostly light-water reactors. What do "light water" and "reactor" mean? What is "heavy water"?
7. List the four essential components of a nuclear power reactor, and explain the function of each.
8. Is the enrichment of U.S. reactor fuel closest to 3%, 10%, 25%, 50%, or 90%? Which of these figures is closest to bomb-grade enrichment?

9. Where might a fusion reactor get its fuel?

ISSUES FOR NUCLEAR AND COAL POWER

10. List five problems with nuclear power and five problems with coal power.
11. Of the four most-used nonrenewable energy resources, which one(s) does the United States possess in greatest abundance (most years remaining at current rate of use)? In least abundance?
12. What is the consensus of most experts concerning the long-term future of coal power? Why?
13. What are high-level nuclear wastes, and where are they stored today? What is the most likely long-term solution to this problem?
14. Describe in general terms what happened at Three Mile Island.
15. Describe in general terms what happened at Chernobyl.
16. What is a breeder reactor? How could breeder reactors affect the uranium resource problem?
17. How could breeder reactors affect the nuclear proliferation problem?

18. Which two renewable energy resources are in widest use today?
19. List three renewable resources that could produce significant amounts of electricity in the future, and describe briefly how each works.
20. What are semiconductors, and what role do they play in photovoltaic cells?
21. What is the difference between active and passive solar heating? Describe some of the techniques used in each.

22. What is the difference between base-load power and peaking power?
23. What is meant by a nation's energy intensity? How does the United States' energy intensity compare with that of the other industrialized nations?
24. How do fluorescent bulbs work?
25. Why are fluorescent bulbs more efficient than incandescent bulbs? How have improvements in fluorescent bulb ballasts made them even better?

Home Projects

1. What color absorbs solar energy best? Get several sheets of construction paper of different colors, place an ice cube on each, and put them in the sun or under a heat lamp. Which melts first? Do you get the same result if the paper is on top of the ice cube? What color should a house roof be in a hot climate? In a cold climate?
2. Measure the "solar angle" (the angle that the sun's noontime rays make with the vertical) on several different dates during this semester. What do your results tell you about the proper design of a passive solar house (Figure 17.21)?
3. Write to your U.S. senator or congressperson about one of the science-related social concerns discussed in this course. Your opinion makes a difference—if you express it! Write to

Senator _____
U.S. Senate
Washington, DC 20510

or to

Congressman (or Congresswoman) _____
U.S. House of Representatives
Washington, DC 20515

Include your name and address, recipient's name and address, and date. Discuss only a single topic. Express your own opinion—your representatives want to know what you think. Back up your opinion with reasons. Keep it to the point—and on one page. Be specific. Try to refer to a specific bill before Congress. Feel free to disagree, but keep it friendly.

For Discussion

1. Make a list of all the energy-consuming devices you used this week that were not available a hundred years ago. How would the world and your life be different without these devices?
2. List several ways by which you and the nation could reduce energy consumption without seriously harming your or its overall quality of life. When considering alternatives such as mass transit or reduced lighting, you will have to decide whether such measures would actually harm the quality of life.

3. Outline two different scenarios for the nation's energy future, both of them plausible. Give arguments in favor of each and opposed to each. Give a reasoned argument for your preferred scenario. Now give a reasoned argument for the other scenario. Graph each of your scenarios by extending (on a blank sheet of paper placed next to the graph) the graph of Figure 17.5 to include the period 1990 to 2050.

Exercises

ENERGY HISTORY AND FUTURE

1. Which are renewables: wood, uranium, trash (as fuel), coal, wind, natural gas, hydroelectric?
2. In reference to Table 17.1, which options contribute to global warming?

3. Use Figure 17.4 to calculate the energy efficiency of the U.S. residential–commercial sector (answer: 71%). From this percentage, what can you plausibly conclude regarding the use of heat engines by this sector?

4. Use Figure 17.2 to estimate the amount of energy the United States got from coal in each of the following years: 1900, 1920, 1940, 1960, 1980. Do the same for oil.

5.*Use Figure 17.2 to estimate the percentage of U.S. energy resource consumption that came from coal in each of the following years: 1900, 1920, 1940, 1960, 1980. Do the same for oil.

6.*Total U.S. energy use was roughly exponential (Section 7.7) from 1880 to 1920. Use Figure 17.2 to estimate its doubling time. Find its annual percentage increase.

7.*From Figure 17.4, find the approximate energy efficiency of the overall U.S. economy.

8.*From Figure 17.5, estimate the equivalent amount of energy services (in exajoules) provided by new (post-1973) conservation measures, in each of the following years: 1975, 1980, 1985, 1990. Now estimate the percentage of the nation's total energy services provided by conservation, in each of these years.

NUCLEAR POWER

9. Every heat engine has a thermal energy input, a work output, and a thermal energy output. At what places in Figure 17.6 does each of these occur?

10. Figure 17.8 shows a tall tower. Is it analogous to a coal-fired plant's stack? What is the function of this tower? Might coal plants have a tower like this?

11. Does a nuclear power plant have a "smoke" stack? Why or why not?

12. Neutrons, created when particles from outer space hit the atmosphere, often hit Earth's surface. Could these neutrons start a fission chain reaction in natural uranium in the ground? Why?

13. Suppose the main water pipe breaks in a nuclear power plant, shutting off the water flow. If the control rods fall immediately into place, stopping the chain reaction, is there still a problem? Why or why not? What safety feature should be used in this case?

ISSUES FOR NUCLEAR AND COAL POWER

14. List two advantages that coal has over nuclear power and two advantages that nuclear power has over coal.

15. Since natural gas creates carbon dioxide, why would it reduce global warming to switch from coal to natural gas power plants?

16. Are radioactive wastes hot? Why?

17. How can a breeder reactor create more fuel than it consumes? Why doesn't this violate the law of conservation of energy?

18. Why might heavy water be considered a strategic material, useful in the production of nuclear weapons?

RENEWABLES

19. Describe how hydroelectric energy is renewed.

20. List one possible disadvantage of each of the eight renewable energy resources listed in Table 17.1.

21. What are some ways that solar energy is routinely used around the home?

22. Is renewable energy used today for any form of transportation? Explain.

23. What physical energy transformation occurs for each of these ways of making electricity: coal, uranium, hydroelectric, biomass, geothermal, wind, solar–thermal, photovoltaic?

24. Suppose that a carbon tax caused coal base-load power to be priced at 10¢/kW·h. According to Figure 17.22, when would wind, solar–thermal, and photovoltaics then compete economically with coal for base loads?

25.*Making estimates. If you covered a football field with photovoltaic cells, for about how many households could it provide electricity? Use the following information: Solar energy hits each square meter of Earth's surface at an average rate (averaged over day and night) of 200 watts; an average household consumes electricity at a rate of 1 kilowatt; and photovoltaic cells are 20% efficient.

26.*Making estimates. Show that about 10,000 square kilometers of land area (about 3% of the land area of a sunny state such as Arizona) would need to be covered with photovoltaic cells in order to provide all of the United States' electricity. See the preceding exercise. The United States consumes electricity at an average rate of about 400 billion watts. What complications would arise if the nation tried to provide all its electricity in this way?

CONSERVATION

27. In what way was the oil crisis, caused by the 1973 Mideast oil embargo, a good thing for the United States?

28. Are there any industrialized nations in which the energy use per person is half, or less than half, what it is in the United States?

18

QUANTUM FIELDS
relativity meets the quantum

What happens when relativity meets quantum theory?

The two theories extend Newtonian physics in different directions. One extends it up to lightspeed, and the other extends it down to the smallest dimensions yet measured, 10^{-19} meters, 10,000 times smaller than an atomic nucleus. But separate relativity and quantum theories cannot be the whole story, because relativity theory doesn't contain the quantum principles so it doesn't work at small sizes, and quantum theory doesn't contain the relativity principles so it doesn't work at high speeds. Is it possible to combine the two into one theory covering all sizes and all speeds?

The answer is yes. The combination is called **quantum field theory.** One part of it, quantum electrodynamics, is the most accurate scientific theory ever invented. Like most post-Newtonian physics, quantum field theory is simple but takes some getting used to.

The next section presents the general idea of quantum field theory, and the remaining sections apply this idea to each of the four known fundamental forces: electromagnetic (Sections 18.2 and 18.3), weak (Section 18.4), strong (Section 18.5), and gravitational (Section 18.6).

18.1 Quantized fields: *the reason there are particles*

Recall (Chapter 9) that a force field, or simply a **field**, is spread out over a region of space. This region needn't contain any matter or any "thing" at all; it can be empty space, a vacuum, containing no matter. A field is a kind of stress in space and can be described as the possibility of a force. For example, an electric field is the possibility of an electric force. Electric fields are created by charged objects and are spread out in space at any point where charged objects would feel a force. Magnetic fields and gravitational fields are other familiar examples of force fields.

At the core of quantum field theory is a view that we have discussed before in connection with mass–energy equivalence. It is the idea that everything is made only of fields, that every thing is "empty," that matter is really just a collection of force fields. Fields have mass because they have energy. Material particles such as electrons and protons are just fields, and the position of a particle is just the center of its field. As poet Gertude Stein put it, there is no "there" there. The table on which this book rests is simply a configuration of quivering force fields, and so is the book. The book doesn't fall through the table, however, because the force fields that we call electrons in the table repel the other electron force fields in the book. And your eye (which is also just fields) sees the book because the book's force fields emit radiation.

There is nothing: no thing. Only fields.

Now consider photons and their relation to quantum theory. Recall that radiation is quantized into separate bundles or quanta called photons, and the psi-field for these photons is the electromagnetic field. The relation between this field and its associated photons is that the field gives the probabilities for finding the photons.

Quantum field theory began in 1926 when the quantum theory of the electromagnetic field was invented by Max Born, Werner Heisenberg, and Pascual Jordan. Its basic idea is to treat the electromagnetic field itself as the fundamental system that is subjected to both quantum and relativity principles, and to develop a mathematical description of this system (the field) that is analogous to the description that Schroedinger developed for material particles. In other words, this theory presents a sort of Schroedinger equation for the motion of the electromagnetic field itself, rather than for the motion of any particle such as an electron or proton or photon. This process—of treating a field as a real physical system and subjecting it to the principles of quantum theory—is called **quantizing the field**. The process is sometimes called *second quantization* because it starts with a quantum entity, a psi-field, and quantizes it. Sounds like gibberish. But this gibberish led to the most accurate theory in scientific history.

The Schroedinger equation for, say, an electron predicts that electron's psi-field; that is, it predicts the probability of finding the electron at each point in space. In the same way, the quantum theory of the electromagnetic field predicts the field's psi-field; that is, it predicts the probability of finding the electromagnetic field to have various strengths at each point in space. Just as the Schroedinger equation for a particle predicts energy quantization for that particle, the quantum theory of the electromagnetic field predicts that the field will be quantized into bundles of energy of various frequencies. The energy of one bundle turns out to be $E = hf$ (compare Chapter 13). These bundles are photons! In other words, the quantum theory of the electromagnetic field predicts that electromagnetic fields will exhibit themselves as photons, thereby explaining why photons exist.

This is typical of all quantized field theories: One starts with a field, quantizes it, and out pop the particles associated with that field. These particles are the **quanta of the field**, the discrete bundles of energy associated with that field. In our previous photon theory of radiation (Chapter 13), photons had to be put into the theory. But if we start by quantizing the electromagnetic field itself, then the theory will predict photons.

In 1929, Heisenberg and Wolfgang Pauli took a new step: They applied the quantized field idea to matter by quantizing the psi-field of the electron. They treated this "electron field" as a real physical system that should itself obey the principles of quantum theory and relativity theory. Note that an electron field (the psi-field of an electron) is different from an electric field: The electron field is a matter field, not a radiation field. Heisenberg and Pauli then discovered that this quantized electron field must exhibit itself as tiny bundles of energy, the quanta of the electron field. But the quanta of the electron field are not photons—they are electrons. So this theory predicts the existence of electrons.

The basic idea of quantum field theory is to quantize all of the psi-fields. We quantize (that is, we subject to quantum principles) the photon's psi-field (the electromagnetic field), the electron's psi-field (the electron field), the proton's psi-field, the neutron's psi-field, and so forth. All of the particles of nature are quanta of their associated fields, just as photons are quanta of the electromagnetic field. So quantum field theory explains why nature exhibits itself as particles. Nature is made of particles because nature is made of fields, and when a field is subjected to the principles of quantum theory, the field's energy turns out to come in tiny bundles—particles. The list of the fundamental ingredients of the world no longer needs to include any particles at all—it needs to include instead a few kinds of quantized fields and the principles of quantum theory.

By viewing nature as made of quantized fields, physicists can combine quantum theory and relativity and put matter and radiation on an equal footing: Both material particles and radiation particles are just bundles of field energy. These particles are subject to the usual quantum uncertainties. The probability of finding a particle at some particular point is determined by the corresponding field's psi-field at that point. All the different fields interact with one another, and these interactions determine everything that happens in nature. To summarize:

> THE QUANTUM THEORY OF FIELDS
> The essential reality is a set of fields. Everything that happens in nature is a result of the interactions and motions of these fields. All fields obey special relativity and quantum theory. Quantum theory requires these fields to exhibit themselves as tiny bundles or quanta of field energy. All of nature's particles of matter and radiation are quanta of this sort. The intensity of a field at a point determines the probability for finding the associated particle.

This theory goes considerably deeper than the quantum theory of matter and radiation (Section 13.7). That earlier quantum theory, which is nonrelativistic, did not explain why nature is made of particles but instead began by assuming that nature is made of particles. The quantum theory of fields shows why particles exist: It is because nature is really made of fields, and when one applies quantum theory to fields, one finds that they exhibit themselves as bundles of field energy.

Quantum field theory resolves—as nearly as it can be resolved—the old wave-versus-particle dilemma. The resolution is that nature is made of fields ("waves") but that these fields are subject to quantum principles and so must exhibit themselves as particles of field energy. Fields and particles are the yin and the yang of relativistic quantum theory.

18.2 Quantum electrodynamics: *the strange theory of electrons and light*

Quantum field theory emerged during the 1930s as the world was marching toward war. Although nuclear physics flourished (Chapter 16), fundamental quantum theory had to wait. After the war, in 1947, two New Yorkers in their twenties, Richard Feynman and Julian Schwinger (Figure 18.1), independently completed quantum field theories of the electromagnetic force. The two theories, known as **quantum electrodynamics**, say the same thing but they look different, perhaps owing to their authors' differing personalities: Whereas Schwinger preferred to work in Einstein-like solitude, Feynman was known to jolt his mind out of a rut by working at a back table in a night club, inspired by the blare of the sound system. We present Feynman's more intuitive version.

Quantum electrodynamics is about photons, the quanta of the electromagnetic field, and electrons, the particle that experiences the electromagnetic force in its purest form. Sounds simple. But the requirement that electrons and photons obey both relativity and quantum theory leads to astonishing results.

Feynman's theory replaces the continuous force between two electrons with a "package" transfer in the form of a photon. Figure 18.2 pictures this

FIGURE 18.1

(a) Richard Feynman in about 1960. Feynman loved to play the drums at parties and while out in the woods. He studied African drumming from experts and kept a set of bongo drums in his office. (b) Julian Schwinger, who independently invented a theory of quantum electrodynamics.

(a)

(b)

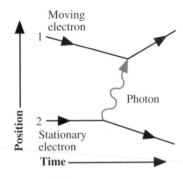

FIGURE 18.2
A schematic diagram showing a single quantum interaction between two electrons. Diagrams like this are known as "Feynman diagrams."

idea. The diagram graphs the positions (upward line) of two electrons at various times (line to the right). It shows a single quantum interaction between the two electrons. Initially, electron 2 is at rest (its position isn't changing), and electron 1 is moving downward. Then electron 2 radiates a photon that travels through space to electron 1, and electron 1 absorbs this photon. When electron 2 emits the photon, electron 2 veers downward, and when electron 1 absorbs the photon, electron 1 veers upward. The electrons repel each other by means of **photon exchange**, much as basketball players interact by passing a basketball back and forth. Surprisingly, however, quantum electrodynamics allows two oppositely charged particles, such as a proton and electron, to veer toward each other when a photon is exchanged.

Each quantum event has quantum uncertainties. In Figure 18.2, the emission and the absorption of the photon are uncertain: That is, it is uncertain whether the emission and absorption will occur in the first place, and if they do, it is uncertain where and when they will occur. Quantum field theory replaces the deterministic electric force law with a formula giving the probability of emission and absorption of a photon. In this theory, for a particle to be **electrically charged** means that it has the ability to emit and absorb photons.

This theory replaces the smooth deterministic Newtonian paths with jerky nondeterministic paths (Figure 18.3). If the individual photons have low energy, quantum theory predicts a fairly smooth, nearly Newtonian, path (Figure 18.3(a)). At higher energies, the quantum predictions are decidedly non-Newtonian (Figure 18.3(b)).

So far, this is the kind of thing we might have expected from quantizing the electric interaction: quantized force packages, and randomness. But something quite unexpected also comes out of this theory. In order for this theory also to obey the special theory of relativity, a new type of material particle must exist in nature.

The argument that leads to this prediction is an interesting one and is typical of the reasoning used in modern physics. It is based on symmetry. In order

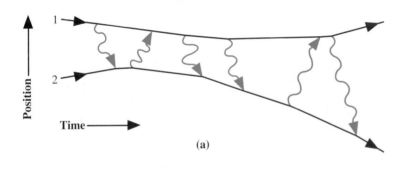

(a)

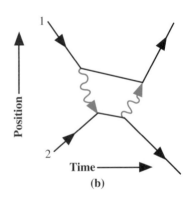

(b)

FIGURE 18.3
(a) A Feynman diagram for a series of interactions between two electrons at lower energy. The electrons' paths approach smooth Newtonian paths. (b) At higher energies, the paths deviate considerably from smooth paths. Individual quantum events and quantum uncertainties become more important at higher energies, and Newtonian physics is longer a good approximation.

Paul Dirac (left) and Werner Heisenberg in 1933. Dirac was an extremely reserved Englishman, and his colleague and good friend Heisenberg was a much more outgoing German.

to obey special relativity, quantum field theory must be "symmetric under time reversal." In other words, if we imagine a universe precisely like ours, only with time running the other way, the laws of quantum field theory must be valid in that universe.* Feynman found that an electron that is imagined to move backward in time would have precisely the same observable effects as would another particle just like the electron, only carrying a positive charge and moving forward in time. In order for the laws of physics to be properly symmetric under time reversal, this positive electron, or **positron**, had to exist.

The positron was first predicted much earlier, in 1928, by the British theorist Paul Dirac (Figure 18.4). Dirac did not use a field-theory approach but instead searched for and found a version of the Schroedinger equation for particles (not fields) that, unlike Schroedinger's original equation, obeyed special relativity. Dirac's new equation turned out to predict a new particle: the positron. It was the first major new prediction to come out of the marriage of quantum theory to special relativity.

HOW DO WE KNOW? THE DISCOVERY OF THE POSITRON AND THE MUON

The path of a subatomic particle can be made visible in a device known as a **cloud chamber.** A container is filled with dust-free air containing "saturated" water vapor, or gaseous H_2O that is just at the point of converting to droplets of liquid water but that lacks the tiny solid "nucleation centers" such as the dust particles that normally start droplets forming from water vapor. When a charged subatomic particle, such as an electron, speeds through the chamber filled with this vapor, it nudges many air molecules along its path. About thirty times in each centimeter of its path, these collisions are strong enough to ionize an air molecule. These ions make excellent nucleation centers for water droplets, so a droplet forms around each ion, and a trail of droplets reveals the particle's path. Jet planes form similar vapor trails in the atmosphere, revealing the plane's path. The cloud chamber was the workhorse of subatomic physics between 1930 and 1960. Its successor is the **bubble chamber** (Figure 18.5), based on the formation of tiny bubbles in a liquid. Its inventor, Donald Glaser, is said to have come up with this innovation in a bar in Ann Arbor, Michigan, while watching the bubbles in a glass of beer.

In 1932, Carl Anderson of the California Institute of Technology generated a strong magnetic field in a cloud chamber. Recall that magnetic fields exert sideways forces on moving charged particles. This sideways force makes electrons curve as they move through magnetic fields. Anderson generated a magnetic field strong enough to make this curvature measurable.

A moving particle's speed and mass can be assessed from the curvature of its path, because faster particles have straighter paths and if two particles move at the same speed, the more massive one will have the

---

* This raises the question of why the forward direction in time is actually different from the backward direction if our most basic physical theory is symmetric in time. For example, why aren't as many people growing younger as are growing older? The answer to this very deep question is not understood, but it is connected to the second law of thermodynamics (Chapter 7).

FIGURE 18.5
*The big bubble chamber at
Fermilab near Chicago.*

FIGURE 18.6
*The photo that won a Nobel Prize.
This photo alone established the
existence of a positive electron.*

straighter path. A particle's speed can also be independently assessed from the thickness of its path, because faster particles ionize fewer air molecules per centimeter and so create fewer bubbles and thinner paths.

In 1932, the only high-energy particles available for experiments came from high-energy particles from space. Anderson allowed these "cosmic rays" to pass through his cloud chamber, and his photographs show a surprising number of thin and fairly straight paths. Electrons and protons were the only charged particles known at that time. The paths indicated a mass too small for the proton. Rather, the paths appeared to be made by fast-moving electrons, but the direction of their curvature was the reverse of what was expected, indicating that the particles carried a positive charge. Anderson's first hypothesis was that these paths were made by electrons that were somehow moving upward through the cloud chamber, despite the expectation that cosmic rays should move downward. He clinched the matter by inserting a thin lead plate across the middle of the chamber. Although the fast-moving particles passed easily through the lead, they slowed down in the process, and so the path's curvature increased after passing through. In Figure 18.6—the photograph that won Anderson a Nobel Prize—the particle is clearly moving from top to bottom, so its curvature shows that it carries a positive charge. Anderson had discovered Dirac's predicted positron.

In order to observe cosmic rays before they interact with much air, Anderson in 1936 built a new magnetic cloud chamber on Pike's Peak in the Colorado Rockies. He found curious tracks that did not seem to fit the behavior of protons or of electrons, even positive ones. The paths were too curved for protons, yet the particles passed easily through lead plates that should have stopped any particle whose mass was as small as the electron's. Anderson found that this new particle was identical to the electron only 200 times more massive, and about one-tenth as massive as a proton. There had been no Dirac to predict this "fat electron," and it was a real surprise. As Columbia University physicist I. I. Rabi put it, "Who ordered that"? Today, we still do not know. This particle is called a **muon**.

18.3 Antimatter

The positron was science's first encounter with **antiparticles**. Relativity's requirement that quantum theory be symmetric under time reversal implies that for every existing type of particle, there must be an antiparticle. A particle and its antiparticle are identical except that they carry opposite charges and so have electromagnetic properties that are mirror images of each other. For example, the electron's existence implies that antielectrons, in other words, positrons, exist. Similarly, the existence of protons and neutrons implies that their antiparticles, the **antiproton** and **antineutron**, exist.

One of the most profound consequences of high-energy physics is the prediction and observation of the **creation and annihilation of matter**. Recall that quantum field theory states that fields come in tiny bundles or quanta of field energy. Photons, for example, are the quanta of the electromagnetic field. Quantum electrodynamics gives the probabilities for this field to exist in the form of various numbers of photons, and it gives the probability that additional photons will be created (increase in number) or annihilated (decrease in number). Such random increases and decreases in the number of photons are called *fluctuations*. Similarly, quantum electrodynamics states that the electron field can exist in the form of various numbers of electrons and allows for the possibility of the creation and annihilation of electrons. However, energy must be conserved, and so electrons must be created from some other form of energy, such as photons. Furthermore, one of the microworld's rules is that the total electric charge must remain unchanged, so it is always **electron–positron pairs** that are created or annihilated.

This is a very non-Newtonian development. As Heisenberg commented:

I believe that the discovery of particles and antiparticles. . .has changed our whole outlook on atomic physics. . . .As soon as one knows that one can create pairs, then one has to consider an elementary particle as a compound system; because virtually it could be this particle plus a pair of particles plus two pairs and so on, and so all of a sudden the whole idea of elementary particles has changed. Up to that time I think every physicist had thought of the elementary particles along the lines of the philosophy of Democritus [Chapter 2], namely by considering [them] as unchangeable units which are just given in nature and are always the same thing, they never change, they never can be transmuted into anything else. They are not dynamical systems, they just exist

Just as it is possible for a particle to be in a quantum state in which it is neither definitely here or there. . .so also it is possible to have a particle in a state in which it is neither definitely an electron nor definitely a neutrino until we measure some property that would distinguish the two, like the electric charge.

Steven Weinberg

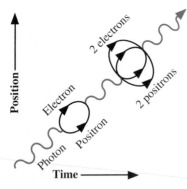

FIGURE 18.7

A few moments in the life history of a photon (which is not a photon but is one electron–positron pair during part of this history and two pairs during another part).

in themselves. After Dirac's discovery everything looked different, because one could ask, why should a photon not sometimes be a photon plus a pair of electron and positron and so on?. . .Thereby the problem of dividing matter had come into a different light.

Quantum electrodynamics predicts that a photon has a certain probability of actually being observed as an electron–positron pair, or as more than one pair, and that such a pair has a certain probability of being observed as one or more photons. Figure 18.7, a Feynman diagram for part of a photon's life history, conveys this notion that a photon can be an electron–positron pair, or two pairs. And conversely, if a particle and its antiparticle are close to each other, they can instead be photons. When this happens, the particle–antiparticle pair are said to "annihilate" each other.

This development implies the possibility of **antimatter**, made of antiprotons, antineutrons, and positrons, instead of the protons, neutrons, and electrons of normal matter. Antimatter is similar to normal matter. Large collections of antimatter, such as antigalaxies, are possible but are thought not to exist, because if they did, we would observe high-energy radiation from annihilation processes when a galaxy and antigalaxy collided. Although we are able to observe many colliding galaxies, we never observe such annihilation processes. The universe is believed to consist almost entirely of matter and very little antimatter. But symmetry seems to suggest that the universe should be made of equal amounts of both. Why so much matter and so little antimatter?

Russian physicist Andrei Sakharov suggested in 1967 that the big bang may have created equal amounts of matter and antimatter and that certain processes during the first second gave rise to a slight excess—less than a part in a billion—of matter, and then the rest of the matter and antimatter annihilated so that the tiny excess formed all the matter in the universe today. It's a good thing for life in the universe, including us, that things worked out this way. If it weren't for that slight excess, the universe would be made nearly entirely of radiation, and we wouldn't be here to think about antimatter!

HOW DO WE KNOW? OBSERVING THE SUBATOMIC WORLD

Matter is routinely created and annihilated in the world's high-energy physics labs (Figure 18.8) when a high-energy particle enters a bubble chamber and collides with the particles of liquid. This creates a shower of new particles, including high-energy photons and particle–antiparticle pairs. Carl Anderson got his high-energy incoming particles from naturally occurring cosmic rays. Today the incoming particles are first accelerated to high energies by electromagnetic forces in **particle accelerators** (Figure 18.9).

We have sought for firm ground and found none. The deeper we penetrate, the more restless becomes the universe; all is rushing about and vibrating in a wild dance.

Max Born

Among the strange phenomena of the microworld, one of the most fascinating is our new view of a **vacuum**—an empty space. As we know, fields exist even in regions that are devoid of material objects, even in regions as empty as the spaces between the galaxies. But if fields do exist, then at any point in a vacuum there is a certain probability that a photon, or a particle–antiparticle pair of any type, will pop into and out of existence. Quantum uncertainties allow the energy present at any point in space to fluctu-

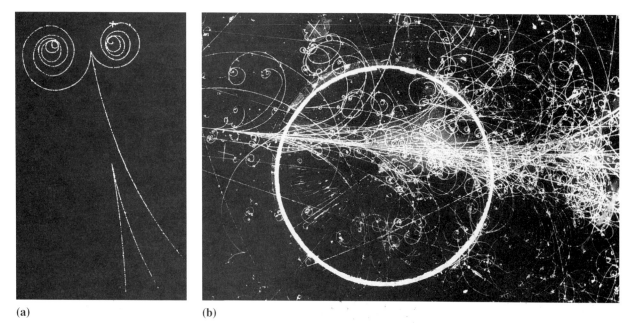

(a)   **(b)**

FIGURE 18.8

(a) A bubble-chamber photograph of electron–positron pair creations, caused by gamma-ray photons. In the event at the top, a photon has struck an atomic electron and knocked it out of its atom (long curving line), and simultaneously created an electron–positron pair (tightly curling spirals). Toward the bottom, a different photon creates an electron–positron pair. How can you tell that each pair has two particles of opposite charge? Of the two pairs, which pair has the highest energy and speed? Why can't you see the path of the photons? (b) A high-energy particle striking a particle in a bubble chamber creates a "spray" of particles of various sorts.

No point is more central than this, that empty space is not empty. It is the seat of the most violent physics.
John Wheeler

ate around its long-term average value. The smaller the region of space is, the larger these **energy fluctuations** can be. Large fluctuations, however, are highly likely to subside in a very short time. So even in empty space, there is always some probability of even very high energy events occurring in small regions during short times. Empty space is not the quiet, uninteresting place we had imagined. Microscopically, a vacuum is a seething soup of creation and annihilation.

It seems that in nothingness, many things are possible.

HOW DO WE KNOW? THE LAMB SHIFT

It is ironic how physics turned out in this century. The 19th and early 20th century was characterized by a materialistic outlook which maintained a sharp distinction between what actually was in the world and what wasn't. Today that distinction still exists, but its meaning has altered. . . .Nothingness contains all of being.
Heinz Pagels, physicist

One consequence of energy fluctuations in a vacuum is a tiny effect on the hydrogen atom's energy levels (Section 14.2). In Schroedinger's non-relativistic treatment of the hydrogen atom, the energies of the quantum states labeled $N = 2$, $L = 0$ and $N = 2$, $L = 1$ (see Figure 14.8) are identical. But when quantum field theory is applied to the hydrogen atom, it is found that vacuum energy fluctuations cause the orbiting electron to jiggle a little and that the energy of this jiggling is slightly different for the $N = 2$, $L = 0$ state than for the $N = 2$, $L = 1$ state. This difference was first noticed experimentally in careful measurements of the hydrogen

FIGURE 18.9

A brief visual history of particle accelerators. The earliest accelerators were electron discharge tubes, similar to today's TV tubes and fluorescent light bulbs. Compare these figures with Figure 1.2: Whether viewing outer space, or the inner subatomic space, large instruments are required at the limits.

(a) Ernest Lawrence invented the "cyclotron" that uses electromagnetic fields to bend charged particles into circular paths while accelerating them. He is shown here holding an early cyclotron in 1930, the year he first proposed the cyclotron idea.

(c) This accelerator, at the University of California at Berkeley, was the first to have enough energy to create proton–antiproton pairs.

(b) Lawrence, next to a 27-inch cyclotron in 1934. In 1937, this machine created the first artificial element, technetium. Although this element has atomic number 43, which puts it in the middle of the periodic table, all of its isotopes are radioactive with short half-lives, so it is not found naturally on Earth.

(d) The Stanford Linear Accelerator (SLAC) at Stanford University in California accelerates electrons down a 3-kilometer straight track. These high speed electrons would radiate much of their energy away if they moved in a circle.

(e) Fermilab near Chicago. The main ring is 6 kilometers around. The small circle in front of the 16-story main lab building is a smaller "booster" accelerator that injects high-energy protons into the main ring.

spectrum by Willis Lamb in 1947. After the experimental discovery of this **Lamb shift**, quantum field theorists calculated it. The theoretically predicted frequency of the radiation absorbed or emitted when a hydrogen atom shifts between these two closely spaced levels is 1057.860 ± 0.009 megahertz. The measured value is 1057.845 ± 0.009 megahertz. This uncanny agreement is testimony to both the accuracy of the theory and the precision of spectral measurements.

Quantum electrodynamics describes not only electrons and positrons but also the electronlike muons along with antimuons. Furthermore, a third type of electron, along with its antiparticle, was discovered in 1976. Called the **tauon**, it is much heavier even than the muon, weighing in at 3500 electron masses, or nearly twice as massive as the proton. Again, nobody knows "who ordered that." These three "generations" of electronlike particles appear today to be among the most fundamental constituents of matter. All three, along with their antiparticles, interact by exchanging photons, and all of their interactions are correctly described by quantum electrodynamics.

The muon and tauon are "unstable"; in other words, they decay spontaneously into lower-energy entities. Today's universe contains muons and tauons only when fleeting pairs of them are created by vacuum fluctuations or in high-energy interactions. However, these two heavy electrons might have played a crucial role during the big bang. Sakharov's process, mentioned earlier, for creating a slight excess of matter over antimatter requires all three generations. Although they seem esoteric and irrelevant today, we might owe our existence to their existence during the first second of the universe.

Are there more generations of still heavier electrons? As we will see, theory combined with astronomical observations indicate that the answer is no.

DIALOGUE 1   Would you like to travel to an antigalaxy? Why?

DIALOGUE 2   A certain gamma-ray source emits photons that have, according to quantum theory, a 20% chance of being found as an electron–positron pair. The source emits 400 photons. How many pairs will be found? Is this an exact figure or only an approximate one? Why?

18.4 Electroweak unification: *neutrinos*

Enrico Fermi (Figure 16.9) hypothesized the weak force in 1933. Wolfgang Pauli had recently suggested that during radioactive beta decay, the nucleus emitted a new kind of particle, a **neutrino**, in addition to the beta particle. Although neutrinos would not be discovered experimentally for another twenty-five years, Fermi took them seriously and argued that they indicated that a new fundamental force was at work. Fermi was aware of the work in progress on the quantum theory of the electric force, and he quickly adapted these ideas

---

1. No. You would be annihilated, for one thing.
2. 20% of 400 is 80 pairs. This is only approximate, because of quantum uncertainties.

FIGURE 18.10
Abdus Salam, coinventor of the electroweak force, was born in Pakistan. He is one of the most prominent scientists of the Islamic faith. He donated his share of the Nobel Prize to the institute with which he is associated in Trieste, Italy, which encourages scientists from developing countries.

FIGURE 18.11
U.S. physicist Steven Weinberg, coinventor of the electroweak force that unified the electromagnetic and weak forces into a single entity. He has written a book for nonscientists about the fundamental forces and other topics, entitled Dreams of a Final Theory.

to the new force. Fermi's quantum theory of the weak force succeeded in predicting the half-lives of radioactive nuclei and the spectrum of energies with which beta particles emerged from the nucleus during beta decay.

The weak nuclear force is the most obscure of nature's four fundamental forces. Gravity and electromagnetism show up all the time in our macroscopic world because they can act over long distances. The strong force holds the nucleus together, is the major actor in nuclear power and nuclear weapons, and is responsible for alpha decay. The most obvious example we have of the weak force is radioactive beta decay. The weak force is elusive because on a microscopic scale, it is both short ranged and weak.

A neutrino barely exists at all. That is, it has almost no properties: It has no charge and perhaps no rest mass (we're still not sure), and it feels neither the electric nor the strong force. Moving at or near lightspeed and feeling only the weak force (and gravity), this "little neutral one" is the most elusive known particle and one of the most fantastic.

Because neutrinos have only weak interactions, they travel through matter without "feeling" it. It would take eight light-years of lead to stop half the neutrinos emitted during beta decay! (Eight light-years is twice the distance to the nearest other star.) No wonder the physicists studying beta decay had so much trouble trapping this thing in their calorimeters (Chapter 6). There are millions of neutrinos from space passing in all directions through your body at any instant, yet it will probably be years before even one of them has even a single quantum interaction within your body. The neutrinos now passing downward through you will exit Earth's far side in less than a tenth of a second and will within 2 seconds be beyond the orbit of the moon.

In 1967 Pakistani physicist Abdus Salam (Figure 18.10) and U.S. physicist Steven Weinberg (Figure 18.11), working independently, uncovered a close connection between the weak force and the electric force. They proposed a new quantum field theory that incorporated both forces into a single **electroweak force**. This unification was comparable to the unification, by Maxwell and others during the nineteenth century, of the electric and magnetic forces into a unified electromagnetic force.

Recall that quantum electrodynamics describes the electric interactions of electrons and positrons and that this interaction occurs via the photons that are exchanged between the charged particles. The Weinberg–Salam theory is a broader version of this picture. It describes the electric and weak interactions of electrons, positrons, neutrinos, and antineutrinos and states that this interaction occurs via the exchange of certain other particles. These **exchange particles** include the photon and three other particles as well. The three new exchange particles are quite unlike the photon, the main difference being that all three have mass—in fact, rather large masses for subatomic particles: Each of them is about 100 times more massive than a proton. They are labeled W^+, W^-, and Z and can be thought of as photons that have, for reasons unknown, acquired a mass. The electroweak theory argues that the weak and electric interactions are really the same thing but only appear to differ because the weak exchange particles (W^+, W^-, Z) have mass and the electric exchange particle (photon) does not. Another difference from the photon is that the two W's are charged, positively and negatively. The Z is, like the photon, not charged. The existence of a massive, charge-neutral exchange particle was a striking new prediction, and its experimental detection six years later in 1973 was a key confirmation of the Weinberg–Salam theory. The theory also correctly predicts the masses of all three exchange particles, although it does not explain why these particles have mass in the first place. After all, the photon performs its exchange function perfectly well without having any mass, so why should the W and Z need it?

Besides the electron there are two other generations of heavier, electronlike particles, the muon and tauon. Since the electroweak force binds the electron and the neutrino together into a single "family," we might guess that there is a second-generation neutrino to go along with the muon and a third-generation neutrino to go along with the tauon. This would be a good guess. The electroweak theory predicts, and experiment confirms, that there is a different kind of neutrino to go along with each kind of electron. All three generations interact by means of the same four exchange particles: the photon, W's, and Z's. The electroweak theory correctly predicts all of the observed interactions among all of these fundamental particles. Table 18.1 summarizes the theory.

Are there more generations of electroweak particles, heavier electrons accompanied by new kinds of neutrinos? In a surprising turn of events, astronomical observations indicate that there are only three generations. The argument comes out of a close connection between the large-scale universe and the microscopic world: Outer space and inner space are connected through a microscopic event that soon expanded and that is still continuing all around us. This event is the big bang.

After the first 4 minutes, the universe was about 75% hydrogen and 25% helium (Section 16.3). These numbers were predicted by theoretical nuclear

At first glance, all of this sounds like medieval mystics discussing the music of the spheres, angels on the head of a pin, or some similar early approach to cosmology. Is it just a mathematical game we are playing, is it just semantics, or is it reality?
Leon Lederman and David Schramm, in *From Quarks to the Cosmos*

TABLE **18.1**

Particles That Interact by Means of the Electroweak Force. The three generations interact among one another by means of the four exchange particles. The second and third generations mostly decayed during the early moments of the big bang and are practically nonexistent today because they are unstable. They may have been important only during the big bang.

Generation	Particle type	Mass (proton = 1)	Charge (proton = +1)
1	electron	0.0005	−1
1	electron neutrino	0?	0
2	muon (mu electron)	0.11	−1
2	muon neutrino	0?	0
3	tau (tau electron)	1.90	−1
3	tau neutrino	0?	0
	Exchange particles		
	photon	0	0
	W^+	86	+1
	W^-	86	−1
	Z	98	0

physics, and they agree with observations of the oldest material in the universe. The theoretically predicted helium fraction depends on the number of generations of electroweak particles: The predicted helium fraction grows larger if the number of generations grows larger. Three generations leads to a predicted helium fraction that agrees with the observed value, while four generations leads to a predicted helium fraction that is too high to agree with the observed helium fraction. So there can be only three generations.

The second and third generations play only a fleeting role in today's universe because they are unstable and vanish soon after they are created. Although only the first generation, the electron and its neutrino, play a role in ordinary matter today, the two higher generations may have played a crucial role during the opening moments of the big bang. Without these phantomlike particles, the universe might be far different from what it is and might be incapable of sustaining life.

Unification is a recurring theme of science (Figure 18.12). For example, Copernicus unified Earth with the other planets; Newton unified Earth-based physics with physics throughout the heavens; and Maxwell found a field theory that unified electricity, magnetism, and light. By the end of the nineteenth century, scientists believed that there were only two fundamen-

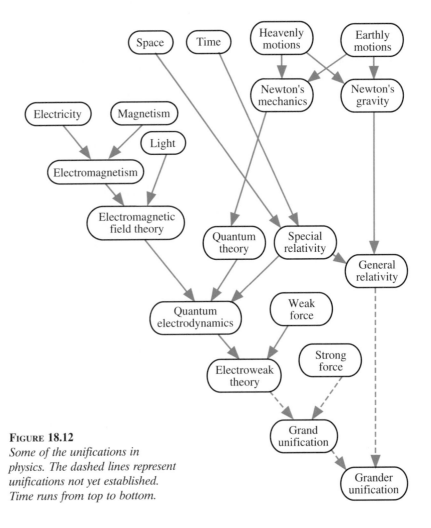

FIGURE 18.12
Some of the unifications in physics. The dashed lines represent unifications not yet established. Time runs from top to bottom.

tal forces, electromagnetism and gravity. Einstein, after fashioning the new theory that explained gravity as a consequence of the geometry of space and time, spent much of his scientific career trying to unify electromagnetism with his gravitational theory, in the hope that a single "unified field theory" would show electricity and gravity to be different aspects of space–time geometry. He was not successful.

Lately we have sought unification at the microscopic level, at the level of quantum field theory. As we have seen, these efforts have achieved significant success by unifying quantum theory, special relativity, and the electric and weak forces. Physicists today are trying to unify the electroweak with the strong force (Section 18.5) and to unify all of these with the gravitational force (Section 18.6).

18.5 Grand unification? *quarks*

As far as the measurements made so far can tell, all of the electroweak particles (Table 18.1) are **point particles**. That is, their force fields appear to be centered on a single point that itself takes up no volume. The electric charge of the electron, for example, appears to be concentrated at a single point.

But protons and neutrons are different. Experiments done in the 1950s showed that their electric and magnetic force centers are spread over a tiny volume about 10^{-15} meters across. Might they be composites that are made of still smaller particles?

There was a time early in this century when protons and electrons were thought to be the only subatomic particles. Then the discovery of both the neutron and the positron in 1932 initiated an era of particle discovery that, by 1960, had produced hundreds of new kinds of supposedly fundamental particles. The collection was frequently referred to as a "zoo." Fermi remarked that "if I could remember the names of all these particles I would have been a botanist." Surely the universe wasn't made of so many kinds of things.

Murray Gell-Mann (Figure 18.13) hoped to bring some order to this zoo. In about 1961 he found ways of grouping the known particles into families that corresponded to significant physical regularities among them. Gell-Mann's work was much like the work of the nineteenth-century chemists who found regularities in the chemical properties of the many known elements and grouped them accordingly into the pattern that we know today as the periodic table. It was only later that this periodic table found its natural explanation in a new model of the atom according to which the more than one hundred elements are built of just three kinds of particles: electrons, protons, and neutrons. In a similar way, Gell-Mann's classification scheme led him to speculate on the existence of a few simpler entities, out of which protons, neutrons, and many other members of the particle zoo could be built. Gell-Mann originally made up a new word, "quork," for them, but changed the spelling to "quark" later when he ran across the line "Three quarks for Master Mark!" in James Joyce's novel, *Finnegan's Wake*.

That set experimentalists on a quark hunt. But despite strenuous searches among bubble-chamber tracks, nobody could come up with direct evidence for quarks, and so they were regarded merely as useful mathematical fictions.

FIGURE 18.13
Murray Gell-Mann (left) talking with Richard Feynman.

HOW DO WE KNOW? THE DISCOVERY OF QUARKS

When Richard Taylor, Jerome Friedman, and Henry Kendall (Figure 18.14), along with twelve coworkers, set out in 1967 to study the proton and the neutron, they weren't looking for quarks. Using the Stanford Linear Accelerator (SLAC, Figure 18.9(d)), they were following up on earlier experiments that had showed both protons and neutrons to be fuzzy balls about 10^{-15} meters across. Hoping to get a clearer picture of the structure of these fuzzballs, they hurled SLAC's high-energy electrons at protons and used huge detectors that they had built specifically to measure the energy and angular deviation of the electrons after they were deflected by the protons (Figure 18.15). At lower electron energies, their "scattered" electrons merely gave them a higher-resolution picture of the same old fuzzballs. But at energies so high that the electrons blew the protons and neutrons to bits, they found a surprise. Some of the electrons were deflected through very large angles, as though they were bouncing off hard little granules buried deep within the fuzzball. The experiment and its outcome paralleled Rutherford's discovery of a tiny hard nucleus deep within what had been supposed to be a fuzz-ball atom (Section 8.6). Only this time there appeared to be not one but three tiny force centers within the proton and within the neutron, a result that became clear only after six years of experimenting and sorting out data. As other labs confirmed these results, it gradually became clear that Taylor, Friedman, and Kendall had discovered Gell-Mann's quarks. The conclusion emerging from the experiments was that all protons and neutrons are made of two kinds of quarks, called the *u* and the *d quarks*. Protons are made of two u's and one d, and neutrons are made of one u and two d's.

  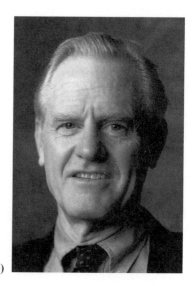

(a) (b) (c)

FIGURE 18.14
*(a) Richard Taylor, (b) Jerome Friedman, and (c) Henry Kendall.
They discovered quarks in 1967.*

FIGURE 18.15

Physicists have found a version of quantum field theory that describes the interactions between quarks and that has so far agreed with all experiments designed to test it. In this theory, the strong force acts directly between quarks, and the force acting between protons and neutrons is created because of the forces between their quarks. The strong force acting between, say, two protons is seen as caused by the interactions among the three quarks in proton 1 and the other three quarks in proton 2.

The force field that is quantized in this new theory is the strong force field, and the matter field that is quantized is the quark field. Think of the strong force field as analogous to the electric field, and the quark field as analogous to the electron psi-field. The quanta of the quark field are quarks of two types, the u quark and d quark (and their antiparticles). They are the material particles of this theory, playing a role similar to that of the electron in quantum electrodynamics. The theory predicts that there are two stable configurations of u and d quarks, namely, u-u-d (the proton) and u-d-d (the neutron).

In addition to feeling and exerting the strong force, quarks must also experience the electric force, because we know that protons experience this force and quarks are supposed to explain protons.

DIALOGUE 3 Surprisingly, quarks turn out to be fractionally charged, the u possessing a positive charge of $+2/3$ of the electron's charge and the d possessing a negative charge of $-1/3$ of the electron's charge. In this case, what is the charge that this theory predicts for the proton? For the neutron?

The quanta of the strong force field are called **gluons** because they are the glue that binds quarks together, and on a larger scale they are the glue that binds the nucleus together. Think of them as the photons of the strong force. Like the photon, they have no mass and no charge.

---

3. $2/3 + 2/3 - 1/3 = +1$ electron charge, $2/3 - 1/3 - 1/3 = 0$.

The quantum field theory of the strong force is the theory of the interactions among quarks and gluons, much as quantum electrodynamics is the theory of the interactions among electrons and photons.

There is an important difference between the way that gluons operate in the strong force and the way that photons operate in the electric force. Gluons themselves directly feel the strong force, unlike photons, which are not electrically charged and do not directly feel the electric force. In quantum electrodynamics, "electric charge" can be thought of as "the ability to emit and absorb photons." In the same way, the property of feeling the strong force can be thought of as the ability to emit and absorb gluons. But gluons themselves feel the strong force, which means that gluons themselves can emit gluons, unlike photons, which cannot emit photons.

This ability of gluons to make more gluons explains one of the most curious features of quarks: The force between quarks grows stronger, not weaker, as they are separated, and as a result it is impossible to isolate single quarks. When a quark within a proton is pulled a short distance from its neighboring quarks, the gluons must fly farther away in order to reach from that quark to its neighbors. This gives these gluons more time to proliferate in flight, which makes more gluons, which makes the force larger as the distance becomes larger. As the quark is pulled farther away, energy quickly builds up in the strong force field, and this energy creates quark–antiquark pairs. Several such pairs are quickly created, and after a brief reshuffling, a new quark is created in the proton from which the first quark had been removed! Furthermore, the removed quark and the newly created quarks team up to form new quark combinations. Figure 18.16 shows in more detail how this process works.

This theory provides a beautifully crazy explanation of why years of looking for isolated quarks in bubble chambers produced no results. Any attempt to pull a quark away from its neighbors just makes more nonisolated quarks. You can't find individual quarks.

As far as anybody can tell, quarks are truly point particles, like electrons and neutrinos. They seem to be good candidates, at last, for the truly fundamental "seeds" in nature's successive seeds within seeds (Figure 18.17).

Recall that there are three generations of electroweak particles (Table 18.1). In just the same way, observation reveals three generations of quarks. The second and third generations each consist of two quarks that are heavier and unstable (short-lived) copies of the u and d, just as the second and third generations of electroweak particles are heavier and unstable copies of the electron and neutrino.

Table 18.2 shows the entire setup for the strong force. The last quark to be discovered experimentally, the t quark, was tentatively confirmed by Fermilab (Figure 18.9(e)) in 1994. The t quark was the most difficult to discover because its mass turned out to be so much larger than the masses of the other five types of quarks, which (because of $E = mc^2$) meant that much more energy was needed to create it. "Weighing" in at an estimated 185 proton masses, the t quark is about as massive as a gold atom! The resemblance between Table 18.1 and 18.2 is striking and points to a close connection between the electroweak and strong forces. Accordingly, physicists believe there should be a "grand unified theory"

FIGURE 18.16
*Here's why you can't separate
quarks:*

Suppose you start with the
three quarks in a proton—

and begin pulling away
one of the quarks. Gluons
travel between the quarks.

As the quark is separated
further, the gluons make
more gluons, which makes
the force between the
quarks stronger.

Finally, all the gluon energy
(strong field energy) creates
a quark–antiquark pair; the
new quark makes the proton
whole again; and the new
antiquark combines with the
old quark to form an unstable
quark–antiquark pair.

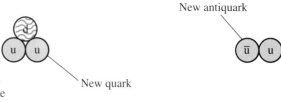

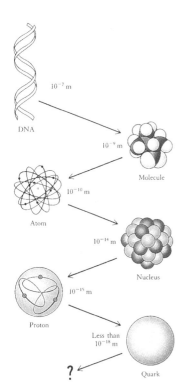

FIGURE 18.17
*Nature's successive seeds within
seeds, from DNA to quarks. Note
the approximate size of each level.*

that would view the electroweak and strong forces as two facets of a single
underlying force. So far, however, nobody has been able to formulate such
a theory.

The particles and organization shown in Tables 18.1 and 18.2 is known as
the **standard model**, a boring title for a theory with such fantastic predictions
as antimatter, neutrinos, and quarks.

To summarize:

THE STANDARD MODEL
Neglecting gravity, there are only two kinds of forces: the
electroweak force and the strong force. Ordinary matter is made
only of electrons, neutrinos, and u and d quarks. Electrons and
neutrinos interact only through the electroweak force, whose
exchange particles are photons, W's, and Z's. The u and d interact
through the electroweak force and also the strong force whose
exchange particles are gluons. In addition, there are two heavier
unstable copies, or "generations," of the four basic particles. In all
there are six kinds of electroweak particles and six kinds of quarks.

TABLE **18.2**
Particles That Interact by Means of the Strong Force. The three generations of quarks interact by means of gluons. Only the first generation is active today, the others being unstable. The other two generations may have been important only during the big bang. The proton is made of u-u-d, and the neutron is made of u-d-d. The force that binds protons and neutrons into the nucleus is actually the strong force acting between their quarks.

Generation	Particle type	Mass (proton = 1)	Charge (proton = +1)
1	u-quark	0.004	+2/3
1	d-quark	0.007	−1/3
2	c-quark	1.6	+2/3
2	s-quark	0.16	−1/3
3	t-quark	185 (unconfirmed)	+2/3
3	b-quark	5.0	−1/3
	Exchange particles gluons	0	0

Think of quarks, electrons, and neutrinos as the bricks of the universe, and exchange particles (the photon and its partners) as the cement.

DIALOGUE 4   According to the standard model, which of the following are "composite" particles and which are "elementary"? Dust grain, electron, positron, hydrogen atom, photon, water molecule, neutrino, neutron, quark, muon, proton, antiproton.

18.6 Grander unification? *gravity and the creation*

As we have seen, two of the four fundamental microscopic forces have been unified into a single quantum field theory of the electroweak force. A similar quantum field theory has been worked out for the strong force, and the similarities between these two theories makes it plausible that a unified theory of the electroweak and the strong force will someday be invented.

That will still leave gravity out of the unified picture. One reason it has been so hard to work gravity into these theories is that so little is known about it at the microscopic level, because it is so weak at this level. For example, the gravitational attraction of a single proton for an electron is 10^{36} (a trillion trillion trillion) times smaller than the electric force. Only if there are large concentrations of matter, as in a planet, are gravitational effects strong enough to be easily observed. In large concentrations of matter, the electric effects of the protons and electrons largely cancel out each other, so that gravity dominates.

---

4. Composite: dust grain, atom, molecule, neutron, proton, antiproton.

Gravity, the first fundamental force to be understood macroscopically, is the last to be understood microscopically.

At the large-scale level, Einstein's general relativity appears to be the correct theory of gravity. It is a field theory whose field is the space–time curvature that is caused by masses (Section 11.6). The most reasonable way to combine this theory with the theories of the other forces would be to quantize the gravitational field to produce a quantum theory of gravity. But this turns out to be very difficult to do in any logically consistent way. General relativity and quantum theory are radically different and difficult to combine.

One particle predicted by all the work on gravity is the **graviton**, the quantum of the gravitational field. Gravitons are similar to photons: They have zero mass, zero charge, and move at lightspeed. From the quantum point of view, the gravitational forces between, say, Earth and the moon, occur by means of the exchange of gravitons between the two bodies. But gravitons have never been observed and perhaps never will be observed in any direct way, because the gravitational force is so weak. For example, if a single proton absorbs a graviton, the proton should recoil, but this recoil is predicted to be so tiny that one cannot hope to observe it.

A marriage, or at least an affectionate relationship, has developed recently between high-energy microscopic physics, on the one hand, and the astrophysics of distant realms of the universe, on the other hand. Inner and outer space have united because the big bang started microscopically. Physicists use high-energy accelerators to study events similar to those that occurred during the creation of the universe, and astronomers study evidence for a big bang that can be regarded as an extremely high energy microscopic physics event. This union has stimulated some remarkable ideas.

In the early days of quantum theory, Max Planck pointed out that nature's fundamental numbers—lightspeed, Planck's constant, and the "gravitational force constant" that specifies the strength of the gravitational force—can be combined in certain ways to obtain what appear to be "natural" units of length, time, and energy. In metric units, these natural **Planck scales** are

$$\text{Planck length} \approx 10^{-35} \text{ meters}$$

$$\text{Planck time} \approx 10^{-43} \text{ seconds}$$

$$\text{Planck energy} \approx 10^{9} \text{ joules}$$

In the 1960s, John Wheeler (Figure 18.18) interpreted Planck's quantities in the light of general relativity and the uncertainty principle. We have seen (Section 18.3) that the uncertainty relations make the energy contained in small regions of space highly uncertain, so that the energy in a small region fluctuates randomly. These energy fluctuations are bigger and briefer in smaller regions of space. Wheeler found that in a sphere whose radius is the Planck length, and during time intervals whose duration is the Planck time, energy fluctuations as large as the Planck energy are likely to occur. The Planck energy is not large by macroscopic standards; in fact it is roughly the chemical energy in one automobile tankful of gasoline. But if this amount of energy is concentrated (because of a quantum fluctuation) in a sphere whose radius is only 10^{-35} meters, the result is remarkable: This energy is

FIGURE 18.18
John Wheeler has headed research groups studying general relativity and the foundations of quantum theory at Princeton University and the University of Texas.

equivalent (because of $E = mc^2$) to a mass of about 20 billionths of a kilogram, and this much mass in such a tiny volume causes space–time to bend back upon itself and form a black hole that is cut off from the rest of the universe.

Wheeler pointed out that this phenomenon would break space and time into tiny pieces at this ultrasmall level, so that the Planck length and time are the smallest lengths and times that have any physical meaning at all. Viewed at this level, space and time are themselves broken up into a "quantum foam" pierced by tiny fluctuating black holes.

It is difficult to investigate such questions experimentally, because the energies going into individual microscopic events in today's large particle accelerators are far smaller than the Planck energy. However, the progress made with existing accelerators already points to a significant trend: The differences among the fundamental forces diminish as the energy rises. The theory of the electroweak force suggests, for example, that at higher energies the differences between the electric force and the weak force diminish, and the two forces attain roughly equal strengths. At still higher energies, it is thought that the strong force becomes similar to the electroweak force, and at even higher energies, namely, the Planck energy, they all become similar to gravity. This might seem surprising, because we pointed out earlier that the gravitational force is normally very weak. But at high energies, that is, at short separation distances (comparable to the Planck length) between different particles, gravity becomes very strong.

All of this leads to a remarkable picture of the big bang, the ultimate high-energy microscopic event (Figure 18.19). Starting from the assumption that the big bang began as a microscopic event, it cannot have initially occurred in a region smaller in diameter than the Planck length or at a time earlier than the Planck time, because smaller volumes and times do not exist. At that time, the universe's total energy was about the Planck energy and was about the Planck length in diameter. The Planck energy, when confined to a region this small, corresponds to an enormous temperature of about 10^{32} degrees Kelvin (degrees above absolute zero). Then something called "inflation" occurred, which allowed the universe to expand rapidly while it cooled.*

Figure 18.19 traces the subsequent history of the universe. By now, the universe has cooled all the way down to about 3 degrees Kelvin, as astronomers have discovered by measuring the temperature of the cosmic background radiation that is left over from the big bang. This radiation was not emitted until 100,000 years after the big bang, because before that time the universe was so hot that electrons could not unite with nuclei to form electrically neutral atoms, and photons were unable to travel through the resulting electrically charged gas of electrons and bare nuclei. Measurements of the cosmic background radiation are seeing this universal first light.

Maybe the universe itself sprang into existence out of nothingness—a gigantic vacuum fluctuation which we know today as the big bang. Remarkably, the laws of modern physics allow for this possibility.
John Wheeler

It is said that there's no such thing as a free lunch. But the universe is the ultimate free lunch.
Alan Guth, originator of the "inflation" idea that explains how the big bang could have created our universe out of a vacuum

---

* But since energy is supposed to be conserved, it is a reasonable question to ask where all of the universe's energy came from if it were only about equal to the energy in a tank of gasoline to begin with. The answer is that the gravitational energy of compact matter is always negative (less than zero), because work must be done on matter in order to pull it apart into separated pieces. The universe's net energy of about 1 billion joules (one "gas tank") could be balanced between an enormous negative gravitational energy and a slightly more enormous but positive kinetic energy.

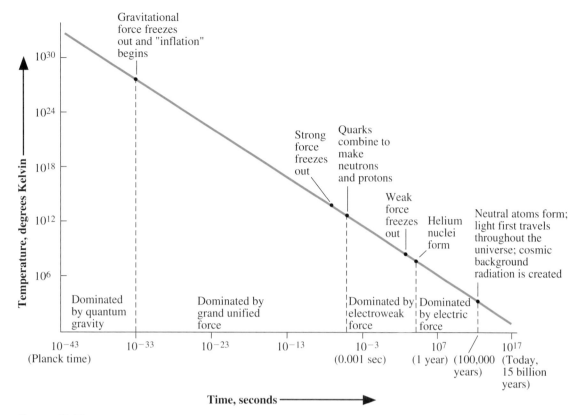

FIGURE 18.19

The thermal history of the universe, starting from the earliest time that physics could exist. Each of the four fundamental forces had an era during which it dominated the physics of the early universe. Once neutral atoms formed, there was light throughout the universe. The cosmic background radiation originated at this time, and stars and galaxies formed after this time. Today, the universe has cooled to 3 degrees Kelvin (3 degrees above absolute zero).

Like a gleam in the darkness, we have appeared for an instant from the black nothingness of ever-unconscious matter, in order to make good the demands of Reason and create a life worthy of ourselves and of the Goal we only dimly perceive.
   Andrei Sakharov, Russian physicist
   and father of the Soviet fusion bomb

There could be many, many other bubble universes existing out there, all out of communication with our universe. Thus, not only is the universe a very big place, but our universe might not be the only one.
Leon Lederman and David Schramm,
   in *From Quarks to the Cosmos*

The cooling of the universe was analogous to the cooling of water. The four fundamental forces were separated by a sort of "freezing out" as the universe cooled. At the Planck energy and temperature, the four forces were similar; physicists say that they had the same "symmetries." Below this temperature, the gravitational force "froze out" of the mixture; it lost the symmetry that had unified it with the other forces. This "symmetry breaking" is analogous to the loss of symmetry when water freezes: All directions are equivalent inside water, but ice crystals line up in specific directions. As the universe continued cooling, the strong force froze out and formed its own unique patterns, and finally the weak force did the same, leaving us with four distinct forces today.

During the early moments of the big bang, the universe was so small that quantum uncertainties should have played a major role. John Wheeler has speculated that a quantum fluctuation in a primordial "quantum foam" could have created our universe. According to this view, it makes no sense to ask "what was before the big bang," because there was no time and so no "before" the big bang. This was where time and space started. To ask what was "before" this event is like asking what is north of Earth's North Pole.

Wheeler expects that if the universe began as a random quantum fluctuation, fundamental numbers such as the masses and charges of the fundamental particles would have been determined randomly. It seems possible that ours could be just one of many universes, each of them born in a new toss of the cosmic dice and each of them characterized by different physical properties.

According to this view, in our universe the numbers happened to have just those lucky values that allowed us to evolve. In any other universe, in which these numbers were very different, life and intelligence might have been physically impossible. Life and intelligence—in other words, our own existence—might turn out to be the best explanation we have for these numbers having the values that they do. This idea—that the universe must be organized in the way that it is because any other organization would not allow intelligent beings to be here to ask the question in the first place—is called the **anthropic principle**.

And this outrageous but plausible connection between the big bang and our lives on Earth is a good place to end our excursion into concepts of physical reality and their connections to our lives.

Summary of Ideas and Terms

Field A physical entity that is spread throughout a region of space. A **force field** exists wherever an object would feel a force.

Quantized field A field that is treated as a real physical entity subject to the laws of quantum theory.

Field quanta The discrete bundles of energy that quantum field theory predicts are associated with any quantized field. Examples: Photons are quanta of the electromagnetic field, and electrons are quanta of the electron field (the psi-field for electrons).

The quantum theory of fields Everything is made of quantized fields that obey the special relativity and quantum theories. All the particles of nature are field quanta. A field's intensity represents the probability for finding the particles that are the quanta of that field.

Quantum electrodynamics The quantum field theory of electrons and photons.

Electric charge Microscopically, charge is the ability to emit and absorb photons. The **electromagnetic force** occurs by means of photon exchanges.

Cloud chamber A device that shows the path of a charged particle as a trail of droplets in water vapor. Its successor, the **bubble chamber**, shows a particle's path as a trail of bubbles in a liquid.

Antiparticle The special relativity theory predicts that for every existing type of particle, there is an antiparticle carrying the opposite charge. Quantum uncertainties allow the **creation and annihilation of particle–antiparticle pairs**.

Positron The electron's antiparticle. Identical to the electron except that it carries a positive charge.

Antimatter is made of antiprotons, antineutrons, and positrons. Today's universe consists overwhelmingly of matter, not antimatter.

Particle accelerator A device to accelerate microscopic particles to high energies.

Energy fluctuations due to quantum uncertainties can cause microscopic events such as pair creation to occur spontaneously during short times, even in a vacuum.

The Lamb shift A small change in the energy levels of the hydrogen atom that is caused by vacuum energy fluctuations in the space surrounding the atom.

Muon, tauon The other two types of electrons that have been found. The only difference among the three is that the muon and tauon are heavier than the electron and are unstable (they have short lifetimes). Because of this instability, muons and tauons have little effect on today's universe, although they were important during the big bang.

Neutrino A particle with no charge and possibly no mass that exerts and feels only the weak force and gravity.

Electroweak force The unified electromagnetic and weak forces. Electrons and neutrinos experience only the electroweak force. The exchange particles, or field quanta, for this force are the **photon, W$^+$, W$^-$**, and **Z**. W's are charged, and the photon and Z are not charged. There are three **generations** of electroweak particles: the electron and its neutrino, the muon and its neutrino, and the tauon and its neutrino.

Strong force One of the four fundamental forces. Its field quanta (or exchange particles) are **gluons**, and the particles that exert and feel this force are **quarks**. Two kinds of quarks, **u** and **d**, are important in ordinary matter. Protons are made of u-u-d, and neutrons are made of u-d-d. Quarks are only fractionally charged and are not found in isolation because any attempt to isolate them creates more quarks. There are two other (unstable) generations of quarks, so there are six kinds of quarks in all.

Standard model The theory of the electroweak and strong force.

Graviton The gravitational field quantum, predicted but not yet observed.

Planck scale According to some theories of quantum gravity, the Planck length and time are the tiny ultimate units into which space–time itself is divided. Smaller dimensions have no physical meaning.

Big bang The creation of our universe in a single event. The consensus today is that this event was initially microscopic and dominated by quantum gravity.

Anthropic principle The idea that any universe that allows the existence of life and conscious observers must be organized similarly to our universe.

Review Questions

QUANTIZED FIELDS

1. What two theories are combined to form quantum field theory?
2. What is a field? What is a quantized field?
3. Name the quanta of the electromagnetic field. Are electrons also "quanta"? Quanta of what?

QUANTUM ELECTRODYNAMICS AND ANTIMATTER

4. What role does the photon play in the electric force between two electrons? Between an electron and a positron?
5. Explain Figure 18.3(a) and (b).

6. What is a muon? A tauon?
7. What is an antiparticle? Name two antiparticles. What is antimatter?
8. Describe the creation of a particle–antiparticle pair.
9. How do scientists observe the subatomic world?
10. Is empty space really empty? What happens there?

ELECTROWEAK UNIFICATION

11. Why is the neutrino so hard to detect? Which of the four fundamental forces does it experience? Is it electrically charged?
12. Name the six types of particles that interact via the electroweak force. Of these, which ones participate significantly in today's universe?
13. Name the exchange particle for the electric force. Name the four exchange particles for the electroweak force.
14. The electroweak particles are laid out in generations. Describe this pattern. How many generations are there?
15. What role might the second and third generations of electroweak particles have once played in our universe?

GRAND UNIFICATION

16. Name the fundamental (not composite) particles that are responsible for the strong force.

17. How were quarks discovered?
18. Are protons fundamental particles? If they are composite particles, of what are they composites? What about neutrons? What about electrons?
19. What force or forces do quarks exert on one another?
20. One property of quarks is that they exert and feel the strong force. List at least two other properties.
21. How many kinds of quarks are there? How many of these are important to today's universe?
22. Name the exchange particles that carry the strong force.
23. Why do we never observe an isolated quark?
24. Which four kinds of particles form the "bricks" of the universe, and which kinds form the "cement"?

GRANDER UNIFICATION

25. Which of the four fundamental forces extend over macroscopic distances?
26. What is a graviton? Has it been discovered experimentally? If so, how? If not, why not?
27. What is the significance of the Planck length and time?
28. What does it mean, in reference to Figure 18.19, to say that a fundamental force "freezes out"?
29. When did the cosmic background radiation originate?
30. What is the anthropic principle?

For Discussion

1. Suppose that our universe had been created in such a way that life could not exist. One possibility, for instance, is that the universe could obey Newtonian physics even at the microscopic level. Chemical reactions would be impossible in such a universe because the existence of the chemical elements depends on the validity of quantum theory. Life would be impossible in a Newtonian universe. Since there is no possibility of any conscious being ever observing such a "dead" universe, would it make sense to say that such a universe "exists" or that it "could exist"?
2. In light of the anthropic principle, might there be a scientific sense in which the universe was made for humans, or at least for conscious beings?

Exercises

QUANTUM ELECTRODYNAMICS AND ANTIMATTER

1. What is the evidence, in reference to Figure 18.8(a), that each pair consists of two oppositely charged particles?
2. In Figure 18.8(a), which of the two pairs has the faster-moving particles? How do you know?
3. In Figure 18.8(a), why don't we see the tracks of the two photons that created the two pairs?
4. Each of the two photons created when an electron–positron pair annihilates has a frequency of about 10^{20} Hertz. In what region of the electromagnetic spectrum does this take place? If the electron and positron were moving instead of being at rest, would it make this photon frequency higher or lower?

5.*Making estimates. Suppose that a proton–antiproton pair, at rest, annihilate and create two photons. Using the data from the preceding exercise, estimate the frequency of each photon. A proton is nearly 2000 times more massive than an electron.

ELECTROWEAK UNIFICATION

6. Of the 10 electroweak particles (Table 18.1), which ones travel at (or near) lightspeed?
7. Of the 10 electroweak particles (Table 18.1), which ones can feel the electric force? Which can exchange photons?
8. Into which one of the "boxes" of Figure 18.12 should the discovery of beta decay be placed?

9. In what ways are the W and Z particles similar to photons? In what ways are they different?

GRAND UNIFICATION

10. In what ways are gluons similar to photons? In what ways are they different?

11. In what ways are quarks similar to electrons? In what ways are they different?

12. Give at least one specific reason (other than a general belief in unity) why scientists believe that there is probably a single theory that can unite the electroweak and the strong force into a single grand unified force.

13. Table 18.2 says that the u quark's mass is only 0.004 proton mass and that the d quark's mass is only 0.007 proton mass. The total mass of the three quarks making up a proton (u-u-d) is so only 0.015 proton mass. Further-more, gluons have no mass. Where, then, does the remaining fraction (0.985 proton mass) of the proton's mass come from?

GRANDER UNIFICATION

14. In what ways was the emergence of the four different fundamental forces during the big bang similar to the change of state of water from liquid to solid?

15. According to the anthropic principle, is it possible that our universe could have been created in such a way that every isotope in the periodic table is unstable (radioactive) with half-lives of, say, less than one year?

16. Is it possible, according to the anthropic principle, that our universe could have been created in such a way that the proton is unstable, with a half-life of less than one year?

EPILOGUE: *summing up*

We have come some distance together, you and I, since we started this book. Now we are almost at journey's end. Let's step back and view the landscape through which we have passed. Although that landscape has ranged from the microscopic to the cosmological and from the ancient to the new, the varying details have been connected by four recurring themes: energy, the social effects of science and technology, Newtonian compared with post-Newtonian physics, and how we know.

Energy unifies all physics, both Newtonian and post-Newtonian, and helps organize our thinking about any physical process. From automobiles to the big bang, the laws and concepts of energy can help clarify the essentials. On a practical level, our culture must get a grip on people's uses of energy if we are ever to fashion a prosperous future. Our most basic energy-related problem, global warming, casts an uncertain shadow over the twenty-first century. Although there are wide uncertainties in our scientific understanding of this problem, enough is known to conclude that the probability of serious global harm is high if we continue escalating our use of fossil fuels. Eventually, the growing use of fossil fuels must be curtailed and probably reversed. Moreover, if global warming is perceived to be damaging the planet, it will take decades to turn our giant energy establishment toward nonfossil fuels: Our energy economy has a lot of inertia. So there are risks in waiting.

We probably face the end of the fossil-fuel age. This is a challenging prospect but also an exciting one, because if we deal intelligently with the transition, we can move toward a technologically sophisticated future in which both humans and their environment can prosper. Your help is needed to fulfill that vision.

On a deeper level, modern physics teaches us that the processes of the universe are not simply mechanical motions of matter but are more appropriately viewed as the organization and transformation of energy. Energy itself is the ability to do work and so bring about change, and it takes the form of motion and interactions. Ultimately, the universe appears to be made of fields, made of the "possibility of an interaction." The material objects of our everyday existence should not be regarded as made of permanent, solid parts but, rather, as made of vibrating fields and energy, bundles of possible interactions. The universe is not made of things; it is made of interactions and change.

Occasionally when I talk with groups about the *social effects of science and technology,* I ask them to name any significant contemporary problems that come to mind. It doesn't take long to accumulate quite a list: overpopulation, urban decay, species extinctions, drug abuse, ozone depletion, AIDS, resource depletion, cults, deforestation, and so forth. After collecting twenty or thirty suggestions, we search for common themes among them. It soon becomes apparent that they all have significant science and technology components.

Without science and technology, we would have other problems, such as early death by disease, but we would not have the particular problems we have today. For instance, because medical science has partly solved the problem of death by disease, we now have the problem of overpopulation. The death problem has been replaced, in a sense, by a birth problem. Because we have accepted science's help in solving the "death" side of the birth-and-death equation but have not simultaneously taken responsibility for the "birth" side, we have burdened the planet with more people than it can handle.

Science gives us great power, and we can use that power in either helpful or harmful ways. Without science and technology, we would not have the automobile or television, for example. When you turn on the switch of either device, you bring great power to bear on yourself, on others, and on Earth.

The problems of science and society come down to this: Humankind is not paying its dues for the fruits of the scientific age. We are quick to accept the speed of the automobile, the fun of television, and the cures of medicine, but we are slow to clean up our exhaust fumes, to maintain intelligent reading habits, or to control our birthrate. The ozone story is a good example: Humankind enjoyed its CFC-powered air conditioning and plastic foams for many decades before anybody took the trouble to consider the effects that all this might be having on the atmosphere, and even then it was fifteen more years before we got serious about eliminating CFCs. Now the ozone is diminished, and it will not soon recover. Earth, including our own health, is paying the price. We dare not accept science's benefits without accepting its responsibilities.

Speaking as a science teacher who is doubtless prejudiced in the matter, my first suggestion is that all of us learn much more science. Scientists, who are usually only narrowly trained in their specialty, need broader scientific knowledge and improved communication skills. Nonscientists need to learn more about the physical universe and our planet. Humankind is using great power today, without knowledge. If we want our technology-based society to succeed, we had better begin to understand what we are doing, for the use of power without knowledge is a prescription for disaster.

The comparison between Newtonian and post-Newtonian physics has implications that run deep into the cultural roots of industrial civilization. Modern culture still assumes that the mechanical, Newtonian worldview is science's view of reality. It is a materialistic worldview that leaves little room for freedom, chance, or creativity. Many people would argue that it leaves little or no room for spiritual values.

But modern physics paints a quite non-Newtonian picture, a more abstract picture made of fields and energy, structured by relativity and quantum theory. Many nonmechanical forms of energy operate in this post-Newtonian universe, and reality emerges not as a predictable clockwork mechanism but as a dynamic and unpredictable network of energy. Material and nonmaterial particles pop unpredictably and briefly in and out of their fields. This is nothing like a clock. In many ways it is the opposite of a clock. Considering the hidden and often distant interactions that seem to be the essence of the physical world, many have suggested that if we are to use images at all (and physicists have used the clockwork image for many centuries now), the universe is more comparable to a live organism than it is to a clock.

It is by no means clear what worldview will emerge from all of this. I feel that the world is just beginning to absorb the impact of relativity and, especially, of quantum theory. After all, more than a century elapsed after Copernicus's death in 1543 before Europe began to absorb the cultural impact of postmedieval science. So it is not surprising that the first post-Newtonian century will be the twenty-first, not the twentieth.

One practical example of the importance of forming a post-Newtonian worldview might be the ongoing reaction, in the United States at least, against the theory of biological evolution. The reaction comes from a perceived threat to religious beliefs. Religious fundamentalists typically view evolutionary theory as mechanical, deterministic, and materialistic, with no room for spiritual values. The real opposition is not so much to evolution itself as it is to evolution *as interpreted through the concepts of Newtonian physics.* A post-Newtonian culture might relieve these old religiously based science anxieties.

More generally, post-Newtonian physics suggests that we cannot control nature in the way that we can control a machine such as a clock, because we live in nature. Nature is a network, and we are in that network. Efforts to subdue nature can react against us, to our own peril. Post-Newtonian physics suggests that our salvation lies not in imposing our own wills on nature but in aiding nature (and so ourselves) in nature's own pursuits.

All this is not to suggest that we should renounce the true benefits of modern science and technology, for if humankind and nature are one, then whatever truly benefits the one will benefit the other. But it is to suggest that we come more humbly to a natural and human world that surpasses our own understanding.

We have often inquired in this book: *How do we know?* The answer is surprisingly simple: We know by experience, as interpreted through intelligence. Science's "method" is really quite simple, but it is fundamental. It is to take nothing for granted, to form one's views on the basis of careful observations

Only by the fusion of science and the humanities can we hope to reach the wisdom appropriate to our day and generation.

I. I. Rabi, physicist

and hard honest thinking, and to be willing to modify those views in the light of new experience.

The twentieth century has been torn by rigidly held and conflicting ideologies. The nationalistic, religious, and economic ideologies seem to come in every imaginable variety, many of them in utter contradiction with one another. Yet all those who believe them are absolutely convinced that they are right. The result has been war, prejudice, fanaticism, and other scourges. Science's view of this is that the danger lies not so much in the beliefs themselves as in their *absolute nature*. Even wrong or harmful beliefs can be corrected if one is willing to trust experience and to be intellectually honest. Even correct and healthy beliefs can become dangerous if accepted uncritically or absolutely.

For the belief in a single truth and in being the possessor thereof is the root cause of all evil in the world.
Max Born, physicist

In thinking about how we might do better in the twenty-first century than we have in the twentieth, we should perhaps ponder science's most basic value: *All ideas are subject to testing by experience and to challenge by critical rational thought.* It is a practical, simple, but demanding code. It is often uncomfortable, even painful, to honestly reevaluate one's beliefs in the light of experience. But it is a code that has worked surprisingly well for science. It might be science's most important benefit.

Answers to odd-numbered exercises

CHAPTER 1

1. Follow its position in the sky for a few weeks. If its position relative to the surrounding stars changes, it is a planet.

3. We are looking northward, and stars rise in the east (left-hand side of photo) and set in the west (right-hand side), so the stars in the photo are circling counterclockwise. If we were looking southward, toward the south pole, the stars would be circling clockwise around a point in the southern sky.

5. Follow a planet every night until it noticeably brightens or dims—an indication that it is closer to or further from Earth.

7. This is possible, if Mars happens to lie a little to the east of the sun in the sky (in other words, close to a line joining Earth to sun in Figure 1.15) and if Venus is below the horizon. This combination of events isn't very likely.

9. Look near the rising or setting sun, just before it rises and just after it sets. If the sun is visible above the horizon, the dim light from Mercury will be obliterated by the light from the sun.

11. The observational imperfections in Ptolemy's and Copernicus's theories that led Kepler to develop his theory were first observed by Brahe. Since these imperfections had not yet been observed in Ptolemy's or Copernicus's times, Kepler's theory would have also agreed with the data known at those earlier times.

13. The "eccentricity" of the elliptical orbits, in other words, the displacement of the focus of the ellipse (the sun) away from the center of the ellipse, is similar to the displaced centers of Ptolemy and Copernicus.

15. Yes. The agreement of the data known during Copernicus's time with both the theory of Copernicus and the theory of Ptolemy is one example.

17. No, it is not testable, because all meter sticks and other measuring devices would double too, so there would be no observable change in anything. This is certainly not good science, and it is not even bad science. It is not science at all, because it doesn't refer to the observable universe.

19. No, some of the material (such as carbon dioxide—see Chapter 2) comes from the air. To test Aristotle's hypothesis, we could put a plant in a pot and carefully weigh it together with any water or fertilizer we put on it as it grows. If the plant's increase in weight is greater than the weight of the water plus fertilizer added to it, then the plant is getting some of its weight from someplace else (namely, the air).

21. This is not a testable hypothesis. So any ESP "theory" that includes this hypothesis should not be considered to be a scientific theory.

23. The answer is 100 billion times 100 billion = 10,000,000,000,000,000,000,000 (a one followed by 22 zeros).

CHAPTER 2

1. No. General scientific principles are never certain. See Chapter 1.

3. $2 + 1 + 4 = 7$.

5. Molecule made of two or more atoms: pure water (H_2O), atmospheric oxygen (O_2), H_2SO_4, carbon dioxide (CO_2), H_2. Single unattached atom: U, He, H.

7. Assume that the coal is pure carbon (C). When the coal burns, each C atom attaches to two O atoms to make CO_2. If we assume, for simplicity, that C and O atoms have the same weight, then a CO_2 molecule would weigh three times as much as a single C atom. So the ton of coal makes three tons of carbon dioxide gas. The more precise answer, based on the weight ratio of 3 to 4 given in exercise 2, is that a ton of coal makes 11/3 (or 3.67) tons of carbon dioxide gas.

9. $10^9 = 1,000,000,000$
$10^{-6} = 0.000\ 001$
$3.6 \times 10^{13} = 36,000,000,000,000$
$5.9 \times 10^{-8} = 0.000\ 000\ 059$

11. A piece of paper is too heavy to respond to being jostled noticeably (as observed by the unaided eye) by atoms.

13. The convict leaves many molecules from his or her body or clothing along the trail. When the dog sniffs these molecules into its nose, it is able to detect them and identify them with the convict, just as your nose is able to identify the odor of violets.

15. Heating causes the air molecules to move faster. This causes them to hit the inner walls of the container harder, which increases the pressure.

17. The air outside the jar pushes downward on the lid more strongly than the air inside the jar pushes upward on the lid. So air pressure holds the lid on the jar.

19. How many times does 10^{-10} m go into 0.1 mm? Since 0.1 mm = 10^{-4} m (because 1 mm = 10^{-3} m), the answer is $10^{-4}/10^{-10} = 10^{-4+10} = 10^6$ atoms, or one million atoms thick.

21. An atom is about 10^{-10} m across (Section 2.3, also exercise 20). Since 0.05 mm = 5×10^{-5} m, the number of atoms needed to stretch across a dust particle is $5 \times 10^{-5}/10^{-10} = 5 \times 10^{-5+10} = 5 \times 10^5$ atoms, or 500,000 atoms (half a million).

23. Hydrogen and oxygen come in the 2-atom form H_2 and O_2. They combine to give water: $H_2 + O_2 \rightarrow H_2O$.

25. Hydrocarbons are made of hydrogen (H) and carbon (C). When these burn in air, containing O_2, the H should combine with O_2 to create H_2O, and the C should combine with O_2 to create CO_2.

27. From the atmosphere, which contains an abundance of N_2 and O_2. These elements don't combine at normal atmospheric temperatures, but at the high temperatures prevailing in automobile engines they do combine.

CHAPTER 3

1. Aristotle: This is the ball's natural motion. Galileo: Friction slowed it to a stop.

3. Nothing keeps them moving. According to the law of inertia, they keep moving simply because there is nothing to stop them.

5. 20 m/s. 20 m/s.

7. They must have different speeds, because she is passing him. Since their speeds are different, their velocities are different, because velocity means speed *and* (not *or*) direction; that is, both the speeds and the directions of motion must be the same before we can say that the velocities are the same.

9. Upper ball: speed = d/t = 0.8 cm/0.2 s = 4 cm/s.
Lower ball: speed = d/t = 1.2 cm/0.2 s = 6 cm/s.

11. Train's track speed = d/t = 330 km/1.5 hr = 220 km/hr.
Plane's flying speed = 330 km/0.5 hr = 660 km/hr.
Total time for train = 1.5 hr + 0.5 hr = 2.0 hr.
Total time for plane = 0.5 hr + 0.5 hr + 0.5 hr + 0.25 hr + 0.75 hr = 2.5 hr.

Overall average speed for train = 330 km/2.0 hr = 165 km/hr.
Overall average speed for plane = 330 km/2.5 hr = 132 km/hr.

13. Zero.

15. Although your foot is on the accelerator, you are not actually accelerating (in the physics sense of the word) when you are driving along a level highway at constant speed; also when you are climbing a straight (unchanging angle, or unchanging "slope") hill at constant speed. When you remove your foot from the accelerator, you slow down; this is an acceleration.

17. The upper ball is accelerated, while the lower ball is not. The lower ball catches up with the upper ball and passes it at time 2, but then the upper ball speeds up and passes the lower ball at time 5. The two balls have the same speed during time 3 to time 4, since they both move the same distance during this time interval.

19. Yes, for example, a drag-racing car just as it starts up from rest. Yes, for instance, a comet; this is speeding up very slightly as it approaches the sun or Earth as it makes one circuit around the sun in one entire year.

21. acceleration = change in speed/time

$$= \frac{(4.5 \text{ m/s} - 3 \text{ m/s})}{5 \text{ s}} = \frac{1.5 \text{ m/s}}{5 \text{ s}} = 0.3 \text{ m/s}^2$$

23. (c)

25. Its acceleration is the same at point B as at A, because objects fall with an unchanging acceleration. Its velocity is larger at point B.

27. 3 times as fast, because speed is proportional to time. 9 times as far, because distance is proportional to the square of the time.

29. In motion with unchanging acceleration, distance is proportional to the square of the time. So the car gets 100 times as far in 10 s as it gets in 1 s. Speed is proportional to the time, so the car is moving 10 times faster after 10 s than it is after 1 s.

CHAPTER 4

1. A force must be exerted on you both when you speed up and when you slow down, in order to accelerate you. Newton's law of motion says so.

3. If the ball accelerates (speeds up, slows down, or changes direction), it must have a force acting on it.

5. Initially the book is at rest; then it quickly speeds up during the fraction of a second that the hammer is actually in contact with the book; after the hammer is no longer touching the book, the book gradually slows down to a stop. There is no net force on the book before the hammer hits it; then there is a large force in the forward direction while the hammer is in contact with the book; then there is a smaller force in the backward direction while the book is slowing down.

7. Since gravity is the only force acting on the ball, the net force on the ball is 8 newtons downward. Thus the ball's acceleration is also downward, opposite to the ball's upward velocity, because Newton's law of motion says that the acceleration is in the direction of the net force. Acceleration and direction are always in the same direction.

9. The floor exerts a force on your feet, upward. You don't accelerate upward because the net force on you is zero.

11. The net force is zero, because the car is not accelerated. The acceleration is zero. The drive force must be 500 newtons, to balance the resistive forces to make the net horizontal force zero.

13. The net force is now $500 - 100 = 400$ newtons in the backward direction. The acceleration is $a = F/m = 400$ N/800 kg = 0.5 m/s^2 backward, i.e., it slows down.

15. Since 1 newton is about 1/4 pound, multiply your weight in pounds by 4 to get your approximate weight in newtons: 100 pounds is roughly 400 newtons, etc.

17. You would be better off having a hunk of gold whose weight is 1 N on the moon, because it would be a more massive hunk (containing more gold) than one whose weight is 1 N on Earth.

19. The net force on the apple at rest is zero. The net force on the falling apple is 2 newtons downward. The net force on the apple moving upward is 2 newtons *downward* because the only force on the apple is gravity.

21. It would be easier to lift a rocket off the moon's surface, because the force of gravity on the rocket (i.e., the rocket's weight) is smaller.

23. False. The boulder has a much larger acceleration than does Earth, because the boulder's mass is much smaller than Earth's mass.

25. The two vehicles exert equally strong forces on each other, and the two vehicles feel equally strong forces from the other vehicle. The car experiences the larger acceleration, because it has the smaller mass.

27. The string pulls upward on the apple, and Earth's gravitational force pulls downward. These do *not* form a force pair. The other members of the two force pairs are (1) the apple pulling downward on the string and (2) the apple's gravitational force pulling upward on Earth.

29. Yes, you push on me too. The two forces are equally strong.

31. The diagram should show a large frictional force, acting backward, in addition to the forces of air resistance (backward), gravity (downward), and the road's perpendicular force (upward). The net force is backward. For a car coasting without braking, the forces are rolling resistance, air resistance, gravity, and the perpendicular force. The net force is strongest in the case of braking.

33. No, the car could be moving with a constant velocity. However, it could be accelerating. The acceleration could be forward (if the car is speeding up) or backward (if the car is slowing down).

35. No; a jet plane could not accelerate (except for the "natural" acceleration of 9.8 m/s^2 downward due to gravity). A rocket-driven airplane could accelerate.

CHAPTER 5

1. Iron. They fall equally fast.

3. The magnitude is your weight (in pounds or in newtons), and the direction is downward.

5. No, you are not in orbit around Earth's center (although you are revolving around that center due to Earth's spinning motion). You are in orbit around the sun.

7. Yes. Yes.

9. One newton, downward. One newton, upward.

11. Yes, this would decrease your weight.

13. Jupiter's radius is much larger than Earth's radius.

15. Yes, you could buy at high altitude (where the gold weighs less) and sell at lower altitude (where the same amount of gold weighs more).

17. 2×10^{-12} newton. 3×10^{-12} newton. 12×10^{-12} newton.

19. Compared with weight on the moon, weight on Earth is multiplied by 100 because of Earth's greater mass, and also divided by $4^2 = 16$ because of Earth's greater radius. The overall effect is to multiply the weight by $100/16 = 6.25$, or about 6.

21. Earth's orbit would be unchanged, because the sun's mass and the distance from Earth to sun would be unchanged. But there would be no sunlight or other radiation from the sun.

23. Because they formed from the flat, pancake-shaped or disk-shaped gas cloud described in Section 5.3.

25. 100 times less than your present weight.

27. Special relativity (because of the high temperature, which implies that particles are moving very rapidly), general relativity (because of the high density, which implies strong gravitational effects), and quantum theory (because of the small size).

CHAPTER 6

1. Yes, because Earth exerts a downward force on you as you walk downward.

3. A meter high, because then you would not have to lift yourself up.

5. A typical person's weight is about 500 newtons. The U.S. population is about 250×10^6, so its total weight is $500 \times 250 \times 10^6$ newtons, or $125,000 \times 10^6$ newtons, or 1.25×10^{11} newtons. To lift this weight by 1 km (1000 meters), the work required is $(1.25 \times 10^{11}$ newtons) $\times$ 1000 meters = 1.25×10^{14} joules, or in round numbers about 10^{14} joules (100 trillion joules).

7. Gravitational, kinetic, chemical, radiant, chemical, thermal.

9. (a) 1000 km high. (b) Increased. (c) At the lower altitude, 6000 km, even though the satellite's *energy* is larger at 12,000 km.

11. Multiplied by 9, because kinetic energy is proportional to the square of the speed. Divided by 4 (i.e., multiplied by 0.25).

13. (a) Chemical energy (of gasoline) → kinetic energy.
(b) Chemical energy (of human body) → kinetic energy.
(c) Electric energy → kinetic energy.
(d) Electric energy → thermal energy.
(e) Electric energy → radiant energy (plus a lot of thermal energy).

15. (a) Chemical energy (of human body) → kinetic energy.
(b) Kinetic energy (of baseball) → thermal energy (of ball and of Jill's hand or baseball mitt).

17. Yes, it must, in both cases, because the energy must come from someplace.

19. No, it would not emit exhaust, and its engine would not be hot because no energy would go into heating anything.

21. (a) 160,000 J. (b) 160,000 J, 160,000 J. (c) One-quarter as much as it had at the high point, because it is one-quarter as high: 40,000 J. So the remaining 120,000 J must be kinetic energy.

23. $(1/2)\ ms^2 = 0.5 \times (1000\ kg) \times (10\ m/s)^2 = 50,000$ J. (b) At twice as fast, it would have four times as much kinetic energy, or 200,000 J.

25. Watts.

27. Power = work/time = 2 J/0.1 s = 20 watt for the 2-joule work source. For the 1000-joule source, power = 1000 J/3600 s = 0.27 watt, which is smaller than the power output of the 2-joule source.

29. 150/1000 = 15%.

31. To answer this, perform a calculation similar to exercise 30 (above) for each appliance used in your home, getting the number of kW·h of electric energy consumed by each appliance during one month. Then add up the consumption of all of the appliances.

CHAPTER 7

1. Using Figure 1.3, it is about 37℃.

3. Thermal energy is removed from the cold inside of the refrigerator, and released into the warmer kitchen. So thermal energy flows from cold to hot, opposite to the "normal" or "natural" direction. This occurs because of the outside assistance, provided by the electric company and by the apparatus of the refrigerator.

5. The efficiency must be less than 100%.

7. You could do it if you had something colder than the ocean to cool the exhaust end of a heat engine that uses the ocean as its warm input material. For example, you could exhaust into the cold stratosphere (but you would need a very long pipe or other device to connect the ship with the stratosphere). In fact, a scheme somewhat like this, using the colder water deeper down in the

ocean, is known as "ocean-thermal conversion" (see Chapter 17).

9. Energy input = 100 J + 400 J = 500 J. Efficiency = 100 J/500 J = 20%.

11. You should get more useful work, because the efficiency would tend to be higher due to the higher input temperature.

13. Transportation.

15. Assuming an efficiency of 13%, about 13 barrels go into getting the car down the road.

17. Hydroelectric, because it is not a heat engine and so is not subject to the inefficiency implied by the second law of thermodynamics.

19. Electricity from a coal-fired generating plant, because such a plant is about 40% efficient while a car engine is only 10–15% efficient (Table 7.1).

21. (15 kg/s) × (3600 s/hr) × (24 hr/day) = 1,300,000 kg/day = 1300 tonnes/day.

23. The football field's area is 100 m × 30 m = 3000 m². So solar energy strikes a football field at an average rate of (200 watts/m²) × (3000 m²) = 600,000 watts. To receive 1 kW, a home would need to use *all* of the solar energy striking an area of 5 m². But since the conversion efficiency is only 10%, the receiving area would need to be 10 times larger, or 50 m². A square-shaped collector would need to be about 7 m on a side.

25. No. Yes.

27. No; for example, it stayed about level during 1930–35. About 150 BkW·h in 1935, about 350 BkW·h in 1945, 600 BkW·h in 1955, 1100 BkW·h in 1965, 2000 BkW·h in 1975. Each of these numbers is roughly twice the preceding number, so the growth is roughly exponential.

29. $T = 70/P = 70/0.8 = 87$ y (U.S.), $70/2.2 = 32$ y (Mexico), $70/4.2 = 17$ y (Kenya).

31. No. The water was not isolated—it had outside help from the colder environment in increasing its order.

CHAPTER 8

1. No, because the water itself actually moves downhill.

3. It must have transformed into thermal energy.

5. Its frequency would halve, because it would take twice as long for each complete wavelength to pass. If you halved the wavelength, the frequency would double.

7. One-half second later, all the ripples will have moved outward by $\frac{1}{2}$ wavelength, so crests will be replaced by valleys and valleys will be replaced by crests. One second later, the surface will look just like it looks in the photo.

9. 3 cm. −3 cm (3 cm downward). 0 cm.

11. Because you cannot see in the dark, even though your eye is open.

13. Firefly, heating element, flashbulb, sun.

15. Because the negative and positive charges are "balanced" (equal amounts of each) in any small region of

space, resulting in no net effect except at the microscopic level.

17. Charge flows from the charged object onto the metal sphere and then down onto the leaves (because metals are good electrical conductors). The charged leaves then repel each other electrically.

19. Both ends of the unmagnetized bar will be attracted to either pole of a magnetized bar. A magnetized bar, on the other hand, has a definite "north" and "south" pole. Thus if you find two ends that *repel* each other, they must belong to the two permanent magnets.

21. Roll a ball into the box to see whether the contents have high mass or low mass. Tie a string around it and pull it; the box's mass can be determined by measurement of the pulling force and the resulting acceleration. Fire bullets through the box from all directions, to see if they mostly pass straight through without hitting anything. Hit the box with a hammer to see if it causes anything inside to roll around.

23. 6 and 2. The ratio is about 12 to 4, or 3 to 1.

25. The ion will carry a positive charge, so it will be repelled by a positively charged transparency, and attracted to a negatively charged transparency. If only one electron was lost, the force on the ion by the transparency would be only half as large, but the directions would still be the same.

27. A typical light wavelength is 0.5×10^{-6} m, which is far bigger than the 10^{-10} m across an atom. The ratio is $0.5 \times 10^{-6}/10^{-10} = 0.5 \times 10^4 = 5000$ times bigger.

CHAPTER 9

1. Yes, because electric currents create magnetic fields. You might be able to detect these fields with a small sensitive compass needle.

3. It will accelerate into motion. An electron would have a larger acceleration (because of its smaller mass) and it would move in a direction opposite to that of the proton (because the electron has a negative charge while the proton has a positive charge).

5. It contains no matter. But it might contain energy, in the form of fields. If so, it is not really "empty" because it contains something that is physically real.

7. Each reply will be delayed by 20 minutes. The telescope would not speed things up.

9. Your ears would "hear" all the radio waves that your radio can receive, all at the same time.

11. The radar signal takes 1.5 s to get to the moon. So the distance to the moon is $(300,000 \text{ km/s}) \times (1.5 \text{ s}) = 450,000$ km.

13. Their wavelengths and frequencies are different. Light's wavelength is much shorter than radio's wavelength, and light's frequency is much higher.

15. Because ultraviolet radiation does not penetrate glass.

17. You would only see ultraviolet and visible light from unnatural sources, such as street lights. But you would see infrared from all over Earth, with warmer places appearing brighter.

19. 92×10^6 hertz. 300,000 km/s. According to Figure 9.7, one wavelength is on the order of 1 meter long (more precisely, about 3 meters long). Your ear is stimulated not by an electromagnetic wave, but by a sound wave.

21. So they will not react chemically inside your body. This inertness means that the gases do not react in the atmosphere, so the gases float up to the upper atmosphere with being broken down along the way.

23. Less than 1 ppb in 1950. It will reach the highest level around the year A.D. 2000. It will return to the 1975 level (about 2 ppb) around A.D. 2050. Without the Ozone Treaty, it would have been about 9 ppb in A.D. 2050.

25. Electricity from solar power, electricity from nuclear power, ethanol from corn (harvesting the corn releases the carbon that was absorbed when the corn was grown, so there is no net carbon increase in the atmosphere).

27. The Ozone Treaty will end the use of CFCs, which not only destroy ozone, but are also greenhouse gases (see Figure 9.24).

29. 2.5% (in buildings) + 5.5% (in industry) = 8%. 1.5% + 3.5% + 9% = 14%.

CHAPTER 10

1. Yes. 8 m/s southward.

3. 6 m/s northward. If Mort had thrown the ball southward, the ball's velocity relative to Velma would be 14 m/s southward.

5. The passenger observes the bullet's velocity to be the same as the gun's muzzle velocity (which is the speed of the bullet relative to the gun), but the sheriff observes the bullet to move faster than the muzzle velocity, because of the train's forward speed. If the gun is fired toward the rear, the passenger still sees the bullet move at the muzzle velocity, but the sheriff sees it move more slowly than this.

7. 560 m/s. 640 m/s.

9. Look out the window; radio to the outside; ask the pilot to look out of the window and to tell you what he sees; stick your hand out of the airplane; etc. All of these involve contact with the outside world.

11. Without specifying the reference frame, this question is meaningless. How fast *relative to what?* Relative to Earth? To the sun? To the center of the galaxy?

13. No. No material object can be speeded all the way up to precisely lightspeed.

15. 300,000 km/s. 300,000 km/s. 300,000.04 km/s.

17. 60 km/s. 60/300,000 = 0.0002, or 2×10^{-4}.

19. The year (or Earth's orbital motion). The vibrations of an atom. The new moon (the moon's orbital motion around Earth). The day (or Earth's spinning motion). An observer moving past these clocks would observe all of them to go slow.

21. No. No. No.

23. 30,000 years. No, it would take more than 30,000 years as measured on Earth. Yes, a person could (theoretically) travel there in as short an amount of their own time as desired; one second, for example. So a person could get there within their own lifetime. This is because of the relativity of time; the person would have to be moving very near lightspeed (relative to Earth), for nearly all of the trip.

25. The speed is 0.1c, so Table 10.1 says that 1 of your seconds is observed on Earth as 1.005 s. So when you return you will have aged by 24 hr while people on Earth have aged by 1.005 × 24 = 24.12 hr = 24 hr and 7.2 min. You have aged less than people on Earth by 7.2 min.

CHAPTER 11

1. According to Figure 11.6, the required speed is about 0.86c. She will observe the United States to be 0.86 × 5000 km = 4300 km wide.

3. According to Figure 11.6, a 10% length contraction (to 90% of the original length) occurs at a speed of about 0.4c.

5. Galileo's relativity predicts 0.75c. Using Table 11.1, Einstein's prediction is 0.667c. Einstein is correct.

7. Using Figure 11.4, Mort measures the mass to be 1.6 × 10,000 kg = 16,000 kg, and the length to be 0.6 × 100 m = 60 m.

9. It is moving at about 0.87c, and is oriented crossways (perpendicularly) to its direction of motion.

11. They are normal; they have not changed at all.

13. It increases. This effect is so small that you wouldn't be able to detect it.

15. mc^2 = (1 kg) × (9 × 10^{16} m^2/s^2) = 9 × 10^{16} J. This much energy goes into lifting. The energy needed to lift a weight through a height is weight × height. The weight of the U.S. population is about (250 × 10^6 persons) × (600 N/person) = 1.5 × 10^{11} N. The height through which this mass could be lifted is 9 × 10^{16} J/1.5 × 10^{11} kg = 6 × 10^5 m = 600 kilometers!

17. Twice as heavy as usual. Half as heavy as usual. Weightless.

19. Latitudinal lines are circles, but only the equator is a "straightest" line—the others are not. They do not meet.

21. The big bang! (The origin of the universe.)

CHAPTER 12

1. No. Each neutron star and black hole must have gone through a supernova explosion, and this explosion would have destroyed any life that had developed previously. Similarly, each white dwarf must have gone through an unstable red giant phase that would destroy any pre-existing life. Our sun will go through such a red giant phase some 5 billion years in the future.

3. 400 × 10^9 × 0.001 × 0.01 × 0.01 = 40,000.

5. There is a rough consensus that there are on the order of a billion "good" places for life in our galaxy, and life probably emerged on many of these, perhaps on millions of these. There is no consensus—not even a rough one—on the remaining questions.

7. The ratio of the distances is the same as the ratio of 11 years to 8 minutes. Convert 11 years to minutes: 11 yr × 365 days/yr × 24 hr/day × 60 min/hr = 5,800,000 min (approximately). Dividing this by 8 min, we find that it is about 700,000 times farther to Tau Ceti than it is to the sun.

9. Any such planet should have a gravitational effect on the other planets, and this effect has not been observed.

11. There were no such reports. There were no stories reporting this fact. Generally, when far-fetched predictions such as this are made, there is little or no attempt to follow up on them.

CHAPTER 13

1. Radiant to electromagnetic.

3. Infrared has greater energy. Radio has longer wavelength. Infrared has larger frequency. Both have the same rest mass (zero). Both move at lightspeed.

5. Ultraviolet photons have more energy, enough to disrupt the chemistry of the cells in your skin.

7. Yes. Each photon's frequency is 1 Hz, so its energy is hf = (6.6 × 10^{-34} J/s) × (1 Hz) = 6.6 × 10^{-34} J.

9. Visible light has a frequency of around 10^{15} Hz, so the energy of the 10 photons is about 10hf = 10 × (6.6 × 10^{-34} J/s) × (10^{15} Hz) = 6.6 × 10^{-18} J.

11. Similarities: both are small particles, both carry energy. Differences: photon is radiation while electron is matter, photon has no rest mass while electron has rest mass, photon moves at lightspeed while electron moves slower than lightspeed.

13. Very small.

15. A slow proton, because slower particles have longer wavelengths.

17. Yes. The wavelength of the protons is about 2000 times smaller than the wavelength of electrons (see Exercise 16), so the "interference lines" (the bright and dark lines on the screen) would be 2000 times smaller.

19. This halves its wavelength. Doubling the mass also halves the wavelength.

21. No, a coin toss is "Newtonian"—the coin obeys Newtonian physics (to a very good approximation), so quantum uncertainties are insignificant. The uncertainties are due to the tosser's lack of detailed information and inability to carry out the calculations needed to predict the outcome.

23. We can predict the overall pattern of electron impacts on the screen. If a large number of electrons go through the experimental apparatus, we can predict

what fraction of them will hit in each portion of the screen.

25. No, this would not be correct. Both electrons come through both slits; electrons coming through slit A or through slit B are inconsistent with the interference pattern.

CHAPTER 14

1. Your radio detects radiation of specific frequencies (namely, the frequency of your radio dial). So you can use a radio to detect the radio-range frequencies that are present. Some of the differences from the spectroscope described in the text: The radio waves do not come from a glowing gas, no thin slit is used, no prism is used.

3. Yes (except for a possible difference in intensities). The sodium and mercury spectra would differ.

5. The wavelength could not be 2.1 m or 1.9 m, because a whole number of these wavelengths cannot be fit onto the string. But a wavelength of 0.5 m would fit; 4 of these wavelengths would just fit onto the string, so this is a possible standing wavelength.

7. Yes. An excited atom has more energy, so it has more mass. When an atom emits a photon, the atom loses energy so it loses mass.

9. 2 to 1. Longest wavelength: 5 to 4.

11. Uncertainties would be smaller. There would be no uncertainties, and Newtonian physics would be correct even at the microscopic level.

13. Yes. Smaller uncertainties would allow atoms to be smaller in size.

15. If the thermometer starts out colder (or warmer) than the water, the thermometer itself will slightly cool (or warm) the water. This is not a quantum effect.

17. It would collapse (or quantum-jump) into a much more spread-out wave, with a larger Δs and a smaller Δx.

19. Before. When the electron hits the screen, the spread out psi-wave collapses to a small point.

21. These correlations have nothing to do with quantum effects.

23. Non-Newtonian: Radioactive decay, the quantum states of an atom, the wave nature of electrons, the particle nature of radiation, quantum jumps, the line spectra of atoms. Newtonian: The path of a baseball, the forces on macroscopic objects, the motions of the planets, the recoil of a rifle, the swinging of a pendulum—macroscopic motions and macroscopic forces in general.

CHAPTER 15

1. The strong force. If the electric force was stronger, the nucleus would fly apart.

3. $^{13}_{6}$C: 6 protons, 7 neutrons
$^{56}_{26}$Fe: 26 protons, 30 neutrons
$^{90}_{38}$Sr: 38 protons, 52 neutrons
$^{3}_{1}$H: 1 proton, 2 neutrons

5. Masses are about the same. Charge is in the ratio of 1 to 2.

7. Hot rock inside Earth; this rock is heated, at least indirectly, by radioactivity inside Earth.

9. Gamma rays, because they are a form of electromagnetic radiation.

11. Yes, in beta decay the daughter nucleus has a higher atomic number.

13. 1/32 (about 0.03) of it will remain.

15. 0.5 grams. Roughly 5 atoms. The prediction that 5 atoms will remain is imprecise—it could easily be 4 or 6, for example. The reason for the imprecision is that radioactive decay is an uncertain process, because of quantum uncertainties. If you start with just one atom, there is a 50–50 chance that you will still have it after 4 days. You cannot predict whether you will have it or not.

17. Table 15.1 tells us that the half-life is 8 days, so 20 days is 2.5 half-lives. Figure 15.11 tells us that in 2.5 half-lives, the fraction remaining is about 0.175. Thus the amount remaining is 0.175×5 g $= 0.875$ g. Table 15.1 tells us that a little more than 4 half-lives are needed for a sample to decline to 5% of its original amount. Thus the time is 4×8 days $= 32$ days.

19. Cosmic rays create the ^{14}C in the atmosphere that is then incorporated into living organisms.

21. According to Figure 15.11, the radioactivity drops to 20% of its original value after about 2.25 half-lives. Table 15.1 says that the half-life of ^{14}C is about 6000 years, so the age of the ax handle is about 13,500 years.

23. The risk from the 4 cigarettes is about 3 times larger than the risk from the wine.

25. The risk has gone down, because the number of deaths has increased by less than 300% (less than 3 times) while the number of people has increased by 400%. More precisely, $315/(750 \times 10^6) = 0.44 \times 10^{-6}$, or 0.44 parts in a million risk in A.D. 1800, $845/(3 \times 10^9) = 0.28 \times 10^{-6}$, or 0.28 parts in a million risk in the mid-20th century.

27. The number of days is $40 \times 365 = 14,600$ days, so the risk is $14,000 \times 10^{-6} = 0.014 = 1.4\%$. If you smoke 20 per day, your risk is multiplied by 20/1.4, or about 14. So the overall risk is now $20 \times 1.4\% = 28\%$. Note: We are assuming here that the risks simply add up, i.e., that previous smoking does not affect the risks involved in present smoking (so that the risk remains 10^{-6} for each 1.4 cigarettes throughout the smoker's life). This assumption is probably not quite correct, because the smoking itself must affect the health (and thus the risks) of the smoker.

CHAPTER 16

1. $^{16}_{8}$O $\rightarrow$ $^{4}_{2}$He $+^{12}_{6}$C. According to the nuclear energy curve, the right side represents the most nuclear energy.

3. Use the nuclear energy curve. Carbon: fusion. Gold: fission. Iron: neither.

5. Warming coffee: the change in mass is so small it cannot be measured. Nuclear reactor: measurable change in mass. TNT explosion: the change in mass is so small it cannot be measured. Fission bomb explosion: measurable change in mass. Fusion bomb explosion: measurable change in mass. Lifting a book: the change in mass is so small it cannot be measured.

7. The sun's energy comes from nuclear fusion.

9. $^{24}_{12}Mg$

11. Lanthanum has atomic number 57, and uranium has atomic number 92. So the atomic number of the other fission product must be $92 - 57 = 35$, which is bromine.

13. 1% of 10 kg is 0.1 kg, or 100 g. $mc^2 = (0.1 \text{ kg}) \times (3 \times 10^8 \text{ m/s})^2 = 9 \times 10^{15}$ J.

15. Yes. All elements having atomic numbers larger than 83 are radioactive.

17. No. Neutrons are required to sustain the chain reaction.

CHAPTER 17

1. Wood, trash, wind, hydroelectric.

3. This sector must not make large use of heat engines, because the efficiency is higher than is normally obtainable from heat engines.

5. Approximate percentages for coal: 80%, 70%, 50%, 20%, 20%. For oil: 0%, 10%, 30%, 40%, 45%.

7. $29/(29 + 56) = 29/85 = 34\%$.

9. Thermal input: in the core of the reactor. Work output: at the turbine. Thermal output: At the condenser.

11. Nuclear power plants do not have a stack, because they do not operate by combustion of fuel.

13. Yes. Because the rods are radioactive, and this generates thermal energy. A supply of cooling water is kept on hand to keep the reactor cool.

15. Gas is a more efficient fuel—more joules of energy can be gotten per ton of carbon.

17. A breeder reactor makes Pu fuel out of the ^{238}U, while the material consumed for energy is ^{235}U. A reactor that consumes more ^{238}U than ^{235}U will create more fuel than it consumes. Two separate reactions occur here: Fission of ^{235}U, and creation of Pu from ^{238}U. Both reactions obey the law of conservation of energy.

19. The sun's radiation warms water, evaporating some of the water. The evaporated water rains down on the land, and is collected behind dams, where it provides the gravitational energy needed for hydroelectric power.

21. Drying clothes, rain watering the lawn, helping to warm the house by day, keeping the house at a comfortable temperature by opening and closing windows at appropriate times, using sunlight for light to see by, sunbathing, growing plants inside and outside the house, burning wood in a fireplace.

23. Coal and biomass: chemical to thermal to electric. Uranium: nuclear to thermal to electric. Hydroelectric: gravitational to electric. Geothermal: thermal to electric. Wind: kinetic to electric. Solar-thermal: radiant to thermal to electric. Photovoltaic: radiant to electric.

25. Since the energy is provided at only 20% efficiency, 5 kW (5000 watts) of power must be provided for each household. The number of square meters required for each household is thus 5000 W/(200 W/m^2) = 25 m^2. A football field is roughly 100 m long (a little less actually) and 30 m wide, so its area is about 3000 m^2. The number of households is thus 3000 m^2/25 m^2 = 120 households.

27. It raised energy prices and thus decreased energy demand, stimulated energy efficiency, and increased the development of alternative energy sources.

CHAPTER 18

1. Both the upper pair of tracks and the lower pair of tracks curve in opposite directions, indicating the forces on the pair of charged particles are in opposite directions, so the particles must have opposite charges.

3. Photons are not charged, and only charged particles make tracks in bubble chambers.

5. Since a proton is 2000 times more massive than an electron, its total energy (mc^2) is 2000 times larger. So the two photons that are created have 2000 times more energy. This means (because of $E = hf$) that each photon has a frequency 2000 times larger than the frequencies of the photons created by electron–positron annihilation. So each photon's frequency is roughly 2000×10^{20} Hz $= 2 \times 10^{23}$ Hz.

7. The charged particles: electron, muon, tau, W^+, and W^-. These are also the particles that can exchange photons, because microscopically "exchanging photons" is what we *mean* by "feeling the electric force."

9. Similarities: the W and Z are exchange particles; the Z is uncharged. Differences: the W and Z have rest mass; the W is charged; the W and Z move at less than lightspeed.

11. Similarities: quarks are material particles (they have rest mass); quarks move at less than lightspeed; quarks are charged (i.e., they feel the electric force). Differences: Quarks feel the strong force while electrons do not; quarks are fractionally charged ($+\frac{2}{3}$ or $-\frac{1}{3}$ of proton charge).

13. Recall the mass–energy equivalence principle ($E = mc^2$). The additional mass is due to the energy of the fields associated with the strong nuclear forces acting between the quarks.

15. No, because then life would not have been possible. (The anthropic principle says that our universe must be consistent with the existence of life and intelligent beings.)

PHOTO CREDITS

All cartoons © Sidney Harris, used with permission.

Figure	Photographer/Source
PO 1	Courtesy of NASA
Fig. 1.1	UPI/The Bettmann Archive
Fig. 1.2a, c	Courtesy of NASA
Fig. 1.2b	Courtesy of European Southern Observatory
Fig. 1.4	Courtesy of National Optical Astronomy Observatories
Fig. 1.7	Courtesy of NASA
Fig. 1.12	Courtesy of New York Public Library Still Pictures Collection
Fig. 1.14	American Institute of Physics
Fig. 1.18	Courtesy of New York Public Library Still Pictures Collection
Fig. 1.19	American Institute of Physics
Fig. 1.20	The Bettmann Archive
Fig. 1.21	American Institute of Physics
Fig. 1.27	Courtesy of NASA
Fig. 1.28	Courtesy of Hale Observatories
Fig. 1.29	Courtesy of P. J. E. Peebles
Fig. 2.8	Macmillan Publishing Company
Fig. 2.10	Courtesy of National Oceanic and Atmospheric Administration
Fig. 2.15	Courtesy of the University of Chicago
Fig. 2.17	Courtesy of International Business Machines Corporation
Fig. 2.18	Courtesy of International Business Machines Corporation
Fig. 2.20	The Bettmann Archive
PO 2	Gregory Dimijian/Photo Researchers, Inc.
Fig. 3.1	T. J. J. See Collection/American Institute of Physics
Fig. 3.4	E. Scott Barr Collection/American Institute of Physics
Fig. 3.6a, b	Courtesy of Kendall/Hunt Publishing
Fig. 3.13	Courtesy of Kendall/Hunt Publishing
Fig. 4.1	American Institute of Physics
Fig. 4.6	Courtesy of National Bureau of Standards

Figure	Photographer/Source
Fig. 5.8a, b, c	Courtesy of NASA
Fig. 5.12	Courtesy of John Wiley & Sons
Fig. 5.14	Courtesy of W. K. Hartmann
Fig. 5.15	Courtesy of the National Optical Astronomy Observatories
Fig. 5.16	Julian Baum/Science Photo Library/Photo Researchers, Inc.
PO 3	Dohrn/Science Photo Library/Photo Researchers, Inc.
Fig. 6.4	American Institute of Physics
Fig. 7.11	Courtesy of General Motors Corporation
Fig. 7.12	Courtesy of Toyota Motor Corporation
Fig. 7.13	Courtesy of Honda North America, Inc.
Fig. 7.15	Courtesy of Honda North America, Inc.
Fig. 7.16	Courtesy of General Motors Corporation
Fig. 7.18	Courtesy of Niagara Mohawk
Fig. 7.20	Grapes-Michaud/Photo Researchers, Inc.
Fig. 8.1	Dohrn/Science Photo Library/Photo Researchers, Inc.
Fig. 8.2	Courtesy of Kendall/Hunt Publishing Company
Fig. 8.12	Courtesy of Kendall/Hunt Publishing Company
Fig. 8.13	Courtesy of Kendall/Hunt Publishing Company
Fig. 8.17	Arthur Hobson
Fig. 8.20	Courtesy of Kendall/Hunt Publishing Company
Fig. 8.21	Courtesy of Kendall/Hunt Publishing Company
Fig. 8.23	Gordon Gore
Fig. 8.25	American Institute of Physics
Fig. 8.26	American Institute of Physics
Fig. 9.2	Courtesy of Kendall/Hunt Publishing Company
Fig. 9.4	American Institute of Physics
Fig. 9.10a, b	Courtesy of National Radio Astronomy Observatory
Fig. 9.10c	John Bova/Photo Researchers, Inc.
Fig. 9.10d, e	Courtesy of NASA
Fig. 9.14	Courtesy of National Science Foundation

Figure	Photographer/Source	Figure	Photographer/Source
PO 4	University of Pennsylvania, Van Pelt Library, Special Collections (E1651M)	Fig. 16.10	Herzfeld Collection/American Institute of Physics
Fig. 10.1	UPI/The Bettmann Archive	Fig. 16.16	Chicago Historical Society
Fig. 10.2	Courtesy of University of New Hampshire	Fig. 16.17	University of Chicago
		Fig. 16.18	Courtesy of Oak Ridge National Laboratory
Fig. 11.11	The Bettmann Archive	Fig. 16.19a	Courtesy of Lawrence-Berkeley Laboratory
		Fig. 16.19b	Bainbridge Collection/American Institute of Physics
Fig. 12.1	Courtesy of Wadsworth Publishing	Fig. 16.23	No. 3381, U. S. Air Force/National Air and Space Museum
Fig. 12.2	Sidney Fox/Visuals Unlimited		
Fig. 12.3	Courtesy of Alexander Oparin	Fig. 16.24	No. 3454, U. S. Air Force/National Air and Space Museum
Fig. 12.4	Michael Abbey/Photo Researchers, Inc.		
Fig. 12.5	Courtesy of NASA		
Fig. 12.6	Courtesy of Ames Space Flight Center, NASA	Fig. 17.7	Nancy J. Pierce/Photo Researchers, Inc.
Fig. 12.7a, b	Courtesy of NASA	Fig. 17.8	Rollin Geppert
		Fig. 17.12	Courtesy of National Renewable Energy Lab
Fig. 13.3	American Institute of Physics	Fig. 17.13	Courtesy of Solar Energy Resource Institute
Fig. 13.7	Dr. Albert Rose/Plenum Press with permission	Fig. 17.14	Courtesy of National Renewable Energy Lab
Fig. 13.8	American Institute of Physics	Fig. 17.17	Courtesy of National Renewable Energy Lab
Fig. 13.14	Courtesy of Claus Jonsson	Fig. 17.18	Courtesy of Solar Energy Resource Institute
Fig. 13.15	Courtesy of A. Tonomura, Advanced Research Laboratory, Hitachi, Ltd.	Fig. 17.20	Michael Godomski/Photo Researchers, Inc.
		Fig. 17.25	Courtesy of Osrom Sylvania, Inc.
Fig. 13.18	Jost Lemmerich/American Institute of Physics		
Fig. 13.20	The Bettmann Archive	Fig. 18.1a	Courtesy of California Institute of Technology
		Fig. 18.1b	Courtesy of Julian Schwinger
Fig. 14.2	Courtesy of Wabash Instrument Company	Fig. 18.4	Max Planck Institute/American Institute of Physics
Fig. 14.5	Courtesy of Educational Development Center		
Fig. 14.13	American Institute of Physics	Fig. 18.5	Courtesy of Fermilab
Fig. 14.19a	Mark Edwards/Still Pictures	Fig. 18.6	Courtesy of Lawrence-Berkeley Laboratory
Fig. 14.19b	Courtesy of CERN	Fig. 18.8a	Courtesy of Lawrence-Berkeley Laboratory
Fig. 14.19c	Courtesy of Lawrence-Berkeley Laboratory	Fig. 18.8b	Courtesy of Fermilab
Fig. 14.19d	Courtesy of Alain Aspect	Fig. 18.9a, c	Courtesy of Lawrence-Berkeley Laboratory
Fig. 14.25	Margarethe Bohr Collection/American Institute of Physics	Fig. 18.9b	Courtesy of Lawrence Radiation Lab/American Institute of Physics
Fig. 14.27	Margarethe Bohr Collection/American Institute of Physics	Fig. 18.9d	Courtesy of Stanford Linear Accelerator/Department of Energy
		Fig. 18.9e	Courtesy of Fermilab
PO 5	CERN/Science Photo Library/Photo Researchers, Inc.	Fig. 18.10	Courtesy of the International Center for Theoretical Physics
Fig. 15.3	W. F. Meggers Collection/American Institute of Physics	Fig. 18.11	American Institute of Physics
		Fig. 18.13	American Institute of Physics
Fig. 15.14	Courtesy of Masao Tsuzuki and Goulichi Kimura/Harper-Collins	Fig. 18.14a	Courtesy of Henry Kendall
		Fig. 18.14b	Courtesy of Donna Conveney/Massachusetts Institute of Technology
Fig. 16.6	Courtesy of NASA	Fig. 18.14c	Courtesy of Henry Kendall and Richard Taylor
Fig. 16.7	National Optical Astronomy Observatories/Science Photo Library/Photo Researchers, Inc.	Fig. 18.15	Courtesy of Stanford Linear Accelerator/Department of Energy
Fig. 16.8	Société Française de Physique/American Institute of Physics	Fig. 18.17	Courtesy of W. H. Freeman
Fig. 16.9	Gouldsmith Collection/American Institute of Physics	Fig. 18.19	Ulli Steltzer/American Institute of Physics

INDEX

526   INDEX

Quick reference list: how things work _____